PHYSICAL CONSTANTS

Quantity	Symbol	Magnitude
Angstrom Unit	Å	$1\,\text{Å} = 10^{-4}\,\mu\text{m} = 10^{-8}\,\text{cm} = 10^{-10}\,\text{m}$
Avogadro's Number	N_{AVO}	6.025×10^{23} molecules/g-mole
Bohr Radius	a_B	$0.53\,\text{Å}$
Boltzmann's Constant	k	1.380×10^{-23} joule/°K $= 8.62 \times 10^{-5}$ eV/°K
Electron Charge	q	1.602×10^{-19} coul
Electron Volt	eV	$1\text{ eV} = 1.602 \times 10^{-19}$ joule $= 23.1$ kcal
Free Electron Mass	m_0	9.108×10^{-31} kg
Micron	μm	$10^{-6}\,\text{m} = 10^{-4}\,\text{cm}$
Permittivity of Free Space	ε_0	8.854×10^{-14} farad/cm $\left(\dfrac{10^9}{4\pi c^2}\right)$
Permeability of Free Space	μ_0	1.257×10^{-8} henry/cm ($4\pi \times 10^{-9}$)
Planck's Constant	h	6.625×10^{-34} joule-sec
Reduced Planck's Constant	$\hbar$	$h/2\pi = 1.054 \times 10^{-34}$ joule-sec
Standard Atmosphere		1.033×10^4 kg/m^2 $= 760$ mm Hg (torr)
Thermal Voltage (T = 300°K)	kT/q	0.0259 volt
Velocity of Light in Free Space	c	2.998×10^{10} cm/sec
Wavelength Associated with 1 eV	λ	$1.239\,\mu\text{m}$

Physics of Semiconductor Devices

Physics of Semiconductor Devices

S. M. SZE

Member of the Technical Staff

Bell Telephone Laboratories, Incorporated

Murray Hill, New Jersey

WILEY-INTERSCIENCE

A Division of John Wiley & Sons

New York · London · Sydney · Toronto

To My Wife and Our Parents

Preface

This book is concerned specifically with physical principles and operational characteristics of semiconductor devices. It is intended as a textbook for first-year graduate students in electrical engineering or applied physics and as a reference for scientists actively involved in solid-state device research and development. It is assumed that the reader has already acquired an introductory understanding of solid-state physics and transistor theory such as is given in the standard texts *Solid-State Physics* by Dekker, *Introduction to Solid-State Physics* by Kittel (Wiley, 1966), *Physics of Semiconductors* by Moll, *Introduction to Semiconductor Physics* by Adler et al. (Wiley, 1964), *Semiconductors* by Smith, and *Physics and Technology of Semiconductor Devices* by Grove (Wiley, 1967). With this as a basis, the present book elaborates on device theory in more detail.

A semiconductor device is defined herein as a unit which consists, partially or wholly, of semiconducting materials and which can perform useful functions in electronic apparatus and solid-state research. Since the invention of the transistor in 1948 the number and variety of semiconductor devices have increased tremendously as advanced technology, new materials, and broadened comprehension have been applied to the creation of new devices. In a book of this length it is not possible to give a complete account of all of them. An attempt, however, has been made to include most of the important devices, in particular, the microwave, interface, and optoelectronic devices; and to present them in a unified and coherent fashion. It is hoped that this book can serve as a basis for the understanding of other devices not included here and perhaps not even conceived of at the present time.

This text began as a set of lecture notes for an out-of-hours course in "Semiconductor Device Physics" given at Bell Laboratories and later presented as a graduate course in "Selected Topics on Solid-State Devices" at Stevens Institute of Technology. For numerical data on semiconductor properties the principal source was the Data Sheets compiled by the Electronic Properties Information Center of Hughes Aircraft Company. More than two thousand papers on semiconductor devices, cataloged at the Murray Hill Library of Bell Laboratories, have provided background. For each chapter a brief historical review, as well as a general outline, is given in its introduction. The physics of the device and its mathematical formulation are then presented in subsequent sections and are generally arranged in logical sequence without heavy reliance on the original papers. The literature and many illustrations used in the book are acknowledged in the reference lists at the close of each chapter. In Chapter 2 extensive data have been compiled and presented for the three most important semiconductors: germanium, silicon, and gallium arsenide. These data are used in analyzing device characteristics throughout this book. In Chapters 3 through 14 each considers a specific device or a group of closely related devices and each is presented in such a way that it is self-contained and essentially independent of the others; for example, if one is interested in bulk-effect devices, he can refer directly to the last chapter without consulting any of those between Chapters 3 and 13.

During the course of the writing I have been deeply grateful to many of my colleagues at Bell Laboratories; without their help this book could not have been written. I wish to express, in particular, my gratitude to Dr. R. M. Ryder for his constant encouragement and constructive criticism and to Drs. W. S. Boyle and F. M. Smits for providing a stimulating and challenging environment in which I have been welcomed, inspired, and abundantly assisted. I am also indebted to Drs. J. M. Andrews, H. J. Boll, J. R. Brews, C. Y. Chang, B. C. DeLoach, J. C. Dyment, R. Edwards, A. Goetzberger, H. K. Gummel, F. Harper, J. E. Iwersen, D. Kahng, S. Knight, M. Kuhn, H. S. Lee, T. P. Lee, M. P. Lepselter, W. T. Lynch, H. Melchior, T. Misawa, E. H. Nicollian, T. Paoli, H. C. Poon, R. J. Powell, J. E. Ripper, W. Rosenzweig, D. L. Scharfetter, L. S. Senhouse, J. Sevick, H. Thim, and T. H. Zachos for their helpful suggestions and discussions on one or more chapters of the book.

I am especially indebted to Mrs. E. S. Blair for her lucid technical editing of the entire manuscript. Thanks are due also to Professor G. J. Herskowitz of Stevens Institute of Technology, who provided the opportunity for me to teach the course on which this text is based, and to Mrs. M. Neuberger of the Electronic Properties Information Center, Hughes Aircraft Company, who provided the up-to-date compilation of data sheets and kindly checked the tables used in Chapter 2. It is a pleasure to acknowledge Mr. G. P. Carey and Mr. A. Loya, who did most of the literature searching on semiconductor devices, and my students at Stevens Institute, in particular Mr. P. P. Bohn and Mr. H. A. Kruegle, who helped to correct many errors in the original class notes. I wish to thank Miss D. A. Williams who typed the final manuscript, Mrs. J. Hendricks, Miss J. McCarthy, and Mrs. E. H. Nevitt who typed various sections of the book in its revision stage, and the members of the Murray Hill Drafting Department of Bell Laboratories who furnished the more than five hundred technical drawings used in this book. The support of C. Y. Tung's Chair Professorship grant and the hospitality of the National Chiao Tung University provided the environment in which to read the proofs of the book. Finally, I wish especially to thank my wife Therese Ling-yi for her assistance in many ways, including the typing of the entire first draft.

S. M. SZE

Contents

Chapter 1 Introduction **1**

    1. General Outline   1
    2. Classification of Semiconductor Devices   3
    3. Specific Remarks   5

PART I SEMICONDUCTOR PHYSICS

Chapter 2 Physics and Properties of Semiconductors—
    A Résumé **11**

    1. Introduction   11
    2. Crystal Structure   12
    3. Energy Bands.   17
    4. Carrier Concentration at Thermal Equilibrium   25
    5. Carrier Transport Phenomena   38
    6. Phonon Spectra and Optical, Thermal, and High-Field
        Properties of Semiconductors   50
    7. Basic Equations for Semiconductor Device Operation   65

PART II p-n JUNCTION DEVICES

Chapter 3 p-n Junction Diodes **77**

    1. Introduction   77
    2. Basic Device Technology   78

3. Depletion Region and Depletion Capacitance   84
4. Current-Voltage Characteristics   96
5. Junction Breakdown   109
6. Transient Behavior and Noise   127
7. Terminal Functions   131
8. Heterojunction   140

Chapter 4   Tunnel Diode and Backward Diode **150**

1. Introduction   150
2. Effects of High Doping   151
3. Tunneling Processes   156
4. Excess Current   169
5. Current-Voltage Characteristics Due to Effects of Doping,
   Temperature, Electron Bombardment, and Pressure   172
6. Equivalent Circuit   190
7. Backward Diode   193

**Chapter 5   Impact-Avalanche Transit-Time Diodes
(IMPATT Diodes)** **200**

1. Introduction   200
2. Static Characteristics   202
3. Basic Dynamic Characteristics   215
4. Generalized Small-Signal Analysis   221
5. Large-Signal Analysis   234
6. Noise   240
7. Experiments   244

Chapter 6   Junction Transistors **261**

1. Introduction   261
2. Static Characteristics   262
3. Microwave Transistor   279
4. Power Transistor   295
5. Switching Transistor   302
6. Unijunction Transistor   310

Chapter 7   p-n-p-n and Junction Field-Effect Devices **319**

1. Introduction   319
2. Shockley Diode and Semiconductor-Controlled Rectifier . . .   320
3. Junction Field-Effect Transistor and Current Limiter   340

PART III   INTERFACE AND THIN-FILM DEVICES

Chapter 8   Metal-Semiconductor Devices **363**

  1. Introduction 363
  2. Schottky Effect 364
  3. Energy Band Relation at Metal-Semiconductor Contact . . . 368
  4. Current Transport Theory in Schottky Barriers 378
  5. Measurement of Schottky Barrier Height 393
  6. Clamped Transistor, Schottky Gate FET, and
     Metal-Semicondutor IMPATT Diode 410
  7. Mott Barrier, Point-Contact Rectifier, and Ohmic Contact . . 414
  8. Space-Charge-Limited Diode 417

Chapter 9   Metal-Insulator-Semiconductor (MIS) Diodes . . . **425**

  1. Introduction 425
  2. Ideal Metal-Insulator-Semiconductor (MIS) Diode 426
  3. Surface States, Surface Charges, and Space Charges 444
  4. Effects of Metal Work Function, Crystal Orientation,
     Temperature, Illumination, and Radiation on
     MIS Characteristics 467
  5. Surface Varactor, Avalanche, Tunneling, and
     Electroluminescent MIS Diodes 479
  6. Carrier Transport in Insulating Films 492

Chapter 10   IGFET and Related Surface Field Effects **505**

  1. Introduction 505
  2. Surface-Space-Charge Region Under Nonequilibrium
     Condition 506
  3. Channel Conductance 512
  4. Basic Device Characteristics 515
  5. General Characteristics 524
  6. IGFET with Schottky Barrier Contacts for Source and Drain . 546
  7. IGFET with a Floating Gate—A Memory Device 550
  8. Surface Field Effects on *p-n* Junctions and Metal-
     Semiconductor Devices 555

Chapter 11   Thin-Film Devices **567**

  1. Introduction 567
  2. Insulated-Gate Thin-Film Transistors (TFT) 568

3. Hot-Electron Transistors   587
4. Metal-Insulator-Metal Structure   614

PART IV   OPTOELECTRONIC DEVICES

Chapter 12   Optoelectronic Devices   625

1. Introduction   625
2. Electroluminescent Devices   626
3. Solar Cell   640
4. Photodetectors   653

Chapter 13   Semiconductor Lasers   687

1. Introduction   687
2. Semiconductor Laser Physics   688
3. Junction Lasers   699
4. Heterostructure and Continuous Room-Temperature
   Operation   723
5. Other Pumping Methods and Laser Materials   725

PART V   BULK-EFFECT DEVICES

Chapter 14   Bulk-Effect Devices   731

1. Introduction   731
2. Bulk Differential Negative Resistance   732
3. Ridley-Watkins-Hilsum (RWH) Mechanism   741
4. Gunn Oscillator and Various Modes of Operation   749
5. Associated Bulk-Effect Devices   778

Author Index   789

Subject Index   799

Physics of Semiconductor Devices

- **GENERAL OUTLINE**
- **CLASSIFICATION OF SEMICONDUCTOR DEVICES**
- **SPECIFIC REMARKS**

I

Introduction

I GENERAL OUTLINE

The contents of this book can be divided into five parts:

Part 1: résumé of physics and properties of semiconductors,
Part 2: p-n junction devices,
Part 3: interface and thin-film devices,
Part 4: optoelectronic devices, and
Part 5: bulk-effect devices.

Part 1 in Chapter 2 is intended as a summary of materials properties, to be used throughout the book as a basis for understanding and calculating device characteristics. Carrier distribution and transport properties are briefly surveyed, with emphasis on the three most important materials: Ge, Si, and GaAs. A compilation of the most recent and most accurate information on these semiconductors is given in the tables and illustrations of Chapter 2.

Part 2, Chapters 3 through 7, treats of devices with one or more interacting p-n junctions. The classic moderate-field p-n junction of Chapter 3 is basic both for itself and as a component of more involved devices. When the junction is doped heavily enough on both sides so that the field becomes high enough for tunneling, one obtains the new features of tunnel diode behavior (Chapter 4). When the junction is operated in avalanche breakdown, under proper conditions one obtains an IMPATT diode which can generate micro-

1

wave radiation (Chapter 5). Chapter 6 treats of the junction transistor, that is, the interaction between two closely-coupled junctions, which is the single most important semiconductor device. Other junction devices are in Chapter 7; among these are the *p-n-p-n* triple-junction devices, and the junction field-effect transistor which utilizes the junction field to control a current flow which is parallel to the junction rather than perpendicular to it.

Part 3, Chapters 8 through 11, deals with interfaces, or surfaces between semiconductors and other materials. Interfaces with metals, in particular the Schottky barrier, are in Chapter 8. The Schottky diode behavior is electrically similar to a one-sided abrupt *p-n* junction, and yet it can be operated as a majority-carrier device with inherent fast response. Metal-insulator-semiconductor devices and the related surface physics of the insulator-semiconductor interface are considered in Chapters 9 and 10. This knowledge of "surface states" is important not only because of the devices themselves but also because of the relevance to stability and reliability of all other semiconductor devices. Chapter 11 considers some thin-film and hot-electron devices which also belong to the interface-device family.

Part 4, Chapters 12 and 13, considers optoelectronic devices which can detect, generate, and convert optical energy to electric energy or vice versa. We shall consider various photodetectors and the solar cell in Chapter 12. Chapter 13 is devoted to one of the most important optoelectronic devices: the semiconductor laser.

Part 5, Chapter 14, considers some so-called "bulk property" devices, primarily those concerned with the intervalley-transfer mechanism (Gunn oscillator and LSA oscillator). These devices do not depend primarily on *p-n* junctions or interfaces, but they do operate at reasonably high fields, so that the velocity-field relationship and various modes of operation are of prime interest in Chapter 14.

In the presentation of each device chapter, the historical events concerning the invention and derivation of a particular device or a group of closely related devices are briefly reviewed. This is then followed by consideration of device characteristics and physical principles. It is intended that each chapter should be more or less independent of the other chapters, so that the instructor or the reader can select or rearrange the device chapters in accordance with his own schedule.

A remark on notation: in order to keep the notations simple, it is necessary to use the simple symbols more than once, with different meanings for different devices. For example, the symbol α is used as the common-base current gain for a junction transistor, as the optical absorption coefficient for a photodetector, and as the electron impact ionization coefficient for an IMPATT diode. This usage is considered preferable to the alternative, which would be to use alpha only once, and then be forced to find more complicated symbols for the other uses. Within each chapter, however, each symbol is used with only one meaning and is defined the first time it appears. Many symbols do have the same or similar meanings consistently throughout this book; they are summarized in Table 1.1 for convenient reference.

2 CLASSIFICATION OF SEMICONDUCTOR DEVICES

In the previous section we have classified semiconductor devices in accordance with material combinations (such as the semiconductor alone, or a combination of metal, semiconductor, and insulator) and material properties (such as the optical property, or the intervalley transfer property). This classification is used because it permits an orderly sequence from one device to another within the book.

TABLE 1.1

LIST OF BASIC SYMBOLS

Symbol	Name	Unit
a	lattice spacing	Å
$\mathscr{B}$	magnetic induction	Weber/m^2
c	velocity of light in free space	cm/sec
C	capacitance	farad
$\mathscr{D}$	electric displacement	coul/cm^2
D	diffusion coefficient	cm^2/sec
E	energy	eV
E_C	bottom of conduction band	eV
E_F	Fermi energy level	eV
E_g	energy gap band	eV
E_V	top of valence band	eV
$\mathscr{E}$	electric field	volt/cm
$\mathscr{E}_c$	critical field	volt/cm
$\mathscr{E}_m$	maximum field	volt/cm
f	frequency	Hz(cps)
$F(E)$	Fermi-Dirac distribution function	
h	Planck's constant	joule-sec
$h\nu$	photon energy	eV
I	current	amp
I_C	collector current	amp
J	current density	amp/cm^2
J_t	threshold current density	amp/cm^2
k	Boltzmann's constant	joule/°K
kT	thermal energy	eV
L	length	cm or μm
m_0	free electron mass	kg
m^*	effective mass	kg

TABLE 1.1 (Cont.)

Symbol	Name	Unit
$\bar{n}$	refractive index	
n	density of free electrons	cm^{-3}
n_i	intrinsic density	cm^{-3}
N	doping concentration	cm^{-3}
N_A	acceptor impurity density	cm^{-3}
N_C	effective density of states in conduction band	cm^{-3}
N_D	donor impurity density	cm^{-3}
N_V	effective density of states in valence band	cm^{-3}
p	density of free holes	cm^{-3}
P	pressure	$dyne/cm^2$
q	magnitude of electronic charge	coulomb
Q_{ss}	surface-state density	$charges/cm^2$
R	resistance	ohm
t	time	sec
T	absolute temperature	°K
v	carrier velocity	cm/sec
v_{sl}	scattering-limited velocity	cm/sec
v_{th}	thermal velocity ($\sqrt{3kT/m}$)	cm/sec
V	voltage	volts
V_{bi}	build-in potential	volts
V_{EB}	emitter-base voltage	volts
V_B	breakdown voltage	volts
W	thickness	cm or μm
W_B	base thickness	cm or μm
x	x-direction	
∇	differential operator	
∇T	temperature gradient	°K/cm
ε_0	free space permittivity	farad/cm
ε_s	semiconductor permittivity	farad/cm
ε_i	insulator permittivity	farad/cm
$\varepsilon_s/\varepsilon_0$ or $\varepsilon_i/\varepsilon_0$	dielectric constant	
τ	lifetime or decay time	second
θ	angle	radian
λ	wavelength	μm or Å
ν	frequency of light	Hz
μ_0	free-space permeability	henry/cm
μ_n	electron mobility	cm^2/V-sec
μ_p	hole mobility	cm^2/V-sec
ρ	resistivity	ohm-cm
ϕ	barrier height or imref	volts
ϕ_{Bn}	Schottky barrier height on n-type semiconductor	volts
ϕ_{Bp}	Schottky barrier height on p-type semiconductor	volts
ϕ_m	metal work function	volts
ω	angular frequency ($2\pi f$ or $2\pi\nu$)	Hz
Ω	ohm	ohm

A more systematic method of classifying semiconductor devices, as proposed by Angello,[1] will be presented in this section. This system can be used to classify all the present and future semiconductor devices, and can serve to organize creative thinking. The system starts by listing a complete set of semiconductor properties and a complete set of external influences (such as applied voltage) which can modify semiconductor attributes. Devices are then classified in the order of the progressive complication of the semiconductor attributes, with external influences applied singly, in pairs, and so on.

We shall start with homogeneous semiconductors (bulk-effect without junction) and apply the external influences one at a time. The devices with the next degree of complication will be the bulk-effect with a pair of external influences, a triple, and so on. A single *p-n* junction will be the simplest departure from bulk-effect devices, and the external influences will be applied singly, in pairs, and so on. Finally, we shall consider multiple junction (or interface) devices, and the external influences will be applied accordingly. Table 1.2 presents some of the bulk-effect devices where the symbols for external influences are defined in Table 1.1. Table 1.3 presents some important single-junction (or interface) devices. Table 1.4 presents some important multiple junction (or interface) devices. These lists are not intended to exhaust all the possibilities. However, by extending the list of effects and influences, other semiconductor devices may be classified similarly.

Table 1.5 shows all the devices that will be considered in this book. They are classified into the aforementioned three groups. We note that a chapter may include devices from all three groups. For example, in Chapter 7, we shall consider the current limiter (bulk-effect), the field-effect diode (single-junction), and the junction field effect transistor (multiple-junction); they are included in this chapter because of the similarities of these devices in configuration and characteristics.

3 SPECIFIC REMARKS

It may be worthwhile to point out some interesting thoughts pertinent to this book.

First of all, the electronics field in general and the semiconductor-device field in particular are so dynamic and so fast-changing that today's concepts may be obsolete tomorrow. Remember these "famous last words"[2]?

"They'll never replace the smoke signal as the fastest means of communication."—Chief White Cloud,

"The telegraph is the ultimate in fast communication."—Engineer, 1850's,

"With the vacuum tube, we've reached the zenith in communication potential."—Engineer, 1920's,

TABLE 1.2

BULK-EFFECT DEVICES

External Influence		Number of Electrodes	Features	Name of Device (where applicable)
No.	Influence			
1	$h\nu$	0	transmission of light over certain frequency	Optical filter
			optical and electron-beam pumping	Lasers
	$\mathscr{E}$	2	$J = \sigma\mathscr{E}$	Resistor
			voltage-controlled negative resistance	Gunn oscillator
			current-controlled negative resistance	Cryosar
			$J = $ const. for $\mathscr{E} > \mathscr{E}_{\text{threshold}}$	Current limiter
	∇T	2	Seeback effect $\mathscr{E} \sim \nabla T$	
2	$h\nu, \mathscr{E}$	2	$J = \sigma(h\nu)\mathscr{E}$	Photoconductor
	$\mathscr{E}, \mathscr{H}$	2	$J = \sigma(\mathscr{H})\mathscr{E}$	Magnetoresistor
		4	Hall effect, $V = f(\mathscr{H}, \mathscr{E})$	Hall generator
	$\mathscr{E}, T$	2	$J = \sigma(T)\mathscr{E}$	Thermistor
	$\mathscr{E}, \nabla T$	2	Thomson heat, $\sim \nabla T$	
	$\mathscr{E}, P$	2	peizoresistance effect	Strain gauge
3	$\mathscr{E}_1, \mathscr{E}_2, \mathscr{H}$	3	Suhl effect	

"Transistors are the final step in the search for speedy, reliable means of communication."—Engineer, 1950's,

"Integrated circuits are IT! They can't possibly go beyond this revolutionary new concept."—Engineer, 1960's, ...

It is thus important for one to understand the fundamental physical processes and to equip himself with sufficient background in physics and mathematics to digest, to appreciate, and to meet the challenge of these dynamic fields.

We should also be aware of the widespread use of digital computers in almost every field. A digital computer is basically a faster, more powerful,

TABLE 1.3
SINGLE-JUNCTION (OR INTERFACE) DEVICES

External Influence		Number of Electrodes	Features	Name of Device (where applicable)
No.	Influence			
1	$h\nu$	2	photovoltaic effect	Photocell, solar cell
	$\mathcal{E}$	2	regular p-n junction (p-n)	Diode, rectifier
			p-n used as variable resistor	Varistor
			p-n used as variable capacitor	Varactor
			p-n used as light source	Electroluminescent diode
			p-n with very high dopings on both sides	Tunnel diode
			with moderately high dopings, direct band-gap, reflection surfaces	Injection laser
			limit voltage by avalanche breakdown or tunneling	Voltage regulator
			microwave generation by impact-avalanche and transit-time effect	IMPATT diode
			unipolar device used as current limiter	Field-effect diode
			junction formed between semiconductors with different band gaps	Heterojunction (n-n, p-n, and p-p)
			contacts between metal and semiconductor	Schottky diode, Mott diode, point-contact
	∇T	2	Seeback effect	Thermocouple, thermoelectric generator
2	$\mathcal{E}_1, \mathcal{E}_2$	3	minority injection into a filament	Unijunction transistor (double-base diode)
	$\mathcal{E}, \nabla T$	2	Peltier effect	Peltier refrigerator
	$\mathcal{E}, h\nu$	2	$J = f(\mathcal{E}, h\nu)$	Photodiode

TABLE 1.4

MULTIPLE-JUNCTION (OR INTERFACE) DEVICES

External Influences		Number of Electrodes	Features	Name of Device (where applicable)
No.	Influence			
1	$\mathscr{E}$	2	four-layer p-n-p-n diode	Shockley diode
		2	metal-insulator-metal device	MIM tunneling diode
		2	metal-insulator-semi-conductor device	MIS diode
1 or 2	$\mathscr{E}_1$ and/or $\mathscr{E}_2$	3	p-n-p and n-p-n	Junction transistors
		3	p-n-p-n with one gate	Semiconductor con-trolled rectifier (SCR)
		3	junction unipolar transistor	JFET
		3	insulated-gate field-effect transistor	IGFET
		3	thin-film transistor with de-posited semiconductor film	TFT
		3	semiconductor-metal-semi-conductor structures and other related structures	Hot-electron transistor
	$\mathscr{E}, h\nu$	3	with incident light	Optical transistor

and more versatile "sliderule." With this "sliderule" we can solve many nonlinear problems, we can handle millions of computations in a short time, and we can simulate dynamic behavior and discover useful results prior to experimental investigations. In this book many theoretial results and illustrations have been obtained with the help of computers. The reader is thus expected to have some basic familiarity with these aids.

It is important to point out that many of the devices, especially the microwave devices such as IMPATT diodes and Gunn oscillators, are still under intensive investigation. Their large-signal behaviors and ultimate performances are by no means fully understood at the present time. The material presented in this book is intended to serve as a foundation. The reader is

TABLE 1.5

CLASSIFICATION OF SEMICONDUCTOR DEVICES

Bulk-Effect Devices	Ch.
• Current limiter	Ch. 7
• Photoconductor	12
• Laser with optical and electron-beam pumping	13
• Bulk current-controlled differential negative resistance devices (cryosar)	14
• Bulk voltage-controlled differential negative resistance devices (Gunn oscillators)	14
• Hall devices	14
• Thermistor	14

Single-Junction (or Interface) Devices	Ch.
• p-n junctions: rectifier, varistor, varactor, voltage regulator	Ch. 3
• Tunnel diode and backward diode	4
• Impact-avalanche transit-time diode (IMPATT diode)	5
• Heterojunction	3
• Unijunction transistor	6
• Field-effect diode	7
• Metal-semiconductor contacts: Schottky, Mott, point-contact	8
• Optoelectronic diodes: solar cell, photodiode, electroluminescent diode	12
• Injection laser	13

Multiple-Junction (or Interface) Devices	Ch.
• p-n-p and n-p-n transistors	Ch. 6
• Shockley diode (p-n-p-n diode)	7
• Semiconductor controlled rectifier (SCR)	7
• Junction field-effect transistor (JFET)	7
• Metal-semiconductor devices	8
• Metal-insulator-semiconductor diode (MIS)	9
• Floating-gate memory device	10
• Insulated-gate field-effect transistor (IGFET)	10
• Thin-film transistors (TFT)	11
• Hot-electron transistors	11

External Influence
electric field ($\mathcal{E}$), magnetic field ($\mathcal{H}$), temperature (T), temperature gradient (∇T), radiation ($h\nu$), pressure (P), etc.

Number of Electrodes
no electrode, two terminals, three terminals, four terminals, etc.

9

expected to enhance his understanding by consulting the original literature with an inquiring mind that will not blindly accept all statements. He is also expected to attend technical conferences to exchange, to listen to, and to argue about technical ideas which may clarify misconcepts, inspire new ideas, and stimulate new thoughts.

Integrated circuitry (IC) is not discussed in this book. This is because IC is basically a technology. We are primarily concerned with the physics and operational principles of the individual devices which form the building blocks of integrated circuits. Only by understanding individual devices can we fully use them either as discrete components or as integrated systems.

REFERENCES

1. S. J. Angello, "Review of Other Semiconductor Devices," Proc. IRE, *46*, 968 (1958).

2. Compiled by H. C. Spencer.

PART I

SEMICONDUCTOR PHYSICS

- Physics and Properties of Semiconductors—
  A Résumé

- INTRODUCTION
- CRYSTAL STRUCTURE
- ENERGY BANDS
- CARRIER CONCENTRATION AT THERMAL EQUILIBRIUM
- CARRIER TRANSPORT PHENOMENA
- PHONON SPECTRA, AND OPTICAL, THERMAL, AND HIGH-FIELD PROPERTIES OF SEMICONDUCTORS
- BASIC EQUATIONS FOR SEMICONDUCTOR DEVICE OPERATION

2

Physics and Properties of Semiconductors—A Résumé

I INTRODUCTION

The physics of semiconductor devices is naturally dependent on the physics of semiconductors themselves. It is the purpose of this chapter to present a summary of the physics and properties of semiconductors. The summary represents only a small cross section of the vast literature on semiconductors; only those subjects pertinent to device operations are included here.

For detailed consideration of semiconductor physics the reader should consult the standard textbooks or reference works by Dunlap,[1] Madelung,[2] Moll,[3] and Smith.[4]

In order to condense a large amount of information into a single chapter, four tables and over thirty illustrations drawn from the experimental data are compiled and presented here. It is well known that, of all semiconductors, the elements germanium and silicon are the most extensively studied. In recent years gallium arsenide has also been intensively investigated because of its many interesting properties which differ from those of Ge and Si, particularly its direct band gap for laser application and its intervalley-carrier transport for generation of microwaves. Hence in this chapter emphasis will be placed on the above three most important semiconductors.

We shall first consider the crystal structure in Section 2, since the electronic properties of solids are intimately related to their lattice structures. We shall review the Miller indices which define crystal planes in real space and designate momenta in energy-momentum space. In Section 3, we shall consider energy bands, and the related parameters such as the effective mass m^* and the energy band gap E_g. It is found that for most semiconductors there is a close resemblance between the motion of an electron in a crystal and the free electron, particularly for an electron with energy near a minimum in energy-momentum space. Most of the electronic behaviors can be described in classical terms as though they were free electrons with an effective mass m^*. The energy gap corresponds to the energy difference between the top of the valence band which is completely filled at $0°K$ and the bottom of the conduction band which is empty at $0°K$. At higher temperatures there are electrons in the conduction band and holes in the valence band. For most semiconductors the transport processes for electrons and holes are considered to be independent of each other; and the total conduction current is simply the sum of the electron and hole current components.

Carrier concentration and carrier transport phenomena, based on the energy band theory, are considered in Sections 4 and 5 respectively. Other related properties of semiconductors under various external influences are discussed in Section 6. Finally, the basic equations for semiconductor device operation are summarized in Section 7.

2 CRYSTAL STRUCTURE

For a crystalline solid there exist three primitive basis vectors, **a**, **b**, **c**, such that the crystal structure remains invariant under translation through any vector which is the sum of integral multiples of these basis vectors. In other words, the direct lattice sites can be defined by the set[5]

$$\mathbf{l} = m\mathbf{a} + n\mathbf{b} + p\mathbf{c} \tag{1}$$

where m, n, and p are integers.

Figure 1 shows some of the important unit cells (direct lattices). A great many of the important semiconductors are of diamond or zincblende lattice structures which belong to the tetrahedral phases, i.e., each atom is surrounded by four equidistant nearest neighbors which lie at the corners of a tetrahedron. The bond between two nearest neighbors is formed by two electrons with opposite spins. The diamond and the zincblende lattices can be considered as two interpenetrating face-centered cubic lattices. For the diamond lattice, such as silicon, all the atoms are silicon; in a zincblende lattice, such as gallium aresnide (GaAs), one of the sublattices is gallium and

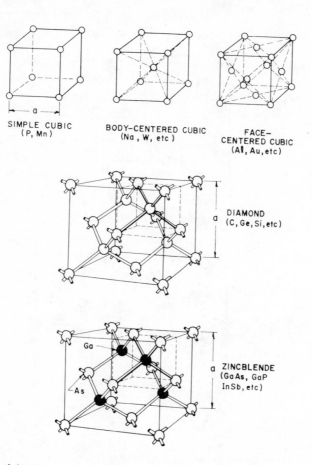

Fig. 1 Some of the important unit cells (direct lattices) and their representative elements or compounds.

the other is arsenic. Gallium arsenide is a III-V compound, since it is formed from elements of Column III and Column V in the Periodic Table. Most III-V compounds crystallize in the zincblende structure, [2,6,7] however, many semiconductors (including some III-V compounds) crystallize in the wurtzite structure which also has a tetrahedral arrangement of nearest neighbors.

Figure 2 shows the hexagonal close-packed lattice as well as the wurtzite lattice which can be considered as two interpenetrating hexagonal close-packed lattices. We note that, as in the zincblende lattice, each atom of the wurtzite lattice is surrounded by four equidistant nearest neighbors at the

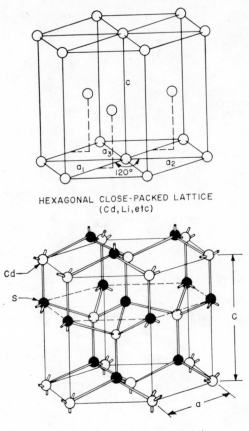

HEXAGONAL CLOSE-PACKED LATTICE
(Cd, Li, etc)

WURTZITE LATTICE (CdS, etc)

Fig. 2 Hexagonal close-packed lattice and wurtzite lattice.

corners of a tetrahedron. The lattice constants of some important semi-conductors are summarized in Table 2.1 along with their crystal structures.[8] It is of interest to note that some compounds, such as zinc sulfide and cad-mium sulfide, can crystallize in both zincblende and wurtzite structures.

For a given set of the direct basis vectors, one can define a set of reciprocal lattice basis vectors, $\mathbf{a}^*$, $\mathbf{b}^*$, $\mathbf{c}^*$ such that

$$\mathbf{a}^* \equiv 2\pi \frac{\mathbf{b} \times \mathbf{c}}{\mathbf{a} \cdot \mathbf{b} \times \mathbf{c}}, \qquad \mathbf{b}^* \equiv 2\pi \frac{\mathbf{c} \times \mathbf{a}}{\mathbf{a} \cdot \mathbf{b} \times \mathbf{c}}, \qquad \mathbf{c}^* \equiv 2\pi \frac{\mathbf{a} \times \mathbf{b}}{\mathbf{a} \cdot \mathbf{b} \times \mathbf{c}}, \qquad (2)$$

TABLE 2.1

	Element or Compound	Name	Crystal Structure	Lattice Constant at 300°K (Å)
Element Semiconductor	C	Carbon (Diamond)	Diamond	3.56679
	Ge	Germanium	Diamond	5.65748
	Si	Silicon	Diamond	5.43086
	Sn	Grey Tin	Diamond	6.4892
IV–IV	SiC	Silicon carbide	Zincblende	4.358
III–V	AlSb	Aluminum antimonide	Zincblende	6.1355
	BN	Boron nitride	Zincblende	3.615
	BP	Boron phosphide	Zincblende	4.538
	GaN	Gallium nitride	Wurtzite	$a = 3.186, c = 5.176$
	GaSb	Gallium antimonide	Zincblende	6.0955
	GaAs	Gallium arsenide	Zincblende	5.6534
	GaP	Gallium phosphide	Zincblende	5.4505
	InSb	Indium antimonide	Zincblende	6.4788
	InAs	Indium arsenide	Zincblende	6.0585
	InP	Indium phosphide	Zincblende	5.8688
II–VI	CdS	Cadmium sulfide	Zincblende	5.832
	CdS	Cadmium sulfide	Wurtzite	$a = 4.16, c = 6.756$
	CdSe	Cadmium selenide	Zincblende	6.05
	ZnO	Zinc oxide	Cubic	4.58
	ZnS	Zinc sulfide	Zincblende	5.42
	ZnS	Zinc sulfide	Wurtzite	$a = 3.82, c = 6.26$
IV–VI	PbS	Lead sulfide	Cubic	5.935
	PbTe	Lead telluride	Cubic	6.460

so that $\mathbf{a} \cdot \mathbf{a}^* = 2\pi$; $\mathbf{a} \cdot \mathbf{b}^* = 0$, etc; and the general reciprocal lattice vector is given by

$$\mathbf{g} = h\mathbf{a}^* + k\mathbf{b}^* + l\mathbf{c}^* \qquad (3).$$

where h, k, and l are integers.

It can be shown that the product $\mathbf{g} \cdot \mathbf{l} = 2\pi \times$ integer; that each vector of the reciprocal lattice is normal to a set of planes in the direct lattice; and that the volume V_c^* of a unit cell of the reciprocal lattice is inversely pro-

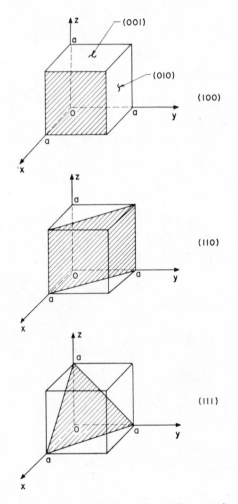

Fig. 3 Miller indices of some important planes in a cubic crystal where a is the lattice constant.

portional to the volume V_c of a unit cell of the direct lattice, i.e., $V_c* = (2\pi)^3/V_c$ where $V_c \equiv \mathbf{a} \cdot \mathbf{b} \times \mathbf{c}$.

A convenient method of defining the various planes in a crystal is by the use of the Miller indices which are determined by first finding the intercepts of the plane with the three basis axes in terms of the lattice constants and then taking the reciprocals of these numbers and reducing them to the smallest three integers having the same ratio. The result is enclosed in parentheses (hkl) as the Miller indices for a single plane or a set of parallel planes. The Miller indices of some important planes in a cubic crystal are shown in Fig. 3. Some other conventions are given as follows:[5]

$(\bar{h}kl)$: for a plane that intercepts the x-axis on the negative side of the origin.

$\{hkl\}$: for planes of equivalent symmetry such as $\{100\}$ for (100), (010), (001), ($\bar{1}$00), ($0\bar{1}0$), and ($00\bar{1}$) in cubic symmetry.

$[hkl]$: for the direction of a crystal such as [100] for the x-axis.

$\langle hkl \rangle$: for a full set of equivalent directions.

$[a_1a_2a_3c]$: for a hexagonal lattice. Here it is customary to use four axes (Fig. 2) with c-axis as the [0001] direction.

For the two semiconductor elements, germanium and silicon, the easiest breakage, or cleavage, planes are the $\{111\}$ planes. In contrast gallium arsenide, which has a similar lattice structure but also has a slight ionic component in the bonds, cleaves on $\{110\}$ planes.

The unit cell of a reciprocal lattice can be generally represented by a Wigner-Seitz cell. The Wigner-Seitz cell is constructed by drawing perpendicular bisector planes in the reciprocal lattice from the chosen center to the nearest equivalent reciprocal lattice sites. A typical example is shown [9, 10] in Fig. 4(a) for a face-centered cubic structure. If one first draws lines from the center point (Γ) to all the eight corners of the cube, then forms the bisector planes, the result is the truncated octahedron within the cube—a Wigner-Seitz cell. It can be shown that[11] a face-centered cubic (fcc) direct lattice with lattice constant "a" has a body-centered cubic (bcc) reciprocal lattice with spacing $4\pi/a$. Thus the Wigner-Seitz cell shown in Fig. 4(a) is the unit cell of the reciprocal lattice of an fcc direct lattice. Similarly we can construct the Wigner-Seitz cell for a hexagonal structure.[12] The result is shown in Fig. 4(b). The symbols used in Fig. 4 are adopted from Group Theory. Some of the symbols will be used in the next section on energy bands.

3 ENERGY BANDS

The band structure of a crystalline solid, i.e., the energy-momentum (E-k) relationship is usually obtained by the solution of the Schrodinger

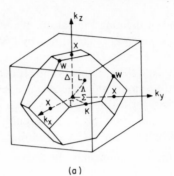

(a)

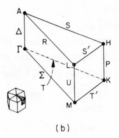

(b)

Fig. 4 (a) Brillouin zone for diamond and zincblende lattices.
(b) Brillouin zone for wurtzite lattice. Also indicated are the most important symmetry points and symmetry lines such as:
Γ: $2\pi/a$ (0, 0, 0), zone center
L: $2\pi/a$ ($\frac{1}{2}$, $\frac{1}{2}$, $\frac{1}{2}$), zone edge along $\langle 111 \rangle$ axes (Λ)
X: $2\pi/a$ (0, 0, 1), zone edge along $\langle 100 \rangle$ axes (Δ)
K: $2\pi/a$ ($\frac{3}{4}$, $\frac{3}{4}$, 0), zone edge along $\langle 110 \rangle$ axes (Σ).
(After Brillouin, Ref. 9; and Cohen, Ref. 12.)

equation of an approximate one-electron problem. One of the most important theorems basic to band structure is the Bloch theorem which states that if a potential energy $PE(\mathbf{r})$ is periodic with the periodicity of the lattice, then the solutions $\phi_\mathbf{k}(\mathbf{r})$ of the Schrodinger equation [11,13]

$$\left[\left(-\frac{\hbar^2}{2m}\nabla^2 + PE(\mathbf{r})\right)\right]\phi_\mathbf{k}(\mathbf{r}) = E_\mathbf{k}\,\phi_\mathbf{k}(\mathbf{r}) \tag{4}$$

are of the form

$$\phi_\mathbf{k}(\mathbf{r}) = e^{j\mathbf{k}\cdot\mathbf{r}}U_\mathbf{k}(\mathbf{r}) = \text{Bloch function} \tag{5}$$

where $U_k(\mathbf{r})$ is periodic in $\mathbf{r}$ with the periodicity of the direct lattice. From the Bloch theorem one can show that the energy E_k is periodic in the reciprocal lattice, i.e., $E_k = E_{k+g}$ where $\mathbf{g}$ is given by Eq. (3). Thus to label the energy uniquely it is sufficient to use only $\mathbf{k}$'s in a primitive cell of the reciprocal lattice. The standard convention is to use the Wigner-Seitz cell in the reciprocal lattice as shown in Fig. 4. This cell is called the Brillouin zone or the first Brillouin zone.[9] It is thus evident that we can reduce any momentum $\mathbf{k}$ in the reciprocal space to a point in the Brillouin zone where any energy state can be given a label in the reduced zone schemes.

The Brillouin zone for the diamond and the zincblende lattices is the same as that of the *fcc* and is shown in Fig. 4(a). The Brillouin zone for the wurtzite lattice is shown in Fig. 4(b). Also indicated are the most important symmetry points and symmetry lines such as the center of the zone $[\Gamma = 2\pi/a(0, 0, 0)]$, the $\langle 111 \rangle$ axes (Λ) and their intersections with the zone edge $[L = 2\pi/a(\frac{1}{2}, \frac{1}{2}, \frac{1}{2})]$, the $\langle 100 \rangle$ axes (Δ) and their intersections $[X = 2\pi/a(0, 0, 1)]$, and the $\langle 110 \rangle$ axes (Σ) and their intersections $[K = 2\pi/a(\frac{3}{4}, \frac{3}{4}, 0)]$.

The energy bands of solids have been studied theoretically using a variety of computer methods. For semiconductors the two methods most frequently used are the orthogonalized plane wave method [14,15] and the pseudopotential method.[16] Recent results[17] of studies of the energy band structures of Ge, Si, and GaAs are shown in Fig. 5. One notices that for any semiconductor there is a forbidden energy region in which no allowed states can exist. Above and below this energy gap are permitted energy regions or energy bands. The upper bands are called the conduction bands; the lower bands, the valence bands. The separation between the energy of the lowest conduction band and that of the highest valence band is called the band gap, E_g, which is the most important parameter in semiconductor physics. Before we discuss the details of the band structure, we shall first consider the simplified band picture as shown in Fig. 6. In this figure the bottom of the conduction band is designated by E_C, and the top of the valence band by E_V. The electron energy is conventionally defined to be positive when measured upwards, and the hole energy is positive when measured downwards. The band gaps of some important semiconductors are listed [18] in Table 2.2.

The valence band in the zincblende structure consists of four subbands when spin is neglected in the Schrodinger equation, and each band is doubled when spin is taken into account. Three of the four bands are degenerate at $k = 0$ (Γ point) and form the upper edge of the band, and the fourth one forms the bottom. Furthermore the spin-orbit interaction causes a splitting of the band at $k = 0$. As shown in Fig. 5 the two top valence bands can be approximately fitted by two parabolic bands with different curvatures: the heavy-hole band (the wider band with smaller $\partial^2 E/\partial k^2$) and the light-hole band (the narrower band with larger $\partial^2 E/\partial k^2$). The effective mass, which is

TABLE 2.2

PROPERTIES OF IMPORTANT SEMICONDUCTORS

	Semiconductor	Band Gap (eV)		Mobility at 300°K (cm²/volt sec)†		Band Structure‡	Effective Mass§ m^*/m_0		Dielectric Constant
		300°K	0°K	Electrons	Holes		Electrons	Holes	
Element	C (Diamond II)	5.47	5.51	1800	1600	Si	0.2	0.25	5.5
	Ge	0.66	0.75	3900	1900	Ge	$m_l^* = 1.6$ $m_t^* = 0.082$	$m_{lh}^* = 0.04$ $m_{hh}^* = 0.3$	16
	Si	1.12	1.16	1500	600	Si	$m_l^* = 0.97$ $m_t^* = 0.19$	$m_{lh}^* = 0.16$ $m_{hh}^* = 0.5$	11.8
	Grey Tin		~0.08			GaAs			
IV-IV	α-SiC	3	3.1	400	50	Si	0.6	1.0	10
III-V	AlSb	1.63	1.75	200	420	GaAs	0.3	0.4	11
	BN (Zincblende)	~7.5				Si			
	BP	6							
	GaN	3.5							
	GaSb	0.67	0.80	4000	1400	GaAs	0.047	0.5	15
	GaAs	1.43	1.52	8500	400	GaAs	0.068	0.5	10.9
	GaP	2.24	2.40	110	75	Si	0.5	0.5	10
	InSb	0.16	0.26	78000	750	GaAs	0.013	0.6	17
	InAs	0.33	0.46	33000	460	GaAs	0.02	0.41	14.5
	InP	1.29	1.34	4600	150	GaAs	0.07	0.4	14

TABLE 2.2 (Cont.)

II–VI	CdS (Wurtzite)	2.42	2.56	300	50	GaAs	0.17	0.6	10
	CdSe	1.7	1.85	800		GaAs	0.13	0.45	10
	ZnO	3.2		200		GaAs	0.27		9
	ZnS	3.6	3.7	165		GaAs	1.1		8
IV–VI	PbS	0.41	0.34	600	700	Ge	0.66	0.5	17
	PbTe	0.32	0.24	6000	4000	Ge	0.22	0.29	30

† The values are for drift mobilities obtained in the purest and most perfect materials available to date.

‡ Ge, Si, and GaAs designate Ge-like, Si-like, and GaAs-like band structure respectively.

§ m_l^* = longitudinal effective mass; m_t^* = transverse effective mass; m_{lh}^* = light hole effective mass; m_{hh}^* = heavy hole effective mass.

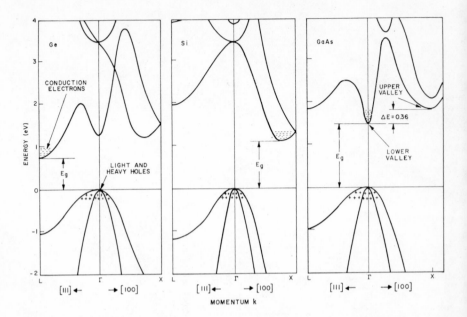

Fig. 5 Energy band structures of Ge, Si, and GaAs where Eg is the energy band gap. $(+)$ signs indicate holes in the valence bands and $(-)$ signs indicate electrons in the conduction bands.
(After Cohen and Bergstresser, Ref. 17.)

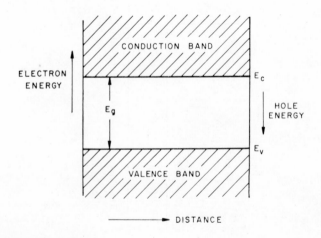

Fig. 6 Simplified band diagram.

defined as

$$m^* \equiv \frac{\hbar^2}{\left(\dfrac{\partial^2 E}{\partial k^2}\right)} , \qquad (6)$$

is listed in Table 2.2 for some important semiconductors.

The conduction band (Fig. 5) consists of a number of subbands. The bottom of the conduction band can appear along the $\langle 111 \rangle$-axes (Λ or L), along the $\langle 100 \rangle$-axes (Δ or X), or at $k = 0$ (Γ). Symmetry considerations alone do not determine the location of the bottom of the conduction band. Experimental results show, however, that in Ge it is along the $\langle 111 \rangle$-axes, in Si the $\langle 100 \rangle$-axes, and in GaAs at $k = 0$. The shapes of the constant energy surfaces [19] are shown in Fig. 7. For Ge there are eight ellipsoids of revolution along the $\langle 111 \rangle$-axes; the Brillouin zone boundaries are at the middle of the ellipsoids. For Si there are six along the $\langle 100 \rangle$-axes with the centers of the ellipsoids located at about three quarters of the distance from the Brillouin zone center. For GaAs the constant energy surface is a sphere at the zone center. By fitting experimental results to parabolic bands, we obtain the electron effective masses; one for GaAs, two for Ge, and two for Si: m_l^* along the symmetry axes and m_t^* transverse to the symmetry axes. These values also are given in Table 2.2.

At room temperature and under normal atmosphere, the values of the band gap are 0.66 eV for Ge, 1.12 eV for Si, and 1.43 eV for GaAs. The above values are for high-purity materials. For highly-doped materials the band gaps become smaller. (The band gaps of degenerate semiconductors are considered in Chapter 4.) Experimental results show that the band gaps of most semiconductors decrease with increasing temperature. The detailed

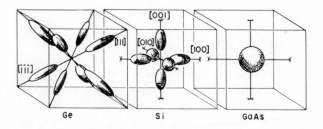

Fig. 7 Shapes of constant energy surfaces in Ge, Si, and GaAs. For Ge there are eight ellipsoids of revolution along $\langle 111 \rangle$ axes, and the Brillouin zone boundaries are at the middle of the ellipsoids. For Si there are six along $\langle 100 \rangle$ axes with the centers of the ellipsoids located at about three quarters of the distance from the Brillouin zone center. For GaAs the constant energy surface is a sphere at the zone center. (After Ziman, Ref. 19.)

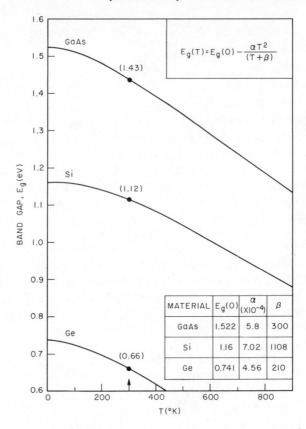

Fig. 8 Energy band gaps of Ge, Si, and GaAs as a function of temperature. (*after* Varshni, Ref. 72; Panish and Casey, Ref. 73).

variations of band gaps as a function of temperature for Ge, Si, and GaAs are shown[4] in Fig. 8. The band gap approaches 0.75, 1.16, and 1.52 eV respectively for the three semiconductors at 0°K. The variation of band gaps with temperature can be expressed by a universal function $E_g(T) = E_g(0) - \alpha T^2/(T + \beta)$, where $E_g(0)$, α and β are given in Fig. 8. The temperature coefficient, dE_g/dT, is negative for the above three semiconductors. There are some semiconductors with positive dE_g/dT, e.g., PbS (Table 2.2), which increases from 0.34 eV at 0°K to 0.41 eV at 300°K. Near room temperature, the band gaps of Ge and GaAs increase with pressure[20] ($dE_g/dP = 5 \times 10^{-6}$ eV/kg/cm² for Ge, and about 9×10^{-6} eV/kg/cm² for GaAs) and that of Si decreases with pressure ($dE_g/dP = -2.4 \times 10^{-6}$ eV/kg/cm²).

4 CARRIER CONCENTRATION AT THERMAL EQUILIBRIUM

The three basic bond pictures of a semiconductor are shown in Fig. 9. Figure 9(a) shows intrinsic silicon which is very pure and contains a negligibly small amount of impurities; each silicon atom shares its four valence electrons with the four neighboring atoms forming four covalent bonds (also see Fig. 1). Figure 9(b) shows schematically an *n*-type silicon where a substitutional phosphorus atom with five valence electrons has replaced a silicon atom, and a negative-charged electron is "donated" to the conduction band. The silicon is *n* type because of the addition of the negative charge carrier, and the phosphorus atom is called a "donor." Similarly as shown in Fig. 9(c), when a boron atom with three valence electrons substitutes for a silicon atom, an additional electron is "accepted" to form four covalent bonds around the boron, and a positive charged "hole" is created in the valence band. This is *p* type, and the boron is an "acceptor."

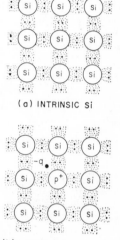

(a) INTRINSIC Si

(b) n-TYPE Si WITH DONOR

(1) Intrinsic Semiconductor

We now consider the intrinsic case. The number of occupied conduction-band levels is given by

$$n = \int_{E_C}^{E_{top}} N(E)F(E)\, dE \tag{7}$$

where E_C is the energy at the bottom of the conduction band and E_{top} is the energy at the top. $N(E)$ is the density of states which for low-enough carrier densities and temperatures can be approximated by the density near the bottom of the conduction band:

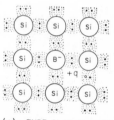

(c) p-TYPE Si WITH ACCEPTOR

$$N(E) = M_c \frac{\sqrt{2}}{\pi^2} \frac{(E - E_c)^{1/2}}{\hbar^3} (m_{de})^{3/2} \tag{8}$$

where M_c is the number of equivalent minima in the conduction band, and m_{de} is the density-of-state effective mass for electrons:[4]

$$m_{de} = (m_1{}^* m_2{}^* m_3{}^*)^{1/3} \tag{9}$$

Fig. 9 Three basic bond pictures of a semiconductor: (a) intrinsic with negligible impurities; (b) n-type with donor (phosphorus); and (c) p-type with acceptor (boron).

where m_1^*, m_2^*, m_3^* are the effective masses along the principal axes of the ellipsoidal energy surface, e.g., in silicon $m_{de} = (m_l^* m_t^{*2})^{1/3}$. $F(E)$ is the Fermi-Dirac distribution function given by

$$F(E) = \frac{1}{1 + \exp\left(\dfrac{E - E_F}{kT}\right)} \tag{10}$$

where k is Boltzmann's constant, T the absolute temperature, and E_F the Fermi energy which can be determined from the charge neutrality condition [see Section 4(3)].

The integral, Eq. (7), can be evaluated to be

$$n = N_C \frac{2}{\sqrt{\pi}} F_{1/2}\left(\frac{E_F - E_C}{kT}\right) \tag{11}$$

where N_C is the effective density of states in the conduction band and is given by

$$N_C \equiv 2\left(\frac{2\pi m_{de} kT}{h^2}\right)^{3/2} M_C, \tag{12}$$

and $F_{1/2}(\eta_f)$ is the Fermi-Dirac integral which is shown[21] in Fig. 10. For the

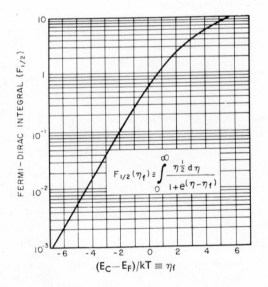

Fig. 10 Fermi-Dirac integral $F_{1/2}$ as a function of Fermi energy. (After Blackmore, Ref. 21.)

case of Boltzmann statistics, i.e., for the Fermi level several kT below E_C in nondegenerate semiconductors, the integral approaches $\sqrt{\pi}\,e^{-\eta_f}/2$ and Eq. (11) becomes

$$n = N_C \exp\left(-\frac{E_C - E_F}{kT}\right). \tag{13}$$

Similarly, we can obtain the hole density near the top of the valence band:

$$p = N_V \frac{2}{\sqrt{\pi}} F_{1/2}\left(\frac{E_V - E_F}{kT}\right) \tag{14}$$

where N_V is the effective density of states in the valence band and is given by

$$N_V \equiv 2\left(\frac{2\pi m_{dh} kT}{h^2}\right)^{3/2} \tag{15}$$

where m_{dh} is the density-of-state effective mass of the valence band:[4]

$$m_{dh} = (m_{lh}^{*3/2} + m_{hh}^{*3/2})^{2/3} \tag{16}$$

where the subscripts refer to "light" and "heavy" hole masses discussed before, Eq. (6). Again under nondegenerate conditions

$$p = N_V \exp\left(-\frac{E_F - E_V}{kT}\right). \tag{17}$$

For intrinsic semiconductors at finite temperatures there is continuous thermal agitation that results in excitation of electrons from the valence band to the conduction band and leaves an equal number of holes in the valence band, i.e., $n = p = n_i$. This process is balanced by recombination of the electrons in the conduction band with holes in the valence band.

The Fermi level for an intrinsic semiconductor (which by definition is nondegenerate) is obtained by equating Eqs. (13) and (17):

$$E_F = E_i = \frac{E_C + E_V}{2} + \frac{kT}{2} \ln \frac{N_V}{N_C} = \frac{E_C + E_V}{2} + \frac{3kT}{4} \ln\left(\frac{m_{dh}}{m_{de}}\right). \tag{18}$$

Hence the Fermi level (E_i) of an intrinsic semiconductor generally lies very close to the middle of the band gap.

The intrinsic carrier concentration is obtained from Eqs. (13), (17), and (18):

$$np = n_i^2 = N_C N_V \exp(-E_g/kT), \tag{19}$$

or

$$n_i = \sqrt{N_C N_V}\, e^{-E_g/2kT}$$

$$= 4.9 \times 10^{15} \left(\frac{m_{de} m_{dh}}{m_0^2}\right)^{3/4} T^{3/2} e^{-E_g/2kT} \tag{19a}$$

where $E_g \equiv (E_C - E_V)$, and m_0 is the free-electron mass. The temperature dependence[22,23] of n_i is shown in Fig. 11 for Ge, Si, and GaAs. As expected, the larger the band gap the smaller the intrinsic carrier concentration.

(2) Donors and Acceptors

When a semiconductor is doped with donor or acceptor impurities, impurity energy levels are introduced. A donor level is defined as being

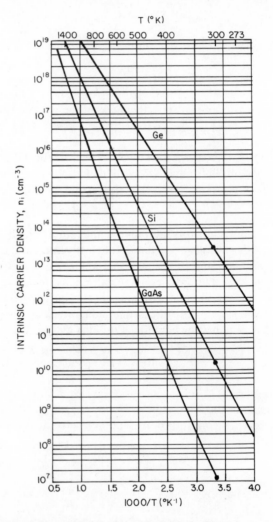

Fig. 11 Intrinsic carrier densities of Ge, Si, and GaAs as a function of reciprocal temperature. (After Morin and Maita, Ref. 22; Hall and Racette, Ref. 23; and Neuberger, Ref. 18.)

neutral if filled by an electron, and positive if empty. An acceptor level is neutral if empty, and negative if filled by an electron.

The simplest calculation of impurity energy levels is based on the hydrogen-atom model. The ionization energy for the hydrogen atom is

$$E_H = \frac{m_0 q^4}{32\pi^2 \varepsilon_0^2 \hbar^2} = 13.6 \, \text{eV} \tag{20}$$

where ε_0 is the free-space permittivity. The ionization energy for the donor E_d, can be obtained by replacing m_0 by the conductivity effective mass[4] of electrons

$$m_{ce} = 3\left(\frac{1}{m_1^*} + \frac{1}{m_2^*} + \frac{1}{m_3^*}\right)^{-1}$$

and ε_0 by the permittivity of the semiconductor, ε_s, in Eq. (20):

$$E_d = \left(\frac{\varepsilon_0}{\varepsilon_s}\right)^2 \left(\frac{m_{ce}}{m_0}\right) E_H. \tag{21}$$

The ionization energy for donors as calculated from Eq. (21) is 0.006 eV for Ge, 0.025 eV for Si, and 0.007 eV for GaAs. The hydrogen-atom calculation for the ionization level for the acceptors is similar to that for the donors. We consider the unfilled valence band as a filled one plus an imaginary hole in the central force field of a negatively charged acceptor. The calculated acceptor ionization energy (measured from the valence band edge) is 0.015 eV for Ge, 0.05 eV for Si, and about 0.05 eV for GaAs.

The above simple hydrogen-atom model certainly cannot account for the details of ionization energy, particularly the deep levels in semiconductors.[24–26] However, the calculated values do predict the correct order of magnitude of the true ionization energies for shallow impurities. Figure 12 shows [27,28] the measured ionization energies for various impurities in Ge, Si, and GaAs. We note that it is possible for a single atom to have many levels, e.g., gold in Ge has three acceptor levels and one donor level in the forbidden energy gap.[29]

Of all the methods of introducing impurities into a host semiconductor, solid-state diffusion[30] is considered, at the present time, to be the most controllable. (The ion-implantation method[31] is still in its infancy.) The diffusion coefficient $D(T)$ in a limited temperature range can be described by

$$D(T) = D_0 \exp(-\Delta E/kT) \tag{22}$$

where D_0 is the diffusion cooefficient extrapolated to infinite temperature, and ΔE is the activation energy of diffusion. Values of $D(T)$ for Ge, Si, and GaAs are plotted in Fig. 13 for various impurities.[32,33] The $D(T)$ for GaAs is for the low-impurity concentration case; for larger concentrations $D(T)$

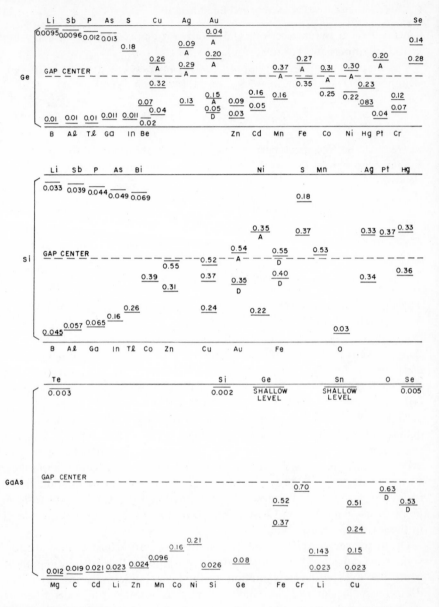

Fig. 12 Measured ionization energies for various impurities in Ge, Si, and GaAs. The levels below the gap centers are measured from the top of the valence band and are acceptor levels unless indicated by D for donor level. The levels above the gap centers are measured from the bottom of the conduction band level and are donor levels unless indicated by A for acceptor level. The band gaps at 300°K are 0.66, 1.12, and 1.43 eV for Ge, Si, and GaAs respectively.

(After Conwell, Ref. 27; and Sze and Irvin, Ref. 28.)

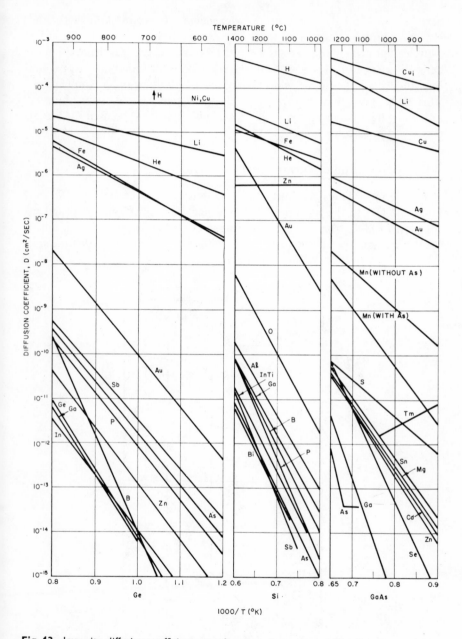

Fig. 13 Impurity diffusion coefficients as a function of temperature for Ge, Si, and GaAs. (After Burger and Donovan, Ref. 32; and Kendall, Ref. 33.)

becomes increasingly concentration-dependent. In connection with the impurity diffusion coefficient there is the solid solubility of the impurity which is the maximum concentration of an impurity that can be accommodated in a solid at any given temperature. The solid solubilities[34] of some important impurities in Si are plotted in Fig. 14 as a function of temperature. This figure shows that arsenic or phosphorus should be used as the impurity in making highly doped n-type silicon while boron should be used for highly doped p-type silicon.

(3) Calculation of Fermi Level

The Fermi level for the intrinsic semiconductor is given by Eq. (18) and lies very close to the middle of the band gap. This situation is depicted in Fig. 15(a) where, from left to right, are shown schematically the simplified band diagram, the density of states $N(E)$, the Fermi-Dirac distribution

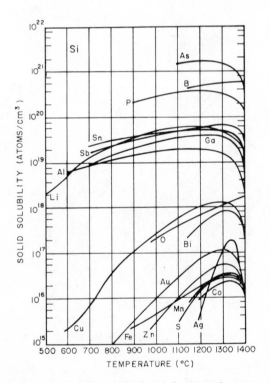

Fig. 14 Solid solubility of various elements in Si as a function of temperature. (After Trumbore, Ref. 34.)

function $F(E)$, and the carrier concentrations. The shaded areas in the conduction band and the valence band are the same indicating the $n = p = n_i$ for the intrinsic case.

When impurity atoms are introduced as shown in Fig. 15(b) and (c), the Fermi level must adjust itself in order to preserve charge neutrality. Consider

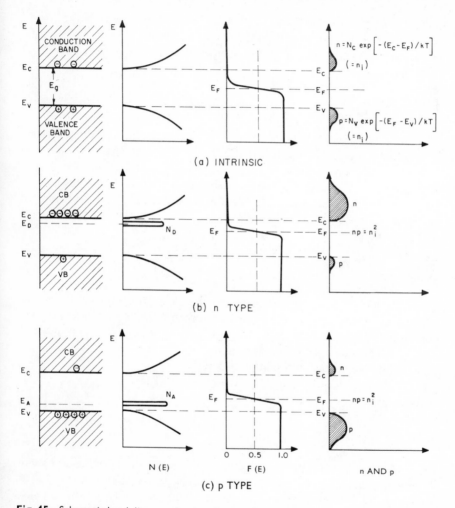

Fig. 15 Schematic band diagram, density of states, Fermi-Dirac distribution, and the carrier concentrations for
(a) intrinsic,
(b) n-type, and
(c) p-type
semiconductors at thermal equilibrium. Note that $pn = n_i^2$ for all three cases.

the case as shown in Fig. 15(b) where donor impurities with the concentration $N_D(\text{cm}^{-3})$ are added to the crystal. To preserve electrical neutrality we must have the total negative charges (electrons and ionized acceptors) equal to the total positive charges (holes and ionized donors), or for the present case

$$n = N_D^+ + p \tag{23}$$

where n is the electron density in the conduction band, p is the hole density in the valence band, and N_D^+ is the number of ionized donors given by[35]

$$N_D^+ = N_D\left[1 - \cfrac{1}{1 + \cfrac{1}{g}\exp\left(\cfrac{E_D - E_F}{kT}\right)}\right] \tag{24}$$

where g is the ground state degeneracy of the donor impurity level which is equal to 2 because of the fact that a donor level can accept one electron with either spin or can have no electron. When acceptor impurities of concentration N_A are added to a semiconductor crystal, a similar expression can be written for the charge neutrality condition, and the expression for ionized acceptors is

$$N_A^- = \cfrac{N_A}{1 + \cfrac{1}{g}\exp\left(\cfrac{E_A - E_F}{kT}\right)} \tag{25}$$

where the ground state degeneracy factor g is 4 for acceptor levels. This is because in Ge, Si, and GaAs each acceptor impurity level can accept one hole of either spin and the impurity level is doubly degenerate as a result of the two degenerate valence bands at $\mathbf{k} = 0$.

Rewriting the neutrality condition of Eq. (23), we obtain

$$N_C \exp\left(-\frac{E_C - E_F}{kT}\right) = N_D \cfrac{1}{1 + 2\exp\left(\cfrac{E_F - E_D}{kT}\right)} + N_V \exp\left(\frac{E_V - E_F}{kT}\right). \tag{26}$$

For a set of given N_C, N_D, N_V, E_C, E_D, E_V, and T, the Fermi level E_F can be uniquely determined from Eq. (26). An elegant graphical method[70] to determine E_F is illustrated in Fig. 16. For this particular solution (with $N_D = 10^{16}$ cm^{-3}, $T = 300°$K) the Fermi level is close to the conduction band edge and adjusts itself so that almost all the donors are ionized. For another temperature, one can first evaluate the values of N_C and N_V which are proportional to $T^{3/2}$, then obtain from Fig. 11 the value of $n_i(T)$ which determines the intercept of the lines $n(E_F)$ and $p(E_F)$; a new Fermi level is thus

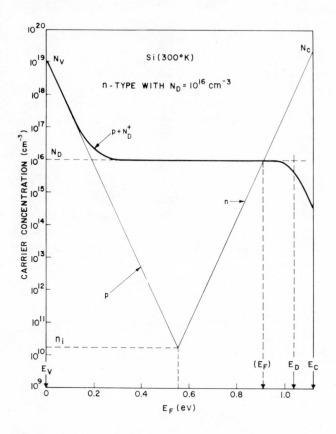

Fig. 16 Graphical method to determine the Fermi energy level E_F. (After Shockley, Ref. 70.)

obtained. As the temperature is lowered sufficiently, the Fermi level rises toward the donor level (for *n*-type semiconductors) and the donor level is partially filled with electrons. The approximate expression for the electron density is then[4]

$$n \simeq \left(\frac{N_D - N_A}{2N_A}\right) N_C \exp(-E_d/kT) \qquad (27)$$

for a partially compensated semiconductor and for

$$N_A \gg \tfrac{1}{2} N_C \exp(-E_d/kT)$$

where $E_d \equiv (E_C - E_D)$, or

$$n \simeq \frac{1}{\sqrt{2}} \tfrac{1}{2}(N_D N_C)^{1/2} \exp(-E_d/2kT) \tag{28}$$

for $N_D \gg \tfrac{1}{2}N_C \exp(-E_d/kT) \gg N_A$. A typical example is shown in Fig. 17 where n is plotted as a function of the reciprocal temperature. At high temperatures we have the intrinsic range since $n \approx p \gg N_D$. At very low temperatures most impurities are frozen out and the slope is given by Eq. (27) or Eq. (28), depending on the compensation conditions. The electron density, however, remains essentially constant over a wide range of temperatures ($\sim 100°$K to 500°K in Fig. 17).

The Fermi level for silicon as a function of temperature and impurity concentration[36] is shown in Fig. 18. The dependence of the band gap on temperature (cf. Fig. 8) is also incorporated in this figure.

When impurity atoms are added, the np product is still given by Eq. (19) which is called the mass-action law and the product is independent of the added impurities. At relatively elevated temperatures, most of the donors and acceptors are ionized, so the neutrality condition can be approximated by

$$n + N_A = p + N_D. \tag{29}$$

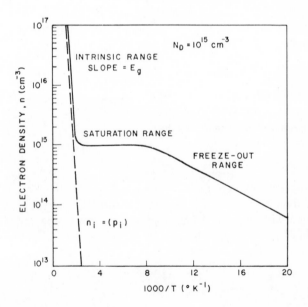

Fig. 17 Electron density as a function of temperature for a Si sample with donor impurity concentration of 10^{15}cm^{-3}.
(After Smith, Ref. 4.)

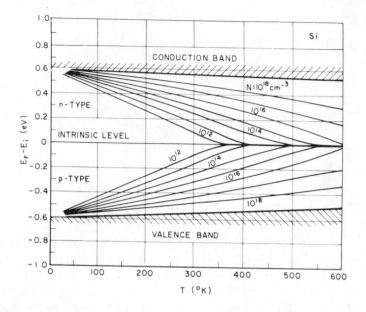

Fig. 18 Fermi level for Si as a function of temperature and impurity concentration. The dependence of the band gap on temperature is also incorporated in the figure. (After Grove, Ref. 36.)

We can combine Eqs. (19) and (29) to give the concentration of electrons and holes in an n-type semiconductor:

$$n_{no} = \tfrac{1}{2}[(N_D - N_A) + \sqrt{(N_D - N_A)^2 + 4n_i^2}]$$

$$\approx N_D \quad \text{if} \quad |N_D - N_A| \gg n_i \quad \text{and} \quad N_D \gg N_A, \tag{30}$$

$$p_{no} = n_i^2/n_{no} \simeq n_i^2/N_D, \tag{31}$$

and

$$E_C - E_F = kT \ln \frac{N_C}{N_D}, \tag{32}$$

or from Eq. (18),

$$E_F - E_i = kT \ln \frac{n_{no}}{n_i}. \tag{33}$$

The concentration of holes and electrons on a p-type semiconductor is given by

$$p_{po} = \tfrac{1}{2}[(N_A - N_D) + \sqrt{(N_A - N_D)^2 + 4n_i^2}]$$
$$\approx N_A \quad \text{if} \quad |N_A - N_D| \gg n_i \quad \text{and} \quad N_A \gg N_D, \tag{34}$$

$$n_{po} = n_i^2/p_{po} \simeq n_i^2/N_A, \tag{35}$$

and

$$E_F - E_V = kT \ln \frac{N_V}{N_A}, \tag{36}$$

or

$$E_i - E_F = kT \ln \frac{p_{po}}{n_i}. \tag{37}$$

In the above formulas the subscripts n and p refer to the type of semiconductors, and the subscripts "o" refer to the thermal equilibrium condition. For n-type semiconductors the electron is referred to as the majority carrier and the hole as the minority carrier, since the electron concentration is the larger of the two. The roles are reversed for p-type semiconductors.

5 CARRIER TRANSPORT PHENOMENA

(1) Mobility

At low electric field the drift velocity (v_d) is proportional to the electric field strength ($\mathscr{E}$), and the proportionality constant is defined as the mobility (μ in cm^2/volt-sec), or

$$v_d = \mu \mathscr{E}. \tag{38}$$

For nonpolar semiconductors such as Ge and Si there are two scattering mechanisms which significantly affect the mobility, namely scattering due to acoustic phonons and to ionized impurities. The mobility due to acoustic phonon, μ_l, is given by[37]

$$\mu_l = \frac{\sqrt{8\pi}q\hbar^4 C_{11}}{3E_{ds}m^{*5/2}(kT)^{3/2}} \sim (m^*)^{-5/2}T^{-3/2} \tag{39}$$

where C_{11} is the average longitudinal elastic constant of the semiconductor, E_{ds} the displacement of the edge of the band per unit dilation of the lattice, and m^* the conductivity effect mass. From the above equation it is expected that the mobility will decrease with the temperature and with the effective mass.

The mobility due to ionized impurities, μ_i, is given by[38]

$$\mu_i = \frac{64\sqrt{\pi}\varepsilon_s^2(2kT)^{3/2}}{N_I q^3 m^{*1/2}} \left\{ \ln\left[1 + \left(\frac{12\pi\varepsilon_s kT}{q^2 N_I^{1/3}} \right)^2 \right] \right\}^{-1}$$

$$\sim (m^*)^{-1/2} N_I^{-1} T^{3/2} \tag{40}$$

where N_I is the ionized impurity density, and ε_s the permittivity. The mobility is expected to decrease with the effective mass but to increase with the temperature. The combined mobility which consists of the above two mechanisms is

$$\mu = \left(\frac{1}{\mu_l} + \frac{1}{\mu_i} \right)^{-1}. \tag{41}$$

For polar semiconductors such as GaAs there is another important scattering mechanism: optical-phonon scattering. The combined mobility can be approximated by[39]

$$\mu \sim (m^*)^{-3/2} T^{1/2}. \tag{42}$$

In addition to the above scattering mechanisms there are other mechanisms which also contribute to the actual mobility: for example (1) the intravalley scattering in which an electron is scattered within an energy ellipsoidal (Fig. 7) and only long-wavelength phonons are involved; and (2) the intervalley scattering in which an electron is scattered from the vicinity of one minimum to another minimum and an energetic phonon is involved.

The measured mobilities of Ge, Si, and GaAs versus impurity concentrations at room temperature are shown[28,25,40] in Fig. 19. We note that, in general, as the impurity concentration increases (at room temperature most impurities are ionized) the mobility decreases as predicted by Eq. (40). Also as m^* increases, μ decreases, thus for a given impurity concentration the electron mobilities for these semiconductors are larger than the hole mobilities (the effect masses are listed in Table 2.2).

The temperature effect on mobility is shown[41] in Fig. 20 for n-type and p-type silicon samples. For lower impurity concentrations the mobility decreases with temperature as predicted by Eq. (39). The measured slopes, however, are different from $(-3/2)$ owing to other scattering mechanisms. It is found that for pure materials near room temperature the mobility varies as $T^{-1.66}$ and $T^{-2.33}$ for n- and p-type Ge respectively; as $T^{-2.5}$ and $T^{-2.7}$ for n- and p-type Si respectively; and as $T^{-1.0}$ and $T^{-2.1}$ for n- and p-type GaAs respectively.

Another important parameter associated with mobility is the carrier diffusion coefficient (D_n for electrons and D_p for holes). In thermal equilibrium the relationship between D_n and μ_n (or D_p and μ_p) is given by[4]

$$D_n = 2\left(\frac{kT}{q}\mu_n\right)F_{1/2}\left(\frac{E_C - E_F}{kT}\right)\bigg/ F_{-1/2}\left(\frac{E_C - E_F}{kT}\right) \qquad (43)$$

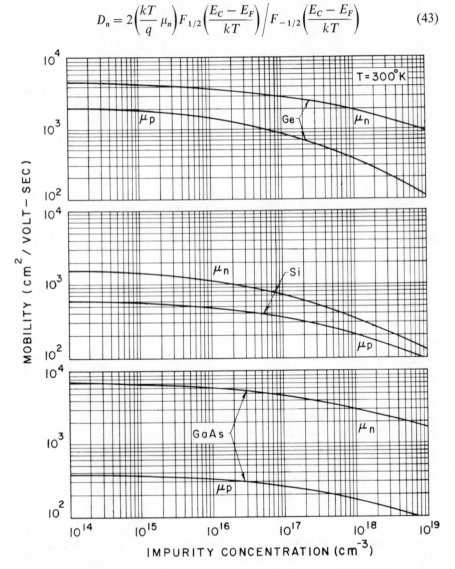

Fig. 19 Drift mobility of Ge and Si and Hall mobility of GaAs at 300°K vs. impurity concentration.

(After Sze and Irvin, Ref. 28; Prince, Ref. 40; and Wolfstirn, Ref. 35.)

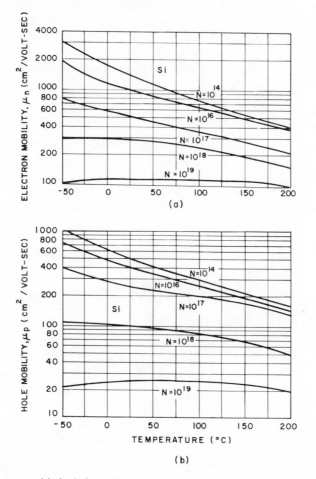

Fig. 20 Electron and hole drift mobilities in Si as a function of temperature and impurity concentration.
(After Gartner, Ref. 41.)

where $F_{1/2}$ and $F_{-1/2}$ are Fermi-Dirac integrals. For nondegenerate semiconductors, Eq. (43) reduces to

$$D_n = \left(\frac{kT}{q}\right)\mu_n,$$ (44a)

and similarly

$$D_p = \left(\frac{kT}{q}\right)\mu_p.$$ (44b)

The above two equations are known as the Einstein relationship. At room temperature $kT/q = 0.0259$ volts, and values of D's are readily obtainable from the mobility results shown in Fig. 19. The mobilities discussed above are the conductivity mobilities, which have been shown to be equal to the drift mobilities.[27] They are, however, different from the Hall mobilities to be considered in the next section.

(2) Resistivity and Hall Effect

The resistivity (ρ) is defined as the proportionality constant between the electric field and the current density (J):

$$\mathscr{E} = \rho J. \tag{45}$$

Its reciprocal value is the conductivity, i.e., $\sigma = 1/\rho$, and

$$J = \sigma \mathscr{E}. \tag{46}$$

For semiconductors with both electrons and holes as carriers, we obtain

$$\rho = \frac{1}{\sigma} = \frac{1}{q(\mu_n n + \mu_p p)}. \tag{47}$$

If $n \gg p$ as in n-type semiconductors,

$$\rho \simeq \frac{1}{q\mu_n n} \tag{48}$$

or

$$\sigma \simeq q\mu_n n. \tag{48a}$$

The most common method for measuring resistivity is the four-point probe method as shown[42,43] in Fig. 21. A small current from a constant

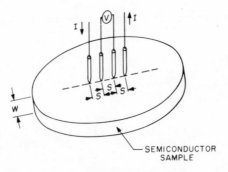

SEMICONDUCTOR
SAMPLE

Fig. 21 Schematic setup of four-point probe method to measure sample resistivity. (After Valdes, Ref. 42.)

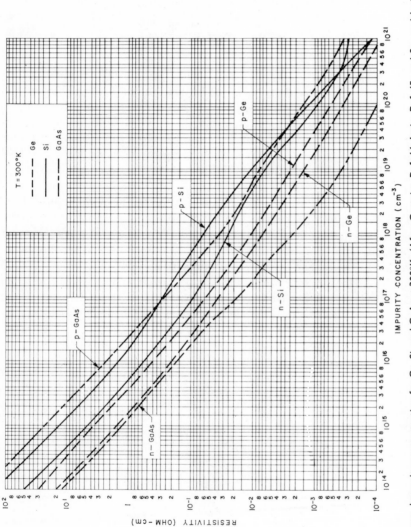

Fig. 22 Resistivity vs. impurity concentration for Ge, Si, and GaAs at 300°K. (After Cuttriss, Ref. 44; Irvin, Ref. 45; and Sze and Irvin, Ref. 28.)

current source is passed through the outer two probes and the voltage is measured between the inner two probes. For a thin slice with its thickness W much smaller than its diameter, and the probe spacing $S \gg W$, the resistivity is given by

$$\rho = \left(\frac{\pi}{\ln 2}\right) \frac{V}{IW} \approx \frac{4.54}{W} \left(\frac{V}{I}\right). \tag{49}$$

The measured resistivity (at 300°K) as a function of the impurity concentration[28,44,45] is shown in Fig. 22 for Ge, Si, and GaAs. Thus we can obtain the impurity concentration of a semiconductor if its resistivity is known. The impurity concentration may be different from the carrier concentration. For example, in a p-type silicon with 10^{17} cm^{-3} gallium acceptor impurities, unionized acceptors at room temperature make up about 23% [from Eq. (25), Figs. 12 and 18]; in other words, the carrier concentration is only 7.7×10^{16} cm^{-3}.

To measure the carrier concentration directly, the most common method uses the Hall effect.[46] The basic setup[47] is shown in Fig. 23 where an electric field is applied along the x-axis and a magnetic field, along the z-axis. Consider a p-type sample. The Lorentz force ($qv_x \times \mathscr{B}_z$) exerts an average downward force on the holes, and the downward-directed current will cause a piling up of holes at the bottom side of the sample, which in turn gives rise to an electric field $\mathscr{E}_y$. Since there is no net current along the y direction in the steady state, the electric field along the y-axis (Hall field) should exactly balance the Lorentz force.

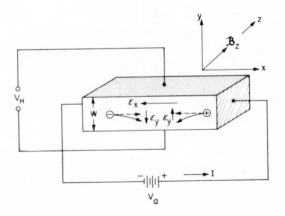

Fig. 23   Basic setup to measure carrier concentration using the Hall effect.

This Hall field can be measured externally and is given by

$$\mathscr{E}_y = (V_y/W) = R_H J_x \mathscr{B}_z \tag{50}$$

where R_H is the Hall coefficient and is given by[4]

$$R_H = r \frac{1}{q} \frac{p - b^2 n}{(p + bn)^2}; \qquad b \equiv \mu_n/\mu_p, \tag{51}$$

$$r \equiv \langle \tau^2 \rangle / \langle \tau \rangle^2. \tag{52}$$

The parameter τ is the mean free time between carrier collisions, which depends on the carrier energy, e.g., for semiconductors with spherical constant-energy surfaces, $\tau \sim E^{-1/2}$ for phonon scattering, $\tau \sim E^{3/2}$ for ionized impurity scattering, and, in general, $\tau = aE^{-s}$ where a and s are constants. From Boltzmann's distribution for nondegenerate semiconductors the average value of the mth power of τ is:

$$\langle \tau^m \rangle = \int_0^\infty \tau^m E^{3/2} \exp(-E/kT)\, dE \Big/ \int_0^\infty E^{3/2} \exp(-E/kT)\, dE \tag{53}$$

so that using the general form of τ

$$\langle \tau^2 \rangle = a^2 (kT)^{-2s} \Gamma(\tfrac{5}{2} - 2s)/\Gamma(\tfrac{5}{2}) \tag{54a}$$

and

$$\langle \tau \rangle^2 = [a(kT)^{-s} \Gamma(\tfrac{5}{2} - s)/\Gamma(\tfrac{5}{2})]^2 \tag{54b}$$

where $\Gamma(n)$ is the gamma function defined as

$$\Gamma(n) \equiv \int_0^\infty x^{n-1} e^{-x}\, dx, \qquad (\Gamma(\tfrac{1}{2}) = \sqrt{\pi}).$$

We obtain from the above expression $r = 3\pi/8 = 1.18$ for phonon scattering and $315\pi/512 = 1.93$ for ionized impurity scattering.

The Hall mobility μ_H is defined as the product of Hall coefficient and conductivity:

$$\mu_H = |R_H \sigma|. \tag{55}$$

The Hall mobility should be distinguished from the drift mobility μ_n (or μ_p) as given in Eq. (48a) which does not contain the factor r. From Eq. (51), if $n \gg p$

$$R_H = r\left(\frac{-1}{qn}\right), \tag{56a}$$

and if $p \gg n$

$$R_H = r\left(\frac{+1}{qp}\right). \tag{56b}$$

Thus the carrier concentration and carrier type (electron or hole) can be obtained directly from the Hall measurement provided that one type of carrier dominates.

In the above consideration the applied magnetic field is assumed to be small enough that there is no change in the resistivity of the sample. However, under strong magnetic fields, significant increase of the resistivity is observed: the so-called magnetoresistance effect. For spherical-energy surfaces the ratio of the incremental resistivity to the bulk resistivity at zero magnetic field is given by[4]

$$\frac{\Delta \rho}{\rho_0} = \left\{ \left[\frac{\Gamma^2(\frac{5}{2})\Gamma(\frac{5}{2} - 3s)}{\Gamma^3(\frac{5}{2} - s)} \right] \left[\frac{\mu_n{}^3 n + \mu_p{}^3 p}{\mu_n n + \mu_p p} \right] \right.$$
$$\left. - \left[\frac{\Gamma(\frac{5}{2})\Gamma(\frac{5}{2} - 2s)}{\Gamma^2(\frac{5}{2} - s)} \right]^2 \left[\frac{\mu_n{}^2 n - \mu_p{}^2 p}{\mu_n n + \mu_p p} \right]^2 \right\} \mathcal{B}_z{}^2. \qquad (57)$$

It is proportional to the square of the magnetic field component perpendicular to the direction of the current flow. For $n \gg p$, $(\Delta \rho / \rho_0) \sim \mu_n{}^2 \mathcal{B}_z{}^2$. A similar result can be obtained for the case $p \gg n$.

(3) Recombination Processes

Whenever the thermal-equilibrium condition of a physical system is distributed, i.e., $pn \neq n_i{}^2$, there are processes by means of which the system can be restored to equilibrium, i.e., $pn = n_i{}^2$. The basic recombination processes are shown in Fig. 24. Figure 24(a) illustrates the band-to-band recombination where an electron-hole pair recombines. This transition of the electron from the conduction band to the valence band is possible by the emission of a photon (radiative process) or by transfer of the energy to another free electron or hole (Auger process). The latter process is the inverse process to impact ionization, and the former is the inverse process to direct optical transitions, which are important for most III-V compounds with direct energy gaps.

Figure 24(b) shows single-level recombination in which only one trapping energy level is present in the band gap, and Fig. 24(c) multiple-level recombination in which more than one trapping level is present in the band gap.

The single-level recombination consists of four steps: namely, electron capture, electron emission, hole capture, and hole emission. The recombination rate, U (in unit of cm^{-3} sec^{-1}), is given by[48]

$$U = \frac{\sigma_p \sigma_n v_{th}(pn - n_i{}^2)N_t}{\sigma_n \left[n + n_i \exp\left(\frac{E_t - E_i}{kT} \right) \right] + \sigma_p \left[p + n_i \exp\left(-\frac{E_t - E_i}{kT} \right) \right]} \qquad (58)$$

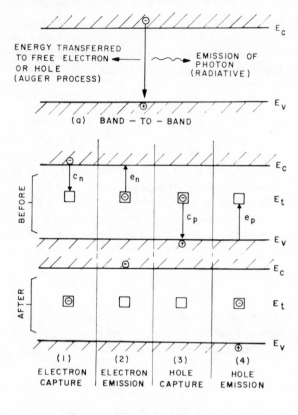

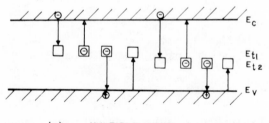

Fig. 24 Recombination processes
(a) band-to-band recombination (radiative or Auger process)
(b) single-level recombination
(c) multiple-level recombination.
(After Sah, Noyce, and Shockley, Ref. 48.)

where σ_p and σ_n are the hole and electron capture cross section respectively, v_{th} the carrier thermal velocity equal to $\sqrt{3kT/m^*}$, N_t the trap density, E_t the trap energy level, E_i the intrinsic Fermi level and n_i the intrinsic carrier density. It is obvious that under a thermal equilibrium condition, $pn = n_i^2$, $U = 0$. Furthermore, under the simplified condition that $\sigma_n = \sigma_p = \sigma$, Eq. (58) reduces to

$$U = \sigma v_{th} N_t \; \frac{(pn - n_i^2)}{n + p + 2n_i \cosh\left(\dfrac{E_t - E_i}{kT}\right)}. \tag{59}$$

The recombination rate approaches a maximum as the energy level of the recombination center approaches mid gap, i.e., $E_t \approx E_i$. Thus the most effective recombination centers are those located near the middle of the band gap.

Under low injection conditions, i.e., when the injected carriers ($\Delta n = \Delta p$) are much fewer in number than the majority carriers, the recombination process may be characterized by the expression

$$U = \frac{p_n - p_{no}}{\tau_p} \tag{60}$$

where p_{no} is the equilibrium minority carrier concentration, $p_n = \Delta p + p_{no}$, and τ_p is the minority carrier lifetime. In an n-type semiconductor, where $n \approx n_{no} =$ the equilibrium majority carrier concentration, and $n \gg n_i$ and p, Eq. (58) becomes

$$U = \sigma_p v_{th} N_t (p_n - p_{no}). \tag{61}$$

Comparison of Eqs. (60) and (61) yields the minority carrier lifetime (hole lifetime) in an n-type semiconductor, and

$$\tau_p = \frac{1}{\sigma_p v_{th} N_t}. \tag{62}$$

Similarly for a p-type semiconductor, the electron lifetime

$$\tau_n = \frac{1}{\sigma_n v_{th} N_t}. \tag{63}$$

For multiple-level traps it has been shown that the recombination processes have gross qualitative features that are similar to those of the single-level case. The details of behavior are, however, different, particularly in the high-injection-level condition (i.e., where $\Delta n = \Delta p \approx$ majority carrier concentration) where the asymptotic lifetime is an average of the lifetimes associated with all the positively charged, negatively charged, and neutral trapping levels.

The expressions, Eqs. (62) and (63), have been verified experimentally by the use of solid-state diffusion and high-energy radiation. It can be seen in Fig. 12 that many impurities have energy levels close to the middle of the band gap. These impurities may serve as efficient recombination centers. A typical example is gold in silicon.[29] It is found that the minority carrier lifetime decreases linearly with the gold concentration over the range of 10^{14} to 10^{17} cm^{-3} where τ decreases from about 2×10^{-7} second to 2×10^{-10} second. This effect is important in some switching device applications when a short lifetime is a desirable feature. Another method of changing the minority carrier lifetime is by high-energy particle irradiation which causes displacement of host atoms and damage to lattices. These, in turn, introduce energy levels in the band gap. For example,[18] in silicon electron irradiation gives rise to an acceptor level at 0.4 eV above the valence band and a donor level at 0.36 eV below the conduction band; neutron irradiation creates an acceptor level at 0.56 eV; deuteron irradiation gives rise to an interstitial state with an energy level 0.25 eV above the valence band. Similar results are obtained for Ge, GaAs, and other semiconductors. Unlike the solid-state diffusion, the radiation-induced trapping centers may be annealed out at relatively low temperatures.

The minority carrier lifetime has generally been measured using the photoconduction effect[49] (*PC*) or the photoelectromagnetic effect[50] (*PEM*). The basic equation for the *PC* effect is given by

$$J_{PC} = q(\mu_n + \mu_p)\Delta n \mathscr{E} \tag{64}$$

where J_{PC} is the incremental current density as a result of illumination and $\mathscr{E}$ is the applied electric field along the sample. The quantity Δn is the incremental carrier density or the number of electron-hole pairs per volume created by the illumination which equals the product of the generation rate of electron-hole pairs resulting from photon (G) and the lifetime (τ) or $\Delta n = \tau G$. The lifetime is thus given by

$$\tau = \frac{\Delta n}{G} = \frac{J_{PC}}{G \mathscr{E} q(\mu_n + \mu_p)} \sim J_{PC}. \tag{65}$$

A measurement setup will be discussed in Section 7. For the *PEM* effect we measure the short-circuit current, which appears when a constant magnetic field $\mathscr{B}_z$ is applied perpendicular to the direction of incoming radiation. The current density is given by

$$J_{PEM} = q(\mu_n + \mu_p)\mathscr{B}_z \frac{D}{L}(\tau G) \tag{66}$$

where $L \equiv \sqrt{D\tau}$ is the diffusion length. The lifetime is given by

$$\tau = \left[\frac{J_{PEM}}{\mathscr{B}_z \sqrt{DGq(\mu_n + \mu_p)}} \right]^2 \sim (J_{PEM}/\mathscr{B}_z)^2. \tag{67}$$

6 PHONON SPECTRA AND OPTICAL, THERMAL, AND HIGH-FIELD PROPERTIES OF SEMICONDUCTORS

In the previous section we have considered the effect of low to moderately high electric fields on the transport of carriers in semiconductors. In this section we shall briefly consider some other effects and properties of semiconductors which are important to the operation of semiconductor devices.

(1) Phonon Spectra

It is well known that for a one-dimensional lattice with only nearest-neighbor coupling and two different masses, m_1 and m_2 placed alternately, the frequencies of oscillation are given by[3]

$$v_{\pm} = \left[\alpha_f \left(\frac{1}{m_1} + \frac{1}{m_2} \right) \pm \alpha_f \sqrt{\left(\frac{1}{m_1} + \frac{1}{m_2} \right)^2 - 4 \sin^2(qa)/m_1 m_2} \right]^{1/2} \tag{68}$$

where α_f is the force constant, q the wave number, and a the lattice spacing. The frequency v_- tends to be proportional to q near $q = 0$. This is the acoustic branch, since it is the analogue of a long-wavelength vibration of the lattice, and since the frequency corresponds to that of sound in such a medium. The frequency v_+ tends to

$$\left[2\alpha_f \left(\frac{1}{m_1} + \frac{1}{m_2} \right) \right]^{1/2}$$

as q approaches zero. This branch is separated considerably from the acoustic mode. This is the optical branch, since the frequency v_+ is generally in the optical range. For the acoustic mode the two sublattices of the atoms with different masses move in the same direction, while for the optical mode, they move in opposite directions.

For a three-dimensional lattice with one atom per unit cell such as a simple cubic, body-centered, or face-centered cubic lattice, there are only three acoustic modes. For a three-dimensional lattice with two atoms per unit cell such as Ge, Si, and GaAs, there are three acoustic modes and three optical modes. Those modes with the displacement vectors of each atom along the direction of the wave vector are called the longitudinally polarized modes; thus we have one longitudinal acoustic mode (LA) and one longitudinal optical mode (LO). Those modes with their atoms moving in the planes normal to the wave vector are called the transversely polarized modes. We have two transverse acoustic modes (TA) and two transverse optical modes (TO).

The measured results[51-53] for Ge, Si, and GaAs are shown in Fig. 25. We note that at small q's, with LA and TA modes, the energies are proportional to q. The Raman phonon energy is that of the longitudinal optical phonon at $q = 0$. Their values are 0.037 eV for Ge, 0.063 eV for Si, and 0.035 eV for GaAs. These results are listed in Table 2.3 along with other important properties of Ge, Si, and GaAs (in alphabetical order).

(2) Optical Property

Optical measurement constitutes the most important means of determining the band structures of semiconductors. Photon-induced electronic

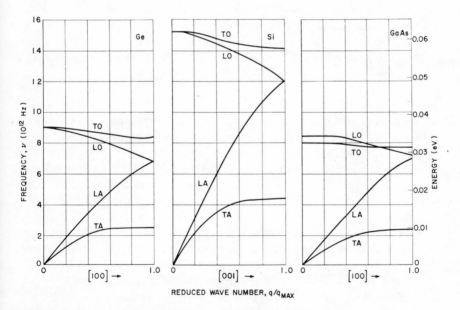

Fig. 25 Measured phonon spectra in Ge, Si, and GaAs. TO for transverse optical modes, LO for longitudinal optical mode, TA for transverse acoustic modes, LA for longitudinal acoustic mode.
(After Brockhouse and Iyengar, Ref. 51; Brockhouse, Ref. 52; and Waugh and Dolling, Ref. 53.)

transitions can occur between different bands which lead to the determination of the energy band gap; or within a single band such as the free-carrier absorption. Optical measurement can also be used to study lattice vibrations.

The transmission coefficient T and the reflection coefficient R are the two important quantities generally measured. For normal incidence they are given by

$$T = \frac{(1 - R^2)\exp(-4\pi x/\lambda)}{1 - R^2 \exp(-8\pi x/\lambda)}, \tag{69}$$

$$R = \frac{(1 - \bar{n})^2 + k^2}{(1 + \bar{n})^2 + k^2} \tag{70}$$

where λ is the wavelength, $\bar{n}$ the refractive index, k the absorption constant, and x the thickness of the sample. The absorption coefficient per unit length α is defined as

$$\alpha \equiv \frac{4\pi k}{\lambda}. \tag{71}$$

By analyzing the $T - \lambda$ or $R - \lambda$ data at normal incidence, or by making observations of R or T for different angles of incidence, both $\bar{n}$ and k can be obtained and related to transition energy between bands.

Near the absorption edge the absorption coefficient can be expressed as[4]

$$\alpha \sim (h\nu - E_g)^\gamma \tag{72}$$

where $h\nu$ is the photon energy, E_g is the band gap, and γ is a constant which equals 1/2 and 3/2 for allowed direct transition and forbidden direct transition respectively [with $k_{min} = k_{max}$ as transitions (a) and (b) shown in Fig. 26]; it equals 2 for indirect transition [transition (c) shown in Fig. 26] where phonons must be incorporated. In addition γ equals 1/2 for allowed indirect transitions to exciton states where an exciton is a bound electron-hole pair with energy levels in the band gap and moving through the crystal lattice as a unit.

Near the absorption edge where the values of $(h\nu - E_g)$ become comparable with the binding energy of an exciton, the Coulomb interaction between the free hole and electron must be taken into account. For $h\nu < E_g$ the absorption merges continuously into that due to the higher excited states of the exciton. When $h\nu \gg E_g$, higher energy bands will participate in the transition processes, and complicated band structures will be reflected in the absorption coefficient.

The experimental absorption coefficients near and above the fundamental absorption edge (band-to-band transition) for Ge, Si, and GaAs are plotted[54-56] in Fig. 27. The shift of the curves towards higher photon

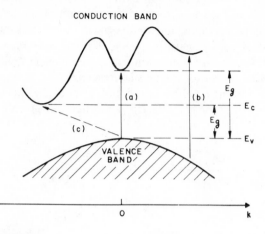

CONDUCTION BAND

Fig. 26 Optical transitions.
(a) and (b) are direct transitions.
(c) is indirect transition involving phonons.

energies at lower temperature is obviously associated with the temperature dependence of the band gap as shown in Fig. 8.

(3) Thermal Property

When a temperature gradient is applied to a semiconductor in addition to an applied electric field, the total current density (in one dimension) is[4]

$$J = \sigma\left(\frac{1}{q}\frac{\partial E_F}{\partial x} - \mathscr{P}\frac{\partial T}{\partial x}\right) \tag{73}$$

where σ is the conductivity, E_F the Fermi energy, and $\mathscr{P}$ the differential thermoelectric power. For a nondegenerate semiconductor with a mean free time $\tau \sim E^{-s}$ as discussed previously, the thermoelectric power is given by

$$\mathscr{P} = -\frac{k}{q}\left(\frac{\left[\frac{5}{2} - s + \ln\left(\frac{N_C}{n}\right)\right]n\mu_n - \left[\frac{5}{2} - s - \ln\left(\frac{N_V}{p}\right)\right]p\mu_p}{n\mu_n + p\mu_p}\right) \tag{74}$$

where k is the Boltzmann constant, and N_C and N_V are the effective density of states in the conduction band and valence band respectively.

This means that the thermoelectric power is negative for n-type semiconductors, and positive for p-type semiconductors. This fact is often used in determining the conduction type of a semiconductor. The thermoelectric power can also be used to determine the position of the Fermi level relative to the band edges. At room temperature the thermoelectric power $\mathscr{P}$ of p-type silicon increases with resistivity: 1 mV/°K for a 0.1-Ω-cm sample and 1.7 mV/°K for a 100-Ω-cm sample. Similar results (except a change of the sign for $\mathscr{P}$) can be obtained for n-type silicon samples.

Another important quantity in thermal effect is the thermal conductivity, κ, which, if $\tau \sim E^{-s}$ for both electrons and holes, is given by

$$\kappa = \kappa_L + \frac{\left(\frac{5}{2} - s\right) k^2 \sigma T}{q^2} + \frac{k^2 \sigma T}{q^2} \frac{\left(5 - 2s + \frac{E_g}{kT}\right)^2 n p \mu_n \mu_p}{(n\mu_n + p\mu_p)^2}. \tag{75}$$

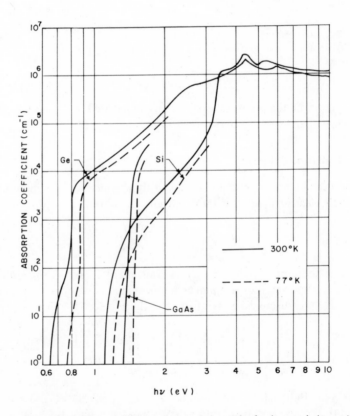

Fig. 27 Measured absorption coefficients near and above the fundamental absorption edge for pure Ge, Si, and GaAs.
(After Dash and Newman, Ref. 54; Philipp and Taft, Ref. 55; and Hill, Ref. 56.)

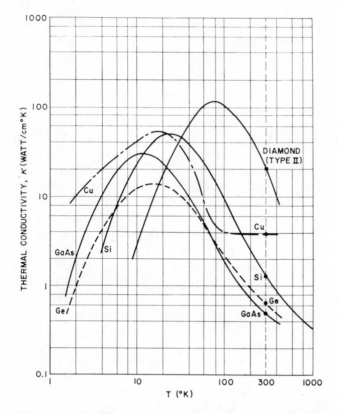

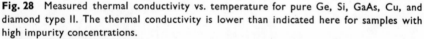

Fig. 28 Measured thermal conductivity vs. temperature for pure Ge, Si, GaAs, Cu, and diamond type II. The thermal conductivity is lower than indicated here for samples with high impurity concentrations.
(After Carruthers et al., Ref. 57; Holland, Ref. 58, 59; White, Ref. 60; and Berman et al., Ref. 61.)

The first, second, and third term on the right-hand side of the above equation represent the lattice contribution, the electronic contribution, and the contributions due to mixed conduction respectively. The contributions of conduction carriers to the thermal conductivity are in general quite small. The third term, however, may be quite large when $E_g \gg kT$. It is expected that the thermal conductivity will first increase with T at low temperatures and then decrease with temperature at higher temperatures.

The measured thermal conductivity[57-59] as a function of lattice temperature for Ge, Si, and GaAs is shown in Fig. 28. Their room temperature values are listed in Table 2.2. Also shown in Fig. 28 are the thermal conductivi-ties[60,61] for Cu and for diamond type II. Copper is the most commonly used metal for thermal conduction in p-n junction devices; diamond type II has the highest room-temperature thermal conductivity known to date and is useful as the thermal sink for junction lasers and IMPATT oscillators to be discussed later.

(4) High-Field Property

As discussed in Section 5(1), at low electric fields, the drift velocity in a semiconductor is proportional to the electric field, and the proportionality constant is called the mobility which is independent of the electric field. When the fields are sufficiently large, however, nonlinearities in mobility and in some cases saturation of drift velocity are observed. At still larger fields, impact ionization will occur. We shall first consider the nonlinear mobility.

At thermal equilibrium the carriers both emit and absorb phonons, and the net rate of exchange of energy is zero. The energy distribution at thermal equilibrium is Maxwellian. In the presence of an electric field the carriers acquire energy from the field and lose it to phonons by emitting more phonons than are absorbed. At reasonably high fields the most frequent scattering event is the emission of optical phonons. The carriers thus, on the average, acquire more energy than they have at thermal equilibrium. As the field increases, the average energy of the carriers also increases, and they acquire an effective temperature T_e which is higher than the lattice tempera-ture T. From the rate equation such that the rate at which energy fed from the field to the carriers must be balanced by an equal rate of loss of energy to the lattice, we obtain for Ge and Si:[3]

$$\frac{T_e}{T} = \frac{1}{2}\left\{1 + \left[1 + \frac{3\pi}{8}\left(\frac{\mu_0 \mathscr{E}}{C_s}\right)^2\right]^{1/2}\right\}, \tag{76}$$

and

$$v_d = \mu_0 \mathscr{E}\sqrt{\frac{T}{T_e}} \tag{77}$$

where μ_0 is the low-field mobility, $\mathscr{E}$ the electric field, and C_s the velocity of sound.

When $\mu_0 \mathscr{E} \ll C_s$, we have the quadratic departure of mobility from con-stant value at low fields, and Eqs. (76) and (77) reduce to

$$T_e \simeq T\left[1 + \frac{3\pi}{32}\left(\frac{\mu_0 \mathscr{E}}{C_s}\right)^2\right], \tag{78}$$

and

$$v_d \simeq \mu_0 \mathscr{E} \left[1 - \frac{3\pi}{64} \left(\frac{\mu_0 \mathscr{E}}{C_s} \right)^2 \right]. \tag{79}$$

When the field is increased until $\mu_0 \mathscr{E} \simeq 8C_s/3$, the carrier temperature has doubled over the crystal temperature and the mobility has dropped by 30%. Finally at sufficiently high fields the drift velocities for Ge and Si approach a scattering-limited velocity:

$$v_{sl} = \sqrt{\frac{8E_p}{3\pi m_0}} \sim 10^7 \quad \text{cm/sec} \tag{79a}$$

where E_p is the optical phonon energy (listed in Table 2.3).

TABLE 2.3

PROPERTIES OF Ge, Si, AND GaAs (AT 300°K IN ALPHABETICAL ORDER)

Properties	Ge	Si	GaAs
Atoms/cm³	4.42×10^{22}	5.0×10^{22}	2.21×10^{22}
Atomic Weight	72.6	28.08	144.63
Breakdown Field (V/cm)	$\sim 10^5$	$\sim 3 \times 10^5$	$\sim 4 \times 10^5$
Crystal Structure	Diamond	Diamond	Zincblende
Density (g/cm³)	5.3267	2.328	5.32
Dielectric Constant	16	11.8	10.9
Effective Density of States in Conduction Band, N_C(cm^{-3})	1.04×10^{19}	2.8×10^{19}	4.7×10^{17}
Effective Density of States in Valence Band, N_V(cm^{-3})	6.1×10^{18}	1.02×10^{19}	7.0×10^{18}
Effective Mass m^*/m_0 Electrons Holes	$m_l^* = 1.6 \quad m_t^* = 0.082$ $m_{lh}^* = 0.04 \; m_{hh}^* = 0.3$	$m_l^* = 0.97 \; m_t^* = 0.19$ $m_{lh}^* = 0.16 \; m_{hh}^* = 0.5$	0.068 0.12, 0.5
Electron Affinity, χ(V)	4.0	4.05	4.07
Energy Gap (eV) at 300°K	0.66	1.12	1.43

TABLE 2.3 (Cont.)

Properties	Ge	Si	GaAs
Intrinsic Carrier Concentration (cm^{-3})	2.4×10^{13}	1.6×10^{10}	1.1×10^{7}
Lattice Constant (Å)	5.65748	5.43086	5.6534
Linear Coefficient of Thermal Expansion $\Delta L/L \Delta T$ (°C^{-1})	5.8×10^{-6}	2.6×10^{-6}	5.9×10^{-6}
Melting Point (°C)	937	1420	1238
Minority Carrier Lifetime (sec)	10^{-3}	2.5×10^{-3}	$\sim 10^{-8}$
Mobility (Drift) (cm^{2}/V-sec) μ_n (electrons) μ_p (holes)	3900 1900	1500 600	8500 400
Raman Phonon Energy (eV)	0.037	0.063	0.035
Specific Heat (Joule/g°C)	0.31	0.7	0.35
Thermal Conductivity at 300°K (watt/cm°C)	0.64	1.45	0.46
Thermal Diffusivity (cm^{2}/sec)	0.36	0.9	0.44
Vapor Pressure (torr)	10^{-3} at 1270°C 10^{-8} at 800°C	10^{-3} at 1600°C 10^{-8} at 930°C	1 at 1050°C 100 at 1220°C
Work Function (V)	4.4	4.8	4.7

For GaAs the velocity-field relationship is more complicated. We must consider the band structure of GaAs as shown in Fig. 5. There is a high-mobility valley ($\mu \approx 4000$ to 8000 cm^{2}/V-sec) located at the Brillouin zone center, and a low-mobility satellite valley ($\mu \approx 100$ cm^{2}/V-sec) along the $\langle 100 \rangle$-axes, about 0.36 eV higher in energy. The effective mass of the electrons is 0.068 m_0 in the lower valley and about 1.2 m_0 in the upper valley; thus the density of states of the upper valley is about 70 times that of the lower valley, from Eq. (12). As the field increases, the electrons in the lower valley can be field-excited to the normally unoccupied upper valley, resulting in a differential negative resistance in GaAs. The intervalley transfer mechanism and the velocity-field relationship will be considered in more detail in Chapter 14.

The experimentally deduced drift velocity is plotted[62,63] in Fig. 29 for Ge, Si, and GaAs. We note that the drift velocities for the three semiconductors all approach 10^7 cm/sec. For GaAs there is a region of differential negative mobility between 3×10^3 to 2×10^4 V/cm.

We next consider impact ionization. When the electric field in a semiconductor is increased above a certain value, the carriers gain enough energy that they can excite electron-hole pairs by impact ionization. The electron-hole pair generation rate (G) due to impact ionization is given by

$$G = \alpha_n n \mu_n + \alpha_p p \mu_p \qquad (80)$$

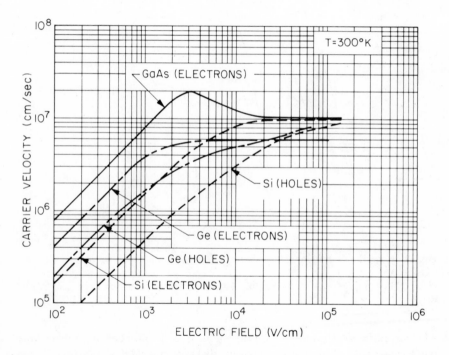

Fig. 29 Measured carrier velocity vs. electric field for high-purity Ge, Si, and GaAs. For highly-doped samples, the initial lines are lower than indicated here. In high-field region, however, the velocity is essentially independent of doping concentration.
(After Seidal and Scharfetter, Ref. 62; Norris and Gibbons, Ref. 62a; Duh and Moll, Ref. 62b; and Ruch and Kino, Ref. 63.)

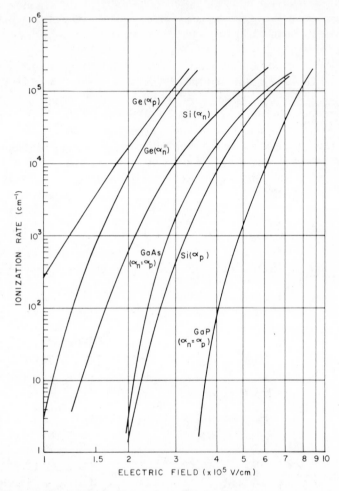

Fig. 30 Measured ionization coefficient for avalanche multiplication vs. electric field for Ge, Si, GaAs. and GaP.
(After Miller, Ref. 64; Lee et al., Ref. 65; Logan and Sze, Ref. 66; and Logan and White, Ref. 67.)

where α_n is the electron ionization rate defined as the number of electron-hole pairs generated by an electron per unit distance traveled. Similarly α_p is the analogously defined ionization rate for holes. Both α_n and α_p are strongly dependent on the electric field. Figure 30 shows the experimental results of the ionization rates[64-67] for Ge, Si, GaAs, and GaP. These results are all obtained by the use of photomultiplication measurements on p-n junctions. It is clear from the figure that over a limited field range one can approximate α_n or α_p as

$$\alpha_n \quad \text{or} \quad \alpha_p = \alpha_0 \left(\frac{\mathscr{E}}{\mathscr{E}_0}\right)^n \tag{81}$$

where α_0 and $\mathscr{E}_0$ are constants, and n, in general, varies from 3 to 9. A better fit to the ionization rates over a wider range of field can be given by the following expression

$$\alpha = A \exp[-(b/\mathscr{E})^m]. \tag{82}$$

The values of A, b, and m are given in Table 2.4.

It has been shown that the dependence of the ionization rate on the electric field and on the lattice temperature can be expressed in terms of a modification of Baraff's three-parameter theory.[68,69] The parameters are: E_I, the ionization threshold energy; λ, the optical phonon mean free path; and $\langle E_p \rangle$, the average energy loss per phonon scattering. The values of E_I for best fit are approximately three-halves the band-gap energy, and λ and $\langle E_p \rangle$ are given by[69]

$$\lambda = \lambda_0 \tanh\left(\frac{E_p}{2kT}\right) \tag{83}$$

$$\langle E_p \rangle = E_p \tanh\left(\frac{E_p}{2kT}\right) \tag{84}$$

and

$$\frac{\lambda}{\lambda_0} = \frac{\langle E_p \rangle}{E_p} \tag{85}$$

where E_p is the optical phonon energy (listed in Table 2.4), and λ_0 is the high-energy low-temperature asymptotic value of the phonon mean free path.

Baraff's result is shown in Fig. 31 where $\alpha\lambda$ is plotted against $E_I/q\mathscr{E}\lambda$, with the ratio of average optical-phonon energy to ionization threshold energy, $\langle E_p \rangle/E_I$, as a parameter. Since for a given set of ionization measurements the values of $E_I(=\frac{3}{2}E_g)$, α, and its field dependence are fixed, one can thus obtain the optical-phonon mean free path, λ, by fitting the ionization data to the Baraff plot. A typical result is shown in Fig. 31 for Ge at 300°K. The value of $\langle E_p \rangle/E_I$ is 0.022. We obtain 64 Å for the room-temperature electron-phonon mean free path, and 69 Å for the hole-phonon mean free path. Similar results have been obtained for Si, GaAs, and GaP. From the room temperature data one can obtain the value of λ_0 from Eq. (84). The average value of λ_0 and λ (at 300°K) are listed in Table 2.4. Once we know the value λ_0 we can predict

TABLE 2.4

Semiconductor	Ge		Si		GaAs	GaP
	Electron	Hole	Electron	Hole		
E_p(eV)	0.037		0.063		0.035	0.050
$\lambda(300°K)$(Å)	65 ± 10		62 ± 5	45 ± 5	35 ± 5	32 ± 5
λ_0(Å)	105 ± 10		76 ± 5	55 ± 5	58 ± 5	42 ± 5
Ionization rate (300°K) $\alpha = Ae^{-(b/\mathscr{E})^m}$ — A(cm^{-1})	1.55×10^7	1.0×10^6	3.8×10^6	2.25×10^7	1.34×10^6	4.0×10^5
b(V/cm)	1.56×10^6	1.28×10^6	1.75×10^6	3.26×10^6	2.03×10^6	1.8×10^6
m	1		1		2	2

$\mathscr{E}$ in V/cm

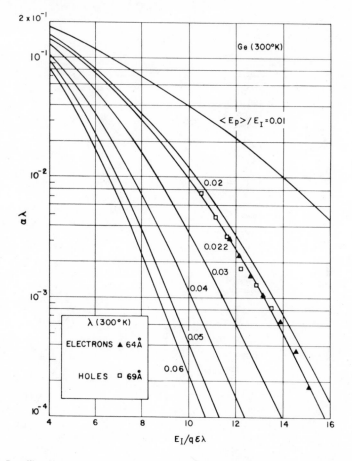

Fig. 31 Baraff's plot—product of ionization rate and optical phonon mean free path $(\alpha\lambda)$ vs. $E_I/q\mathscr{E}\lambda$ where $E_I = \tfrac{3}{2}Eg$ and $\mathscr{E}$ is the electric field. The running parameter is the ratio $\langle E_p \rangle/E_I$ where $\langle E_p \rangle$ is the average optical phonon energy. The solid curves are theoretical results.
(After Baraff, Ref. 68.)
The experimental data are obtained from Ge p-n junctions with $\lambda = 64\text{Å}$ for electrons and $\lambda = 69\text{Å}$ for holes.
(After Logan and Sze, Ref. 66.)

the values of λ at various temperatures; and from the temperature dependence of $\langle E_p \rangle$, Eq. (84), we can choose the correct Baraff plot. The theoretical predicted electron ionization rates in silicon as obtained from the above approach are shown in Fig. 32. Also shown are the experimental results at

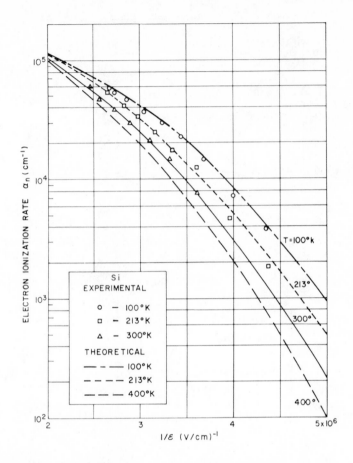

Fig. 32 Electron ionization rate vs. reciprocal electric field for Si. (After Crowell and Sze, Ref. 69.)

three different temperatures. The agreement is satisfactory. To facilitate numerical analysis, the curves as shown in Fig. 31 can be represented by the following approximation:[69]

$$\alpha\lambda = \exp\left\{ \begin{array}{l} (11.5r^2 - 1.17r + 3.9 \times 10^{-4})x^2 \\ + (46r^2 - 11.9r + 1.75 \times 10^{-2})x \\ + (-757r^2 + 75.5r - 1.92) \end{array} \right\} \quad (86)$$

where

$$r \equiv \langle E_p \rangle / E_I, \quad \text{and} \quad x \equiv E_I / q \mathscr{E} \lambda.$$

The errors are within $\pm 2\%$ over the range $0.01 < r < 0.06$ and $5 < x < 16$.

7 BASIC EQUATIONS FOR SEMICONDUCTOR DEVICE OPERATION

(1) Basic Equations[70]

The basic equations for semiconductor device operation are those which describe the static and dynamic behaviors of carriers in semiconductors under the influence of external fields which cause deviation from the thermal-equilibrium conditions. The basic equations can be classified in three groups: the Maxwell equations, the current density equations, and the continuity equations.

A. Maxwell Equations for Homogeneous and Isotropic Materials

$$\nabla \times \mathscr{E} = -\frac{\partial \mathscr{B}}{\partial t}, \tag{87}$$

$$\nabla \times \mathscr{H} = \frac{\partial \mathscr{D}}{\partial t} + \mathbf{J}_{\text{cond}} = \mathbf{J}_{\text{tot}}, \tag{88}$$

$$\nabla \cdot \mathscr{D} = \rho(x, y, z), \tag{89}$$

$$\nabla \cdot \mathscr{B} = 0, \tag{90}$$

$$\mathscr{B} = \mu_0 \mathscr{H}, \tag{91}$$

$$\mathscr{D}(\mathbf{r}, t) = \int_{-\infty}^{t} \varepsilon_s(t - t') \mathscr{E}(\mathbf{r}, t') \, dt' \tag{92}$$

where Eq. (92) reduces to $\mathscr{D} = \varepsilon_s \mathscr{E}$ under static or very low frequency conditions, $\mathscr{E}$ and $\mathscr{D}$ are the electric field and displacement vector respectively, $\mathscr{H}$ and $\mathscr{B}$ are the magnetic field and induction vector respectively, ε_s and μ_0 are the permittivity and permeability respectively, $\rho(x, y, z)$ the total electric charge density, $\mathbf{J}_{\text{cond}}$ the conduction current density, and $\mathbf{J}_{\text{tot}}$ the total current density (including both conduction and convection current components and $\nabla \cdot \mathbf{J}_{\text{tot}} = 0$). Among the above six equations the most important is the Poisson equation, Eq. (89), which determines the properties of the *p-n* junction depletion layer (to be discussed in the next chapter).

B. Current Density Equations

$$\mathbf{J}_n = q\mu_n n\mathscr{E} + qD_n\nabla n, \tag{93}$$

$$\mathbf{J}_p = q\mu_p p\mathscr{E} - qD_p\nabla p, \tag{94}$$

$$\mathbf{J}_{\text{cond}} = \mathbf{J}_n + \mathbf{J}_p. \tag{95}$$

Where $\mathbf{J}_n$ and $\mathbf{J}_p$ are the electron current density and the hole current density respectively, they consist of the drift component due to field and the diffusion component due to carrier concentration gradient. The values of the electron and hole mobilities (μ_n and μ_p) have been given in Section 5(1). For non-degenerate semiconductors the carrier diffusion constants (D_n and D_p) and the mobilities are given by the Einstein relationship ($D_n = (kT/q)\mu_n$, etc.).

For a one-dimensional case, Eqs. (93) and (94) reduce to

$$J_n = q\mu_n n\mathscr{E} + qD_n\frac{\partial n}{\partial x} = q\mu_n\left(n\mathscr{E} + \frac{kT}{q}\frac{\partial n}{\partial x}\right), \tag{93a}$$

$$J_p = q\mu_p p\mathscr{E} - qD_p\frac{\partial p}{\partial x} = q\mu_p\left(p\mathscr{E} - \frac{kT}{q}\frac{\partial p}{\partial x}\right). \tag{94a}$$

The above equations do not include the effect due to externally applied magnetic field. With an applied magnetic field, another current of $\mathbf{J}_{n\perp}\tan\theta_n$ and $\mathbf{J}_{p\perp}\tan\theta_p$ should be added to Eqs. (93) and (94) respectively, where $\mathbf{J}_{n\perp}$ is the current component of $\mathbf{J}_n$ perpendicular to the magnetic field and $\tan\theta_n \equiv q\mu_n n R_H|\mathscr{H}|$ (which has negative value because the Hall coefficient R_H is negative for electrons); similar results are obtained for the hole current.

C. Continuity Equations

$$\frac{\partial n}{\partial t} = G_n - U_n + \frac{1}{q}\nabla\cdot\mathbf{J}_n \tag{96}$$

$$\frac{\partial p}{\partial t} = G_p - U_p - \frac{1}{q}\nabla\cdot\mathbf{J}_p \tag{97}$$

where G_n and G_p are the electron and hole generation rate respectively ($\text{cm}^{-3}\ \text{sec}^{-1}$) due to external influence such as the optical excitation with high energy photons or impact ionization under large electric fields. U_n is the electron recombination rate in p-type semiconductors. Under low injection conditions (i.e., when the injected carrier density is much less than the equilibrium majority carrier density) U_n can be approximated by the expressions $(n_p - n_{po})/\tau_n$ where n_p is the minority carrier density, n_{po} the thermal-equilibrium minority carrier density, and τ_n the electron (minority) lifetime. There is a similar expression for the hole recombination rate with lifetime τ_p.

If the electrons and holes are generated and recombined in pairs with no trapping or other effects, $\tau_n = \tau_p$.

For the one-dimensional case under a low-injection condition, Eqs. (96) and (97) reduce to

$$\frac{\partial n_p}{\partial t} = G_n - \frac{n_p - n_{po}}{\tau_n} + n_p \mu_n \frac{\partial \mathscr{E}}{\partial x} + \mu_n \mathscr{E} \frac{\partial n_p}{\partial x} + D_n \frac{\partial^2 n_p}{\partial x^2}, \qquad (96a)$$

$$\frac{\partial p_n}{\partial t} = G_p - \frac{p_n - p_{no}}{\tau_p} - p_n \mu_p \frac{\partial \mathscr{E}}{\partial x} - \mu_p \mathscr{E} \frac{\partial p_n}{\partial x} + D_p \frac{\partial^2 p_n}{\partial x^2}. \qquad (97a)$$

(2) Simple Examples

A.   Decay of Photoexcited Carriers.   Consider an n-type sample, as shown in Fig. 33(a), which is illuminated with light and where the electron-hole pairs are generated uniformly throughout the sample with a generation rate G. The boundary conditions are $\mathscr{E} = 0$, and $\partial p_n / \partial x = 0$. We have from Eq. (97a):

$$\frac{\partial p_n}{\partial t} = G - \frac{p_n - p_{no}}{\tau_p}. \qquad (98)$$

At steady state, $\partial p_n / \partial t = 0$, and

$$p_n = p_{no} + \tau_p G = \text{constant}. \qquad (99)$$

If at an arbitrary time, say $t = 0$, the light is suddenly turned off, the boundary conditions are $p_n(0) = p_{no} + \tau_p G$ as given in Eq. (99) and $p_n(t \to \infty) = p_{no}$. The differential equation is now

$$\frac{\partial p_n}{\partial t} = -\frac{p_n - p_{no}}{\tau_p} \qquad (100)$$

and the solution is

$$p_n(t) = p_{no} + \tau_p G e^{-t/\tau_p}. \qquad (101)$$

The variation of p_n with time is shown in Fig. 33(b).

The above example presents the main idea of the Stevenson-Keyes method for measuring minority carrier lifetime.[49] A schematic setup is shown in Fig. 33(c). The excess carriers generated uniformly throughout the sample by the light pulses cause a momentary increase in the conductivity. The increase manifests itself by a drop in voltage across the sample when a constant current is passed through it. The decay of this photoconductivity can be observed on an oscilloscope and is a measure of the lifetime. (The pulse width must be much less than the lifetime.)

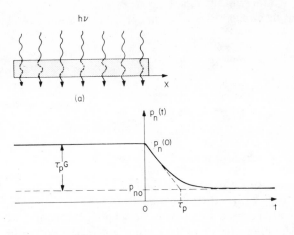

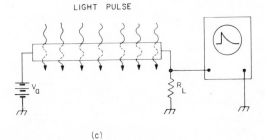

Fig. 33 Decay of photoexcited carriers
(a) *n*-type sample under constant illumination
(b) decay of minority carriers (holes) with time
(c) Stevenson-Keyes experiment.
(After Stevenson and Keyes, Ref. 49.)

B. Steady-State Injection From One Side. Figure 34(a) shows another simple example where excess carriers are injected from one side (e.g., by high-energy photons which create electron-hole pairs at the surface only). Referring to Fig. 27, we note that for $hv = 3.5$ eV the absorption coefficient is about 10^6 cm^{-1}, in other words, the light intensity decreases by $1/e$ in a distance of 100 Å.

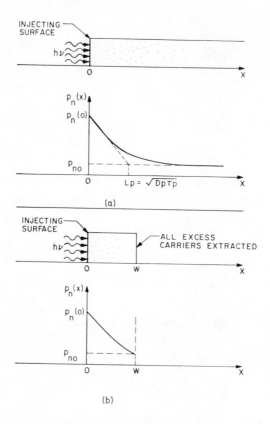

Fig. 34 Steady-state carrier injection from one side.
(a) semi-infinite sample
(b) sample with length W

At steady state there is a concentration gradient near the surface. The differential equation is, from Eq. (97a),

$$\frac{\partial p_n}{\partial t} = 0 = -\frac{p_n - p_{no}}{\tau_p} + D_p \frac{\partial^2 p_n}{\partial x^2}. \tag{102}$$

The boundary conditions are $p_n(x = 0) = p_n(0) = $ constant value, and $p_n(x \to \infty) = p_{no}$. The solution of $p_n(x)$ which is shown in Fig. 34(a), is

$$p_n(x) = p_{no} + [p_n(0) - p_{no}]e^{-x/L_p} \tag{103}$$

where $L_p \equiv \sqrt{D_p \tau_p}$, the diffusion length. The maximum values of L_p and

$L_n(\equiv \sqrt{D_n \tau_n})$ are of the order of 1 cm in germanium and silicon, but only of the order of 10^{-2} cm in gallium arsenide.

If the second boundary condition is changed so that all excess carriers at $x = W$ are extracted or $p_n(W) = p_{no}$, then we obtain from Eq. (102) a new solution,

$$p_n(x) = p_{no} + [p_n(0) - p_{no}] \left[\frac{\sinh\left(\dfrac{W - x}{L_p}\right)}{\sinh\left(\dfrac{W}{L_p}\right)} \right]. \qquad (104)$$

The above result is shown in Fig. 34(b). The current density at $x = W$ is given by Eq. (94a):

$$J_p = -qD_p \left. \frac{\partial p}{\partial x} \right|_W = q[p_n(0) - p_{no}] \frac{D_p}{L_p} \frac{1}{\sinh\left(\dfrac{W}{L_p}\right)}. \qquad (105)$$

It will be shown later that the above equation is related to the current-gain in junction transistors (Chapter 6).

C. Transient and Steady-State Diffusion. When localized light pulses generate excess carriers in a semiconductor, Fig. 35(a), the transport equation after the pulse is given by Eq. (97a).

$$\frac{\partial p_n}{\partial t} = -\frac{p_n - p_{no}}{\tau_p} - \mu_p \mathscr{E} \frac{\partial p_n}{\partial x} + D_p \frac{\partial^2 p_n}{\partial x^2}. \qquad (106)$$

If there is no field applied along the sample, $\mathscr{E} = 0$, the solution is given by

$$p_n(x, t) = \frac{N}{\sqrt{4\pi D_p t}} \exp\left[-\frac{x^2}{4D_p t} - \frac{t}{\tau_p} \right] + p_{no} \qquad (107)$$

where N is the number of electrons or holes generated per unit area. The solution is shown in Fig. 35(b), from which it can be seen that the carriers diffused away from the point of injection, and that they also recombined.

If an electric field is applied along the sample, the solution is in the form of Eq. (107) but with x replaced by $(x - \mu_p \mathscr{E} t)$ as shown in Fig. 35(c). This means that the whole "package" of excess carrier moves toward the negative end of the sample with the drift velocity $\mu_p \mathscr{E}$. At the same time, the carriers diffuse outward and recombine as in the field-free case.

The above example is essentially the celebrated Haynes-Shockley experiment[71] for the measurement of carrier drift mobility in semiconductors. With known sample length, applied field, and the time delay between the applied

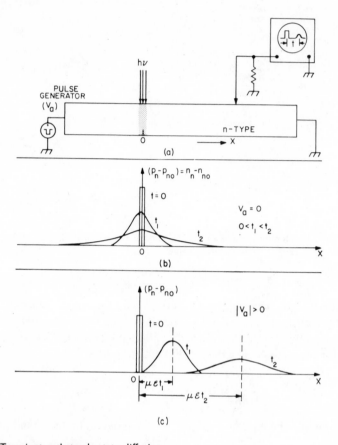

Fig. 35 Transient and steady-state diffusion
(a) Haynes-Shockley experiments
(After Haynes and Shockley, Ref. 71.)
(b) Without applied field
(c) With applied field

electric pulse and the detected pulse (both displayed on the oscilloscope), one can calculate the drift mobility ($\mu = x/\mathscr{E}t$).

D. Surface Recombination.[36] When one introduces surface recombination at one end of a semiconductor sample (Fig. 36), the boundary condition at $x = 0$ is given by

$$qD_n \frac{\partial p_n}{\partial x}\bigg|_{x=0} = qS_p[p_n(0) - p_{no}] \tag{108}$$

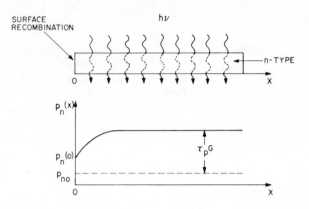

Fig. 36 Surface recombination at $x = 0$. The minority carrier distribution near the surface is affected by the surface recombination velocity.

which states that the minority carriers which reach the surface recombine there; the constant S_p which has the dimension cm/sec is defined as the surface recombination velocity. The boundary condition at $x = \infty$ is the same as that for example A. The differential equation is

$$\frac{\partial p_n}{\partial t} = G - \frac{p_n - p_{no}}{\tau_p} + D_p \frac{\partial^2 p_n}{\partial x^2}. \tag{109}$$

The solution of the equation subject to the above boundary conditions is

$$p_n(x) = p_{no} + \tau_p G \left[1 - \frac{\tau_p S_p \exp(-x/L_p)}{L_p + \tau_p S_p} \right] \tag{110}$$

which is plotted in Fig. 36 for a finite S_p. When $S_p \to 0$, $p_n(x) \to p_{no} + \tau_p G$, the same as obtained previously (example A); when $S_p \to \infty$, $p_n(x) \to p_{no} + \tau_p G[1 - \exp(-x/L_p)]$, and the minority carrier density at the surface approaches its thermal equilibrium value p_{no}. Analogous to the low-injection bulk recombination process, in which the reciprocal of the minority carrier lifetime $(1/\tau)$ is equal to $\sigma_p v_{th} N_t$, the surface recombination velocity is given by

$$S_p = \sigma_p v_{th} N_{st} \tag{111}$$

where N_{st} is the number of surface trapping centers per unit area at the boundary region.

REFERENCES

1. W. C. Dunlap, *An Introduction to Semiconductors*, John Wiley & Sons, New York (1957).

2. O. Madelung, *Physics of III-V Compounds*, John Wiley & Sons (1964).

3. J. L. Moll, *Physics of Semiconductors*, McGraw-Hill Book Co. (1964).

4. R. A. Smith, *Semiconductors*, Cambridge at the University Press (1959).

5. See for example, C. Kittel, *Introduction to Solid State Physics*, John Wiley & Sons, New York, 2nd ed. (1956), 3rd ed. (1966).

6. C. Hilsum and A. C. Rose-Innes, *Semiconducting III-V Compounds*, Pergamon Press, New York (1961).

7. R. K. Willardson and A. C. Beer, Ed. *Semiconductors and Semimetals*, Vol. 2, Physics of III-V Compound, Academic Press, New York (1966).

8. W. B. Pearson, *Handbook of Lattice Spacings and Structure of Metals and Alloys*, Pergamon Press (1967).

9. L. Brillouin, *Wave Propagation in Periodic Structures*, Dover Publication, 2nd ed., New York (1953).

10. F. Seitz, *The Modern Theory of Solids*, McGraw-Hill Book Co., New York (1940).

11. J. M. Ziman, *Principles of the Theory of Solids*, Cambridge University Press (1964).

12. M. L. Cohen, "Pseudopotential Calculations for II-VI Compounds," a chapter of *II-VI Semiconducting Compounds*, Edited by D. G. Thomas, W. A. Benjamin, Inc., New York, p. 462 (1967).

13. C. Kittel, *Quantum Theory of Solids*, John Wiley & Sons, New York (1963).

14. C. Herring, "A New Method for Calculating Wave Functions in Crystals," Phys. Rev., *57*, 1169 (1940).

15. F. Herman, "The Electronic Energy Band Structure of Silicon & Germanium," Proc. IRE, *43*, 1703 (1955).

15a. L. C. Allen, "Interpolation Scheme for Energy Bands in Solids," Phys. Rev., *98*, 993 (1955).

16. J. C. Phillips, "Energy-Band Interpolation Scheme Based on a Pseudopotential," Phys. Rev., *112*, 685 (1958).

17. M. L. Cohen and T. K. Bergstresser, "Band Structures and Pseudopotential Form Factors for Fourteen Semiconductors of the Diamond and Zinc-blende Structures," Phys. Rev., *141*, 789 (1966).

18. Compiled from Data Sheets of Electronic Properties Information Center (EPIC), Hughes Aircraft Co., Culver City, California.

18a. M. Neuberger, *Germanium Data Sheets*, DS-143 (Feb. 1965) and *Germanium Bibliographic Supplement*, DS-143 (Supplement 1) (Oct. 1966).

18b. M. Neuberger, *Silicon Data Sheets* DS-137 (May 1964) and *Silicon Bibliographic Supplement*, DS-137 (Supplement) (July 1968).

18c. M. Neuberger, *Gallium Arsenide Data Sheets*, DS-144 (April 1965) and *Gallium Arsenide Bibilographic Supplement*, DS-144 (Supplement 1) (Sept. 1967).

19. J. M. Ziman, *Electrons and Phonons*, Oxford at the Clarendon Press (1960).

20. W. Paul and D. M. Warschauer, Ed., *Solids Under Pressure*, McGraw–Hill Book Co., Inc., New York (1963).

21. J. S. Blakemore, "Carrier Concentrations and Fermi Levels in Sem.conductors," Elec. Commun., *29*, 131 (1952).

22. F. J. Morin and J. P. Maita, "Electrical Properties of Silicon Containing Arsenic and Boron," Phys. Rev., *96*, 28 (1954); also "Conductivity and Hall Effect in the Intrinsic Range of Germanium," Phys. Rev., *94*, 1525 (1954).

23. R. N. Hall and J. H. Racette, "Diffusion and Solubility of Copper in Extrinsic and Intrinsic Germanium, Silicon, and Gallium Arsenide," J. Appl. Phys., *35*, 379 (1964).

24. P. N. Lebeder, "Deep Levels in Semiconductors," Soviet Physics JETP, *18*, 253 (1964).

25. J. Hermanson and J. C. Phillips, "Pseudopotential Theory of Exciton and Impurity States," Phys. Rev., *150*, 652 (1966).

26. J. Callaway and A. J. Hughes, "Localized Defects in Semiconductors," Phys. Rev., *156*, 860 (1967).

27. E. M. Conwell, "Properties of Silicon and Germanium, Part II," Proc. IRE, *46*, 1281 (1958).

28. S. M. Sze and J. C. Irvin, "Resistivity, Mobility, and Impurity Levels in GaAs, Ge, and Si at 300°K," Solid State Electron., *11*, 599 (1968).

29. W. M. Bullis, "Properties of Gold in Silicon," Solid State Electron., *9*, 143 (1966).

30. W. Jost, *Diffusion in Solids, Liquids, Gases*, Academic Press (1952).

30a. P. G. Shewmon, *Diffusion in Solids*, McGraw-Hill Book Co., New York (1963).

30b. B. I. Boltaks, *Diffusion in Semiconductors*, Academic Press (1963).

30c. H. S. Carslaw and J. C. Jaeger, *Conduction of Heat in Solids*, Oxford at the Clarendon Press (1959).

31. For a review see J. F. Gibbons, "Ion Implantation in Semiconductors Part I, Range Distribution Theory and Experiments," Proc. IEEE, *56*, 295 (1968).

32. R. M. Burger and R. P. Donovan, Ed., *Fundamentals of Silicon Integrated Device Technology*, Vol. 1, Prentice-Hall Inc., Englewood Cliffs, New Jersey (1967).

33. D. L. Kendall, *Diffusion in III-V Compounds with Particular Reference to Self-Diffusion in InSb*, Rep. No. 65-29, Dept. of Material Science, Stanford University, Stanford (August 1965).

34. F. A. Trumbore, "Solid Solubilities of Impurity Elements in Germanium and Silicon," Bell System Tech. J., *39*, 205 (1960).

35. K. B. Wolfstirn, "Holes and Electron Mobilities in Doped Silicon From Radio Chemical and Conductivity Measurements," J. Phys. Chem. Solids, *16*, 279 (1960).

36. A. S. Grove, *Physics and Technology of Semiconductor Devices*, John Wiley and Sons, New York (1967).

37. J. Bardeen and W. Shockley, "Deformation Potentials and Mobilities in Nonpolar Crystals," Phys. Rev., *80*, 72 (1950).

38. E. Conwell and V. F. Weisskopf, "Theory of Impurity Scattering in Semiconductors," Phys. Rev., *77*, 388 (1950).

39. H. Ehrenreich, "Band Structure and Electron Transport in GaAs," Phys. Rev. *120*, 1951 (1960).

40. M. B. Prince, "Drift Mobility in Semiconductors I, Germanium," Phys. Rev., *92*, 681 (1953).

41. W. W. Gartner, *Transistors, Principles, Design, and Applications*, D. Van Nostrand Co. Inc., Princeton (1960).

42. L. B. Valdes, "Resistivity Measurement on Germanium for Transistors," Proc. IRE, *42*, 420 (1954).

43. F. M. Smits, "Measurement of Sheet Resistivities with the Four-Point Probe," Bell Syst. Tech. J., *37*, 711 (1958).

44. D. B. Cuttriss, "Relation Between Surface Concentration and Average Conductivity in Diffused Layers in Ge," Bell Syst. Tech. J., *40*, 509 (1961).

45. J. C. Irvin, "Resistivity of Bulk Silicon and of Diffused Layers in Silicon," Bell Syst. Tech. J., *41*, 387 (1962).

46. E. H. Hall, "On a New Action of the Magnet on Electric Currents," Am. J. Math., *2*, 287 (1879).

47. L. J. Van der Pauw, "A Method of Measuring Specific Resistivity and Hall Effect of Disc or Arbitrary Shape," Philips Research Reports, *13*, No. 1, p. 1-9 (Feb. 1958).

48. C. T. Sah, R. N. Noyce, and W. Shockley, "Carrier Generation and Recombination in p-n Junction and p-n Junction Characteristics," Proc. IRE, *45*, 1228 (1957).

48a. R. N. Hall, "Electron-Hole Recombination in Germanium," Phys. Rev., *87*, 387 (1952).

48b. W. Shockley and W. T. Read, "Statistics of the Recombination of Holes and Electrons," Phys. Rev., *87*, 835 (1952).

49. D. T. Stevenson and R. J. Keyes, "Measurement of Carrier Lifetime in Germanium and Silicon," J. Appl. Phys., *26*, 190 (1955).

50. W. W. Gartner, "Spectral Distribution of the Photomagnetic Electric Effect," Phys. Rev., *105*, 823 (1957). And also see other references listed in Ref. 41.

51. B. N. Brockhouse and P. K. Iyengar, "Normal Modes of Germanium by Neutron Spectrometry," Phys. Rev., *111*, 747 (1958).

52. B. N. Brockhouse, "Lattice Vibrations in Silicon and Germanium," Phys. Rev. Letters, *2*, 256 (1959).

53. J. L. T. Waugh and G. Dolling, "Crystal Dynamics of Gallium Arsenide," Phys. Rev., *132*, 2410 (1963).

54. W. C. Dash and R. Newman, "Intrinsic Optical Absorption in Single-Crystal Germanium and Silicon at 77°K and 300°K," Phys. Rev., *99*, 1151 (1955).

55. H. R. Philipp and E. A. Taft, "Optical Constants of Germanium in the Region 1 to 10 eV," Phys. Rev., *113*, 1002 (1959), also "Optical Constants of Silicon in the Region 1 to 10 eV," Phys. Rev. Letters, *8*, 13 (1962).

56. D. E. Hill, "Infrared Transmission and Fluorescence of Doped Gallium Arsenide," Phys. Rev., *133*, A866 (1964).

57. J. A. Carruthers, T. H. Geballe, H. M. Rosenberg, and J. M. Ziman, "The Thermal Conductivity of Germanium and Silicon between 2 and 300 K," Proc. Roy. Soc., London, *238*, 502 (1957).

58. M. G. Holland, "Low Temperature Thermal Conductivity in Silicon," Proc. International Conf. Semiconductor Physics, Prague (1960), Academic Press, New York (1961).

59. M. G. Holland, "Phonon Scattering in Semiconductors from Thermal Conductivity Studies," Phys. Rev., *134*, A471 (1964).

60. G. K. White, "The Thermal and Electrical Conductivity of Copper at Low Temperatures," Austr. J. Phys., *6*, 397 (1953).

61. R. Berman, "The Thermal Conductivity of Diamond at Low Temperatures," Proc. Roy. Soc. (London), Ser. A, *200*, 171 (1953).

62. T. E. Seidel and D. L. Scharfetter, "Dependence of Hole Velocity upon Electric Field and Hole Density for p-type Silicon," J. Phys. Chem. Solids, *28*, 2563 (1967).

62a. C. B. Norris and J. F. Gibbons, "Measurement of High-Field Carrier Drift Velocities in Si by a Time-of-Flight Technique," IEEE Trans. Electron Devices, *ED-14*, 38 (1967).

62b. C. Y. Duh and J. L. Moll, "Electron Drift Velocity in Avalanching Silicon Diodes," IEEE Trans. Electron Devices, *ED-14*, 38 (1967).

63. J. G. Ruch and G. S. Kino, "Measurement of the Velocity-Field Characteristics of Gallium Arsenide," Appl. Phys. Letters, *10*, 40 (1967).

64. S. L. Miller, "Avalanche Breakdown on Germanium," Phys. Rev., *99*, 1234 (1955).

65. C. A. Lee, R. A. Logan, R. L. Batdorf, J. J. Kleimack, and W. Wiegmann, "Ionization Rates of Holes and Electrons in Silicon," Phys. Rev., *134*, A761 (1964).

66. R. A. Logan and S. M. Sze, "Avalanche Multiplication in Ge and GaAs *p-n* Junctions," Proc. International Conference on the Physics of Semiconductors, Kyoto, and J. Phys. Soc. Japan Supplement Vol. 21, p. 434 (1966).

67. R. A. Logan and H. G. White, "Charge Multiplication in GaP *p-n* Junctions," J. Appl. Phys., *36*, 3945 (1965).

68. G. A. Baraff, "Distribution Junctions and Ionization Rates for Hot Electrons in Semiconductors," Phys. Rev., *128*, 2507 (1962).

69. C. R. Crowell and S. M. Sze, "Temperature Dependence of Avalanche Multiplication in Semiconductors," Appl. Phys. Letters, *9*, 242 (1966).

70. W. Shockley, *Electrons and Holes in Semiconductors*, D. Van Nostrand Co. Inc., Princeton (1950).

71. J. R. Haynes and W. Shockley, "The Mobility and Life of Injected Holes and Electrons in Germanium," Phys. Rev., *81*, 835 (1951).

72. Y. P. Varshni, "Temperature Dependence of the Energy Gap in Semiconductors," Physica, *34*, 149 (1967).

73. M. B. Panish and H. C. Casey, Jr., "Temperature Dependence of the Energy Gap in GaAs and GaP," J. Appl. Phys., *40*, 163 (1969).

PART II

p-n JUNCTION DEVICES

- *p-n* Junction Diodes
- Tunnel Diode and Backward Diode
- Impact-Avalanche Transit-Time Diodes (IMPATT Diodes)
- Junction Transistors
- *p-n-p-n* and Junction Field-Effect Devices

- INTRODUCTION
- BASIC DEVICE TECHNOLOGY
- DEPLETION REGION
  AND DEPLETION CAPACITANCE
- CURRENT-VOLTAGE CHARACTERISTICS
- JUNCTION BREAKDOWN
- TRANSIENT BEHAVIOR AND NOISE
- TERMINAL FUNCTIONS
- HETEROJUNCTION

3

p-n Junction Diodes

I INTRODUCTION

The *p-n* junctions are of supreme importance both in the modern electronic applications and in the understanding of other semiconductor devices. The *p-n* junction theory serves as the foundation of the physics of semiconductor devices. The basic theory of current-voltage characteristics of *p-n* junction was established by Shockley.[1] This theory was then extended by Sah et al.[2] and by Moll.[3]

In this chapter we shall first briefly discuss the basic device technology which is pertinent not only to *p-n* junctions but also to most semiconductor devices. Then the basic equations presented in Chapter 2 will be used to develop the ideal static and dynamic characteristics of *p-n* junctions. Departures from the ideal characteristics due to generation and recombination in the depletion layer, to high-injection, and to series resistance effects are then discussed. Junction breakdown (especially that due to avalanche multiplication) is considered in detail after which transient behaviors and noise performance in *p-n* junctions are presented.

A *p-n* junction is a two-terminal device. Depending on doping profile, device geometry, and biasing condition, a *p-n* junction can perform various terminal functions which are briefly considered in Section 7. The chapter closes with attention to an interesting group of devices, the heterojunctions, which are junctions formed between dissimilar semiconductors, e.g., *n*-type Ge on *p*-type GaAs or *n*-type Ge on *n*-type GaAs.

2   BASIC DEVICE TECHNOLOGY[4]

Some of the important junction fabrication methods are shown in Fig. 1. In the alloy method,[5] Fig. 1(a), a small pellet of aluminum is placed on an *n*-type $\langle 111 \rangle$ oriented silicon wafer. The system is then heated to a temperature slightly higher than the eutectic temperature ($\sim 580°C$ for the Al-Si system) so that a small puddle of molten Al-Si mixture is formed. The temperature is then lowered and the puddle begins to solidify. A recrystallized portion, which is saturated with the acceptor impurities and with the same crystal orientation, forms the heavily doped *p*-type region (p^+) on the *n*-type substrate. The aluminum button on top can be used as an ohmic contact for the *p*-type region. For the ohmic contact on the *n*-type wafer, a Au-Sb alloy (with $\sim 0.1\%$ Sb) can be evaporated onto the wafer and alloyed at about 400°C

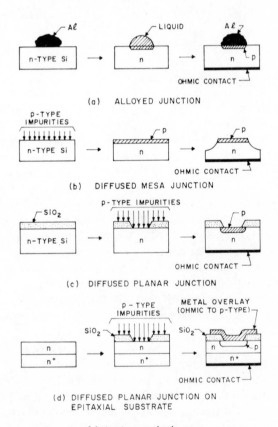

Fig. 1   Some important junction fabrication methods.

to form a heavily doped n-type region (n^+). One can interchange the roles of the aluminum and the Au-Sb alloy on a p-type wafer in order to form the n^+ junction on top and a p^+ ohmic contact on the bottom of the wafer. The junction location obtained by the alloy method depends critically on the temperature-time alloying cycle and is difficult to control precisely.

The solid-state diffusion method[6] was later developed to give precise control of the impurity profile. A diffused mesa junction method, where p-type impurities (e.g., boron in the form of BBr_3) are diffused into the n-type substrate, is shown in Fig. 1(b). After the diffusion, portions of the surface are protected (e.g., by wax or metal contacts) and the rest are etched out to form the mesa structures.

A new degree of control over the geometry of the diffused junction is achieved by the use of an insulating layer that can prevent most donor and acceptor impurities from diffusing through it.[7] A typical example is shown in Fig. 1(c). A thin layer of silicon dioxide (~ 1 µm) is thermally grown on silicon. With the help of photoresist techniques (e.g., the KPR process— Kodak photoresist process) portions of the oxide can be removed and windows (or patterns) cut in the oxide. The impurities will diffuse only through the exposed silicon surface, and p-n junctions will form in the oxide windows. This is the celebrated planar process[8, 54] which since 1960 has become the principal method of fabricating semiconductor devices and integrated circuits.

To reduce series resistance, an epitaxial substrate[9] is generally used in the planar process, Fig. 1(d). Epitaxy, derived from the Greek word EPI—meaning on, and TAXIS—meaning arrangement, describes a technique of crystal growth by chemical reaction used to form, on the surface of a crystal, thin layers of semiconductor materials with lattice structures identical to those of the crystal. In this method lightly doped high-resistivity epitaxial layers are grown on and supported by a heavily doped low-resistivity substrate thus ensuring both the desired electrical properties and mechanical strength. A typical impurity distribution of an epitaxial wafer is shown in Fig. 2. The doping of the original n^+ substrate is about 10^{19} cm^{-3}. The gradual transition from the n^+ substrate to the epitaxial n layer ($\sim 10^{15}$ cm^{-3}) is mainly due to the out-diffusion from the n^+ substrate during growth.

We shall now briefly discuss the two main processes of planar technology, namely the formation of the insulating layer and the diffusion of impurities.

The most important insulator for silicon is the silicon dioxide which can be formed by a vapor-phase reaction,[10] by anodization,[11] or by a plasma reaction.[12] The most frequently used method is, however, by the thermal oxidation[13] of silicon through the chemical reaction: Si(solid) + O_2(dry oxygen) $\rightarrow SiO_2$(solid) or Si(solid) + $2H_2O$(steam) $\rightarrow SiO_2$(solid) + $2H_2$. It can be shown that for short reaction times the oxide thickness increases linearly with time, and for prolonged oxidation the thickness varies as the square root of

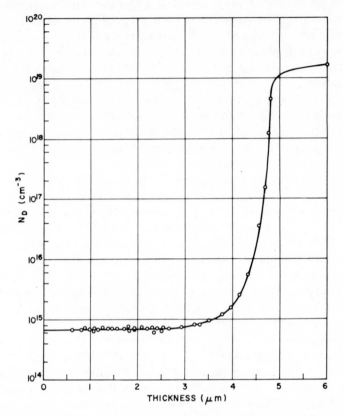

Fig. 2 Typical impurity distribution in an epitaxial layer. The substrate doping is about 2×10^{19} cm^{-3}, and the expitaxial layer doping is about 6.5×10^{14} cm^{-3}.

the time—the so-called parabolic relationship.[14] When a silicon dioxide film of thickness W_0 is formed, a layer of silicon of thickness 0.45 W_0 is consumed. Figure 3 shows the experimental results of the oxide thickness as a function of the reaction time and temperature for both the dry oxygen growth and the steam growth. Unlike silicon oxide, the germanium oxides formed on germanium are water-soluble. For most germanium as well as gallium arsenide planar devices, one uses silicon oxide deposited by thermal decomposition of ethyl orthosilicate (EOS) in a nitrogen ambient.[15]

The simple one-dimensional diffusion process can be given by the Fick equation[16]

$$\frac{\partial C(x, t)}{\partial t} = D \frac{\partial^2 C(x, t)}{\partial x^2} \qquad (1)$$

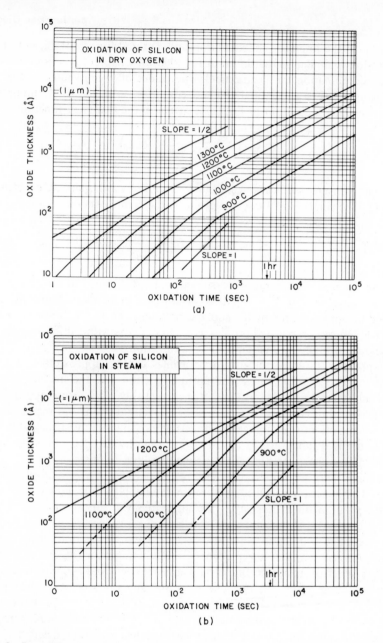

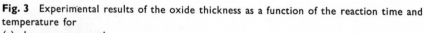

Fig. 3 Experimental results of the oxide thickness as a function of the reaction time and temperature for
(a) dry oxygen growth
(b) steam growth.
(After Deal and Grove, Ref. 14.)

where C is the impurity concentration and D the diffusion coefficient. This expression is similar to that given in Eq. (96a) of Chapter 2, without generation, recombination, or electric field. The values of D are shown in Fig. 13 of Chapter 2. For a "limited source" condition, with the total amount of impurities S, the solution of Eq. (1) is given by the Gaussian function

$$C(x, t) = \frac{S}{\sqrt{\pi D t}} \exp\left(-\frac{x^2}{4Dt}\right). \tag{2}$$

For the "constant surface concentration" condition with a surface concentration C_s, the solution of Eq. (1) is given by the error function compliment

$$C(x, t) = C_s \, \mathrm{erfc}\left(\frac{x}{2\sqrt{Dt}}\right). \tag{3}$$

The normalized concentration versus normalized distance to the above two solutions is shown in Fig. 4. The diffusion profiles of many impurities can indeed be approximated by the above expressions. There are, however, many that have more complicated profiles, e.g., Zn in GaAs with a diffusion process depending strongly on the impurity concentration.[17]

In practice, most of the diffusion profiles can be approximated by the following two limiting cases: the abrupt junction and the linearly graded junction as shown in Fig. 5(a) and 5(b) respectively. The abrupt approximation provides an adequate description for alloyed junctions and for shallowly diffused junctions. The linearly graded approximation is reasonable for deeply diffused junctions. In the following sections we shall study the static and dynamic characteristics of the above two limiting cases.

There is another important effect resulting from the planar processes. When a *p-n* junction is formed by diffusion into a bulk semiconductor through a window in an insulating layer, the impurities will diffuse downward and also sideway. Hence the junction consists of a plane (or flat) region with approximately cylindrical edges, as shown in Fig. 6(a). In addition, if the diffusion mask contains sharp corners, the junction near the corner will be roughly spherical in shape, Fig. 6(b). These spherical and cylindrical regions have profound effects on the junction especially for the avalanche multiplication process[18] which will be discussed in Section 5. At microwave frequencies, the junction size must be kept small. In the limit[19] the junction formed by diffusion through a narrow-stripe window approaches a semicylindrical junction if the width of the stripe approximates the junction depth, Fig. 6(c); and the junction formed by diffusion through a small circular window approaches a hemispherical junction if the radius of the window is equal to or smaller than the junction depth, Fig. 6(d).

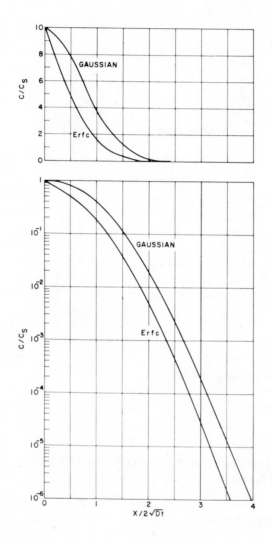

Fig. 4 Normalized concentration versus normalized distance for Gaussian and error function compliment (erfc) distributions which are plotted in both linear and semilog scales. (After Carslaw and Jaeger, Ref. 16.)

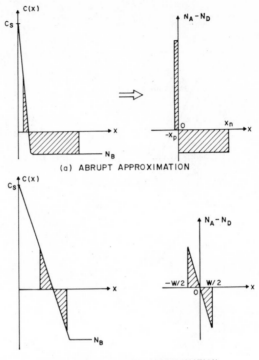

(a) ABRUPT APPROXIMATION

(b) LINEARLY GRADED APPROXIMATION

Fig. 5 Approximate doping profiles
(a) abrupt junction
(b) linearly graded junction.

3 DEPLETION REGION AND DEPLETION CAPACITANCE

(1) Abrupt Junction

A. Diffusion Potential and Depletion-Layer Width. When the impurity concentration in a semiconductor changes abruptly from acceptor impurities (N_A) to donor impurities (N_D), as shown in Fig. 7(a), one obtains an abrupt junction. In particular, if $N_A \gg N_D$, one obtains a onesided abrupt junction or $p^+ n$ junction.

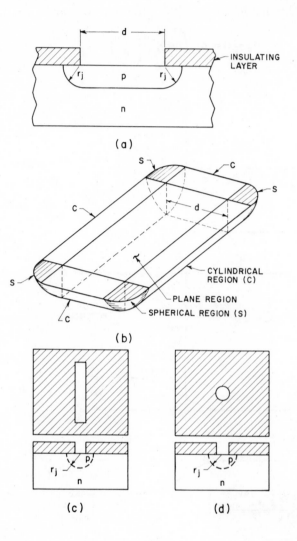

Fig. 6

(a) Planar diffusion process which forms junction curvature near the edges of the diffusion mask. r_j is the radius of curvature.

(b) The formation of approximately cylindrical and spherical regions by diffusion through rectangular mask.

(c) Diffusion through a narrow-stripe window.

(d) Diffusion through a small circular window.

(After Lee and Sze, Ref. 19.)

We first consider the thermal equilibrium condition, i.e., one with no applied voltage and no current flow. From Eqs. (33) and (93a) in Chapter 2

$$J_n = 0 = q\mu_n\left(n\mathscr{E} + \frac{kT}{q}\frac{\partial n}{\partial x}\right) = \mu_n n\frac{\partial E_F}{\partial x}, \tag{4}$$

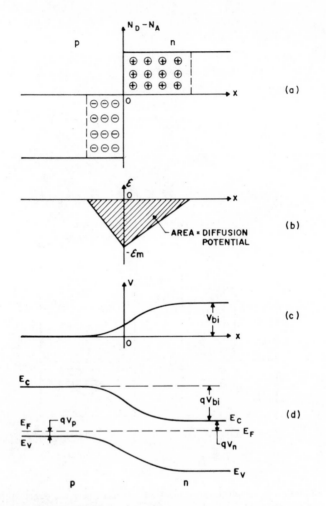

Fig. 7 Abrupt *p-n* junction in thermal equilibrium
(a) impurity distribution
(b) field distribution
(c) potential variation with distance
(d) energy band diagram,
where V_{bi} is the built-in potential.

or

$$\frac{\partial E_F}{\partial x} = 0. \tag{4a}$$

Similarly,

$$J_p = 0 = \mu_p \, p \, \frac{\partial E_F}{\partial x}. \tag{5}$$

Thus the condition of zero net electron and hole currents requires that the Fermi level must be constant throughout the sample. The diffusion potential, or built-in potential V_{bi}, as shown in Fig. 7(c) is equal to

$$qV_{bi} = E_g - (qV_n + qV_p)$$

$$= kT \ln\left(\frac{N_C N_V}{n_i^2}\right) - \left[kT \ln\left(\frac{N_C}{n_{no}}\right) + kT \ln\left(\frac{N_V}{p_{po}}\right) \right]$$

$$= kT \ln\left(\frac{n_{no} \, p_{po}}{n_i^2}\right) \simeq kT \ln\left(\frac{N_A N_D}{n_i^2}\right). \tag{6}$$

Since at equilibrium $n_{no} p_{no} = n_{po} p_{po} = n_i^2$,

$$V_{bi} = \frac{kT}{q} \ln\left(\frac{p_{po}}{p_{no}}\right) = \frac{kT}{q} \ln\left(\frac{n_{no}}{n_{po}}\right). \tag{7}$$

Equation (7) gives the relationship between the hole and electron densities on either side of the junction:

$$p_{no} = p_{po} \exp\left(-\frac{qV_{bi}}{kT}\right) \tag{8a}$$

$$n_{po} = n_{no} \exp\left(-\frac{qV_{bi}}{kT}\right). \tag{8b}$$

The approximate values of V_{bi} for one-sided abrupt p-n junctions in Ge, Si, and GaAs are shown in Fig. 8.

Since in equilibrium the electric field in the neutral regions (far from the junction at either side) of the semiconductor must be zero, the total negative charge per unit area in the p side must be precisely equal to the total positive charge per unit area in the n side:

$$N_D x_n = N_A x_p. \tag{9}$$

From Poisson's equation we obtain (for abrupt approximation):

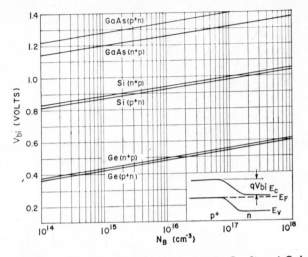

Fig. 8 Built-in potential for one-sided abrupt junctions in Ge, Si, and GaAs where p^+ is for heavily doped p side and n^+ is for heavily doped n side. The background doping N_B is for the impurity concentration of the lightly doped side.

$$-\frac{\partial^2 V}{\partial x^2} \equiv \frac{\partial \mathscr{E}}{\partial x} = \frac{\rho(x)}{\varepsilon_s} = \frac{q}{\varepsilon_s} [p(x) - n(x) + N_D^+(x) - N_A^-(x)]$$

$$\approx \frac{q}{\varepsilon_s} N_D \qquad \text{for} \qquad 0 < x \le x_n$$

$$\approx \frac{-q}{\varepsilon_s} N_A \qquad \text{for} \qquad -x_p \le x < 0. \tag{10}$$

The electric field is then obtained by integrating Eq. (10) as shown in Fig. 7(b) to be

$$\mathscr{E}(x) = -\frac{qN_A(x + x_p)}{\varepsilon_s} \qquad \text{for} \qquad -x_p \le x < 0 \tag{11a}$$

and

$$\mathscr{E}(x) = -\mathscr{E}_m + \frac{qN_D x}{\varepsilon_s}$$

$$= \frac{qN_D}{\varepsilon_s}(x - x_n) \qquad \text{for} \qquad 0 < x \le x_n \tag{11b}$$

where $\mathscr{E}_m$ is the maximum field which exists at $x = 0$ and is given by

$$|\mathscr{E}_m| = \frac{qN_D x_n}{\varepsilon_s} = \frac{qN_A x_p}{\varepsilon_s}. \tag{12}$$

Integration once again of Eq. (10), Fig. 7(c), gives the potential distribution $V(x)$ and the built-in potential V_{bi}:

$$V(x) = \mathscr{E}_m\left(x - \frac{x^2}{2W}\right) \tag{13}$$

$$V_{bi} = \tfrac{1}{2}\mathscr{E}_m W \equiv \tfrac{1}{2}\mathscr{E}_m(x_n + x_p) \tag{14}$$

where W is the total depletion width. Elimination of $\mathscr{E}_m$ from Eqs. (12) and (14) yields

$$W = \sqrt{\frac{2\varepsilon_s}{q}\left(\frac{N_A + N_D}{N_A N_D}\right)V_{bi}} \tag{15}$$

for a two-sided abrupt junction. For a one-sided abrupt junction, Eq. (15) reduces to

$$W = \sqrt{\frac{2\varepsilon_s V_{bi}}{qN_B}} \tag{15a}$$

where $N_B = N_D$ or N_A depending on whether $N_A \gg N_D$ or vice versa. The values of W as a function of the impurity concentration for one-sided abrupt junctions in silicon are shown in Fig. 9 (dashed line for zero bias).

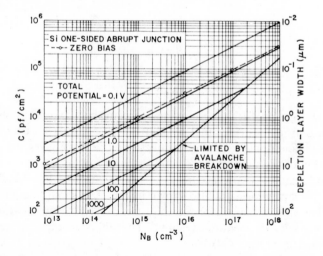

Fig. 9 Depletion-layer capacitance per unit area and depletion-layer width as a function of doping for one-sided abrupt junction in Si. The dashed line is for the case of zero bias voltage.

When a voltage V is applied to the junction, the total electrostatic potential variation across the junction is given by $(V_{bi} + V)$ for reverse bias (positive voltage on n region with respect to the p region) and by $(V_{bi} - V)$ for forward bias. Substitution of these values of voltage in Eqs. (15) or (15a) yields the depletion layer width as a function of the applied voltage. The results for one-sided abrupt junctions in silicon are shown in Fig. 9. The values above the zero-bias line (dashed line) are for the forward-biased condition; and below, for the reverse-biased condition.

These results can also be used for GaAs since both Si and GaAs have approximately the same static dielectric constants. To obtain the depletion-layer width for Ge, one must multiply the results of Si by the factor

$$\sqrt{\varepsilon_s(\text{Ge})/\varepsilon_s(\text{Si})} = 1.17.$$

B. Depletion-Layer Capacitance. The depletion-layer capacitance per unit area is defined as $C \equiv dQ_c/dV$ where dQ_c is the incremental increase in charge per unit area upon an incremental change of the applied voltage dV.

For one-sided abrupt junctions the capacitance per unit area is given by

$$C \equiv \frac{dQ_c}{dV} = \frac{d(qN_B W)}{d\left(\frac{qN_B}{2\varepsilon_s} W^2\right)} = \frac{\varepsilon_s}{W} = \sqrt{\frac{q\varepsilon_s N_B}{2(V_{bi} \pm V)}} \qquad \text{pf/cm}^2, \qquad (16)$$

or

$$\frac{1}{C^2} = \frac{2}{q\varepsilon_s N_B}(V_{bi} \pm V), \qquad (16a)$$

$$\frac{d(1/C^2)}{dV} = \frac{2}{q\varepsilon_s N_B} \qquad (16b)$$

where the $\pm$ signs are for the reverse- and forward-bias conditions respectively. It is clear from Eq. (16a) that by plotting $1/C^2$ versus V, a straight line should result for a one-sided abrupt junction. The slope gives the impurity concentration of the substrate (N_B), and the intercept (at $1/C^2 = 0$) gives the built-in potential V_{bi} (more accurate consideration gives $V_{bi} - 2kT/q$). The results of the capacitance are also shown in Fig. 9. It should be pointed out that, for the forward bias, there is a diffusion capacitance in addition to the depletion capacitance mentioned above. The diffusion capacitance will be discussed later in Section 4(4).

For the cases of cylindrical *p-n* junctions, Fig. 6(c), the capacitance per unit length is equivalent to the capacitance of a coaxial transmission line and is given by[19]

$$C_c = \frac{2\pi\varepsilon_s}{\ln\dfrac{r_2}{r_1}} \quad \text{pf/cm} \tag{17}$$

where r_1 and r_2 are the radii of inner and outer boundaries of the depletion layer respectively. The capacitance for a spherical junction, Fig. 6(d), is equivalent to that of two concentric spheres and is given by

$$C_s = \frac{4\pi\varepsilon_s}{\dfrac{1}{r_1} - \dfrac{1}{r_2}} \quad \text{pf} \tag{18}$$

The results are shown in Fig. 10 for C_c and C_s versus normalized reverse voltages where

$$V_{AC} \equiv \frac{qN_B r_j^2}{4\varepsilon_s} \tag{18a}$$

is for the cylindrical junction, with r_j the radius of curvature at the metallurgical junction; and

$$V_{AS} \equiv \frac{qN_B r_j^2}{6\varepsilon_s} \tag{18b}$$

is for the spherical junction. We note that for large values of V, the slopes are less than $\frac{1}{2}$; and as V decreases, the slope approaches the ideal value of $\frac{1}{2}$ for plane abrupt junctions.

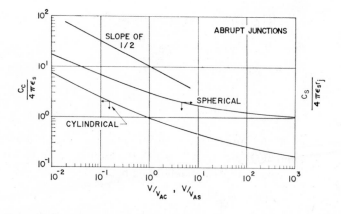

Fig. 10 Normalized capacitance versus normalized voltage for abrupt cylindrical junction and abrupt spherical junction.
(After Lee and Sze, Ref. 19.)

Exact calculation[20] shows that for one-sided or two-sided junctions the relation shown in Eq. (16b) is well obeyed. However, for one-sided abrupt junctions with large doping ratios (e.g., $N_A \gg N_D$), the voltage intercept in a $1/C^2$ versus V plot is nearly independent of the doping but increases approximately as $\ln V$. This results from the fact that, when one side is much more heavily doped than the other, mobile charges in the immediate neighborhood of the metallurgical junction cannot be neglected. The mobile charges result from carriers spilled over from the heavily doped side.

(2) Linearly Graded Junction

Consider the thermal equilibrium case first. The impurity distribution for linearly graded junctions is shown in Fig. 11(a). The Poisson equation for this case is

$$-\frac{\partial^2 V}{\partial x^2} \equiv \frac{\partial \mathscr{E}}{\partial x} = \frac{\rho(x)}{\varepsilon_s} = \frac{q}{\varepsilon_s}[p - n + ax] \approx \frac{q}{\varepsilon_s}ax \qquad -\frac{W}{2} \le x \le \frac{W}{2} \qquad (19)$$

where a is the impurity gradient in cm^{-4}. By integration of Eq. (19) once, we obtain field distribution as shown in Fig. 11(b):

$$\mathscr{E}(x) = -\frac{qa}{\varepsilon_s}\frac{\left(\frac{W}{2}\right)^2 - x^2}{2} \qquad (20)$$

with the maximum field $\mathscr{E}_m$ at $x = 0$,

$$|\mathscr{E}_m| = \frac{qaW^2}{8\varepsilon_s}. \qquad (20a)$$

Integration of Eq. (19) once again gives the built-in potential as shown in Fig. 11(c).

$$V_{bi} = \frac{qaW^3}{12\varepsilon_s}, \qquad (21)$$

or

$$W = \left(\frac{12\varepsilon_s V_{bi}}{qa}\right)^{1/3}. \qquad (21a)$$

Since the values of the impurity concentrations at the edges of the depletion region ($-W/2$ and $W/2$) are equal to $aW/2$, the built-in potential for linearly graded junctions can be approximately given by an expression similar to Eq. (6):

$$V_{bi} \simeq \frac{kT}{q}\ln\left[\frac{\left(\frac{aW}{2}\right)\left(\frac{aW}{2}\right)}{n_i^2}\right] = \frac{kT}{q}\ln\left(\frac{aW}{2n_i}\right)^2. \qquad (22)$$

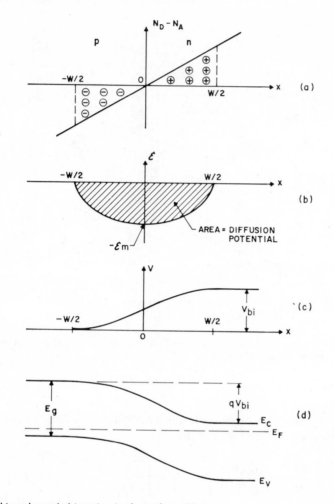

Fig. 11 Linearly graded junction in thermal equilibrium
(a) impurity distribution
(b) field distribution
(c) potential variation with distance
(d) energy band diagram.

From Eqs. (21) and (22) we can determine both V_{bi} and W. The values of the built-in potential V_{bi} for Ge, Si, and GaAs are in Fig. 12. The values of W at zero bias for Si and GaAs are shown in Fig. 13 (dashed line). For Ge the silicon data should be multiplied by the factor $[\varepsilon_s(\text{Ge})/\varepsilon_s(\text{Si})]^{1/3} = 1.1$.

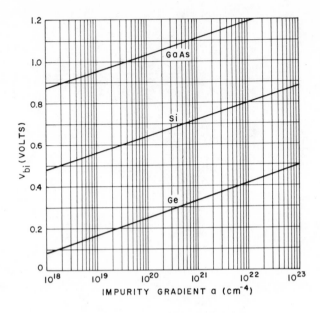

Fig. 12 Built-in potential for linearly graded junctions in Ge, Si, and GaAs. The potential is plotted against the impurity gradient "*a*."

The depletion-layer capacitance can be obtained similarly to that for the abrupt junction case and is given by

$$C \equiv \frac{dQ_c}{dV} = \frac{d(qaW^2/8)}{d(qaW^3/12\varepsilon_s)} = \frac{\varepsilon_s}{W} = \left[\frac{qa\varepsilon_s^2}{12(V_{bi} \pm V)}\right]^{1/3} \qquad \text{pf/cm}^2 \qquad (23)$$

where the signs + and − are for the reverse and forward bias respectively Typical values of C versus the impurity gradient for the silicon junction are shown in Fig. 13.

The above discussions are for plane junction. For cylindrical and spherical linearly graded junctions, similar results are obtained.[19] The normalized capacitances versus normalized reverse voltage are plotted in Fig. 14, where

$$V_{LC} \text{ (cylindrical junctions)} \equiv \frac{9qar_j^3}{8\varepsilon_s}, \qquad (24a)$$

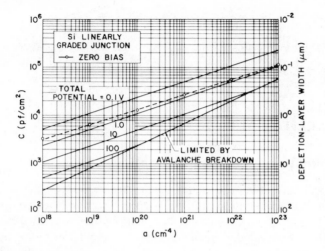

Fig. 13 Depletion-layer capacitance per unit area and depletion-layer width as a function of impurity gradient for linearly graded junctions in Si. The dashed line is for the case of zero bias voltage.

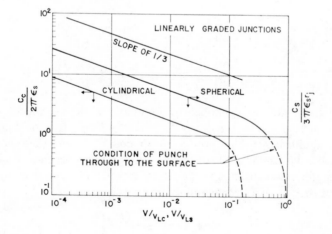

Fig. 14 Normalized capacitance versus normalized voltage for linearly graded cylindrical and spherical junctions.
(Ref. 19.)

and

$$V_{LS} \text{ (spherical junctions)} \equiv \frac{8qar_j{}^3}{81\varepsilon_s}. \tag{24b}$$

We note that over most of the voltage range, as long as the doping profiles are linearly graded, the slopes are the same ($\frac{1}{3}$) independent of junction geometries.

4 CURRENT-VOLTAGE CHARACTERISTICS

(1) Ideal Case—Shockley Equation[1]

The ideal current-voltage characteristics are derived on the basis of the following four assumptions: (1) the abrupt depletion-layer approximation, i.e., the built-in potential and applied voltages are supported by a dipole layer with abrupt boundaries, and outside the boundaries the semiconductor is assumed to be neutral; (2) the Boltzmann approximation, i.e., throughout the depletion layer, the Boltzmann relations similar to Eqs. (33) and (37) of Chapter 2 are valid; (3) the low injection assumption, i.e., the injected minority carrier densities are small compared with the majority-carrier densities; and (4) the facts that no generation current exists in the depletion layer, and the electron and hole currents are constant through the depletion layer.

We first consider the Boltzmann relation. At thermal equilibrium this relation is given by

$$n = n_i \exp\left[\frac{E_F - E_i}{kT}\right] \equiv n_i \exp\left[\frac{q(\psi - \phi)}{kT}\right] \tag{25a}$$

$$p = n_i \exp\left[\frac{E_i - E_F}{kT}\right] \equiv n_i \exp\left[\frac{q(\phi - \psi)}{kT}\right] \tag{25b}$$

where ψ and ϕ are the potentials corresponding to the intrinsic level and the Fermi level respectively (or $\psi \equiv -E_i/q$, $\phi \equiv -E_F/q$). It is obvious that at thermal equilibrium the *pn* product from Eqs. (25a) and (25b) is equal to $n_i{}^2$. When a voltage is applied, the minority carrier densities on both sides of a junction are changed, and the *pn* product is no longer given by $n_i{}^2$. We shall now define the imrefs as follows:

$$n \equiv n_i \exp\left[\frac{q(\psi - \phi_n)}{kT}\right] \tag{26a}$$

$$p = n_i \exp\left[\frac{q(\phi_p - \psi)}{kT}\right] \tag{26b}$$

where ϕ_n and ϕ_p are the imrefs or quasi-Fermi levels for electrons and holes respectively. From Eqs. (26a) and (26b) we obtain

$$\phi_n \equiv \psi - \frac{kT}{q} \ln\left(\frac{n}{n_i}\right) \tag{27a}$$

$$\phi_p \equiv \psi + \frac{kT}{q} \ln\left(\frac{p}{n_i}\right) \tag{27b}$$

the pn product becomes

$$pn = n_i^2 \exp\left[\frac{q(\phi_p - \phi_n)}{kT}\right]. \tag{28}$$

For a forward bias, $(\phi_p - \phi_n) > 0$, and $pn > n_i^2$; on the other hand, for a reversed bias, $(\phi_p - \phi_n) < 0$, and $pn < n_i^2$.

From Eq. (93) of Chapter 2, Eq. (26a), and the fact that $\mathscr{E} \equiv -\nabla\psi$ we obtain

$$\mathbf{J}_n = q\mu_n\left(n\mathscr{E} + \frac{kT}{q}\nabla n\right) = q\mu_n n(-\nabla\psi) + q\mu_n \frac{kT}{q}\left[\frac{qn}{kT}(\nabla\psi - \nabla\phi_n)\right]$$

$$= -q\mu_n n\nabla\phi_n. \tag{29}$$

Similarly, we obtain

$$\mathbf{J}_p = -q\mu_p p\nabla\phi_p. \tag{30}$$

Thus the electron and hole current densities are proportional to the gradients of the electron and hole imref respectively. If $\phi_n = \phi_p = \phi = $ constant (at thermal equilibrium), then $\mathbf{J}_n = \mathbf{J}_p = 0$.

The idealized potential distributions and the carrier concentrations in a p-n junction under forward-bias and reverse-bias conditions are shown in Fig. 15. The variations of ϕ_n and ϕ_p with distance are related to the carrier concentrations as given in Eq. (27). Since the electron density n varies in the junction from the n side to the p side by many orders of magnitude, whereas the electron current J_n is almost constant, it follows that ϕ_n must also be almost constant over the depletion layer. The electrostatic potential difference across the junction is given by

$$V = \phi_p - \phi_n. \tag{31}$$

Equations (28) and (31) can be combined to give the electron density at the boundary of the depletion-layer region on the p side ($x = -x_p$)

$$n_p = \frac{n_i^2}{p_p}\exp\left(\frac{qV}{kT}\right) = n_{po}\exp\left(\frac{qV}{kT}\right) \tag{32}$$

where n_{po} is the equilibrium electron density on the p side. Similarly,

$$p_n = p_{no}\exp\left(\frac{qV}{kT}\right) \tag{33}$$

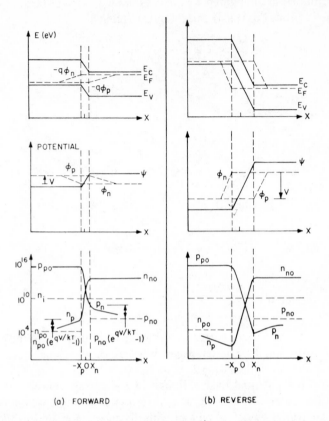

(a) FORWARD   (b) REVERSE

Fig. 15 Energy band diagram, intrinsic Fermi level (ψ), quasi-Fermi level also referred to as imref (ϕ_n for electrons, ϕ_p for holes), and carrier distributions under
(a) forward bias
(b) reverse bias condition.
(After Shockley, Ref. 1.)

at $x = x_n$ for the n type boundary. The above equations serve as the most important boundary conditions for the ideal current-voltage equation.

From the continuity equations we obtain for the steady-state:

$$-\frac{n_n - n_{no}}{\tau_n} + \mu_n \mathscr{E} \frac{\partial n_n}{\partial x} + \mu_n n_n \frac{\partial \mathscr{E}}{\partial x} + D_n \frac{\partial^2 n_n}{\partial x^2} = 0, \qquad (34a)$$

$$-\frac{p_n - p_{no}}{\tau_p} - \mu_p \mathscr{E} \frac{\partial p_n}{\partial x} - \mu_p p_n \frac{\partial \mathscr{E}}{\partial x} + D_p \frac{\partial^2 p_n}{\partial x^2} = 0. \qquad (34b)$$

We can eliminate the term in $\partial\mathscr{E}/\partial x$ from the above equations with the condition that $(p_n - p_{no})/\tau_p$ equals $(n_n - n_{no})/\tau_n$. This gives

$$-\frac{p_n - p_{no}}{\tau_p} + \mu_a \mathscr{E}\left(p_n \frac{\partial n_n}{\partial x} - n_n \frac{\partial p_n}{\partial x}\right)\bigg/(p_n - n_n)$$

$$+ D_a\left(p_n \frac{\partial^2 n_n}{\partial x^2} + n_n \frac{\partial^2 p_n}{\partial x^2}\right)\bigg/(p_n + n_n) = 0 \quad (35)$$

where

$$D_a \equiv D_n D_p(p_n + n_n)/(n_n D_n + p_n D_p) \equiv \text{ambipolar diffusion coeff.} \quad (36)$$

$$\mu_a \equiv \mu_n \mu_p(p_n - n_n)/(n_n \mu_n + p_n \mu_p) \equiv \text{ambipolar mobility.} \quad (37)$$

It can be shown that from the low-injection assumption (e.g., $p_n \ll n_n \approx n_{no}$ in the n-type semiconductor) Eq. (35) reduces to

$$-\frac{p_n - p_{no}}{\tau_p} - \mu_p \mathscr{E} \frac{\partial p_n}{\partial x} + D_p \frac{\partial^2 p_n}{\partial x^2} = 0 \quad (38)$$

which is Eq. (34b) with the exception that the term $\mu_p p_n \partial\mathscr{E}/\partial x$ is missing; under the low-injection assumption this term is of the same order as the neglected terms.

In the neutral region where there is no electric field, Eq. (38) further reduces to

$$\frac{\partial^2 p_n}{\partial x^2} - \frac{p_n - p_{no}}{D_p \tau_p} = 0. \quad (39)$$

The solution of Eq. (39) with the boundary condition Eq. (33) and $p_n(x = \infty) = p_{no}$, gives

$$p_n - p_{no} = p_{no}(e^{qV/kT} - 1)e^{-(x - x_n)/L_p} \quad (40)$$

where

$$L_p \equiv \sqrt{D_p \tau_p}. \quad (41)$$

And at $x = x_n$

$$J_p = -qD_p \frac{\partial p_n}{\partial x}\bigg|_{x_n} = \frac{qD_p p_{no}}{L_p}(e^{qV/kT} - 1). \quad (42)$$

Similarly we obtain for the p side

$$J_n = qD_n \frac{\partial n_p}{\partial x}\bigg|_{-x_p} = \frac{qD_n n_{po}}{L_n}(e^{qV/kT} - 1). \quad (43)$$

The minority carrier densities and the current densities for the forward-bias and reverse-bias condition are shown in Fig. 16.

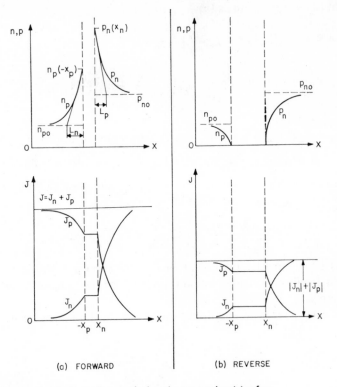

Fig. 16 Carrier distributions (linear plot) and current densities for
(a) forward bias
(b) reverse bias condition.
(After Shockley, Ref. 1.)

The total current is given by the sum of Eqs. (42) and (43):

$$J = J_p + J_n = J_s(e^{qV/kT} - 1), \tag{44}$$

$$J_s \equiv \frac{qD_p p_{no}}{L_p} + \frac{qD_n n_{po}}{L_n}. \tag{45}$$

This is the celebrated Shockley equation[1] which is the ideal diode law. The ideal current-voltage relation is shown in Fig. 17(a) and (b) in the linear and semilog plots respectively. In the forward direction (positive bias on *p*) for $V > 3kT/q$, the rate of rise is constant, Fig. 17(b); at 300°K for every decade change of current there is a 59.5-mV ($= 2.3 \, kT/q$) change in voltage. In the reverse direction the current density saturates at $-J_s$.

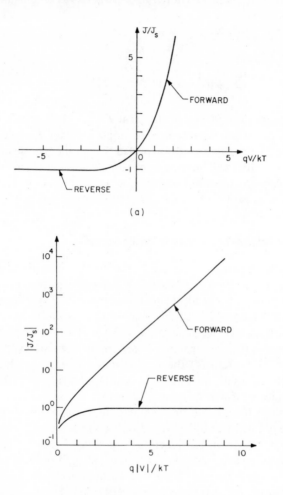

Fig. 17 Ideal current-voltage characteristics
(a) linear plot
(b) semilog plot.

We shall now briefly consider the temperature effect on the saturation current density J_s. We shall consider only the first term in Eq. (45), since the second term will behave similarly to the first one. For the one-sided p^+n abrupt junction (with donor concentration N_D), $p_{no} \gg n_{po}$, the second term also can be neglected. The quantities D_p, p_{no}, and L_p ($\equiv \sqrt{D_p \tau_p}$) are all temperature-dependent. If D_p/τ_p is proportional to T^γ, where γ is a constant, then

$$J_s \simeq \frac{qD_p p_{no}}{L_p} \simeq q\sqrt{\frac{D_p}{\tau_p}} \frac{n_i^2}{N_D}$$

$$\sim \left[T^3 \exp\left(-\frac{E_g}{kT}\right)\right] T^{\gamma/2} = T^{(3+\gamma/2)} \exp\left(-\frac{E_g}{kT}\right). \tag{46}$$

The temperature dependence of the term $T^{(3+\gamma/2)}$ is not important in comparison with the exponential term. The slope of a plot J_s versus $1/T$ is determined by the energy gap E_g. It is expected that in the reverse direction where $|J_R| \sim J_s$, the current will increase approximately as $e^{-E_g/kT}$ with temperature; and in the forward direction where $J_F \sim J_s e^{qV/kT}$, the current will increase approximately as $\exp[-(E_g - qV)/kT]$.

(2) Generation-Recombination Process[2]

The Shockley equation adequately predicts the current-voltage characteristics of germanium p-n junctions at low-current densities. For Si and GaAs p-n junctions, however, the ideal equation can give only qualitative agreement. The departures from the ideal are mainly due to the following effects: (1) the surface effect, (2) the generation and recombination of carriers in the depletion layer, (3) the tunneling of carriers between states in the band gap (particularly for GaAs), (4) the high-injection condition which may occur even at relatively small forward bias, and (5) the series resistance effect. In addition, under sufficiently larger field in the reverse direction, the junction will break down (as a result, for example, of avalanche multiplication). The junction breakdown will be discussed in Section 5.

The surface effects on p-n junctions are mainly due to the ionic charges on or outside the semiconductor surface which induce image charges in the semiconductor and thereby cause the formation of the so-called surface channels or surface depletion-layer regions. Once a channel is formed, it modifies the junction depletion region and gives rise to surface leakage current. The details of the surface effect will be discussed in Chapters 9 and 10 of metal-insulator-semiconductor (MIS) devices. For Si planar p-n junctions the surface leakage current is in general much smaller than the generation current in the depletion region.

Consider first the generation current under the reverse-bias condition. Because of the reduction in carrier concentration under reverse bias ($pn \ll n_i^2$), the dominant recombination-generation processes as discussed in Chapter 2 are those of emission. The rate of generation of electron-hole pairs can be obtained from Eq. (58) of Chapter 2 with the condition $p < n_i$ and $n < n_i$:

$$U = - \left[\frac{\sigma_p \sigma_n v_{th} N_t}{\sigma_n \exp\left(\dfrac{E_t - E_i}{kT}\right) + \sigma_p \exp\left(\dfrac{E_i - E_t}{kT}\right)} \right] n_i \equiv - \frac{n_i}{\tau_e} \qquad (47)$$

where τ_e is the effective lifetime and is defined as the reciprocal of the expression in the square bracket. The current due to the generation in the depletion region is thus given by

$$J_{gen} = \int_0^W q|U| \, dx \simeq q|U|W = \frac{qn_i W}{\tau_e} \qquad (48)$$

where W is the depletion-layer width. If the effective lifetime is a slowly varying function of temperature, the generation current will then have the same temperature dependence as n_i. At a given temperature, J_{gen} is proportional to the depletion-layer width which, in turn, is dependent on the applied reverse bias. It is thus expected that

$$J_{gen} \sim (V_{bi} + V)^{1/2} \qquad (49a)$$

for abrupt junctions, and

$$J_{gen} \sim (V_{bi} + V)^{1/3} \qquad (49b)$$

for linearly graded junctions.

The total reverse current (for $p_{no} \gg n_{po}$ and $|V| > 3kT/q$) can be approximately given by the sum of the diffusion components in the neutral region and the generation current in the depletion region:

$$J_R = q \sqrt{\frac{D_p}{\tau_p}} \frac{n_i^2}{N_D} + \frac{qn_i W}{\tau_e}. \qquad (50)$$

For semiconductors with large values of n_i (such as Ge) the diffusion component will dominate at room temperature and the reverse current will follow the Shockley equation; but if n_i is small (such as for Si), the generation current may dominate. A typical result[3] for Si is shown in Fig. 18, curve (e). At sufficiently high temperatures, however, the diffusion current will dominate.

At forward bias, where the major recombination-generation processes in the depletion region are the capture processes, we have a recombination current J_{rec} in addition to the diffusion current. Substitution of Eq. (28) in Eq. (58) of Chapter 2 yields

$$U = \frac{\sigma_p \sigma_n v_{th} N_t n_i^2 (e^{qV/kT} - 1)}{\sigma_n \left[n + n_i \exp\left(\dfrac{E_t - E_i}{kT}\right) \right] + \sigma_p \left[p + n_i \exp\left(\dfrac{E_i - E_t}{kT}\right) \right]}. \qquad (51)$$

Under the assumptions that $E_i = E_t$ and $\sigma_n = \sigma_p = \sigma$, Eq. (51) reduces to

$$U = \frac{\sigma v_{th} N_t n_i^2 (e^{qV/kT} - 1)}{n + p + 2n_i} \tag{52}$$

$$= \frac{\sigma v_{th} N_t n_i^2 (e^{qV/kT} - 1)}{n_i \left\{ \exp\left[\dfrac{q(\psi - \phi_n)}{kT}\right] + \exp\left[\dfrac{q(\phi_p - \psi)}{kT}\right] + 2 \right\}}. \tag{52a}$$

The maximum value of U exists in the depletion region where ψ is halfway between ϕ_p and ϕ_n, or $\psi = (\phi_n + \phi_p)/2$, and the denominator of Eq. (52a) becomes $2n_i[\exp(qV/2kT) + 1]$. We obtain for $V > kT/q$

$$U \simeq \frac{1}{2} \sigma v_{th} N_t n_i \exp\left(\frac{qV}{2kT}\right), \tag{53}$$

and

$$J_{rec} = \int_0^W qU \, dx \approx \frac{qW}{2} \sigma v_{th} N_t n_i \exp\left(\frac{qV}{2kT}\right) \sim n_i N_t. \tag{54}$$

Similar to the generation current in reverse bias, the recombination current in forward bias is also proportional to n_i. The total forward current can be approximated by the sum of Eqs. (44) and (54) (for $p_{no} \gg n_{po}$ and $V > kT/q$):

$$J_F = q\sqrt{\frac{D_p}{\tau_p}} \frac{n_i^2}{N_D} \exp\left(\frac{qV}{kT}\right) + \frac{qW}{2} \sigma v_{th} N_t n_i \exp\left(\frac{qV}{2kT}\right). \tag{55}$$

The experimental results in general can be represented by the following empirical form

$$J_F \sim \exp\left(\frac{qV}{nkT}\right) \tag{56}$$

where the factor $n = 2$ when the recombination current dominates, Fig. 18 curve (a), and $n = 1$ when the diffusion current dominates, Fig. 18, curve (b). When both currents are comparable, n has a value between 1 and 2.

(3) High-Injection Condition

At high current densities (under the forward-bias condition) such that the injected minority carrier density is comparable with the majority concentration, both drift and diffusion current components must be considered. The individual conduction current densities can always be given by Eqs. (29) and (30) and are repeated here:

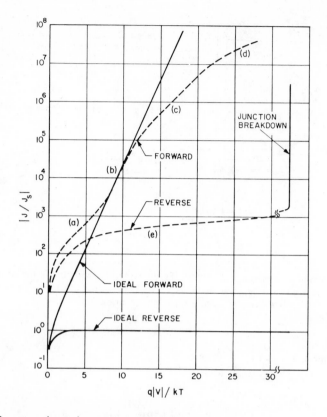

Fig. 18 Current-voltage characteristics of a practical Si diode
(a) generation-recombination current region
(b) diffusion current region
(c) high-injection region
(d) series resistance effect
(e) reverse leakage current due to generation-recombination and surface effect.
(After Moll, Ref. 3.)

$$J_p = -q\mu_p p \nabla \phi_p$$

$$J_n = -q\mu_n n \nabla \phi_n.$$

Since J_p, q, μ_p, and p are positive, the hole imref decreases monotonically to the right as shown in Fig. 15(a). Likewise the electron imref increases monotonically to the left. Thus, everywhere the separation of the imrefs must be less than, or equal to, the applied voltage and therefore[21]

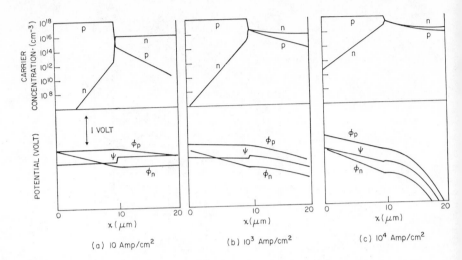

Fig. 19 Carrier concentrations, intrinsic Fermi level (ψ), and imrefs for a Si p-n step junction with the following parameters: $N_A = 10^{18}$ cm^{-3}, $N_D = 10^{16}$ cm^{-3}, $\tau_n = 3 \times 10^{-10}$ sec and $\tau_p = 8.4 \times 10^{-10}$ sec for various injection conditions
(a) 10 amp/cm^2
(b) 10^3 amp/cm^2
(c) 10^4 amp/cm^2.
(After Gummel, Ref. 21.)

$$pn \leq n_i^2 \exp\left(\frac{qV}{kT}\right). \tag{57}$$

This is true even under the high-injection condition. Note also that the above argument does not depend on recombination in the depletion region. As long as recombination takes place somewhere, currents will flow.

As an illustration of the high-injection case, we present in Fig. 19 plots of numerical results for intrinsic Fermi level (ψ), imrefs (ϕ_n and ϕ_p), and carrier concentrations for a silicon p-n step junction with the following parameters: $N_A = 10^{18}$ cm^{-3}, $N_D = 10^{16}$ cm^{-3}, $\tau_n = 3 \times 10^{-10}$ sec, and $\tau_p = 8.4 \times 10^{-10}$ sec. The current densities in Fig. 19(a), (b), and (c) are 10, 10^3, and 10^4 amp/cm^2. At 10 amp/cm^2 the diode is in the low-injection regime. Almost all of the potential drop occurs across the junction. The hole concentration on the n-side is small compared to the electron concentration. At 10^3 amp/cm^2 the electron concentration near the junction exceeds the donor concentration appreciably. An ohmic potential drop appears on the n-side. At 10^4 amp/cm^2 we have very high injection; the potential drop across the junction is insignificant in comparison to ohmic drops on both sides. Even though only the center region of the diode is shown in Fig. 19, it is apparent that the separation of the imrefs at the junction is less than or equal to the difference in the hole imref to the left of the junction and the electron imref to the right of the junction for all forward-bias levels.

From Fig. 19(b) and (c) the carrier densities at the n side of the junction are comparable or $n \approx p$. Substituting this condition in Eq. (57), we obtain $p_n(x = x_n) \approx n_i \exp(qV/2kT)$. The current then becomes roughly proportional to $\exp(qV/2kT)$ as shown in Fig. 18, curve (c).

At high-injection levels we should consider another effect associated with the finite resistivity in the quasi-neutral regions of the junction. This resistance absorbs an appreciable amount of the voltage drop between the diode terminals. This effect is shown in Fig. 18 curve (d). The series resistance effect can be substantially reduced by the use of epitaxial materials.

(4) Diffusion Capacitance

The depletion-layer capacitance considered previously accounts for most of the junction capacitance when the junction is reverse-biased. When forward-biased, there is in addition a significant contribution to junction capacitance from the rearrangement of minority carrier density, the so-called diffusion capacitance. When a small ac signal is applied to a junction which is forward-biased to a voltage V_0 and current density J_0, the total voltage and current are defined by

$$V(t) = V_0 + V_1 e^{j\omega t}$$

$$J(t) = J_0 + J_1 e^{j\omega t} \tag{58}$$

where V_1 and J_1 are the small-signal amplitude of the voltage and current density respectively. The electron and hole densities at the depletion region boundaries can be obtained from Eqs. (32) and (33) by using $(V_0 + V_1 e^{j\omega t})$ instead of V. The small-signal ac component of the hole density is given by

$$\tilde{p}_n(x, t) = p_{n1}(x)e^{j\omega t}; \tag{59}$$

we obtain for $V_1 \ll V_0$

$$p_n = p_{no} \exp\left[\frac{q(V_0 + V_1 e^{j\omega t})}{kT}\right]$$

$$\simeq p_{no} \exp\left(\frac{qV_0}{kT}\right) + \frac{p_{no}qV_1}{kT} \exp\left(\frac{qV_0}{kT}\right)e^{j\omega t}. \tag{60}$$

Similar expression is obtained for the electron density. The first term in Eq. (60) is the dc component, and the second term is the small-signal ac component at the depletion layer boundary $[p_{n1}(x_n)e^{j\omega t}]$. Substitution of p_n into the continuity equation [Eq. (97) of Chapter 2 with $G = 0$] yields

$$j\omega\tilde{p}_n = -\frac{\tilde{p}_n}{\tau_p} + D_p \frac{\partial^2 \tilde{p}_n}{\partial x^2}$$

or

$$\frac{\partial^2 \tilde{p}_n}{\partial x^2} - \frac{\tilde{p}_n}{D_p \tau_p/(1 + j\omega\tau_p)} = 0. \tag{61}$$

The above equation is identical to Eq. (39) if the carrier lifetime is expressed as

$$\tau_p^* = \frac{\tau_p}{1 + j\omega\tau_p}. \tag{62}$$

We can then obtain the alternating current density from Eq. (44) by making the appropriate substitutions:

$$J_1 = \frac{qV_1}{kT} \left[\frac{qD_p p_{no}}{L_p/\sqrt{1 + j\omega\tau_p}} + \frac{qD_n n_{po}}{L_n/\sqrt{1 + j\omega\tau_n}} \right] \exp\left(\frac{qV_0}{kT}\right). \tag{63}$$

Equation (63) leads directly to the ac admittance:

$$Y \equiv \frac{J_1}{V_1} = G_d + j\omega C_d. \tag{64}$$

For relatively low frequencies ($\omega\tau_p$, $\omega\tau_n \ll 1$), the diffusion conductance G_{d0} is given by

$$G_{d0} = \frac{q}{kT} \left(\frac{qD_p p_{no}}{L_p} + \frac{qD_n n_{po}}{L_n} \right) e^{qV_0/kT} \qquad \text{mho/cm}^2 \tag{65}$$

which has exactly the same value obtained by differentiating Eq. (44). The low-frequency diffusion capacitance C_{d0} is given by

$$C_{d0} = \frac{q}{kT} \left(\frac{qL_p p_{no}}{2} + \frac{qL_n n_{po}}{2} \right) e^{qV_0/kT} \qquad \text{farad/cm}^2. \tag{66}$$

The frequency dependence of the conductance and capacitance is shown in Fig. 20 as a function of the normalized frequency $\omega\tau$ where only one term in Eq. (63) is considered (for example the term contains p_{no} if $p_{no} \gg n_{po}$). The insert shows the equivalent circuit of the ac admittance. It is clear from Fig. 20 that the diffusion capacitance decreases with increasing frequency. For

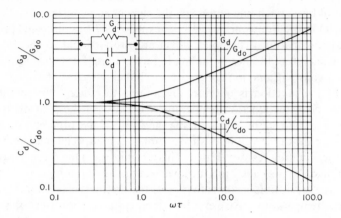

Fig. 20 Normalized diffusion conductance and diffusion capacitance versus $\omega\tau$. Insert shows the equivalent circuit of a p-n junction under forward bias.

large frequencies, $C_d \sim (\omega)^{-1/2}$. The diffusion capacitance, however, increases with the direct current level ($\sim e^{qV_0/kT}$). This is the reason that C_d is especially important at low frequencies and under forward-bias conditions.

5 JUNCTION BREAKDOWN[22]

When a sufficiently high field is applied to a p-n junction, the junction "breaks down" and conducts a very large current. There are basically three breakdown mechanisms: the thermal instability, the tunneling effect, and the avalanche multiplication. We shall consider briefly the first two mechanisms, and discuss in detail the avalanche multiplication.

(1) Thermal Instability

The breakdown due to thermal instability is responsible for the maximum dielectric strength in most insulators at room temperature, and is also a major effect in semiconductors with relatively small band gaps (e.g., Ge). Because of the heat dissipation which is caused by the reverse current at high reverse voltage, the junction temperature increases. This, in turn, increases the reverse current in comparison with its value at lower voltages. The temperature effect[23] on reverse current-voltage characteristics is shown in Fig. 21. In this figure the reverse currents, J_s, are represented by a family of horizontal lines. Each line represents the current at a constant junction temperature, and the current varies as $T^{3+\gamma/2} \exp(-E_g/kT)$, as discussed previously. The heat dissipation hyperbolas which are proportional to the IV product are shown as straight lines in the log-log plot. These lines also correspond to curves of constant junction temperature. The reverse current-voltage characteristic of the junction is obtained by joining the intersection points of the curves of constant junction temperature. Because of the heat dissipation at high reverse voltage, the characteristic shows a negative differential resistance. In this case the diode will be destroyed unless some special measure such as a large series-limiting resistor is used. This effect is called the thermal instability. The voltage, V_U, is called the turnover voltage. For *p-n* junctions with relatively large saturation currents (e.g., Ge) the thermal instability is important at room temperature. At very low temperatures, however, the thermal instability becomes less important in comparison with other mechanisms.

(2) Tunneling Effect

We next consider the tunneling effect. It is well known that for a one-dimensional square energy barrier with barrier height E_0 and thickness W, the quantum-mechanical transmission probability, T_t, is given by[24]

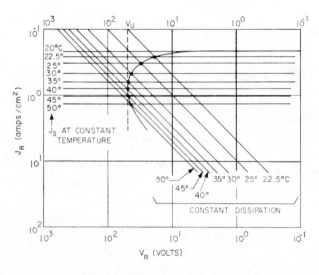

Fig. 21 Reverse current-voltage characteristics of thermal breakdown where V_u is the turnover voltage. (Note: Direction of coordinate increases are opposite to usual conventions.) (After Strutt, Ref. 23.)

$$T_t = \left[1 + \frac{E_0{}^2 \sinh^2 \beta W}{4E(E_0 - E)} \right]^{-1} \tag{67}$$

with

$$\beta \equiv \sqrt{\frac{2m(E_0 - E)}{\hbar^2}}$$

where E is the energy of the carrier. The probability decreases monotonically with decreasing E. When $\beta W \gg 1$, the probability becomes

$$T_t \approx \frac{16E(E_0 - E)}{E_0{}^2} \exp(-2\beta W). \tag{67a}$$

A similar expression has been obtained for p-n junctions. The detailed mathematical treatment will be given in Chapter 4. The tunneling current density is given by[22]

$$J_t = \frac{\sqrt{2m^*} q^3 \mathscr{E} V}{4\pi^2 \hbar^2 E_g^{1/2}} \exp\left(- \frac{4\sqrt{2m^*} E_g^{3/2}}{3q\mathscr{E}\hbar} \right) \tag{68}$$

where $\mathscr{E}$ is the electric field at the junction, E_g the band gap, V the applied voltage, and m^* the effective mass.

When the field approaches 10^6 V/cm in Ge and Si, significant current begins to flow by means of the band-to-band tunneling process. In order to obtain such a high field, the junction must have relatively high impurity concentrations on both the p and n sides. The mechanism of breakdown for Si and Ge junctions with breakdown voltages less than about $4E_g/q$ is found to be due to the tunneling effect. For junctions with breakdown voltages in excess of $6E_g/q$, the mechanism is caused by the avalanche multiplication. At voltages between 4 and 6 E_g/q the breakdown is due to a mixture of both avalanche and tunneling. Since the energy band gaps, E_g, in Ge, Si, and GaAs decrease with increasing temperature (refer to Chapter 2) the breakdown voltage in these semiconductors due to the tunneling effect has a negative temperature coefficient, i.e., the voltage decreases with increasing temperature. This is because a given breakdown current, J_t, can be reached at smaller reverse voltages (or fields) at higher temperatures, Eq. (68). A typical example is shown in Fig. 22. This temperature effect is generally used to distinguish the tunneling mechanism from the avalanche mechanism which has a positive temperature coefficient, i.e., the breakdown voltage increases with increasing temperature.

(3) Avalanche Multiplication

The avalanche multiplication (or impact ionization) is the most important mechanism in junction breakdown, since the avalanche breakdown voltage

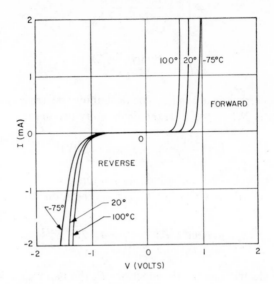

Fig. 22 Current-voltage characteristics of tunneling breakdown. Note that for a given reverse current the voltage decreases with temperature, i.e., negative temperature coefficient for breakdown due to tunneling. (After Strutt, Ref. 23.)

imposes an upper limit on the reverse voltage for most diodes and on the collector voltage of all transistors, in addition, the impact ionization mechanism can be used to generate microwave power as in IMPATT devices (*Impact-Avalanche and Transit-Time* Devices) to be discussed in Chapter 5. The electron and hole ionization rates (α_n and α_p) have been considered in Chapter 2.

We shall first derive the basic ionization integral which determines the breakdown condition. Assume a current I_{po} is incident at the left-hand side of the depletion region with width W. If the electric field in the depletion region is high enough that electron-hole pairs are generated by the impact ionization process, the hole current (I_p) will increase with distance through the depletion region and reaches a value $M_p I_{po}$ at W. Similarly, the electron current (I_n) will increase from $x = W$ to $x = 0$. The total current $I(=I_p + I_n)$ is constant at steady state. The incremental hole current at x is equal to the number of electron-hole pairs generated per second in the distance dx,

$$d(I_p/q) = (I_p/q)(\alpha_p \, dx) + (I_n/q)(\alpha_n \, dx) \tag{69}$$

or

$$dI_p/dx - (\alpha_p - \alpha_n)I_p = \alpha_n I \tag{70}$$

The solution† of Eq. (70) with the boundary condition that $I = I_p(W) = M_p I_{po}$ is given by

$$I_p(x) = I\left\{\frac{1}{M_p} + \int_0^x \alpha_n \exp\left[-\int_0^x (\alpha_p - \alpha_n)\, dx'\right] dx\right\} \Big/ \exp\left[-\int_0^x (\alpha_p - \alpha_n)\, dx'\right] \tag{71}$$

where M_p is the multiplication factor of holes and is defined as

$$M_p \equiv \frac{I_p(W)}{I_p(0)}. \tag{72}$$

Equation (71) can be written as

$$1 - \frac{1}{M_p} = \int_0^W \alpha_p \exp\left[-\int_0^x (\alpha_p - \alpha_n)\, dx'\right] dx. \tag{73}$$

The avalanche breakdown voltage is defined as the voltage where M_p approaches infinity. Hence the breakdown condition is given by the ionization integral

$$\int_0^W \alpha_p \exp\left[-\int_0^x (\alpha_p - \alpha_n)\, dx'\right] dx = 1. \tag{74}$$

If the avalanche process is initiated by electrons instead of holes, the ionization integral is given by

$$\int_0^W \alpha_n \exp\left[-\int_x^W (\alpha_n - \alpha_p)\, dx'\right] dx = 1. \tag{75}$$

If the avalanche process is initiated by both electrons and holes, the breakdown condition is given by the simultaneous solution of both Eqs. (74) and (75). For semiconductors with equal ionization rates ($\alpha_n = \alpha_p = \alpha$) such as GaAs and GaP, Eqs. (74) and (75) reduce to the simple expression

$$\int_0^W \alpha\, dx = 1. \tag{76}$$

† Equation (70) has the form $y' + Py = Q$ where $y \equiv I_p$. The standard solution is

$$y = \left[\int_0^x Q e^{\int_0^x P\, dx'}\, dx + C\right] \Big/ e^{\int_0^x P\, dx'}$$

where C is the constant of integration.

From the above breakdown conditions and the field dependence of the ionization rates, the breakdown voltages, the maximum electric field, and the depletion layer width can be calculated. As discussed previously, the electric field and potential in the depletion layer are determined from the solutions of Poisson's equation. The depletion layer boundaries, such that Eq. (74) is satisfied, can be obtained numerically with a computer using the iteration method. With known boundaries we obtain

$$V_B \text{ (breakdown voltage)} = \frac{\mathscr{E}_m W}{2} = \frac{\varepsilon_s \mathscr{E}_m^{\;2}}{2q} (N_B)^{-1} \qquad (77)$$

for one-sided abrupt junctions, and

$$V_B = \frac{2\mathscr{E}_m W}{3} = \frac{4\mathscr{E}_m^{3/2}}{3} \left(\frac{2\varepsilon_s}{q}\right)^{1/2} (a)^{-1/2} \qquad (78)$$

for linearly graded junctions where N_B is the ionized background impurity concentration of the lightly doped side, ε_s the semiconductor permittivity, "a" the impurity gradient, and $\mathscr{E}_m$ the maximum field.

Figure 23 shows the calculated breakdown voltage[25] as a function of N_B for abrupt junctions in Ge, Si, GaAs, and GaP. Figure 24 shows the calculated voltage versus the impurity gradient for linearly graded junctions in these semiconductors. The experimental results of breakdown voltages in p-n junctions are generally in good agreement with the calculated values.[26] The dashed lines in the above figures indicate the upper limits of N_B or "a" for which the avalanche breakdown calculation is valid. This limitation is based on the criteria $6E_g/q$. Above these values the tunneling mechanism will also contribute to the breakdown process and eventually dominate.

The calculated values of the maximum field $\mathscr{E}_m$ and the depletion layer width at breakdown for the above four semiconductors are shown[25] in Fig. 25 for the abrupt junctions and in Fig. 26 for the linearly graded junctions. Because of the strong dependence of the ionization rates on the field, the maximum field varies very slowly with either N_B or a. Thus, as a first approximation, we can assume that, for a given semiconductor, $\mathscr{E}_m$ has a fixed value. Then from Eqs. (77) and (78) we obtain $V_B \sim N_B^{-1.0}$ for abrupt junctions and $V_B \sim a^{-0.5}$ for linearly graded junctions. Figures 23 and 24 show that the above patterns are generally followed. Also as expected, for a given N_B or a, the breakdown voltage increases with the energy band gap, since the avalanche process requires band-to-band excitations.

An approximate universal expression can be given as follows for the above results comprising all of the semiconductors studied:

$$V_B \cong 60(E_g/1.1)^{3/2}(N_B/10^{16})^{-3/4} \qquad \text{volts} \qquad (79)$$

for abrupt junctions where E_g is the band gap in eV, and N_B is the background

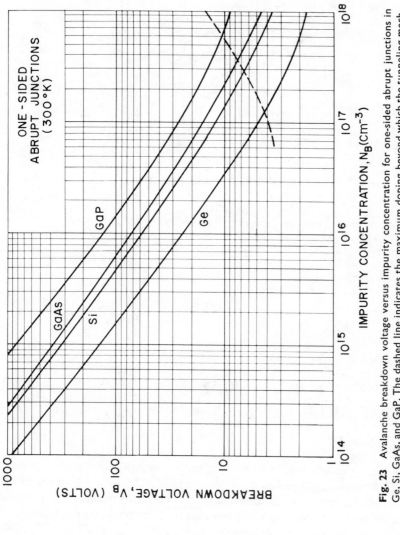

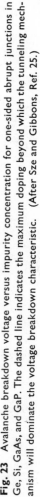

Fig. 23 Avalanche breakdown voltage versus impurity concentration for one-sided abrupt junctions in Ge, Si, GaAs, and GaP. The dashed line indicates the maximum doping beyond which the tunneling mechanism will dominate the voltage breakdown characteristic. (After Sze and Gibbons, Ref. 25.)

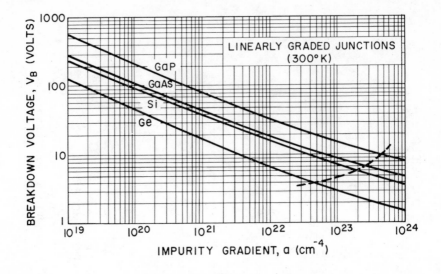

Fig. 24 Avalanche breakdown voltage versus impurity gradient for linearly graded junctions in Ge, Si, GaAs, and GaP. The dashed line indicates the maximum gradient beyond which the tunneling mechanism will set in. (Ref. 25.)

doping in cm^{-3}; and

$$V_B \cong 60(E_g/1.1)^{6/5}(a/3 \times 10^{20})^{-2/5} \qquad \text{volts} \qquad (80)$$

for linearly graded junctions where a is the impurity gradient in cm^{-4}.

The above results (Fig. 23 through 26) are for avalanche breakdowns at room temperature. At higher temperatures the breakdown voltage increases. A simple explanation for this increase is that the hot carriers passing through the depletion layer under high field lose part of their energy to optical phonons after traveling each electron-phonon mean free path, λ. The value of λ decreases with increasing temperature, Eq. (83) of Chapter 2. Therefore the carriers lose more energy to the crystal lattice along a given distance at constant field. Hence the carriers must pass through a greater potential

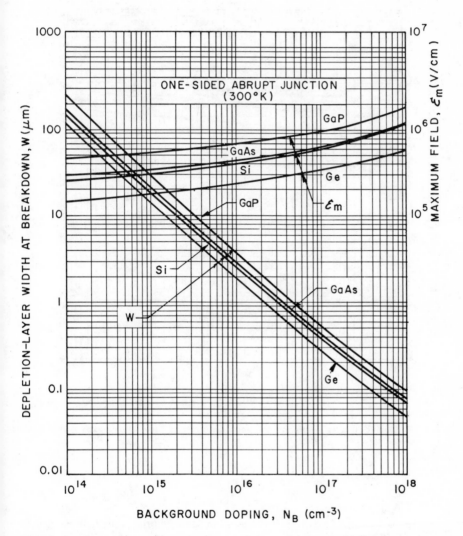

Fig. 25 Depletion-layer width and maximum field at breakdown for one-sided abrupt junctions in Ge, Si, GaAs, and GaP. (Ref. 25.)

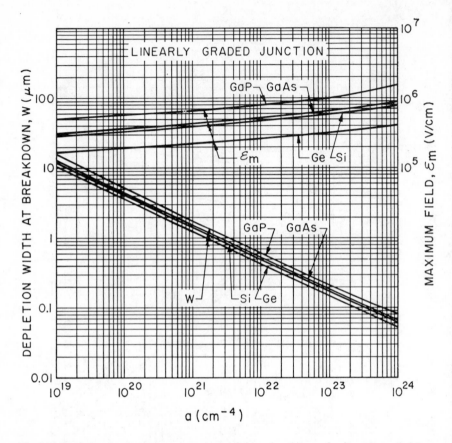

Fig. 26 Depletion-layer width and maximum field at breakdown for linearly graded junctions in Ge, Si, GaAs, and GaP.
(Ref. 25.)

difference (or higher voltage) before they can acquire sufficient energy to generate an electron-hole pair. The detailed calculations have been done by the use of a modification of Baraff's theory [26a] as discussed in Chapter 2. The predicted values of V_B normalized to the room-temperature value are shown in Fig. 27 for Ge and Si. For the same doping profile the predicted percentage change on V_B with temperature is about the same for GaAs as it is for Ge and for GaP as it is for Si junctions. We note that there are substantial increases of the breakdown voltage especially for lower dopings (or small gradient) at higher temperatures. Figure 28 shows the measured results,[27] which agree quite well with this theory.

For planar junctions there is an important effect which should be considered—the junction curvature effect. A schematic diagram of a planar junction has been shown in Fig. 6(b). Since the cylindrical and/or spherical regions of the junction have a higher field intensity, the avalanche breakdown voltage is determined by these regions. The potential $V(r)$ and the electric field $\mathscr{E}(r)$ in a cylindrical or spherical p-n junction can be calculated from Poisson's equation:

$$\frac{1}{r}\frac{d}{dr}[r\mathscr{E}(r)] = \frac{\rho(r)}{\varepsilon_s} \tag{81a}$$

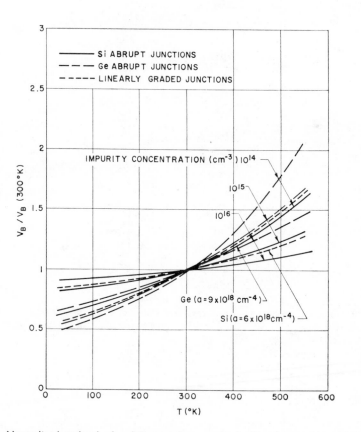

Fig. 27 Normalized avalanche breakdown voltage versus lattice temperature. The breakdown voltage increases with increasing temperature, i.e., positive temperature coefficient for breakdown due to avalanche multiplication.
(After Crowell and Sze, Ref. 26a.)

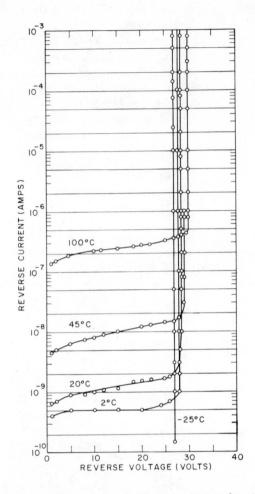

Fig. 28 Temperature dependence of reverse *I-V* characteristics of a microplasma-free n^+p Si diode with $N_B = 2.5 \times 10^{16}$ cm^{-3} and an *n*-type guard ring. The temperature coefficient is 0.024 V/C°.
(After Goetzberger et al., Ref. 27.)

for the cylindrical junction, and

$$\frac{1}{r^2}\frac{d}{dr}\left[r^2\mathscr{E}(r)\right] = \frac{\rho(r)}{\varepsilon_s} \tag{81b}$$

for the spherical junction. The solution for $\mathscr{E}(r)$ can be obtained from Eq.

(81) and is given by

$$\mathscr{E}(r) = \frac{1}{\varepsilon_s r^n} \int_{r_j}^{r} r^n \rho(r)\, dr + \frac{\text{constant}}{r^n} \qquad (82)$$

where n equals 1 for the cylindrical junction and 2 for the spherical junction, r_j the radius of curvature of the metallurgical junction; and the constant must be adjusted so that the breakdown condition Eq. (74) or (75) is satisfied. The calculated results[18] for Si one-sided abrupt junctions at 300°K are shown in Fig. 29. We note that the plane junction result is the limiting case when $r_j \to \infty$. With finite r_j, and for a given concentration, the breakdown voltage of the spherical junction is always smaller than that of the cylindrical junction. Typically, for a doping of 10^{15} cm^{-3}, V_B is 330 V for a plane junction, 80 V for a cylindrical junction with $r_j = 1$ μm, and 39 V for a spherical junction with the same junction radius. Similar results are obtained for Ge, GaAs, and GaP. A general expression can be given for V_B (in units of volts) as follows:

$$V_B \cong 60\left(\frac{E_g}{1.1}\right)^{3/2}\left(\frac{N_B}{10^{16}}\right)^{-3/4} \{[(n + 1 + \gamma)\gamma^n]^{1/(n+1)} - \gamma\} \qquad (83)$$

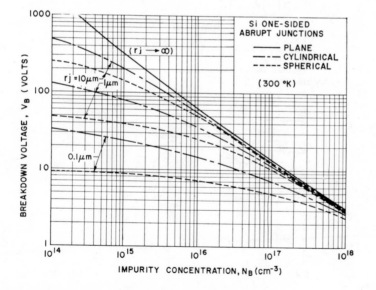

Fig. 29 Avalanche breakdown voltage versus impurity concentration for one-sided abrupt doping profile with cylindrical and spherical junction geometries where r_j is the radius of curvature as indicated in Fig. 6.
(After Sze and Gibbons, Ref. 18.)

where $n = 1$ and 2 for cylindrical and spherical junctions respectively, and $\gamma \equiv r_j/W$ where W is the depletion-layer width (in microns) shown in Fig. 25. When $r_j \rightarrow \infty$, the expression in { } approaches unity and Eq. (83) reduces to Eq. (79). The maximum electric fields at avalanche breakdown in abrupt Si junctions are shown in Fig. 30 as a function of background doping.[18] The parameter r_j is the radius of curvature of the cylindrical junction. The corresponding depletion-layer widths at breakdown for various N_B and r_j are shown in Fig. 31.

For linearly graded junctions, it is found that the breakdown voltage is essentially independent of r_j, and the results obtained for plane, cylindrical, and spherical junctions differ by only about 5% or less, and the plots shown in Fig. 24 are valid for the above three cases. In Fig. 32 are the results for idealized Si composite junctions in which the space charge terminates in a graded region on one side of the junction and in a uniformly doped region on the other. In the limits of large and small impurity gradients, the same results are obtained as for the abrupt junctions and linearly graded junctions respectively. Figure 32 has been plotted for two values of r_j (1 μm in dashed

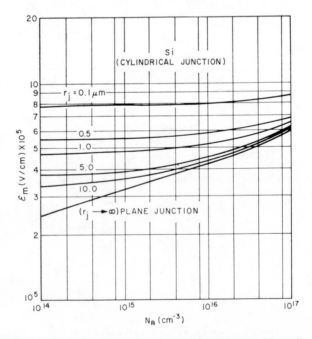

Fig. 30 Maximum field at breakdown for Si cylindrical junctions. Note that for a given background doping, N_B, the maximum field increases with decreasing radius of curvature. (Ref. 18.)

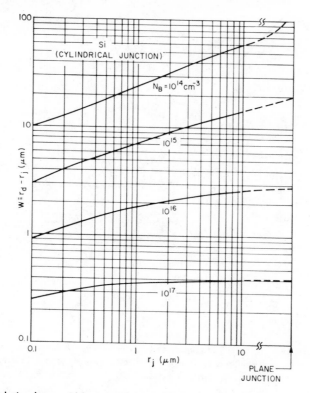

Fig. 31 Depletion-layer width at breakdown for Si cylindrical junction. (Ref. 18.)

lines and 10 μm in solid lines). The vertical dotted lines delineate the sweep-out condition which occurs when the electric field reaches the surface and the space-charge between $r = 0$ and $r = r_j$ is not enough to terminate the field.

The reason that the junction curvature has great effect on the breakdown of the abrupt junctions but has virtually no effect on the linearly graded junction is as follows: in linearly graded junctions Eq. (82) can be expressed as

$$\mathscr{E}(r) = -\mathscr{E}_1(r) + \mathscr{E}_2(r)$$

$$\mathscr{E}_1(r) \equiv \mathscr{E}_1(x) = \frac{qax^2}{2\varepsilon_s} \frac{(1 + 2x/3r_j)}{(1 + x/r_j)}$$

$$\mathscr{E}_2(r) \equiv \mathscr{E}_2(x) = \frac{C}{(x + r_j)}$$

(84)

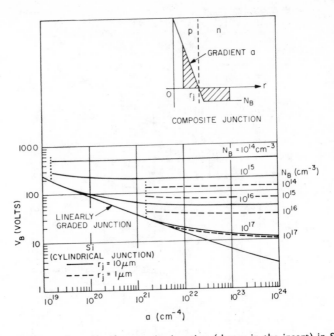

Fig. 32 Breakdown voltage V_B of composite junction (shown in the insert) in Si for two values of r_j. For small gradient, V_B approaches that for linearly graded junction; for large gradient, V_B approaches that for abrupt junction. (Ref. 18.)

where $x \equiv r - r_j$ and C is the constant of integration. Figure 33 shows plots of $\mathscr{E}_1$ and $\mathscr{E}_2$ versus x for Si linearly graded junctions with $r_j = 1$ μm (dashed lines), and 10 μm (solid lines), and the proper values of C chosen to satisfy the breakdown condition. One notes that the $\mathscr{E}_2$ field profile is approximately symmetrical with respect to the metallurgical junction. This symmetry causes the area which lies between $\mathscr{E}_2$ and $\mathscr{E}_1$ for $r - r_j < 0$, to be approximately equal to the area for $r - r_j > 0$. The breakdown voltage, which is represented by the area enclosed by $\mathscr{E}_1(x)$ and $\mathscr{E}_2(x)$, is therefore virtually independent of r_j. For abrupt junctions, however, the field profile is unsymmetrical with respect to the metallurgical junction (as shown in Fig. 34). Therefore there is no compensation in areas, and the breakdown voltage, which is equal to the area enclosed by $\mathscr{E}_1(x)$, $\mathscr{E}_2(x)$, increases with r_j resulting in the strong dependence of V_B on r_j in abrupt junctions.

Aside from the junction curvature effect on breakdown, there are many other effects such as punch-through, microplasma, and surface-field effect

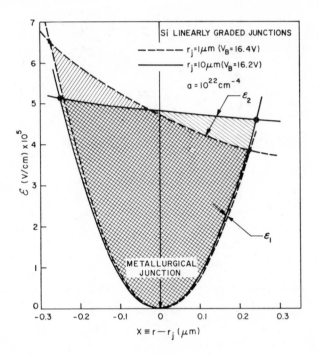

Fig. 33 Field distribution at breakdown for linearly graded junctions with two different values of r_j.
(Ref. 18.)

which also influence the breakdown characteristics. The punch-through (or sweep-out) effect is shown in Fig. 35 in which the breakdown voltage is plotted against the background doping for Si one-sided abrupt junctions formed on epitaxial substrates (e.g., n on n^+) with the epitaxial-layer thickness, W, as a parameter. For a given thickness the breakdown voltage approaches a constant value as the doping decreases corresponding to a complete sweep-out of the epitaxial layer. The microplasma effect, which occurs at lattice defects or around metallic precipitates, is a localized breakdown in small high-field regions. With reduced crystal imperfections (e.g., etch pits with a density less than $100 \ cm^{-2}$), the microplasma effect is expected to be reduced. The surface field effect also has profound influence on breakdown characteristics. We shall, however, discuss this effect in detail in Chapter 10.

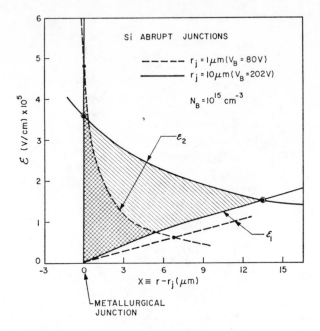

Fig. 34 Field distribution at breakdown for one-sided abrupt junction with two different values of r_j.
(Ref. 18.)

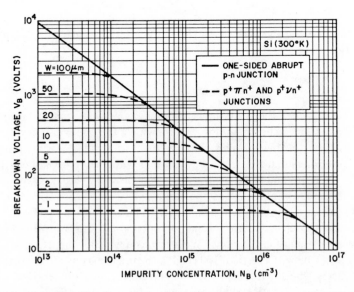

Fig. 35 Breakdown voltage for $p^+\pi n^+$ and $p^+\nu n^+$ structure where π is for lightly doped p type and ν for lightly doped n type. W is the thickness of the π or ν region.

6 TRANSIENT BEHAVIOR AND NOISE

(1) Transient Behavior

For switching applications it is required that the transition from forward bias to reverse bias be nearly abrupt and the transient time be short. In Fig. 36(a) a simple circuit is shown where a forward current I_F is flowing in the

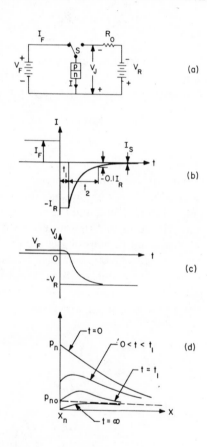

Fig. 36 Transient behavior of a p-n junction
(a) basic switching circuit
(b) transient response where t_1 is the time interval for the constant-current phase and t_2 is that for the decay phase
(c) junction voltage as a function of time
(d) minority carrier distribution for various time intervals.
(After Kingston, Ref. 28.)

p-n junction; at time $t = 0$, the switch S is suddenly thrown to the right, and initial reverse current $I_R \simeq V/R$ flows. The transient time is defined as the time in which the current reaches 10 percent of the initial current I_R, and is equal to the sum of t_1 and t_2 as shown in Fig. 36(b), where t_1 and t_2 are the time intervals for the constant-current phase and the decay phase respectively. Consider the constant-current phase (also called storage phase) first. The continuity equation as given in Chapter 2 can be written for the *n*-type side (assume $p_{po} \ll n_{no}$) as

$$\frac{\partial p_n(x, t)}{\partial t} = D_p \frac{\partial^2 p_n(x, t)}{\partial x^2} - \frac{p_n(x, t) - p_{no}}{\tau_p} \tag{85}$$

where τ_p is the minority-carrier lifetime. The boundary conditions are that at $t = 0$ the initial distribution of holes is a steady-state solution to the diffusion equation, and that the voltage across the junction is given from Eq. (33) as

$$V_j = \frac{kT}{q} \ln \frac{p_n(0, t)}{p_{no}}. \tag{86}$$

The distribution of the minority carrier density p_n with time is shown[28] in Fig. 36(d). From Eq. (86) it can be calculated that, as long as $p_n(0, t)$ is greater than p_{no} (in the time interval $0 < t < t_1$), the junction voltage V_J remains of the order of kT/q, as shown in Fig. 36(c), and the current I_R is approximately given by $V/R = $ constant. Hence in this time interval the reverse current is constant and we have the constant-current phase. However, at or near t_1 the hole density approaches zero, the junction voltage tends to minus infinity, and a new boundary condition now holds. This is the decay phase with the boundary condition $p(0, t) = p_{no} = $ constant. The solutions have been given by Kingston,[28] and the times t_1 and t_2 are given by the transcendental equations

$$\mathrm{erf} \sqrt{\frac{t_1}{\tau_p}} = \frac{1}{1 + \dfrac{I_R}{I_F}} \tag{87}$$

$$\mathrm{erf} \sqrt{\frac{t_2}{\tau_p}} + \frac{\exp\left(-\dfrac{t_2}{\tau_p}\right)}{\sqrt{\pi \dfrac{t_2}{\tau_p}}} = 1 + 0.1 \left(\frac{I_R}{I_F}\right). \tag{88}$$

The results are shown in Fig. 37 where the solid lines are for the plane junction with the length of the *n*-type material W much greater than the diffusion length ($W \gg L_p$), and the dashed lines are for the narrow-base

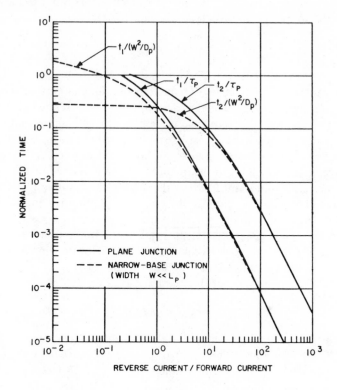

Fig. 37 Normalized time versus the ratio of reverse current to forward current. (Ref. 28.)

junction with $W \ll L_p$. For a large ratio I_R/I_F, the transit time can be approximated by

$$(t_1 + t_2) \simeq \frac{\tau_p}{2} \left(\frac{I_R}{I_F}\right)^{-1} \tag{89a}$$

for $W \gg L_p$, or

$$(t_1 + t_2) \simeq \frac{W^2}{2D_p} \left(\frac{I_R}{I_F}\right)^{-1} \tag{89b}$$

for $W \ll L_p$. If one switches a junction $(W \gg L_p)$ from forward 10 ma to reverse 10 ma $(I_R/I_F = 1)$, the time for the constant-current phase is 0.3 τ_p, and that for the decay phase is about 0.6 τ_p. Total transient time then is 0.9 τ_p. For a fast switch, one thus requires that τ_p be small. The lifetime τ_p

can be substantially reduced by the introduction of impurities with deep levels in the forbidden gap (such as gold in silicon).

(2) Noise[29]

The term "noise" refers to the spontaneous fluctuations in the current passing through, or the voltage developed across, semiconductor bulk materials or devices. A study of the noise phenomena is important for semiconductor devices. Since the devices are mainly used to measure small physical quantities or to amplify small signals, the spontaneous fluctuations in current or voltage set a lower limit to the quantities to be measured or the signals to be amplified. It is important to know the factors contributing to these limits, to use this knowledge to optimize the operating conditions, and to find new methods and new technology to reduce the noise.

Observed noise is generally classified into thermal noise, flicker noise, and shot noise. The thermal noise occurs in any conductor or semiconductor and is caused by the random motion of the current carriers. The open-circuit mean square voltage $\langle V_n^2 \rangle$ of thermal noise is given by

$$\langle V_n^2 \rangle = 4kTBR \tag{90}$$

where k is the Boltzmann constant, T the absolute temperature in $°K$, B the the bandwidth in Hz, and R the real part of the impedance between terminals. At room temperature, for a semiconductor material with 1 kΩ resistance, the root mean square voltage, $\sqrt{\langle V_n^2 \rangle}$, measured with a 1-Hz bandwidth is only about 4 nV (1 nV = 10^{-9} volt).

The flicker noise is distinguished by its peculiar spectral distribution which is proportional to $1/f^\alpha$ with α generally close to unity (the so-called $1/f$ noise). Flicker noise is important at lower frequencies. It has been shown that, for most semiconductor devices, the origin of the flicker noise is due to the surface effect. The $1/f$ noise-power spectrum has been correlated both qualitatively and quantitatively with the lossy part of the metal-insulator-semiconductor (MIS) gate impedance due to carrier recombination at the interface surface states. We shall defer the discussion of the flicker noise until the chapter on MIS devices (Chapter 9).

The shot noise constitutes the major noise in most semiconductor devices. It is independent of frequency (white spectrum) at low and intermediate frequencies. At higher frequencies the shot-noise spectrum also becomes frequency dependent. The mean square noise current of shot noise for a *p-n* junction is given by

$$\langle i_n^2 \rangle = 2qBI \tag{91}$$

where I is the current which is positive in the forward and negative in the reverse direction. For low injection the total mean square noise current (neglecting $1/f$ noise) is given by

$$\langle i_n^2 \rangle = 4kTBG - 2qBI. \tag{92}$$

From the Shockley equation we obtain

$$G = \frac{\partial I}{\partial V} = \frac{\partial}{\partial V} [I_s(e^{qV/kT} - 1)] = \frac{qI_s}{kT} e^{qV/kT}. \tag{93}$$

Substitution of Eq. (93) into Eq. (92) yields for the forward-bias condition

$$\langle i_n^2 \rangle = 2qI_s Be^{qV/kT} + 2qBI_s. \tag{94}$$

Experimental measurements indeed confirm that the mean square noise current is proportional to the saturation current I_S which can be varied by irradiation.

It has been suggested by Van Der Ziel that a better terminology should be used to classify the noises according to the physical processes: generation-recombination noise (GR), diffusion noise, and modulation noise. The GR noise is caused by spontaneous fluctuations in the generation rates, recombination rates, and trapping rates of the carriers, thus causing fluctuations in the free carrier density. For junction devices the GR noise shows close resemblance to shot noise. The diffusion noise is caused by the fact that diffusion is a random process; consequently fluctuations in the diffusion rate give rise to localized fluctuations in the carrier density. In bulk material it is the cause of thermal noise; in junction devices, however, it is a major contributor to shot noise. The modulation noise refers to noise not directly caused by fluctuations in the drift or diffusion rates but due, instead, to carrier density fluctuations or current fluctuations caused by some modulation mechanisms such as the surface field effect in MIS devices.

7 TERMINAL FUNCTIONS

A p-n junction is a two-terminal device which can perform various terminal functions depending upon its biasing conditions as well as its doping profile and device geometry. In this section we shall briefly discuss some of the interesting device performances based on the current-voltage, capacitance-voltage, and breakdown characteristics discussed in the previous sections. Two important cases which will not be considered here are the tunnel diode and IMPATT diode. We shall consider these two devices in detail in the two subsequent chapters.

(1) Rectifier

A rectifier is a *p-n* junction diode which is specifically designed to rectify alternating current, i.e., to give a very low resistance to current flow in one direction and a very high resistance in the other direction. The forward and reverse resistances of a rectifier can be easily derived from the current-voltage relationship of a practical diode,

$$I = I_s(e^{qV/nkT} - 1) \tag{95}$$

where I_S is the saturation current and the factor n generally has a value between 1 and 2 ($n = 1$ for diffusion current and $n = 2$ for recombination current). The forward dc (or static) resistance R_F and small-signal (or dynamic) resistance r_F are obtainable from Eq. (95):

$$R_F \equiv \frac{V_F}{I_F} \left(\simeq \frac{V_F}{I_s} e^{-qV_F/nkT} \quad \text{for} \quad V \geq 3kT/q \right) \tag{96a}$$

$$r_F = \frac{dV_F}{dI_F} = \frac{nkT}{qI_F}. \tag{96b}$$

The reverse dc resistance R_R and small-signal resistance r_R are given by

$$R_R \equiv \frac{V_R}{I_R} \simeq \left(\frac{V_R}{I_s} \quad \text{for} \quad |V_R| \geq 3kT/q \right) \tag{97a}$$

$$r_R \equiv \frac{\partial V_R}{\partial I_R} = \frac{nkT}{qI_s} e^{q|V_R|/kT} \tag{97b}$$

Comparison of Eqs. (96) and (97) shows that the dc rectification ratio, R_R/R_F, varies with $\exp(qV_F/nkT)$; while the ac rectification ratio, r_R/r_F, varies with $I_F/[I_s \exp(-q|V_R|/kT)]$.

Rectifiers generally have slow switching speeds, i.e., a significant time delay is necessary to obtain high impedance after switching from the forward-conduction state to the reverse-blocking state. This time delay (proportional to the minority carrier lifetime as shown in Fig. 37) is of little consequence in rectifying 60-Hz currents. For high-frequency applications the lifetime should be sufficiently reduced to maintain rectification efficiency. The majority of rectifiers has power-dissipation capabilities from 0.1 to 10 watts, reverse breakdown voltages from 50 to 2500 volts (for high-voltage rectifier two or more *p-n* junctions are connected in series), and switching times from 50 ns (1 ns = 10^{-9} sec) for low-power diodes to about 500 ns for high-power diodes.

(2) Voltage Regulator

A voltage regulator is a p-n junction diode operated in the reverse direction up to its breakdown voltage. Prior to the breakdown, the diode has a very high resistance, and after breakdown the diode has a very small dynamic resistance. The voltage is thus limited (or regulated) by the breakdown voltage.

Most voltage regulators are made of Si. This is mainly because of the low saturation current in Si diodes and the advanced Si technology. As discussed in Section 5, for breakdown voltage, V_B, larger than $6E_g/q$ (~ 8 volts for Si), the breakdown mechanism is mainly avalanche multiplication, and the temperature coefficient of V_B is positive. For $V_B < 4E_g/q$ (~ 5 volts for Si) the breakdown mechanism is band-to-band tunneling, and the temperature coefficient of V_B is negative. For $4E_g/q < V_B < 6E_g/q$, the breakdown is due to a combination of these two mechanisms. One can connect, for example, a negative-temperature-coefficient diode in series with a positive-temperature-coefficient diode to produce a low-temperature-coefficient regulator (with a temperature coefficient of the order of $0.002\%/°C$) which is suitable as a voltage reference.

(3) Varistor

A varistor (*vari*able resi*stor*) is a two-terminal device which shows nonohmic behavior. We have shown in Eqs. (91) and (92) the nonohmic characteristics of a p-n junction diode. Similar nonohmic characteristics are obtainable from metal-semiconductor diodes which will be considered in Chapter 8. An interesting application of varistors is their use as symmetrical fractional-voltage (~ 0.5 volt) limiter by connecting two diodes in parallel, oppositely poled. The two-diode unit will exhibit the forward I-V characteristics in either direction.

(4) Varactor[30]

The term "varactor" comes from the words *var*iable re*actor* and means a device whose reactance can be varied in a controlled manner with a bias voltage. Varactor diodes are widely used in parametric amplification, harmonic generation, mixing, detection, and voltage variable tuning.[31,32]

The basic capacitance-voltage relationships have already been derived in Section 3. We shall now extend the previous derivations of abrupt and linearly graded doping distributions to a more general case. The one-dimensional

Poisson equation is given as

$$\frac{\partial^2 V}{\partial x^2} = -\frac{qN}{\varepsilon_s} \tag{98}$$

where N is the generalized doping distribution as shown in Fig. 38 (assuming one side is very heavily doped):

$$N = Bx^m \quad \text{for} \quad x \geq 0. \tag{99}$$

For $m = 0$ we have $B = N_B$ corresponding to the uniformly doped (or one-sided abrupt junction) case. For $m = 1$, the doping profile corresponds to a one-sided linearly graded case. For $m < 0$, the device is called a "hyper-abrupt" junction.[33] The hyperabrupt doping profile can be achieved by epitaxial process. The boundary conditions are $V(x = 0) = 0$ and $V(x = W) = V + V_{bi}$ where V is the applied voltage and V_{bi} is the built-in voltage.

Integrating Poisson's equation with the boundary conditions, we obtain for the depletion-layer width and the differential capacitance per unit area

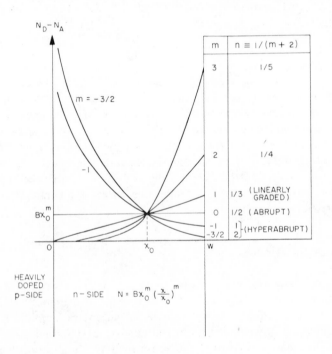

Fig. 38 Various impurity distributions for varactors.
(After Norwood and Shatz, Ref. 30.)

$$W = \left[\frac{\varepsilon_s(m + 2)(V + V_{bi})}{qB}\right]^{1/(m+2)} \tag{100}$$

$$C \equiv \frac{\partial Q_c}{\partial V} = \left[\frac{qB(\varepsilon_s)^{m+1}}{(m+2)(V + V_{bi})}\right]^{1/(m+2)} \sim (V + V_{bi})^{-n} \tag{101}$$

$$n \equiv \frac{1}{(m+2)}$$

where Q_c is the charge per unit area and is equal to the product of ε_s and the maximum electric field (at $x = 0$). It is apparent that, when $m = 0$ or 1, the expressions of Eqs. (100) and (101) reduce to the abrupt and linearly graded junction cases derived previously. Of particular interest is the hyperabrupt case where $m = -3/2$ and $n = 2$. For this impurity distribution the resonant frequency is linearly dependent upon bias voltage or

$$f_r = \frac{1}{2\pi\sqrt{LC}} \sim \frac{1}{\sqrt{C}} \sim (V + V_{bi})^{n/2} = (V + V_{bi})^1. \tag{102}$$

This device behavior is useful in frequency modulation and in elimination of distortion.

The simplified equivalent circuit of a varactor is shown[32] in Fig. 39(a), where C_J is the junction capacitance, R_S is the series resistance, and R_P is the parallel equivalent resistance of generation-recombination current, diffusion current, and surface leakage current. Both C_J and R_S decrease with the reverse-bias voltage, while R_P generally increases with voltage. The efficiency of a varactor is expressed as a quality factor Q, which is the ratio of energy stored to energy dissipation:

$$Q \simeq \frac{\omega C_J R_P}{(1 + \omega^2 C_J{}^2 R_P R_S)}. \tag{103}$$

The above expression can be differentiated to obtain the angular frequency of maximum Q, ω_0, and the value of this maximum, Q_{max}. These expressions are shown as

$$\omega_0 \simeq \frac{1}{C_J(R_P R_S)^{1/2}} \tag{104}$$

$$Q_{max} \approx \left(\frac{R_P}{4R_S}\right)^{1/2}. \tag{105}$$

Figure 39(b) shows a qualitative graph of the relationship between Q, frequency, and bias voltage. For a given bias, Q varies as $\omega C_J R_P$ at low frequencies and as $1/\omega C_J R_S$ at high frequencies. The maximum bias voltage is limited by the breakdown voltage V_B.

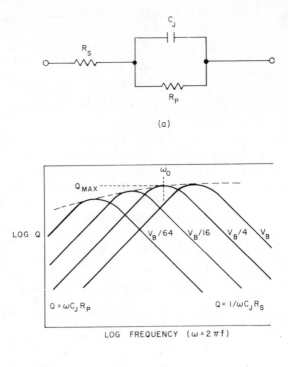

Fig 39
(a) Simplified equivalent circuit of a varactor.
(After Penfield and Rafuse, Ref. 32.)
(b) Quality factor Q versus frequency for various bias voltages.
(After Norwood and Shatz, Ref. 30.)

(5)   Fast-Recovery Diode

Fast-recovery diodes are designed to give ultrahigh switching speed. The devices can be classified into two types: diffused *p-n* junction diodes and metal-semiconductor diodes. The equivalent circuit of both types can be represented by that of the varactor diode, Fig. 39(a); and the general switching behavior of both types can be described by that shown in Fig. 36(b).

The total recovery time $(t_1 + t_2)$ for a *p-n* junction diode can be substantially reduced by introducing recombination centers such as Au in Si. Although the recovery time is directly proportional to the lifetime τ, as shown in Fig. 37, unfortunately it is not possible to reduce recovery times to zero by introducing an extremely large number of recombination centers, N_t, since the reverse generation current of a *p-n* junction is proportional to N_t [Eqs. (47) and (48)]. For direct band gap semiconductors such as GaAs, the minority-carrier lifetimes are generally much smaller than that of Si. This results in ultrahigh-speed for GaAs *p-n* junction diodes with recovery times of the order of 0.1 ns or less. For Si the practical recovery time is in the range of 1 to 5 ns.

The metal-semiconductor diode (Schottky diode) also exhibits ultrahigh-speed characteristics. This is because most Schottky diodes are majority-carrier devices and there is negligible minority-carrier storage effect. We shall discuss metal-semiconductor contacts in detail in Chapter 8.

(6) Charge-Storage Diode

In contrast to fast-recovery diodes a charge-storage diode is designed to store charge while conducting in the forward direction and upon switching to conduct for a short period in the reverse direction. A particularly interesting charge-storage diode is the step-recovery diode (also called the snapback diode) which conducts in the reverse direction for a short period then abruptly cuts off the current as the stored charges have been dispelled. This cutoff occurs in the range of picoseconds and results in a fast-rising wavefront rich in harmonics. Because of these characteristics step-recovery diodes are used as harmonic generators and pulse formers. Most charge-storage diodes are made from Si with relatively long minority-carrier lifetimes ranging from 0.5 to 5 μs. Note that the lifetimes are about one thousand times longer than that for fast-recovery diodes.

(7) p-i-n Diode[34,35]

A *p-i-n* diode is a *p-n* junction with a doping profile tailored in such a way that an intrinsic layer, "*i* region," is sandwiched in between a *p* layer and an *n* layer, Fig. 40(a). In practice, however, the idealized *i* region is approximated by either a high-resistivity *p* layer (referred to as π layer) or a high-resistivity *n* layer (*v* layer). The impurity distribution, space charge density, and field distribution in *p-i-n* and *p-π-n* diodes are shown[36] in Fig. 40(b), (c), (d) respectively. Because of low doping in the *i* region, most of the potential will drop across this region. For a practical *p-i-n* diode the impurity distribution in the *p* and *n* layers varies more gradually than that shown in Fig. 40. It can be fabricated, for example, using (1) the epitaxial process, (2) the diffusion of *p* and *n* regions into a high-resistivity semiconductor substrate, and (3) the ion-drift (e.g., lithium) method to introduce the highly compensated intrinsic region.[37]

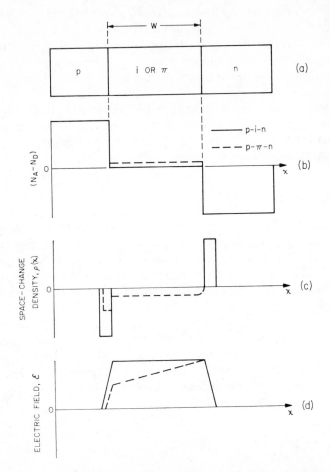

Fig. 40 Impurity distribution, space charge density, and field distribution in *p-i-n* and *p-π-n* junctions.
(After Veloric and Prince, Ref. 36.)

The typical impedance-voltage relationships[38,39,40] of a *p-i-n* diode are shown in Fig. 41. Figure 41(a) shows the reverse characteristics where the depletion-layer capacitance remains essentially at a constant value and the series resistance decreases with increasing reverse bias. The simplified equivalent circuit is shown in the insert. These results can be explained by referring to the simple *p-n* junction theory. For a *p-i-n* or *p-π-n* diode such that the *p* and *n* regions are heavily doped, the depletion capacitance can be calculated similarly to that of a one-sided abrupt junction. Because of the low doping concentration, the entire π region is usually depleted even at zero bias. The

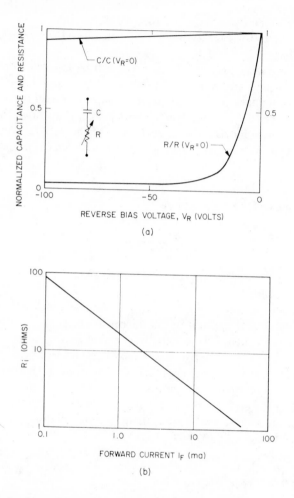

Fig. 41 Normalized impedance-voltage relationship of a *p-i-n* diode. (After Senhouse, Ref. 38; Larrabee, Ref. 39; and Olsen, Ref. 40.)

capacitance per unit area is then given by ε_s/W where W is the width of the π region. With increasing reverse bias the depletion layer will widen slightly into the p and n regions and result in a slight decrease in capacitance. The zero-bias resistance is mainly due to a slightly unswept-out π region. As the reverse bias increases, the series resistance decreases rapidly toward an asymtotic value corresponding to the contact resistance and resistance of the substrate. The breakdown voltage V_B of a *p-i-n* diode is approximately equal to $\mathscr{E}_m W$ where $\mathscr{E}_m$ is the maximum electric field at breakdown. For Si $\mathscr{E}_m \approx 2 \times 10^5$ V/cm, so that $V_B \approx 1000$ volts for an i region 50 μm thick. Computed results for breakdown voltages in Si and Ge p-i-n diodes[41] are shown in Chapter 5.

Under the forward bias condition, holes will be injected from the p layer to the i region and electrons will be injected from the n layer. The current crossing the boundary between the p and i layers will be entirely hole current given by $I_F = AqpW/\tau_p$ where A is the device area, p the hole density, and τ_p the hole lifetime. Similarly, the current crossing the n-"i" boundary will be entirely electron current. The conductivity of the i layer is given by

$$\sigma = q(\mu_p p + \mu_n n) \sim I_F. \tag{106}$$

The resistance of the i layer will be

$$R_i = \frac{W}{\sigma A} \sim \frac{1}{I_F}. \tag{107}$$

The above prediction is in reasonable agreement with the experimental result shown in Fig. 41(b).

Because of these interesting forward and reverse characteristics, the p-i-n diode has found wide applications in microwave circuits.[42,43] It can be used as a microwave switch with essentially constant depletion-layer capacitance and high power-handling capability. The switching speed[44] is approximately given by $W/2v_{sl}$ where v_{sl} is the scattering-limited velocity across the i region. In addition, a p-i-n diode can be used as a variolosser (variable attenuator) by controlling the device resistance which varies approximately linearly with the forward current. It can also be used as a modulator to perform signal modulating functions up to the GHz range.

8   HETEROJUNCTION

A heterojunction is a junction formed between two semiconductors having different energy band gaps. In 1951 Shockley[45] proposed the abrupt heterojunction to be used as an efficient emitter-base junction in a transistor. Kroemer[46] later analyzed a similar, though graded, heterojunction as a wide-gap emitter. Since then heterojunctions have been extensively studied, and many applications have been proposed, among them are the majority-carrier rectifier, the high-speed band-pass photodetector, the beam-of-light transistor, and the indirect-gap injection lasers.[47] A heterojunction can be formed by various methods, among them are (1) the interface-alloy technique[47,48,49] which utilizes the difference in melting points between the two semiconductors for selective melting and regrowth, (2) the epitaxial vapor growth technique,[50] and (3) the vacuum deposition of one material on another.

A perfect match of lattice constants (refer to Table 2.1) and thermal expansion coefficients, however, are not normally possible in heterojunctions and, therefore, defects such as interfacial dislocations are generally present at the heterojunction interface.[51] These interface states can act as trapping centers and severely limit the device potential. Nevertheless, the heterojunction is a useful device in the sense that it can further understanding of the carrier transport processes and of the variation of band gap across the interface; it will also further the study of lattice mismatch and epitaxial processes, as well as the investigation of many interesting physical phenomena associated with discontinuity at the interface.

The energy band model of an ideal abrupt heterojunction without interface states was proposed by Anderson[50] based on the previous work of Shockley. We shall now consider this model, since it can adequately explain most of the transport processes, and only slight modification of the model is needed to account for nonideal cases such as interface states. Figure 42(a) shows the energy-band diagram of two isolated pieces of semiconductors. The two semiconductors are assumed to have different band gaps (E_g), different permittivities (ε), different work functions (ϕ_m), and different electron affinities (χ). Work function and electron affinity are defined, respectively, as that energy required to remove an electron from the Fermi level (E_F) and from the bottom of the conduction band (E_C) to a position just outside the material (vacuum level). The difference in energy of the conduction-band edges in the two semiconductors is represented by ΔE_C and that in the valence band edges by ΔE_V.

When a junction is formed between these semiconductors, the energy-band profile at equilibrium is shown in Fig. 42(b). This is called an n-p heterojunction a junction formed between an n-type narrow-gap semiconductor and a p-type wide-gap semiconductor. Since the Fermi level must coincide on both sides in equilibrium and the vacuum level is everywhere parallel to the band edges and is continuous, the discontinuity in conduction-band edges (ΔE_C) and valence-band edges (ΔE_V) is invariant with doping in those cases where E_g and χ are not functions of doping (i.e., nondegenerate semiconductors). The total built-in potential V_{bi} is equal to the sum of the partial built-in voltages ($V_{b1} + V_{b2}$) where V_{b1} and V_{b2} are the electrostatic potentials supported at equilibrium by semiconductors 1 and 2 respectively.

The depletion widths and capacitance can be obtained by solving Poisson's equation for the step junction on either side of the interface.[50,52] One of the boundary conditions is the continuity of electric displacement, i.e., $\varepsilon_1 \mathscr{E}_1 = \varepsilon_2 \mathscr{E}_2$ at the interface. We obtain

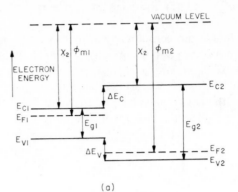

(a)

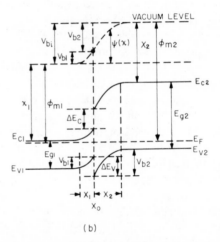

(b)

Fig. 42
(a) Energy band diagram for two isolated semiconductors in which space charge neutrality is assumed to exist in each region.
(b) Energy band diagram of an ideal *n-p* heterojunction at equilibrium.
(After Anderson, Ref. 50.)

$$x_1 = \left[\frac{2N_{A2}\,\varepsilon_1\varepsilon_2(V_{bi} - V)}{qN_{D1}(\varepsilon_1 N_{D1} + \varepsilon_2 N_{A2})} \right]^{1/2} \tag{108}$$

$$x_2 = \left[\frac{2N_{D1}\varepsilon_1\varepsilon_2(V_{bi} - V)}{qN_{A2}(\varepsilon_1 N_{D1} + \varepsilon_2 N_{A2})} \right]^{1/2} \tag{109}$$

and

$$C = \left[\frac{q N_{D1} N_{A2} \varepsilon_1 \varepsilon_2}{2(\varepsilon_1 N_{D1} + \varepsilon_2 N_{A2})(V_{bi} - V)} \right]^{1/2} \tag{110}$$

The relative voltage supported in each of the semiconductors is

$$\frac{V_{b1} - V_1}{V_{b2} - V_2} = \frac{N_{A2} \varepsilon_2}{N_{D1} \varepsilon_1}. \tag{111}$$

where $V = V_1 + V_2$. It is apparent the above expressions will reduce to that for the p-n junction (homojunction) discussed in Section 3, where both sides of the heterojunction have the same materials.

The case of an n-n heterojunction of the above two semiconductors is somewhat different. Since the work function of the wide-gap semiconductor is the smaller, the energy bands will be bent oppositely to the n-p case [see Fig. 43(a)].[53] The relation between $V_{b1} - V_1$ and $V_{b2} - V_2$ can be found from the boundary condition of continuity of electric displacement at the interface. For an accumulation in region 1 governed by Boltzmann statistics, [for detailed derivation see Section 2 of Chapter 9] the electric displacement $\mathcal{D}_1$ at x_0 is given by

$$\mathcal{D}_1 = \varepsilon_1 \mathscr{E}_1(x_0) = \left\{ 2\varepsilon_1 q N_{D1} \left[\frac{kT}{q} \left(\exp \frac{q(V_{b1} - V_1)}{kT} - 1 \right) - (V_{b1} - V_1) \right] \right\}^{1/2}. \tag{112}$$

The electric displacement at the interface for a depletion in region 2 is given by

$$\mathcal{D}_2 = \varepsilon_2 \mathscr{E}_2(x_0) = [2\varepsilon_2 q N_{D2}(V_{b2} - V_2)]^{1/2}. \tag{113}$$

Equating Eqs. (112) and (113) gives a relation between $(V_{b1} - V_1)$ and $(V_{b2} - V_2)$ which is quite complicated. However, if the ratio $\varepsilon_1 N_{D1}/\varepsilon_2 N_{D2}$ is of the order of unity and $V_{bi}(\equiv V_{b1} + V_{b2}) \gg kT/q$, we obtain[53]

$$\exp\left[\frac{q(V_{b1} - V_1)}{kT} \right] \simeq \frac{q}{kT}(V_{bi} - V) \tag{114}$$

where V is the total applied voltage and is equal to $(V_1 + V_2)$. Also shown in Fig. 43 are the idealized equilibrium energy band diagrams for p-n (narrow-gap p-type and wide-gap n-type) and p-p heterojunctions.

If interface states are present, the above idealized conditions should be modified. The energy band at the interface is free to move up or down with the necessary charge being supplied by electrons (or their absence) in the interface states. The discontinuity in the conduction band is still equal to the difference in electron affinities, however, the height of the conduction band edge above the Fermi level at the interface is determined primarily by the interface states. Figure 44(a) shows the energy band diagram of a Ge-Si *n-n* heterojunction (with about 4% lattice mismatch).[51] It is observed that both sides of the junction are depleted, a situation made possible by the acceptor nature of the interface states. A magnified energy band diagram of the *n-n* heterojunction with interface states is shown in Fig. 44(b). The interface states

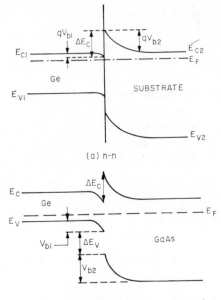

(a) n-n

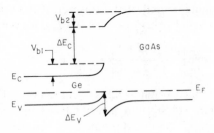

(b) p-n

(c) p-p

Fig. 43
(a) Energy band diagram for an ideal abrupt *n-n* heterojunction.
(After Chang, Ref. 53.)
(b) and (c) Energy band diagrams for ideal *p-n* and *p-p* heterojunctions respectively.
(After Anderson, Ref. 50.)

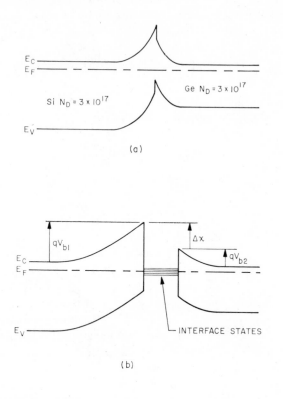

Fig. 44
(a) Ge-Si n-n heterojunction with interface states.
(b Magnified interface region.
(After Oldham and Milnes, Ref. 51.)

are assumed to lie in a thin layer sandwiched between the two depletion regions; and these states can act as generation-recombination centers.

The current-voltage characteristics in heterojunctions are influenced by various mechanisms depending on the band discontinuities at the interface and the density of interface states. For example, if the barrier to holes is much higher than that for electrons, then the current will consist almost entirely of electrons; or if the density of interface states is very high then the dominant current will be generation-recombination current from the interface. The dominant current can also be due to tunneling if the barrier width is very thin, or to thermionic emission if the interface acts as a metal-semiconductor contact.

We shall consider an interesting case as shown in Fig. 43(a). The conduction mechanism is governed by thermionic emission (refer to Chapter 8 for details) and the current density is given by[53]

$$J = A^* T^2 \exp\left(-\frac{qV_{b2}}{kT}\right)\left[\exp\left(\frac{qV_2}{kT}\right) - \exp\left(\frac{qV_1}{kT}\right)\right] \qquad (115)$$

where A^* is the effective Richardson constant. Substitution of Eq. (114) into Eq. (115) yields the current-voltage relationship:

$$J = J_0\left(1 - \frac{V}{V_{bi}}\right)\left[\exp\left(\frac{qV}{kT}\right) - 1\right] \qquad (116)$$

where

$$J_0 \equiv \frac{qA^*TV_{bi}}{k}\exp\left(-\frac{qV_{bi}}{kT}\right).$$

The above expression is somewhat different from that for metal-semiconductor contact. The value of J_0 is different and so is its temperature dependence. The reverse current never saturates but increases linearly with voltage at large V. In the forward, the dependence of J on qV/kT can be approximated by an exponential function or $J \sim \exp qV/nkT$. The above predicted results have been observed in Ge-GaAs$_{1-x}$P$_x$ n-n heterojunctions for $x \simeq 0.1$ or less. For substrates with $x = 0.3$, however, the lattice mismatch is greater than 1% and the effect of interface states causes the bands to bend, as is shown in Fig. 44. The device then behaves as a structure with two diodes connected back to back in which the interface states cause the depletion of electrons from both semiconductors.

REFERENCES

1. W. Shockley, "The Theory of *p-n* Junctions in Semiconductors and *p-n* Junction Transistors," Bell Syst. Tech. J., *28*, 435 (1949); also *Electrons and Holes in Semiconductors*, Van Nostrand Book Co. (1950).

2. C. T. Sah, R. N. Noyce, and W. Shockley, "Carrier Generation and Recombination in *p-n* Junction and *p-n* Junction Characteristics," Proc. IRE, *45*, 1228 (1957).

3. J. L. Moll, "The Evolution of the Theory of the Current-Voltage Characteristics of *p-n* Junctions," Proc. IRE, *46*, 1076 (1958).

4. For example see A. G. Grove, *Physics and Technology of Semiconductor Devices*, John Wiley and Sons, Inc., New York (1967).

5. R. N. Hall and W. C. Dunlap, "*p-n* Junctions Prepared by Impurity Diffusion," Phys. Rev., *80*, 467 (1950).

6. M. Tanenbaum and D. E. Thomas, "Diffused Emitter and Base Silicon Transistors," Bell Syst. Tech. J., *35*, 1 (1956).

7. C. J. Frosch and L. Derrick, "Surface Protection and Selective Masking During Diffusion in Silicon," J. Electrochem. Soc., *104*, 547 (1957).

8. J. A. Hoerni, "Planar Silicon Transistor and Diodes," IRE Electron Devices Meeting, Washington, D.C. (1960).

9. H. C. Theuerer, J. J. Kleimack, H. H. Loar, and H. Christenson, "Epitaxial Diffused Transistors," Proc. IRE, *48*, 1642 (1960).

10. E. L. Jordan, "A Diffusion Mask for Germanium," J. Electrochem. Soc., *108*, 478 (1961).

11. P. F. Schmidt and W. Michel, "Anodic Formation of Oxide Films on Silicon," J. Electrochem. Soc., *104*, 230 (1957).

12. J. R. Ligenza, "Silicon Oxidation in an Oxygen Plasma Excited by Microwaves," J. Appl. Phys., *36* 2703 (1965).

13. M. M. Atalla, "Semiconductor Surfaces and Films; the Silicon-Silicon Dioxide System," in *Properties of Elemental and Compound Semiconductors*, H. Gatos, Ed., Vol. 5, pp. 163-181, Interscience (1960).

14. B. E. Deal and A. S. Grove, "General Relationship for the Thermal Oxidation of Silicon," J. Appl. Phys., *36*, 3770 (1965).

15. D. Flatley, N. Goldsmith, and J. Scott, "A Zinc Diffusion Mask," presented at the Electrochemical Society Meeting, Toronto, Canada (May 1964).

16. For a general reference, see H. S. Carslaw and J. C. Jaeger, *Conduction of Heat in Solids*, Oxford University Press, 2nd Ed. (1959).

17. D. L. Kendall, *Diffusion in III-V Compounds with Particular Reference to Self-Diffusion in InSb*, Report No. 65-29, Dept. of Material Science, Stanford University, Stanford (August 1965).

18. S. M. Sze and G. Gibbons, "Effect of Junction Curvature on Breakdown Voltages in Semiconductors," Solid State Electron., *9*, 831 (1966).

19. T. P. Lee and S. M. Sze, "Depletion Layer Capacitance of Cylindrical and Spherical p-n Junctions," Solid State Electron., *10*, 1105 (1967).

20. H. K. Gummel and D. L. Scharfetter, "Depletion-Layer Capacitance of p^+n Step Junction," J. Appl. Phys., *38*, 2148 (1967).

21. H. K. Gummel, "Hole-Electron Product of p-n Junctions," Solid State Electron., *10*, 209 (1967).

22. For a general discussion, see J. L. Moll, *Physics of Semiconductors*, McGraw-Hill Book Co., New York (1964).

23. M. J. O. Strutt, *Semiconductor Devices*, Vol. 1, Semiconductor and Semiconductor Diodes, Academic Press, New York, Ch. 2, (1966).

24. L. J. Schiff, *Quantum Mechanics*, 2nd Ed., McGraw-Hill Book Co., New York (1955)

25. S. M. Sze and G. Gibbons, "Avalanche Breakdown Voltages of Abrupt and Linearly Graded p-n Junctions in Ge, Si, GaAs, and GaP," Appl. Phys. Letters, *8*, 111 (1966).

26. H. Kressel, "A Review of the Effect of Imperfections on the Electrical Breakdown of *p-n* Junctions," RCA Review, Vol. 28, pp. 175-207 (1967).

26a. C. R. Crowell and S. M. Sze, "Temperature Dependence of Avalanche Multiplication in Semiconductors," Appl. Phys. Letters, *9*, 242 (1966).

27. A. Goetzberger, B. McDonald, R. H. Haitz, and R. M. Scarlet, "Avalanche Effects in Silicon *p-n* Junction. II. Structurally Perfect Junctions," J. Appl. Phys., *34*, 1591 (1963).

28. R. H. Kingston, "Switching Time in Junction Diodes and Junction Transistors," Proc. IRE, *42*, 829 (1954).

29. A. Van Der Ziel, *Fluctuation Phenomena in Semiconductors*, Academic Press, New York (1959).

30. For a review, see M. H. Norwood and E. Shatz, "Voltage Variable Capacitor Tuning —A Review," Proc. IEEE, *56*, 788 (1968).

31. J. H. Forster and R. M. Ryder, "Diodes Can Do Almost Anything," Bell Labs Rec., *39*, 2 (1961).

32. P. Penfield, Jr. and R. P. Rafuse, *Varactor Applications*, Cambridge, Mass., MIT Press (1962).

33. M. E. McMahon and G. F. Straube, "Voltage Sensitive Semiconductor Capacitor," IRE WESCON Conv. Rec., Pt. 3, pp. 72-82 (August 1958).

34. M. B. Prince, "Diffused *p-n* Junction Silicon Rectifiers," Bell Syst. Tech. J., *35*, 661 (1956).

35. R. N. Hall, "Power Rectifiers and Transistors," Proc. IRE, *40*, 1512 (1952).

36. H. S. Veloric and M. B. Prince, "High Voltage Conductivity-Modulated Silicon Rectifier," Bell Syst. Tech. J., *36*, 975 (1957).

37. E. M. Pell, "Ion Drift in an *n-p* Junction," J. Appl. Phys., *31*, 291 (1960); also J. W. Mager, "Characteristics of *p-i-n* Junction Produced by Ion-Drift Techniques in Silicon," J. Appl. Phys., *33*, 2894 (1962).

38. L. S. Senhouse, "Reverse Biased *p-i-n* Diode Equivalent Circuit Parameters at Microwave Frequencies," IEEE Trans. Electron Devices, *ED-13*, 314 (1966).

39. R. D. Larrabee, "Current Voltage Characteristics of Forward Biased Long *p-i-n* Structures," Phys. Rev., *121*, 37 (1961).

40. H. M. Olsen, "Design Calculation of Reverse Bias Characteristics for Microwave *p-i-n* Diodes," IEEE Trans. Electron Devices, *ED-14*, 418 (1967).

41. G. Gibbons and S. M. Sze, "Avalanche Breakdown in Read Diodes and *p-i-n* Diodes," Solid State Electron., *11*, 225 (1968).

42. H. Benda, A. Hoffman, and E. Spenke, "Switching Processes in Alloyed *p-i-n* Rectifiers," Solid State Electron., *8*, 887 (1965).

43. D. Leenov, "The Silicon *p-i-n* Diode as a Microwave Radar Protector at Megawatt Levels," IEEE Trans. Electron Devices, *ED-11*, 53 (1964).

44. G. Lucovsky, R. F. Schwarz, and R. B. Emmons, "Transit-Time Considerations in *p-i-n* Diodes," J. Appl. Phys., *35*, 622 (1964).

45. W. Shockley, U.S. Patent 2,569,347 (1951).

46. H. Kroemer, "Theory of a Wide-Gap Emitter for Transistors," Proc. IRE, *45*, 1535 (1957).

47. For a summary of the proposed applications, see for example, R. H. Rediker, S. Stopek, and J. H. R. Ward, "Interface-Alloy Epitaxial Heterojunctions," Solid State Electron., *7*, 621 (1964).

48. E. D. Hinkley and R. H. Rediker, "GaAs-InSb Graded-Gap Heterojunction," Solid State Electron., *10*, 671 (1967).

49. J. Shewchun and L. Y. Wei, "Germanium-Silicon Alloy Heterojunction," J. Electrochemical Soc., *111*, 1145 (1964).

50. R. L. Anderson, "Experiments on Ge-GaAs Heterojunctions," Solid State Electron., *5*, 341 (1962).

51. W. G. Oldham and A. G. Milnes, "Interface States in Abrupt Semiconductor Heterojunctions," Solid State Electron., *7*, 153 (1964).

52. L. L. Chang, "Comments on the Junction Boundary Conditions for Heterojunctions," J. Appl. Phys., *37*, 3908 (1966).

53. L. L. Chang, "The Conduction Properties of Ge-GaAs$_{1-x}$P$_x$ *n-n* Heterojunctions," Solid State Electron., *8*, 721 (1965).

54. The "planar" process uses techniques all of which were previously known, such as oxide masking against diffusion. The distinguishing feature is their use in combination, permitting an unprecedented fineness of control of the sizes and shapes of electrodes and diffused regions. The name "planar" comes from the requirement that the wafer surface must be approximately flat. If the surface is rough, the liquid photoresist does not coat it evenly, and imperfections result.

■ INTRODUCTION

■ EFFECTS OF HIGH DOPING

■ TUNNELING PROCESSES

■ EXCESS CURRENT

■ CURRENT-VOLTAGE CHARACTERISTICS
DUE TO EFFECTS OF DOPING,
TEMPERATURE, ELECTRON BOMBARDMENT,
AND PRESSURE

■ EQUIVALENT CIRCUIT

■ BACKWARD DIODE

4

Tunnel Diode and Backward Diode

I INTRODUCTION

The first acknowledged paper[1] on the tunnel diode (also referred to as the Esaki diode) was reported by Leo Esaki in 1958. There he described, in the course of studying the internal field emission in degenerate germanium p-n junctions, an "anomalous" current-voltage characteristic observed in the forward direction, i.e., a negative resistance region over part of the forward characteristic. It was said that before 1958 this anomalous characteristic of some p-n junctions was observed by many solid-state research scientists, but these p-n junctions were rejected immediately because they did not follow the "classic" diode equation. Esaki, however, explained this anomalous characteristic by the use of a quantum tunneling concept and obtained reasonable agreement between his tunneling theory and the experimental results. This work quickly led others to more sophisticated approaches in both theories and experiments. The ultrahigh-speed, low-power operation, and particularly low-noise characteristics of the tunnel diode are useable for many circuit applications including microwave amplification, high-speed switching, and binary memory.[2]

The tunneling phenomenon is a majority carrier effect. In addition the tunneling time of carriers through the potential energy barrier is not governed by the conventional transit time concept ($\tau = W/v$ where W is the barrier width and v is the carrier velocity) but rather by the quantum transition probability per unit time which is proportional to $\exp[-2\bar{k}(0)W]$ where $\bar{k}(0)$ is the average value of momentum encountered in the tunneling path corresponding to an incident carrier with zero transverse momentum and energy equal to the Fermi energy.[3] Reciprocating gives the tunneling time proportional to $\exp[2\bar{k}(0)W]$. This tunneling time is very short, permitting the use of tunnel diodes well into the millimeter-wave region.

In this chapter we shall first discuss the effect of high doping concentrations. In Section 3 the tunneling processes and the derivation of the tunneling current will be considered. Section 4 discusses the role of excess current. Section 5 presents detailed consideration of the current-voltage character-istics due to effects of doping, temperature, electron bombardment, and pres-sure. Also included is a brief discussion of the degradation phenomenon. In Section 6 we shall discuss the equivalent circuit and the basic characterization of a tunnel diode. Finally, Section 7 considers the backward diode which is akin to the tunnel diode and is useful because of its large reverse current and its nonlinearity near the zero-bias condition.

2 EFFECTS OF HIGH DOPING

A tunnel diode consists of a simple *p-n* junction in which both the *p* and the *n* sides are very heavily doped with impurities. There are three major effects due to high doping densities: (1) the Fermi level is located within the allowed bands themselves, (2) the impurity states broaden into bands, and (3) the intrinsic band gap is reduced (so-called band-edge tailing). In addition the impurity bands give rise to the valley current in tunnel diodes which will be discussed in the section on excess current.

For a semiconductor with an impurity concentration much less than the effective density of states (N_C or N_V), the semiconductor is nondegenerate, and the Fermi level lies in the band gap. When the impurity concentration is approaching or greater than the effective density of states, the Fermi level is located within the conduction or valence band, and the semiconductor is defined as degenerate. One necessary condition for obtaining a tunnel diode is that both *p*-type and *n*-type semiconductor materials must be degenerate. We can calculate the doping density which is necessary to make the Fermi level just coincide with the edge of band. Let us consider an *n*-type semicon-ductor in which we assume that the donor level E_D is a discrete level. From Eqs. (11) and (24) of Chapter 2 and the condition $n \approx N_D^+$ we obtain by setting $E_F - E_C = 0$ and $E_F - E_D = E_d$:

$$N_D = N_C \frac{2}{\sqrt{\pi}} F_{1/2}(0) \left[1 + 2 \exp\left(\frac{E_d}{kT}\right) \right]$$

$$\approx 0.68 \, N_C \left[1 + 2 \exp\left(\frac{E_d}{kT}\right) \right]. \tag{1}$$

The concentrations calculated from Eq. (1) for shallow donor levels are about 2×10^{19} cm^{-3} for Ge and about 6×10^{19} cm^{-3} for Si. An n-type semiconductor with density of N_D equal to or greater than that of Eq. (1) is a degenerate semiconductor. A similar calculation can be made for p-type semiconductors.

The assumption of a discrete impurity level is actually violated at very high doping levels, since the donor and acceptor states broaden out into bands themselves. The result of Eq. (1), nevertheless, serves as a first-order approximation and gives a reasonable prediction of the order of magnitude for the concentrations required to give degeneracy. Figure 1 compares the density distribution functions of nondegenerate and degenerate semiconductors.

We note that, as the doping increases, the distribution of impurity semiconductor states changes from a delta function to an impurity band. Four basic models[4] for the impurity band are shown in Fig. 2. Figure 2(a) and (b) are for impurity band calculations based on a regular close-packed lattice structure of hydrogen-like impurities. Figure 2(a) shows the density of states and the electron population at room temperature with $N_D = 2 \times 10^{19}$/cm^3. The density of states of the impurity band is given by

$$n(E) = \frac{4N_D}{\Delta E} \exp[-4\pi(E/\Delta E)^2] \quad \text{states}/eV$$

$$= 3.2 \times 10^{20} \exp[-4\pi(E/\Delta E)^2] \tag{2}$$

where the Fermi level is taken as zero, and ΔE is the bandwidth parameter of the Gaussian distribution. For $N_D = 2 \times 10^{19}$ cm^{-3}, ΔE is found to be 0.25 eV. Figure 2(b) is similar to Fig. 2(a) except that the maximum value of $n(E)$ is reduced. In order to give the same total number of states, the function $n(E)$ is given by

$$n(E) = 2.2 \times 10^{20} \exp[-6(E/\Delta E)^2] \quad \text{states}/eV. \tag{2a}$$

Figures 2(c) and (d) show the impurity bands which are calculated from a one-dimensional crystal consisting of randomly distributed delta-function potentials. Figure 2(c) shows the density of states and the electron population at room temperature with $N_D = 1.7 \times 10^{19}$ cm^{-3}. Figure 2(d) is similar to 2(c) except that the impurity band contains $2N_D$ states instead of N_D states as in Fig. 2(c). In all cases we note that, at sufficiently high impurity densities, the impurity bands merge with the main bands.

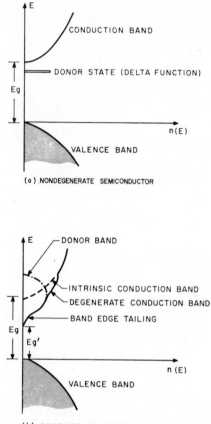

Fig. 1 Density-of-states distribution functions $n(E)$ vs E,
(a) for a nondegenerate semiconductor,
(b) for a degenerate semiconductor.

Figure 1(b) and Fig. 2 also reveal the third consequence of high doping densities. The intrinsic band gap E_g is reduced to E_g'. From emission spectra it is found[5] that the band gap of a degenerate germanium with a doping density of $10^{19}/cm^3$ is only 0.5 eV (in comparison with the intrinsic band gap of 0.66 eV). Figure 3 shows the results of the density of states of p-type gallium arsenide, metal-semiconductor tunneling experiments.[6] It should be noted that in degenerate semiconductors, the conduction-band or the valence-band edge is not sharp; it tails off gradually into the band gap. This phenomenon is

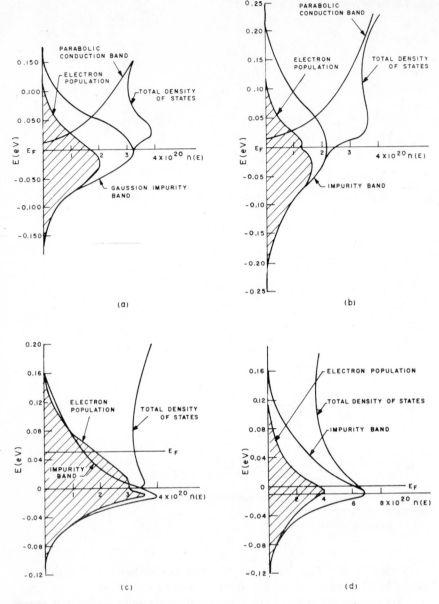

Fig. 2 Impurity band broadening and the merger of impurity band and the main band.
(a) and (b) for impurity band calculations based on a regular close-packed lattice structure with donor density 2×10^{19} cm^{-3}.
(c) and (d) for the impurity bands which are calculated from a one-dimensional crystal consisting of randomly distributed delta-function potentials with donor density 1.7×10^{19} cm^{-3} and 3.4×10^{19} cm^{-3} respectively.
(After Brody, Ref. 4.)

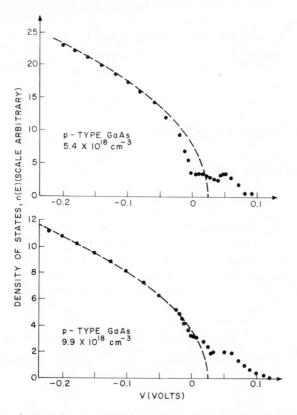

Fig. 3 Density of states of p-type GaAs obtained experimentally. The dashed lines are the best fits of the parabolic band. Note the large band-edge tailing effect at higher doping levels. (After Mahan and Conley, Ref. 6.)

referred to as the band-edge tailing. As the doping increases, the band-edge tailing also increases. The theoretical band-edge tailing has been studied by Kane[7] using a semiclassical or Thomas-Fermi approach. The density of states is found to be:

$$n(E) = m^{*3/2}(2\eta)^{1/2}\pi^{-2}\hbar^{-3}y(E/\eta) \text{ Vol}$$

$$Y(x) \equiv \pi^{-1/2} \int_{-\infty}^{\infty} (x - \zeta)^{1/2} \exp(-\zeta^2) \, d\zeta \tag{3}$$

where $x \equiv E/\eta$, Vol. is the volume, and η is a parameter proportional to $(N_D)^{5/12}$ where N_D is the average impurity density. We can express the above result in terms of the function $Y(x)$ with no parameters. $Y(x)$ is plotted in

Fig. 4 where the dotted curve is for the nondegenerate case which at high energies emerges into the solid curve for the degenerate case. This result compares favorably with the experimental results of Fig. 3. It is also clear from Eq. (3) that $n(E) \sim \exp{(-E^2/\eta^2)}$ at low energies and $n(E) \sim E^{1/2}$ at high energies.

3 TUNNELING PROCESSES

(1) Qualitative Consideration

A typical static current-voltage characteristic of a tunnel diode is shown in Fig. 5(a). In the reverse direction (p-side negative with respect to n-side) the current increases monotonically. In the forward direction the current first increases to a maximum value (peak current or I_P) at a voltage V_P, then decreases to a minimum value I_V at a voltage V_V. For voltage larger than V_V, the current increases exponentially with the voltage. The static characteristic is the result of many current components: namely the band-to-band tunnel

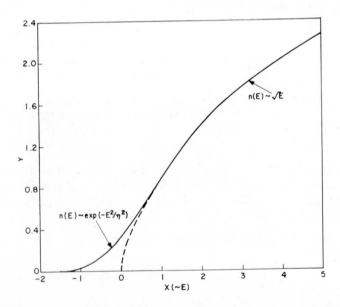

Fig. 4 Theoretical band-edge tailing plot. Y is proportional to the density of states and x is proportional to the energy.
(After Kane, Ref. 7.)

current, the valley current, the exponential excess current, and the thermal current, Fig. 5(b). We shall consider the tunnel current component in this section. The other current components will be considered in the next section.

We shall first discuss qualitatively the tunneling processes at absolute zero temperature using the simplified band structures as shown in Fig. 6. We note that the Fermi levels are within the bands of the semiconductor, and at thermal

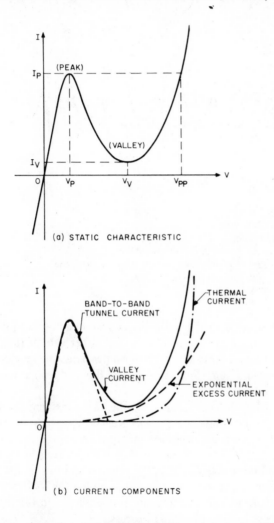

(a) STATIC CHARACTERISTIC

(b) CURRENT COMPONENTS

Fig. 5
(a) Static current-voltage characteristic of a typical tunnel diode. I_P and V_P are the peak current and peak voltage respectively. I_V and V_V are the valley current and valley voltage respectively. V_{PP} is the voltage at which the current again reaches I_P.
(b) The static current-voltage characteristic is decomposed into the current components.

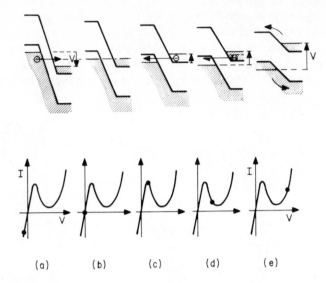

Fig 6 Simplified energy-band diagrams of tunnel diode at
(a) reverse bias,
(b) thermal equilibrium,
(c) forward bias such that peak current is obtained,
(d) forward bias such that the valley current is approached, and
(e) forward bias with thermal current flowing.
(After Hall, Ref. 48.)

equilibrium, Fig. 6(b), the Fermi level is a constant across the junction. Above the Fermi level there are no filled states on either side of the junction, and below the Fermi level there are no empty states available on either side of the junction. Hence no tunneling current can flow at zero externally applied voltage.

When a biasing voltage is applied, the electrons may tunnel from the valence band to the conduction or vice versa. The necessary conditions for tunneling are: (1) there are occupied energy states on the side from which the electron tunnels, (2) there are unoccupied energy states at the same energy levels as in (1) on the side to which the electron can tunnel, (3) the tunneling potential barrier height should be low and the barrier width should be small enough that there is a finite tunneling probability, and (4) the momentum must be conserved in the tunneling process.

Figure 6(a) shows electron tunneling from the valence band into the conduction band when a reversed bias is applied. The corresponding current is also designated by the dot on the I-V curve. When a foward bias is applied,

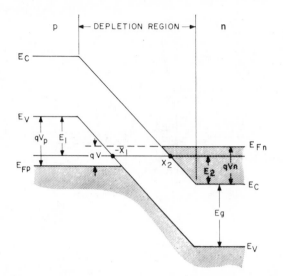

Fig. 7 Simplified energy band diagram with constant electric field in the depletion region. The amounts of degeneracy are qV_n and qV_p on the n-side and p-side respectively. V is the applied voltage. E_1 and E_2 are the electron energies measured for the n and p band edges respectively.

Fig. 6(c), a band of energies exists for which there are filled states on the n side corresponding to states which are available and unoccupied on the p side. The electrons can thus tunnel from the n side to the p side. When the forward voltage is further increased, there are fewer available but unoccupied states on the p side, Fig. 6(d). If forward voltage is applied such that the band is " uncrossed," i.e., the edge of the conduction band is exactly opposite the top of the valence band, there are no available states opposite filled states. Thus at this point the tunneling current can no longer flow. With still further increases of the voltage the normal thermal current will flow, Fig. 6(e), and will increase exponentially with the applied voltage. One thus expects that as the forward voltage increases, the tunneling current increases from zero to a maximum I_P and then decreases to zero when $V = V_n + V_p$ where V is the applied forward voltage, V_n the amount of degeneracy on the n-side ($V_n \equiv (E_{Fn} - E_C)/q$), and V_p is the amount of degeneracy on the p side ($V_p \equiv (E_V - E_{Fp})/q$) as shown in Fig. 7. It is the decreasing portion after the peak current that gives rise to the negative resistance region.

The tunneling process can be either direct or indirect. Figure 8(a) shows direct tunneling where the E-k relationships at the classical turning points are

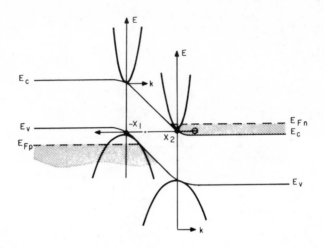

(a) DIRECT TUNNELING ($k_{min} = k_{max}$)

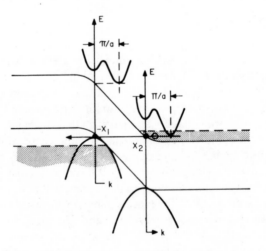

(b) INDIRECT TUNNELING ($k_{min} \neq k_{max}$)

Fig. 8
(a) Direct tunneling process with E-k relationship at the classical turning points ($-x_1$ and x_2) superimposed on the E-x relationship of the tunnel junction.
(b) Indirect tunneling process where $k_{min} \neq k_{max}$.

superimposed on the E-x relationship of the tunnel junction. The electrons can tunnel from the vicinity of the minimum of the conduction-band energy-momentum surface to a corresponding value of momentum in the vicinity of the valence-band maximum of the energy-momentum surface. For direct tunneling to occur, the conduction-band minimum and the valence-band maximum must have the same momentum. This condition can be fulfilled by semiconductors such as GaAs and GaSb that have a direct band gap. This condition can also be fulfilled by semiconductors with indirect band gap (such as Ge) when the applied voltage is sufficiently large that the valence-band maximum (Γ point)[8] is in line with the indirect conduction-band minimum (Γ point). For indirect tunneling, the conduction-band minimum does not occur at the same momentum as the valence band maximum, Fig. 8(b). In order to conserve momentum, the difference in momentum between the conduction-band minimum and the valence-band maximum must be supplied by scattering agents such as phonons or impurities. For phonon-assisted tunneling, it is required that both the energy and the momentum should be conserved; i.e., the sum of the phonon energy and the initial electron energy tunneling from the n to the p side is equal to the final electron energy after it has tunneled to the p side; and the sum of the initial electron momentum and the phonon momentum ($\hbar k_p$) is equal to the final electron momentum after it has tunneled. In general the probability for indirect tunneling is much lower than the probability for direct tunneling when direct tunneling is possible. Also, indirect tunneling involving several phonons has a much lower probability than that with only a single phonon.

(2) Tunneling Probability and Tunneling Current

When the electric field in a semiconductor is sufficiently high (of the order of 10^6 V/cm), there is a finite probability of quantum tunneling or direct excitation of electrons from the valence band into the conduction band. The tunneling probability, T_t, can be given by the WKB approximation (Wentzel-Kramers-Brillouin method):[9]

$$T_t \simeq \exp\left[-2 \int_{-x_1}^{x_2} |k(x)|\, dx \right] \tag{4}$$

where $|k(x)|$ is the absolute value of the wave vector of the carrier in the barrier; $-x_1$ and x_2 are the classical turning points as shown in Figs. 7 and 8.

The tunneling of an electron through a forbidden band is formally the same as a particle tunneling through a barrier. At the present time the detailed form of the potential barrier for an electron in the forbidden gap is not known. Nevertheless it will be shown that the tunneling exponent is not very sensitive to the choice of the potential barriers.

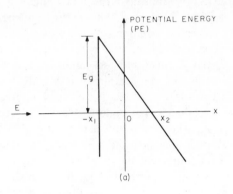

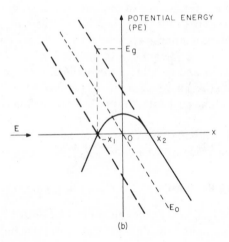

Fig. 9
(a) Triangular potential barrier.
(b) Parabolic potential barrier.

We shall consider two special potential barriers, namely the triangular and the parabolic barriers as shown in Fig. 9(a) and 9(b) respectively. For the triangular barrier the wave vector is given by

$$k(x) = \sqrt{\frac{2m^*}{\hbar^2}(PE - E)} = \sqrt{\frac{2m^*}{\hbar^2}\left(\frac{E_g}{2} - q\mathscr{E}x\right)} \tag{5}$$

where PE is the potential energy, E the incoming electron energy, E_g the band gap of the semiconductor, and $\mathscr{E}$ the electric field. Substitution of Eq. (5) into Eq. (4) yields

$$T_t \simeq \exp\left[-2\int_{-x_1}^{x_2}\sqrt{\frac{2m^*}{\hbar^2}\left(\frac{E_g}{2}-q\mathscr{E}x\right)}\,dx\right]$$

$$= \exp\left[+\frac{4}{3}\frac{\sqrt{2m^*}}{q\mathscr{E}\hbar}\left(\frac{E_g}{2}-q\mathscr{E}x\right)^{3/2}\right]\Bigg|_{-x_1}^{x_2}. \tag{6}$$

Since at

$$x = x_2,\quad \left(\frac{E_g}{2}-q\mathscr{E}x\right)=0;$$

and at

$$x = -x_1,\quad \left(\frac{E_g}{2}-q\mathscr{E}x\right)=E_g,$$

we have[9a]

$$T_t = \exp\left(-\frac{4\sqrt{2m^*}E_g^{3/2}}{3q\hbar\mathscr{E}}\right). \tag{7}$$

For the parabolic energy barrier, E_0 is defined as the energy measured from the electron energy to the center of the band, and the form of $(PE-E)$ is

$$PE - E = \frac{(E_g/2)^2 - E_0^2}{E_g} = \frac{(E_g/2)^2 - (q\mathscr{E}x)^2}{E_g}. \tag{8}$$

This form is also the simplest algebraic function that has the correct behavior at the band edges.[10] The probability is then given by

$$T_t = \exp\left[-2\int_{-x_1}^{x_2}\sqrt{\frac{2m^*}{\hbar^2}\left(\frac{E_g^2/4 - q^2\mathscr{E}^2x^2}{E_g}\right)}\,dx\right]$$

$$= \exp\left[-\frac{m^{*1/2}E_g^{3/2}}{2\sqrt{2}\,q\hbar\mathscr{E}}\int_{-1}^{1}(1-y^2)^{1/2}\,dy\right]\Bigg|_{y\equiv 2q\mathscr{E}x/E_g}$$

$$= \exp\left(-\frac{\pi m^{*1/2}E_g^{3/2}}{2\sqrt{2}\,q\hbar\mathscr{E}}\right). \tag{9}$$

The above expression is virtually identical to Eq. (7) except for the numerical constant which is $\pi/2\sqrt{2} = 1.11$ for the parabolic barrier, and $4\sqrt{2}/3 = 1.88$ for the triangular barrier.

Since the total momentum must be conserved in the tunneling process the transverse momentum must be included in the calculation of the tunneling probability, we divide the total energy into E_x and $E_\perp$ where $E_\perp$ is the energy associated with the momentum perpendicular to the direction of tunneling (or the transverse momentum), and E_x is the energy associated with momentum in the tunneling direction. Then for $E_\perp > 0$ as shown in Fig. 9(c)

$$PE - E_x = \frac{E_g^2/4 - E_0^2}{E_g} + E_\perp, \tag{10}$$

and the classical turning points are at

$$-x_1', x_2' = \mp \frac{1}{q\mathscr{E}} \sqrt{E_g^2/4 + E_g E_\perp}. \tag{11}$$

We obtain from Eqs. (4), (10), and (11) the tunneling probability

$$T_t = \exp\left(-\frac{\pi m^{*1/2} E_g^{3/2}}{2\sqrt{2}q\hbar\mathscr{E}}\right)\exp\left(-\frac{2E_\perp}{\bar{E}}\right) \tag{12}$$

where the first exponent is the same as in Eq. (9), corresponding to the probability of tunneling with zero transverse momentum, and $\bar{E}$ is given by

$$\bar{E} \equiv \frac{\sqrt{2}q\hbar\mathscr{E}}{\pi m^{*1/2} E_g^{1/2}} \tag{13}$$

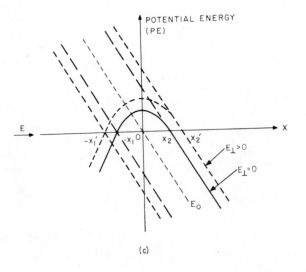

(c)

Fig. 9
(c) Parabolic potential barrier with transverse energy component $E_\perp$.
(After Moll, Ref. 12.)

which is a measure of the significant range of transverse momentum. Thus if $\bar{E}$ is very small, only the electron with small transverse momentum can tunnel. In other words, the effect of perpendicular energy is to further reduce the transmission by the factor $\exp(-2E_\perp/\bar{E})$. Transverse momentum must be conserved in direct tunneling. From the above results it is clear that to obtain large tunneling probability, both the effective mass and the band gap should be small and the electric field should be large.

We next consider the tunneling current and shall present the first-order approach using the density of states in the conduction and valence bands.[1] We shall then discuss a more rigorous theory of tunneling[10] in which the momentum is conserved.

At thermal equilibrium the tunneling current $I_{V \to C}$ from the valence band to the empty state of the conduction band and the current $I_{C \to V}$ from the conduction band to the empty state of the valence band should be detail-balanced. Expressions for $I_{C \to V}$ and $I_{V \to C}$ are formulated as follows:

$$I_{C \to V} = A \int_{E_C}^{E_V} F_C(E) n_C(E) T_t \{1 - F_V(E)\} n_V(E)\, dE \tag{14a}$$

$$I_{V \to C} = A \int_{E_C}^{E_V} F_V(E) n_V(E) T_t \{1 - F_C(E)\} n_C(E)\, dE \tag{14b}$$

where A is a constant and the tunneling probability T_t is assumed to be equal for both directions; $F_C(E)$ and $F_V(E)$ are the Fermi-Dirac distribution functions; $n_C(E)$ and $n_V(E)$ are the density of states in the conduction band and valence bands respectively. When the junction is biased, the observed current I is given by

$$I = I_{C \to V} - I_{V \to C} = A \int_{E_C}^{E_V} [F_C(E) - F_V(E)] T_t\, n_C(E) n_V(E)\, dE. \tag{15}$$

A closed form of the above expression can be obtained under the following assumptions: (1) the tunneling probability is almost a constant in the small voltage range involved, (2) the densities of states in the conduction band and valence band vary as $(E - E_C)^{1/2}$ and $(E_V - E)^{1/2}$ respectively, and (3) both V_n and V_p as shown in Fig. 7 are equal or less than $2kT$. With the above assumptions, the Fermi-Dirac distributions can be approximated by linear functions of E, i.e., $F_C(E) \simeq \frac{1}{2} - (E - E_{Fn})/4kT$, and $F_V(E) \simeq \frac{1}{2} + (E_{Fp} - E)/4kT$. Equation (15) can then be expressed as[11]

$$I = A' T_t \frac{qV}{kT} (V_n + V_p - qV)^2 \tag{15a}$$

where A' is a constant, V is the applied voltage, and V_n and V_p are respectively the penetrations of Fermi level into the conduction band and valence band.

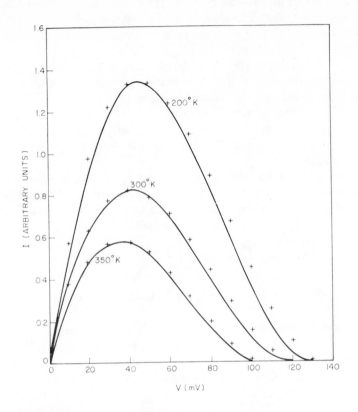

Fig. 10 Comparison of experimental curves and the theoretical results. (After Karlovsky, Ref. 11.)

Figure 10 shows a comparison of the experimental curves and the theoretical result calculated from Eq. (15a). The constant in Eq. (15a) is chosen to fit the peak current values. We note that there is reasonable agreement, especially at room and elevated temperatures.

It should be pointed out, however, that the above approach does not take into account the conservation of momentum. In addition the expression for the density of states is not exactly correct for degenerate semiconductors as discussed in the previous section. More accurate results are given by Kane[10] for both direct tunneling and phonon-assisted indirect tunneling.

For direct tunneling, a simple model of a tunnel diode with constant field in the depletion region is used (Fig. 7). The tunneling probability is given by

Eq. (12) in which the transverse momentum is included. The incident current per unit area in the energy range $dE_x\, dE_\perp$ is given by:[10,12]

$$dJ_x = \frac{q(m_y^* m_z^*)^{1/2}}{2\pi^2\hbar^3}\, dE_x\, dE_\perp \tag{16}$$

where

$$E = E_x + E_\perp, \quad E_x = \hbar^2 k_x^2/2m_x^* \tag{17}$$

By taking m^* as isotropic and equal for the n and p sides, and by using Eqs. (12) and (17), we obtain the tunneling current per unit area, J_t

$$J_t = \frac{qm^*}{2\pi^2\hbar^3}\exp\left(-\frac{\pi m^{*1/2}E_g^{3/2}}{2\sqrt{2}\,q\hbar\mathscr{E}}\right)\int [F_C(E) - F_V(E)]\exp(-2E_\perp/\bar{E})\, dE\, dE_\perp. \tag{18}$$

We have used Eq. (16) to give E and $E_\perp$ as the variables of integration. The limits of integration are determined by the condition $0 \le E_\perp \le E_1, 0 \le E_\perp \le E_2$ where E_1 and E_2 are the electron energies measured from the n and p band edges, respectively, as shown in Fig. 7. The limits on E are given by the band edges.

The integral over $E_\perp$ can be carried out immediately with the result

$$J_t = \frac{qm^*}{2\pi^2\hbar^2}\exp\left(-\frac{\pi m^{*1/2}E_g^{3/2}}{2\sqrt{2}\,\hbar q\mathscr{E}}\right)\left(\frac{\bar{E}}{2}\right)D \tag{19}$$

$$D \equiv \int [F_C(E) - F_V(E)][1 - \exp(-2E_S/\bar{E})]\, dE \tag{20}$$

where E_S is the smaller of E_1 and E_2 and $\bar{E}$ is given by Eq. (13). The quantity D is an overlap integral which determines the shape of the I-V characteristic. It has the dimensions of energy (in units of eV) and depends on the temperature and the depth of penetration of the Fermi levels into the energy bands qV_n and qV_p. At $T = 0°K$, both F_C and F_V are step functions, and the quantity D versus forward voltage is shown in Fig. 11 (dotted lines) for the cases $V_n = V_p$ and $V_n = 3V_p$. The maximum value of D occurs for $V = V_{\min}$ if $V_{\max} \ge 2V_{\min}$ and for $V = (V_n + V_p)/3$ if $V_{\max} \le 2V_{\min}$ where $V_{\max}$ and $V_{\min}$ refer to the larger and the smaller of V_n, V_p, respectively.

Also shown in Fig. 11 are the corresponding results of indirect tunneling (solid lines). For phonon-assisted indirect tunneling the tunneling probability is given by[10,12a]

$$T_t \simeq \exp\left[\frac{-4\sqrt{2}\,m_{rx}^{*1/2}(E_g - E_p)^{3/2}}{3q\hbar\mathscr{E}}\right] \tag{21}$$

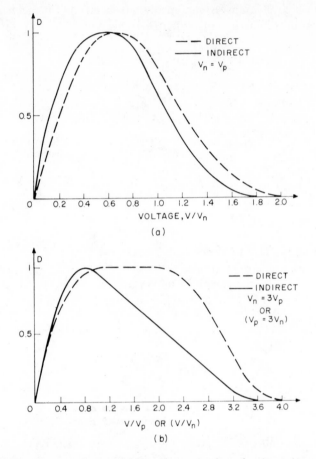

Fig. 11 Effective density of states "D" vs. forward voltage for direct (dotted lines) and indirect tunneling (solid lines) with $\bar{E}$ very large.
(a) $V_n = V_p$.
(b) $V_n = 3V_p$ (or $V_p = 3V_n$).
(After Kane, Ref. 10.)

where m_{rx}^* is the reduced effective mass in the tunneling direction, and E_p is the phonon energy. The expression for the tunneling current is very similar to Eq. (19) where the maximum in the forward characteristic occurs at

$$V = V_n + V_p - (V_n{}^2 + V_p{}^2)^{1/2}. \tag{22}$$

The maximum value of D is

$$D_{\max} = \frac{q^3}{3\bar{E}^2} \left[(V_n{}^2 + V_p{}^2)^{3/2} - (V_n{}^3 + V_p{}^3) \right]. \tag{23}$$

The current-voltage characteristics of phonon-assisted tunneling will be considered in Section 5(2).

4 EXCESS CURRENT

From the previous discussion for an ideal tunnel diode, the tunneling current should decrease to zero at biases where $V \geq (V_n + V_p)$. In other words, tunneling of electrons from the conduction band to the valence band in a single energy-conserving transition should then be impossible, and only normal diode current due to the forward injection of minority carriers should flow. In practice, however, the actual current at such biases is considerably in excess of the normal diode current; hence the term, excess current.

There are two major components of the excess current: (1) valley current due to band-tail tunneling and (2) exponential excess current due to carrier tunneling by way of energy states within the forbidden gap. Other possible mechanisms leading to excess current such as those due to photon, phonon, or plasmon have been considered and found to be too small to be important.[10]

The effect of the density-of-states tails on the current-voltage characteristic of a tunnel junction has been studied by Kane.[7] Figure 12 shows a comparison of the theoretical I-V curves with and without the band-tail tunneling for a silicon tunnel diode with $N_A = 2.3 \times 10^{19}$ cm^{-3} and $N_D = 4.8 \times 10^{19}$ cm^{-3}.

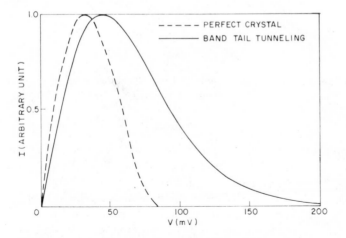

Fig. 12 Comparison of the theoretical current-voltage characteristics with and without the band-tail tunneling for a Si tunnel diode with $N_A = 2.3 \times 10^{19}$ cm^{-3} and $N_D = 4.8 \times 10^{19}$ cm^{-3}.
(After Kane, Ref. 7.)

We note that for the case of "perfect crystal," i.e., without band-tail tunneling, the tunneling current decreases to zero at about 85 mV. However with band-tail tunneling, the tunneling current extends beyond 200 mV. A comparison between the experimental results[13] and the theoretical curves, using band-edge tail theory[7] is shown in Fig. 13 for three silicon tunnel diodes. For each junction the pairs of curves have been normalized to have the same peak current but the relative magnitudes for the three junctions have no significance. It can be seen that the agreement is very good.

The exponential excess current has been derived by Chynoweth et al.,[14] with the help of Fig. 14 where some examples of possible tunneling routes are shown. An electron starting at C in the conduction band might tunnel to an

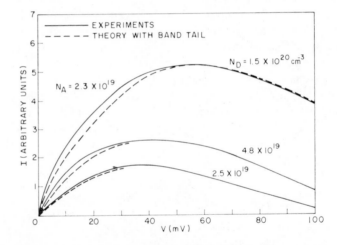

Fig. 13 Comparison between experimental results and theoretical curves using band-tail tunneling theory for three silicon tunnel diodes.
(After Logan and Chynoweth, Ref. 13.)

appropriate local level at A from which it could then drop down to the valence band D. Alternatively, the electron could drop down from C to an empty level at B from which it could tunnel to D. A third variant is a route such as $CABD$, where the electron dissipates its excess energy in a process which could be called impurity band conduction between A and B. A fourth route which should also be included is a staircase from C to D which consists of a series of tunneling transitions between local levels together with a series of vertical steps in which the electron loses energy by transferring from one level to another, a process made possible when the concentration of intermediate levels is sufficiently high. The route CBD can be regarded as the basic mechanism, the other routes being simply more complicated modifications. Let the junction be at a bias V, and consider an electron making a tunneling transition from B to D. The energy E_x through which it must tunnel is given by

$$E_x \approx E_g - qV + q(V_n + V_p) \approx q(V_{bi} - V) \qquad (24)$$

where V_{bi} is the built-in potential (assuming that the electron ends up near the top of the valence band). The tunneling probability, T_t, for the electron on the level at B can be given by

$$T_t = \exp\left(\frac{-4\sqrt{2}\,m_x^{*1/2}E_x^{3/2}}{3qh\mathscr{E}}\right) = \exp(-\alpha_x E_x^{3/2}/\mathscr{E}). \qquad (25)$$

This expression is the same as Eq. (7) except that E_g is replaced by E_x. The

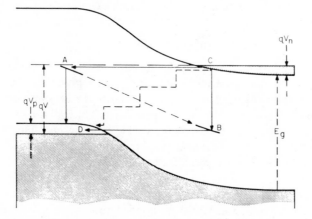

Fig. 14 Band diagram illustrating proposed mechanisms of tunneling via states in the forbidden gap for the excess current flow. (After Chynoweth et al., Ref. 14.)

maximum electric field for a step junction is given by

$$\mathscr{E} = 2(V_{bi} - V)/W \qquad (26)$$

where W is the depletion layer width given by

$$W = \left[\frac{2\varepsilon_s}{q}\left(\frac{N_A + N_D}{N_A N_D}\right)(V_{bi} - V)\right]^{1/2}. \qquad (27)$$

Let the volume density of the occupied levels at energy E_x above the top of the valence band be D_x. Then the excess current will be given by

$$I_x \simeq A D_x T_t \qquad (28)$$

where A is a constant and a reasonable assumption is made that the excess current will vary predominantly with the parameters in the exponent of T_t rather than with those in the factor D_x. Substitution of Eqs. (24) through (27) into (28) yields the expression for the excess current:[14]

$$I_x \simeq A D_x \exp\{-\alpha'_x[E_g - qV + q(V_n + V_p)]\} \tag{29}$$

where α'_x is a constant. The above expression predicts that the excess current will increase with the volume density of band-gap levels (through D_x), and also increase exponentially with the applied voltage V (provided $E_g \gg qV$). For temperatures within a limited range the band gaps of Ge, Si, and GaAs vary linearly with temperature

$$E_g(T) = E_g(0) - \beta T \tag{30}$$

where $E_g(0)$ is the extrapolated band gap to $0°K$ and β is the temperature coefficient (refer to Section 3 of Chapter 2). Under the assumption that $[E_g(T) - qV] \gg q(V_n + V_p)$, Eq. (29) becomes

$$I_x \simeq A D_x \exp\{-\alpha'_x[E_g(0) - qV] + \alpha'_x \beta T\}. \tag{31}$$

In Eq. (31), the factor α'_x involves the effective masses which are found to be only weakly dependent on temperature. Consequently, Eq. (31) predicts a simple temperature dependence for I_x proportional to $\exp(\text{const} \times T)$. By contrast injection current varies as $\exp(qV/kT)$. Thus, if Eq. (31) applies, a plot of $\ln I_x$ versus T for fixed voltage should be a straight line.

Experimental results do confirm these predictions. Figure 15(a) shows the nearly straight line for a typical GaAs tunnel diode.[15] Figure 15(b) shows $\ln I_x$ versus V for the same GaAs diode. The linear relationship is in good agreement with Eq. (29).

The above discussion can also be applied to cases where "hump" current[16-18] or a second negative resistance region exists, Fig. 16(b). Certain chemical impurities or lattice defects can give rise to a narrow band of energies in the band gap as shown in Fig. 16(a). The electrons can tunnel into this band as they can into the route of CAD shown in Fig. 14, and a secondary peak region is seen to result. The location of the second peak indicates the position of the band-gap energy levels.

5 CURRENT-VOLTAGE CHARACTERISTICS DUE TO EFFECTS OF DOPING, TEMPERATURE, ELECTRON BOMBARDMENT, AND PRESSURE

In the previous sections we have considered the basic current-voltage characteristics of tunnel diodes. We have concluded that at reverse bias and small forward bias, the current is due to band-to-band tunneling, and the

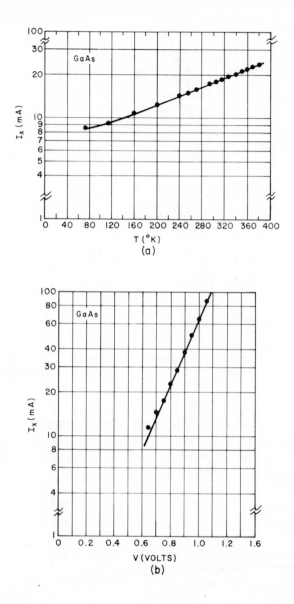

Fig. 15

(a) Experimental results of excess current vs. temperature of GaAs tunnel diode.

(b) Experimental results of excess current vs. forward voltage of the same GaAs tunnel diode as in (a).

(After Nanavati and DeAndrade, Ref. 15.)

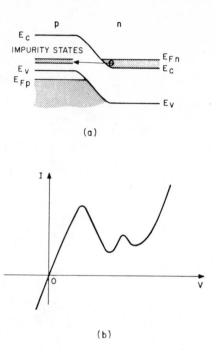

(a)

(b)

Fig. 16
(a) Tunneling into the narrow band of energies in the band gap.
(b) Current-voltage characteristic which has a hump current or a second negative resistance region.

band-edge tailing effect should be included. The basic expression for the above region is given by Eq. (19). For large forward biases, however, the current is mainly given by the thermal diffusion current, i.e., $J \simeq J_s \exp(qV/nkT)$ where all the symbols have their usual meanings. For forward biases intermediate to the tunneling region and the diffusion current region, the dominant current component is the excess current due to tunneling via band-gap states and the basic expression is given by Eq. (29). In this section more detailed considerations will be given to the current-voltage characteristics resulting from the effects of doping concentrations, temperature, electron bombardment, and pressure. In addition, the degradation of the I-V characteristics under normal operating conditions will be discussed.

In practice, most tunnel diodes are made using one of the following techniques: (1) ball alloy: a small metal alloy pellet containing the counter-dopant of high solid solubility is alloyed to the surface of a mounted semiconductor substrate (with high doping) in a precisely controlled temperature-time cycle under inert or hydrogen gas, the desired peak current level I_p is obtained by an etching process; (2) pulse bond: the contact and the junction are made simultaneously when the junction is pulse-formed between the semiconductor substrate and the metal alloy containing the counterdopant; (3) planar processes: planar tunnel diode fabrication consists basically of using planar technology including solution growth, diffusion, and controlled alloy. A typical planar Ge tunnel diode configuration[19] is shown in Fig. 17. The structure is designed to be self-supporting (the gold beam leads serve as mechanical support for the diode and also as electrical connections) thereby eliminating the need for a conventional package with its associated parasitic impedance.

(1) Doping Effect

Because of the requirement of narrow junction width, tunnel diodes as formed by the above techniques have doping profiles which can be approximated as two-sided abrupt junctions. The junction capacitance per unit area, C, according to the equation for an abrupt junction is given by

$$C = \sqrt{\frac{q\varepsilon_s N^*}{2(V_{bi} - V)}} \quad \text{farad/cm}^2 \tag{32}$$

where V_{bi} is the built-in potential and is equal to the voltage intercept, V_i, at

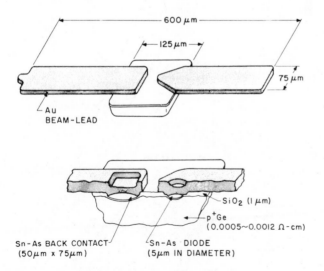

Fig. 17 Planar Ge tunnel diode configuration. The beam leads serve as mechanical support for the diode as well as electrical connections.
(After Davis and Gibbons, Ref. 19.)

$1/C^2 = 0$ of a $1/C^2$ versus V plot, and N^* is the effective concentration given by

$$N^* = N_A N_D/(N_A + N_D). \tag{33}$$

Typical examples[20] of the $1/C^2$ versus voltage plots are shown in Fig. 18. It is clear that the abrupt approximation for tunnel diodes is valid. The average field corresponding to the peak voltage V_P for an abrupt junction is given by

$$\mathscr{E} = (V_{bi} - V_P)/W \approx (N^* E_g/2\varepsilon_s)^{1/2} \tag{34}$$

where an approximation has been made that $q(V_{bi} - V_P) \approx E_g$.

From Eq. (34) and Eq. (19) it is expected that the peak current I_P (or peak current density J_P) should decrease as N^* or $\mathscr{E}$ decreases. Figure 19 shows the measured results[21] for Ge tunnel diodes at 4.2°K. The broken line is calculated from Eq. (19). Apparently, the agreement is very good. It should be pointed out, however, that germanium is an indirect semiconductor, the tunneling

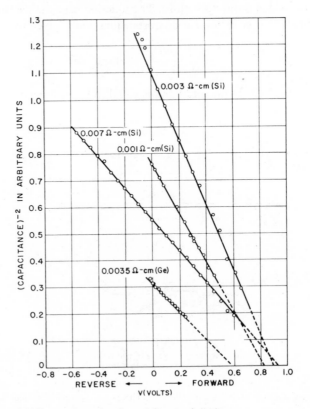

Fig. 18 Plots of $1/C^2$ vs. voltage for Ge and Si tunnel diodes. (After Chynoweth et al., Ref. 20.)

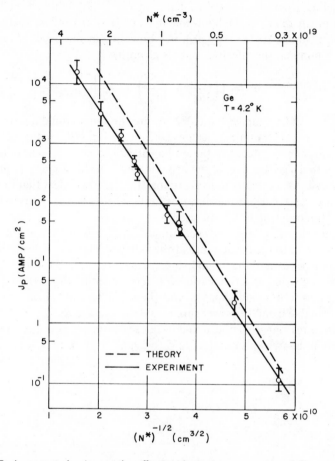

Fig. 19 Peak current density vs. the effective doping concentration of Ge tunnel diodes at 4.2°K. The broken curve is calculated from Eq. (19). (After Meyerhofer et al., Ref. 21.)

takes place between the light-hole band and the ⟨111⟩ conduction band minima. Phonon participation or impurity scattering is needed to conserve momentum. It has been found that for As-doped diodes, the phonon-assisted tunneling is not the dominant mechanism. However, whatever process is operative, the variation of J_P with N^* will be dominated by the tunneling exponential of Eq. (19). The appropriate effective mass is given by[22]

$$m^* = \left(\frac{1}{m_e^*} + \frac{1}{m_{lh}^*} \right)^{-1} \tag{35}$$

for tunneling from the light hole band to the $\langle 000 \rangle$ conduction band of germanium where m_{lh}^* is the light-hole mass ($=0.04\ m_o$) and m_e^* is the $\langle 000 \rangle$ conduction band mass ($=0.036\ m_o$). For tunneling in the $\langle 100 \rangle$ direction to the $\langle 111 \rangle$ minima, the effective mass is given by

$$m^* = \left[\left(\frac{1}{3m_l^*} + \frac{2}{3m_t^*} \right) + \frac{1}{m_{lh}^*} \right]^{-1} \tag{36}$$

where $m_l^* = 1.6\ m_o$ and $m_t^* = 0.082\ m_o$ are the longitudinal and transverse masses of the $\langle 111 \rangle$ minima. The exponents in Eq. (19) or Eq. (25) differ, however, only by 5 percent in these two cases. The peak current density can also be plotted as a function of the depletion layer width. Figure 20 shows the averaged value of J_P versus W in Ge tunnel diodes.[19] It is clear that in order to obtain large peak current density, it is required that a very narrow junction width, of the order of 100 Å, be used.

Also shown in Fig. 20 is the speed index[19] which is defined as the ratio of the peak current to the capacitance at the valley voltage V_V. The speed index, which is independent of the junction area, is a figure of merit for tunnel diodes. The switching speed of a tunnel diode is determined by the current available for charging the junction capacity and therefore depends on the amount of current available from the power supply and the average RC product. Since R, the negative resistance of the tunnel diode, is inversely proportional to the peak current, a large speed index, I_P/C (or small RC product), is required for

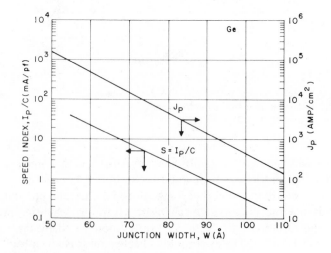

Fig. 20 Averaged value of peak current density and the speed index ($=I_P/C$) vs. depletion layer width of Ge tunnel diodes at 300°K.
(After Davis and Gibbons, Ref. 19.)

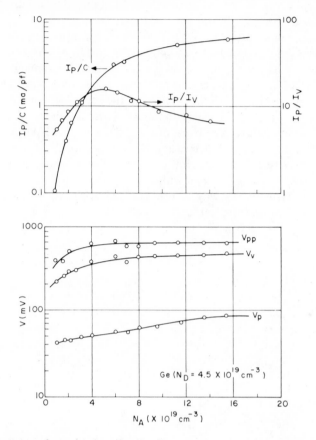

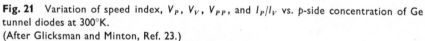

Fig. 21 Variation of speed index, V_P, V_V, V_{PP}, and I_P/I_V vs. p-side concentration of Ge tunnel diodes at 300°K.
(After Glicksman and Minton, Ref. 23.)

fast switching. One thus must use narrow junction width or large effective concentration in order to obtain large speed index (Fig. 20). Similar results[23] are shown in Fig. 21 where the p-side concentration is varied from 1×10^{19} to about 1.5×10^{20} cm^{-3}. We note that, as expected, the speed index, the peak voltage (V_P), the valley voltage (V_V), and V_{PP} all increase with increasing N_A or (N^*). The leveling off which occurs at high p-region doping is due to the fact that both I_P and C are limited by the less heavily doped n-region. Thus in order to obtain a high speed index it is necessary to increase N_D as well as N_A.

Another important tunnel diode parameter is the peak-current to valley-current ratio, I_P/I_V, which affects maximum current gain. To maximize current gain, high I_P/I_V values are required. Figure 21 shows the variation[23] of the ratio as a function of N_A. In this particular example a maximum value of 16/1 is obtained at a concentration of 5×10^{19} cm^{-3}. It is also seen that the ratio decreases markedly at both lower and higher doping levels. This is because, for lower values of N_A, the depletion width is too large for tunneling, and for larger values of N_A, the excess current component increases; so does the valley current. A clearer demonstration of the effect of doping on excess current is shown in Fig. 22 where the excess current density, J_x, is plotted against the depletion layer width, W, at the valley voltage.[24] We note that the excess current increases with decreasing W. In other words, J_x increases with increasing effective doping concentration. The linear relationship obtained between J_x and W is in agreement with the conclusion that the excess current is the result of tunneling by way of local energy levels in the forbidden gap.[14] A comparison of the typical current-voltage characteristics of Ge,[21,24]

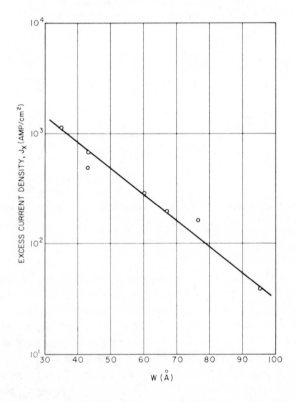

Fig. 22 Excess current at the valley voltage vs. depletion layer width of Ge tunnel diodes. (After Minton and Glicksman, Ref. 24.)

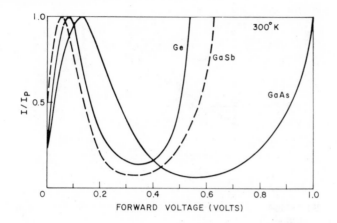

Fig. 23 Typical current-voltage characteristics of Ge, GaSb, and GaAs tunnel diodes at 300°K. The ratio I_P/I_V is 8/1 for Ge and 12/1 for GaSb and GaAs tunnel diodes.

GaSb,[25] and GaAs[15,17,26] tunnel diodes at 300°K is shown in Fig. 23. The current ratios of I_p/I_V are 8/1 for Ge and 12/1 for GaSb and GaAs. Tunnel diodes have been made in many other semiconductors such as Si[26,27,28] with a current ratio of about 4/1; InAs[29] with a ratio of 2/1 at room temperature and 10/1 at 4°K (this is because of its small band gap), and InP[17,30] with a ratio of 5/1. The current ratio for a given semiconductor, in general, can be increased by increasing the doping concentrations on both n and p sides. The ultimate limitation on the current ratio depends on (1) the peak current which in turn depends on the effective tunneling mass and the band gap, and (2) the valley current which in turn depends on the distribution and concentration of energy levels in the forbidden gap.

The position of the peak voltage[21] as a function of the sum of the Fermi level penetration V_n and V_p is shown in Fig. 24 for Ge tunnel diodes. We note that the peak voltage shifts toward higher values as the doping increases. This result is in qualitative agreement with both the simple model proposed by Esaki[1] and the tunneling theory by Kane.[10]

(2) Temperature Effect

Figure 25 shows the percentage change in the peak current with temperature within the range of $-55°C$ to $150°C$ for various doping concentrations in Ge tunnel diodes.[24] At concentrations of 1.45×10^{19} cm^{-3}, the temperature coefficient is negative over the full range. However, as the doping is increased,

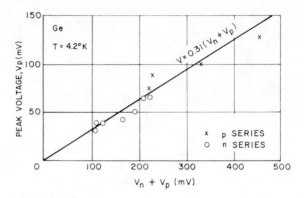

Fig. 24 Variation of peak voltage of Ge tunnel diodes as a function of the sum of V_n and V_p. (After Meyerhofer et al., Ref. 21.)

the temperature coefficient is changed from negative to positive. The negative and positive temperature coefficients of tunnel diodes can be explained by the variation of the factor D and the energy gap, E_g, in Eq. (19). At high concentrations the effect of temperature on the factor D is small, and the negative value of $\partial E_g/\partial T$ is primarily responsible for the change in tunneling probability.

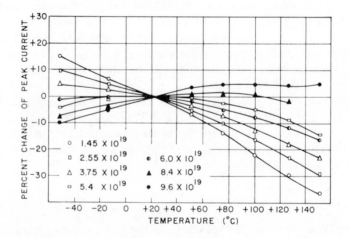

Fig. 25 Percentage change in the peak current with temperature within the range $-55°C$ to $150°C$ for various doping concentrations in Ge tunnel diodes. (After Minton and Glicksman, Ref. 24.)

As a result, the peak current increases with temperature. In the more lightly doped tunnel diodes, the decrease of D with temperature dominates, and the temperature coefficient is negative. Typical current-voltage characteristics as a function of temperature are shown[2] in Fig. 26. Although the peak current may increase or decrease with temperature as discussed previously, the valley current generally increases with increasing temperature. This is in agreement with Eq. (31) and Fig. 15(a). To show more clearly the variation of the current components as a function of temperature, the forward characteristics of a typical tunnel diode are decomposed into different conduction processes[21] and are shown in Fig. 27.

As mentioned in the Introduction, the tunnel diode is a useful device not only for electronic applications but also for the study of fundamental physical parameters. An important example is shown in Fig. 28 for the I-V characteristics of a Si tunnel diode.[27] As the temperature is reduced to 4.2°K, the curve reveals definitely some fine structure and has two bending points A and B. These bending points actually correspond to phonon-assisted tunneling processes; and the energies (or voltages) at A and B correspond to the acoustic and optical phonons respectively. Similar observations are made in Group III-V semiconductor junctions. Figure 29 shows the plots of the conductance (dI/dV) versus V at 4.2°K for GaP, InAs, and InSb.[31] The arrows indicate the corresponding optical phonon energies in these semiconductors. The phonon-assisted indirect tunneling can also be studied in more detail

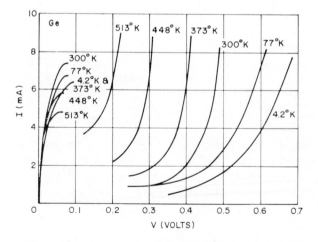

Fig. 26 Typical current-voltage characteristics of Ge tunnel diodes as a function of temperature.
(After Chow, Ref. 2; and Hall, Ref. 48.)

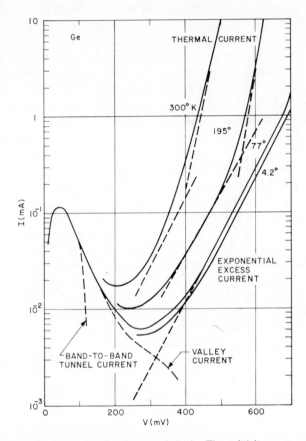

Fig. 27 Forward characteristic of a Ge tunnel diode. The solid lines are measured. The dotted lines are proposed decomposition into conduction processes. The diode is made of an In-$\frac{1}{2}$% Ga dot alloyed onto As-doped Ge wafer with 1.8×10^{19} cm^{-3}. (After Meyerhofer et al., Ref. 21.)

by the second-derivative technique. Studies have shown that there are twelve phonon and phonon-combination energies in Si junctions and seven in Ge junctions.[27a] The energies are mostly due to combinations of the transverse acoustic or optical phonons with intervalley scattering phonons and optical phonons of zero wave number.

(3) Electron Bombardment, Pressure, and Other Effects

Figure 30(a) shows the electron bombardment-induced changes in the *I-V* characteristics of a silicon tunnel diode.[32] Curve 0 refers to the junction

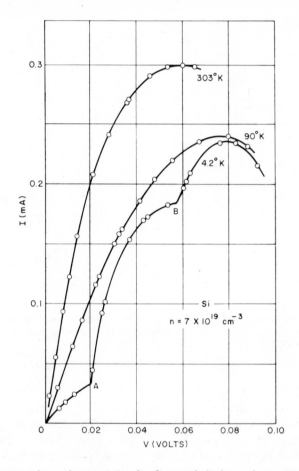

Fig. 28 Current-voltage characteristic of a Si tunnel diode at three temperatures. At 4.2°K the bending points A and B correspond to phonon-assisted tunneling processes. (After Esaki and Miyahara, Ref. 27.)

before bombardment, and the other curves show the effects of three successive bombardments with 1-MeV electrons. Figure 30(b) shows the results obtained by interrupting the annealing at 380°C at the times shown. It is seen that the major effect of the bombardment is to increase the excess current, and the increased excess current apparently can be gradually annealed out. The excess current is shown to be a sensitive indicator of the density and distribution of states introduced into the forbidden gap by bombardment. Similar results are expected to be observed for other radiations such as γ-ray.

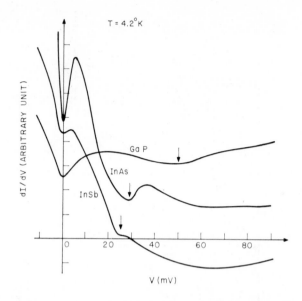

Fig. 29 Conductance (dI/dV) in arbitrary units vs. voltage for tunnel diodes made from Group III–V compounds. The arrows indicate positions of optical phonon energies. (After Hall et al., Ref. 31.)

Figure 31(a) and (b) show stress effect on the current-voltage characteristics of Ge and Si tunnel diodes respectively.[33] The most noticeable change is the increase of excess current with increasing stress. The changes are found to be reversible. It is suggested that this effect arises from deep-lying energy states associated with the strain-induced lattice defects in the junction region.[33,33a] In other words, as the stress increases, more states are introduced in the forbidden gap. These give rise to larger excess current. In addition the variation of tunnel diode I-V characteristics under hydrostatic pressure can lead to the study of the change of the energy band structure.[34,34a]

It is observed that the room-temperature peak current of an As-doped Ge tunnel diode annealed at 300°C decreases steadily with increasing annealing time.[35] On the other hand, the room-temperature valley current of the same Ge diode increases steadily with the annealing time. These results are explained by the fact that the impurity arsenic ions diffuse and drift at 300°C, causing the widening of the tunneling junction width which reduces the peak current, and at the same time introduces additional states in the energy gap which increases the valley current. In order to avoid the change of the I-V characteristics, a tunnel diode should be operated below a certain temperature (150°C

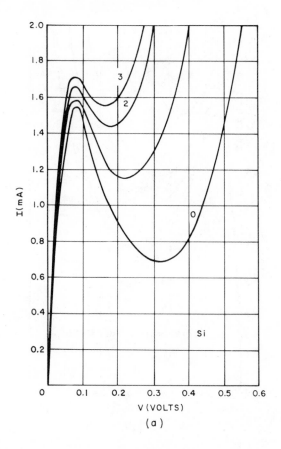

Fig. 30

(a) Electron bombardment-induced changes in the *I–V* characteristics of a silicon tunnel diode. Curve O refers to the junction before bombardment. The other curves show the effects of three successive bombardments with 1-MeV electrons.

for As-doped Ge diodes) such that the effects of impurity diffusion and drift can be neglected.

Permanent degradation of GaAs tunnel diodes is also observed during normal operation at room temperature.[36] This degradation is characterized by a large decrease in peak current and is correlated with a widening of the depletion layer width. Since degradation results when the GaAs diode is operated in the forward region, and since this region can be described by

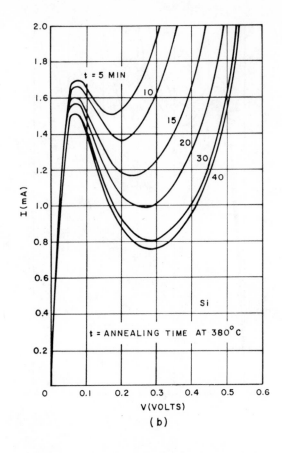

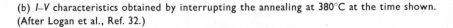

(b) *I–V* characteristics obtained by interrupting the annealing at 380°C at the time shown. (After Logan et al., Ref. 32.)

a recombination current, it is suggested that the hole-electron recombinations may occur at the recombination centers with energies of recombination sufficient to cause displacement of zinc ions which subsequently diffuse to the edge of junction giving rise to space-charge widening.[37] It has also been observed that surface effects are responsible for some of the degradation in GaAs and Ge diodes.[38,39] For instance exposure of GaAs diodes to a temperature of 455°C in air for two minutes produces drastic degradation. The original

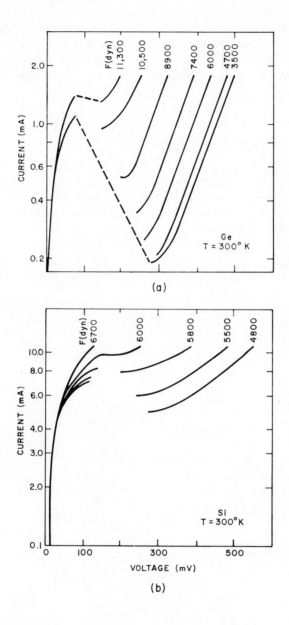

Fig. 31 Stress effect on the current-voltage characteristics of Ge and Si tunnel diodes. (After Bernard et al., Ref. 33.)

characteristics can be restored, however, by deep etching or by cleaving off the exposed surfaces. Hence the degradation is attributed to the diffusion of oxygen into the surface of the junction, resulting in band bending and a lowering of the junction barrier at the exposed edges of the junction. It is also suggested that the diffusion and drift of impurities with large diffusion coefficients such as Cu and Li may also cause permanent degradation of the GaAs tunnel diodes.

6 EQUIVALENT CIRCUIT

The symbol[40] of a tunnel diode is shown in Fig. 32(a), and the basic equivalent circuit is shown in Fig. 32(b) which consists of four elements: the series inductance L_S, the series resistance R_S, the diode capacitance C, and the negative diode resistance $-R$.

The series resistance R_S includes the lead resistance, the ohmic contacts, and the spreading resistance in the wafer which is given by $\rho/2d$ where ρ is the resistivity of the semiconductor and d is the diameter of the diode area. The

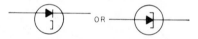

(a) SYMBOL OF TUNNEL DIODE

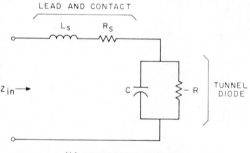

(b) EQUIVALENT CIRCUIT

Fig. 32
(a) Symbol of tunnel diode
(b) Equivalent circuit of tunnel diode.
(Ref. 40.)

series inductance L_S in a coaxial cavity is given by[41]

$$L_S = \frac{2.303\mu_0 l}{2\pi} \ln \frac{r_2}{r_1} \tag{37}$$

where μ_0 is the permeability of the medium, l is the length, and r_1 and r_2 are the inner and outer radii of the coaxial-line respectively. For the planar tunnel diode shown in Fig. 17, the small size of the unit allows the circuit parasitic series inductance to be reduced. We shall see that these parasitic elements establish important limits on the performance of the tunnel diode.

To consider the diode capacitance and negative resistance, we refer to Fig. 33 where a typical current-voltage characteristic of a GaAs tunnel diode is shown in (a). The differential resistance which is defined as $(dI/dV)^{-1}$ is plotted in Fig. 33(b). The value of the negative resistance at the inflection point, which is the minimum negative resistance in the region, is designated by R_{min}. This resistance can be approximated by

$$R_{min} \approx 2V_P/I_P \tag{38}$$

where V_P and I_P are the peak voltage and peak current respectively, and the above relationship has been used in defining the speed index as shown in Fig. 20 and 21. Also shown in Fig. 33(c) is the conductance plot, (dI/dV) versus V. At the peak and valley voltages, the conductance becomes zero; the diode capacitance is usually measured at the valley voltage, and is designated by C_j.

The input impedance Z_{in} of the equivalent circuit of Fig. 32 is given by

$$Z_{in} = \left[R_S + \frac{-R}{1 + (\omega RC)^2} \right] + j\left[\omega L_S + \frac{-\omega CR^2}{1 + (\omega RC)^2} \right]. \tag{39}$$

From Eq. (39) we see that the resistive (real) part of the impedance will be zero at a certain frequency, and the reactive (imaginary) part of the impedance will also be at a second frequency. We denote these frequencies by the resistive cutoff frequency f_r and the reactive cutoff frequency f_x respectively; these frequencies are given by

$$f_r \equiv \frac{1}{2\pi RC} \sqrt{\frac{R}{R_S} - 1}, \tag{40}$$

$$f_x \equiv \frac{1}{2\pi} \sqrt{\frac{1}{L_S C} - \frac{1}{(RC)^2}}. \tag{41}$$

For cutoff frequencies specified at the minimum resistance and valley

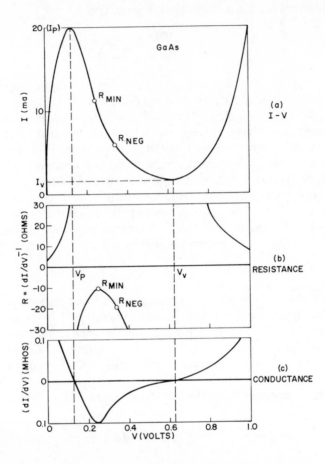

Fig. 33

(a) Current-voltage characteristics of a typical GaAs tunnel diode at 300°K.

(b) Differential resistance $(dI/dV)^{-1}$ vs. voltage. R_{min} is the minimum resistance. R_{neg} is the resistance corresponding to the minimum noise figure.

(c) Differential conductance, $G \equiv (dI/dV)$, vs. voltage. At peak and valley currents, $G = 0$.

capacitance, we have

$$f_{r0} \equiv \frac{1}{2R_{min}C_j}\sqrt{\frac{R_{min}}{R_S} - 1} \geq f_r \tag{42}$$

$$f_{x0} \equiv \frac{1}{2\pi}\sqrt{\frac{1}{L_S C_j} - \frac{1}{(R_{min}C_j)^2}} \leq f_x \tag{43}$$

where f_{r0} is the maximum resistive cutoff frequency, at which the diode will no longer exhibit negative resistance; and f_{x0} is the minimum reactive cutoff frequency or the self-resonant frequency, at which the reactance of diode is zero and at which the diode would oscillate if $f_{r0} > f_{x0}$. In most applications where the diode is operated into the negative resistance region, it is desirable to have $f_{x0} > f_{r0}$ and $f_{r0} \gg f_0$, the operating frequency. It can be shown from Eqs. (42) and (43) that in order to fulfill the requirement that $f_{x0} > f_{r0}$, the series inductance L_S must be lowered.

In addition to the cutoff frequencies we will consider one more important quantity associated with the equivalent circuit, the noise figure which is defined as

$$NF \equiv 1 + \frac{q}{2kT} |RI|_{min} \tag{44}$$

where $|RI|_{min}$ is the minimum value of the negative resistance-current product on the current-voltage characteristic. The corresponding value of R (designated by R_{neg}) is shown in Fig. 33. The product $q|RI|_{min}/2kT$ is called the noise constant K and is a material constant. Typical values of K at room temperature are 1.2 for Ge, 2.4 for GaAs, and 0.9 for GaSb. The GaSb tunnel diode has the lowest noise figure (or the most quiet performance) because of its small effective mass (0.047 m_o compared with 0.082 m_o for Ge) and its small energy gap (0.67 eV compared with 1.43 eV for GaAs).

7 BACKWARD DIODE

In connection with the tunnel diode, when the doping concentrations on the p and n sides of a p-n junction are nearly or not quite degenerate, the current in the "reverse" direction for small bias, as shown in Fig. 34, is larger than the current in the "forward" direction—hence, the name "backward diode."

The energy band diagram of the backward diode is shown in Fig. 35 where at thermal equilibrium, Fig. 35(a), the Fermi level is very close to the band edges. When a small reverse bias (p side negative with respect to n side) is applied, electrons can readily tunnel from the valence band into the conduction band and give rise to a tunneling current given by Eq. (19) which can be written in the form

$$J \simeq A_1 \exp(+|V|/A_2) \tag{45}$$

where A_1 and A_2 are positive quantities and are slowly varying functions of the applied voltage V. Equation (45) indicates that the reverse current increases approximately exponentially with the voltage.

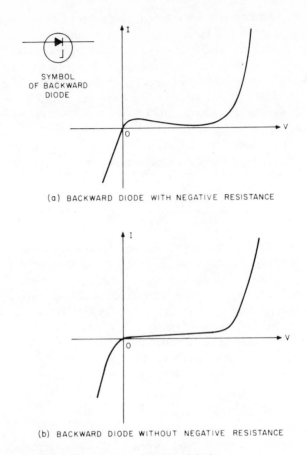

(a) BACKWARD DIODE WITH NEGATIVE RESISTANCE

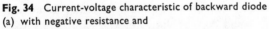

(b) BACKWARD DIODE WITHOUT NEGATIVE RESISTANCE

Fig. 34 Current-voltage characteristic of backward diode
(a) with negative resistance and
(b) without negative resistance.

The backward diode can be used for rectification of small signals, and for microwave detection and mixing. Similar to the tunnel diode, the backward diode has a good frequency response because there is no minority carrier storage effect. In addition, the current-voltage characteristic is insensitive to temperature and to radiation effect; and the backward diode has very low $1/f$ noise.[42-44]

For nonlinear applications such as high-speed switching, a device figure of merit is γ, the ratio of the second derivative to the first derivation of the current-voltage characteristic. It is also referred to as the curvature coefficient:[45]

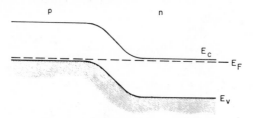

(a) THERMAL EQUILIBRIUM (BACKWARD DIODE)

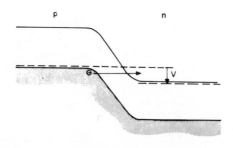

(b) REVERSE BIAS

Fig. 35 Energy band diagram of backward diode
(a) at thermal equilibrium and
(b) with reverse bias.

$$\gamma \equiv \left[\left(\frac{d^2 I}{dV^2} \right) \middle/ \left(\frac{dI}{dV} \right) \right].$$ (46)

The value of γ is a measure of the degree of nonlinearity normalized to the operating admittance level. For a forward-biased p-n junction or a Schottky barrier (refer to Chapter 8) the value of γ is simply given by q/nkT. Thus γ varies inversely as T and at room temperature, the value of γ for an ideal p-n junction ($n = 1$) is about 40 V^{-1} independent of bias. For a reverse-biased p-n junction, however, the value of γ is very small at low voltages and increases linearly with the avalanche multiplication factor near breakdown voltage. Although the ideal reverse breakdown characteristic would give a value of γ greater than 40 V^{-1}, because of the statistical distribution of impurities and the effect of space-charge resistance much lower values of γ are expected.[46]

For a backward diode the value of γ can be obtained from Eqs. (7), (15a) and (26); and is given by[47]

$$\gamma(\text{for } V = 0) = \frac{4}{V_n + V_p} + \frac{2}{\hbar}\sqrt{\frac{2\varepsilon_s m^*}{N^*}} \tag{47}$$

where m^* is the average effective mass of the carriers

$$[m^* \simeq m_e^* m_h^*/(m_e^* + m_h^*)],$$

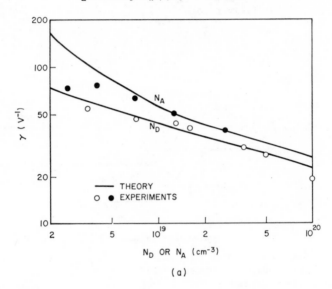

(a)

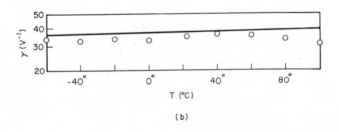

(b)

Fig. 36
(a) Curvature coefficient at room temperature for $V \approx 0$ vs. acceptor concentration in Ge (for a fixed $N_D = 2 \times 10^{19}$ cm^{-3}) or donor concentration (for a fixed $N_A = 10^{19}$ cm^{-3}). Solid lines are computed results. Data points are from experiment measures.
(b) Curvature coefficient vs. temperature. Solid line and data points are computed and measured respectively.
(After Karlovsky, Ref. 47.)

and N^* is the effective doping concentration given by Eq. (33). It is clear that the curvature coefficient γ depends upon the impurity concentrations on both sides of the junction and the effective masses. It is also expected that in contrast to Schottky barriers, the value of γ is relatively insensitive to temperature variation since the parameters in Eq. (47) are slowly varying functions of temperature.

Figure 36 shows a comparison between theoretical and experimental values of γ for Ge backward diodes. The solid lines are computed from Eq. (47) using $m_e^* = 0.22\ m_o$ and $m_h^* = 0.39\ m_o$. The agreement is generally good over the doping range considered. We also note that there are two interesting features of γ for backward diodes: (1) γ can exceed 40 V^{-1}, and (2) it is insensitive to temperature variation.

REFERENCES

1. L. Esaki, "New Phenomenon in Narrow Germanium *p-n* Junctions," Phys. Rev., *109*, 603 (1958).

2. For a general discussion, see W. F. Chow, *Principles of Tunnel Diode Circuits*, John Wiley and Sons, Inc., New York (1964).

3. K. K. Thornber, Thomas C. McGill, and C. A. Mead, "The Tunneling Time of an Electron," J. Appl. Phys., *38*, 2384 (1967).

4. T. P. Brody, "Nature of the Valley Current in Tunnel Diodes," J. Appl. Phys., *33*, 100 (1962).

5. J. I. Pankove, "Influence of Degeneracy on Recombination Radiation in Germanium," Phys. Rev. Letters, *4*, 20 (1960); and "Optical Absorption by Degenerate Germanium," Phys. Rev. Letters, *4*, 454 (1960).

6. G. D. Mahan and J. W. Conley, "The Density of State in Metal-Semiconductor Tunneling," Appl. Phys. Letters, *11*, 29 (1967).

7. E. O. Kane, "Thomas-Fermi Approach to Impure Semiconductor Band Structure," Phys. Rev., *131*, 79 (1963).

8. J. V. Morgan and E. O. Kane, "Observation of Direct Tunneling in Germanium," Phys. Rev. Letters, *3*, 466 (1959).

9. L. D. Landau and E. M. Lifshitz, *Quantum Mechanics*, Ch. VII, p. 174, Addison-Wesley Book Co., 1958.

9a. R. H. Fowler and L. Nordheim, "Electron Emission in Intense Electric Fields," Proc. Roy. Soc. (London), *119*, 173 (1928).

10. E. O. Kane, "Theory of Tunneling," J. Appl. Phys., *32*, 83 (1961); and also "Tunneling in InSb," Phys. Chem. Solids, *2*, 181 (1960).

11. J. Karlovsky, "Simple Method for Calculating the Tunneling Current of an Esaki Diode," Phys. Rev., *127*, 419 (1962).

12. J. L. Moll, *Physics of Semiconductors*, Ch. 12, p. 252, McGraw-Hill Book Co. (1964).

12a. L. V. Keldysh, "Behavior of Non-Metallic Crystals in Strong Electric Fields," Soviet J. Exptl. Theoret. Phys., *6*, 763 (1958).

13. R. A. Logan and A. G. Chynoweth, "Effect of Degenerate Semiconductor Band Structure on Current-Voltage Characteristics of Silicon Tunnel Diodes," Phys. Rev., *131*, 89 (1963).

14. A. G. Chynoweth, W. L. Feldmann, and R. A. Logan, "Excess Tunnel Current in Silicon Esaki Junctions," Phys. Rev., *121*, 684 (1961).

15. R. P. Nanavati and C. A. Morato De Andrade, "Excess Current in Gallium Arsenide Tunnel Diodes," Proc. IEEE, *52*, 869 (1964).

16. R. S. Claassen, "Excess and Hump Current in Esaki Diodes, " J. Appl. Phys., *32*, 2372 (1961).

17. N. Holonyak, Jr., "Evidence of States in the Forbidden Gap of Degenerate GaAs and InP—Secondary Tunnel Current and Negative Resistance," J. Appl. Phys., *31*, 130 (1960).

18. A. S. Epstein and J. F. Caldwell, "Lithium-Doped GaAs Tunnel Diodes," J. Appl. Phys., *35*, 3050 (1964).

19. R. E. Davis and G. Gibbons, "Design Principles and Construction of Planar Ge Esaki Diodes," Solid-State Electron., *10*, 461 (1967).

20. A. G. Chynoweth, W. L. Feldmann, C. A. Lee, R. A. Logan, G. L. Pearson, and P. Aigrain, "Internal Field Emission at Narrow Silicon and Germanium *p-n* Junctions," Phys. Rev., *118*, 425 (1960).

21. D. Meyerhofer, G. A. Brown, and H. S. Sommers, Jr., "Degenerate Germanium I, Tunnel, Excess, and Thermal Current in Tunnel Diodes," Phys. Rev., *126*, 1329 (1962).

22. P. N. Butcher, K. F. Hulme, and J. R. Morgan, "Dependence of Peak Current Density on Acceptor Concentration in Germanium Tunnel Diodes," Solid-State Electron., *3*, 358 (1962).

23. R. Glicksman and R. M. Minton, "The Effect of *p*-Region Carrier Concentration on the Electric Characteristics of Germanium Epitaxial Tunnel Diodes," Solid-State Electron., *8*, 517 (1965).

24. R. M. Minton and R. Glicksman, "Theoretical and Experimental Analysis of Germanium Tunnel Diode Characteristics," Solid-State Electron., *7*, 491 (1964).

25. W. N. Carr, "Reversible Degradation Effects in GaSb Tunnel Diodes," Solid-State Electron., *5*, 261 (1962).

26. N. Holonyak, Jr., and I. A. Lesk, "GaAs Tunnel Diodes," Proc. IRE, *48*, 1405 (1960).

27. L. Esaki and Y. Miyahara, "A New Device Using the Tunneling Process in Narrow *p-n* Junctions," Solid-State Electron., *1*, 13 (1960).

27a. A. G. Chynoweth, R. A. Logan, and D. E. Thomas, "Phonon-Assisted Tunneling in Silicon and Germanium Esaki Junctions," Phys. Rev., *125*, 877 (1962).

28. V. M. Franks, K. F. Hulme, and J. R. Morgan, "An Alloy Process for Making High Current Density Silicon Tunnel Diode Junctions," Solid-State Electron., *8*, 343 (1965).

29. H. P. Kleinknecht, "Indium Arsenide Tunnel Diodes," Solid-State Electron., *2*, 133 (1961).

30. C. A. Burrus, "Indium Phosphide Esaki Diodes," Solid-State Electron., *3*, 357 (1962).

31. R. N. Hall, J. H. Racette, and H. Ehrenreich, "Direct Observation of Polarons and Phonons During Tunneling In Group 3-5 Semiconductor Junctions," Phys. Rev. Letters, *4*, 456 (1960).

32. R. A. Logan, W. M. Augustyniak, and J. F. Gilbert, "Electron Bombardment Damage in Silicon Esaki Diodes," J. Appl. Phys., *32*, 1201 (1961).

33. W. Bernard, W. Rindner, and H. Roth, "Anisotropic Stress Effect on the Excess Current in Tunnel Diodes," J. Appl. Phys., *35*, 1860 (1964).

33a. B. Bazin, "Pressure Effects on Silicon Tunnel Diode," J. Phys. Chem. Solids, *26*, 2075 (1965).

34. H. Fritzsche and J. J. Tiemann, "Effect of Elastic Strain on Interband Tunneling in Sb-doped Germanium," Phys. Rev., *130*, 617 (1963).

34a. For a general discussion of pressure effect of solids, see W. Paul and D. M. Warschauer, Ed., *Solids Under Pressure*, McGraw-Hill Book Co. Inc., New York (1963).

35. J. H. Buckingham, K. F. Hulme, and J. R. Morgan, "Impurity Diffusion and Drift in Germanium Tunnel-Diode Junctions," Solid-State Electron., *6*, 233 (1963).

36. R. D. Gold and L. R. Weisberg, "Permanent Degradation of GaAs Tunnel Diodes," Solid-State Electron., *7*, 811 (1964).

37. A. S. Epstein and J. F. Caldwell, "Degradation in Zn-doped GaAs Tunnel Diodes," J. Appl. Phys., *35*, 2481 (1964).

38. H. Kessler and N. N. Winogradoff, "Surface Aspects of the Thermal Degradation of GaAs *p-n* Junction Lasers and Tunnel Diodes," IEEE Trans. Electron Devices, *ED-13*, 688 (1966).

39. J. E. Alberghini and R. M. Brondy, "Surface Excess Conductance in Ge Tunnel Junction via Surface States," Appl. Phys. Letters, *9*, 362 (1966).

40. "Standards on Definitions, Symbols, and Methods of Test for Semiconductor Tunnel (Esaki) Diodes and Backward Diodes," IEEE Trans. Electron Devices, *12*, 374 (1965).

41. W. B. Hauer, "Definition and Determination of the Series Inductance of Tunnel Diodes," IRE Trans. on Electron Devices, *8*, 470 (1961).

42. H. V. Shurmer, "Backward Diodes As Microwave Detectors," Proc. Inst. Elec. Eng., London, *111*, 1511 (1964).

43. S. T. Eng, "Low-Noise Properties of Microwave Backward Diodes," IRE Trans. on MTT, *8*, 419 (1961).

44. J. B. Hopkins, "Alloyed InAs Microwave Backward Diodes," IEEE International Solid State Device Conference, Paper 13.5, Washington, D.C. (Nov. 1967).

45. H. C. Torrey and C. A. Whitmer, *Crystal Rectifiers*, Ch. 8, McGraw-Hill Book Co. (1948).

46. S. M. Sze and R. M. Ryder, "The Nonlinearity of the Reverse Current-Voltage Characteristics of a *p-n* Junction Near Avalanche Breakdown," Bell Syst. Tech. J., *46*, 1135 (1967).

47. J. Karlovsky, "The Curvature Coefficient of Germanium Tunnel and Backward Diodes," Solid-State Electron., *10*, 1109 (1967).

48. R. N. Hall, "Tunnel Diodes," IRE Trans. Electron Devices, *ED-7*, 1 (1960).

- **INTRODUCTION**
- **STATIC CHARACTERISTICS**
- **BASIC DYNAMIC CHARACTERISTICS**
- **GENERALIZED SMALL-SIGNAL ANALYSIS**
- **LARGE-SIGNAL ANALYSIS**
- **NOISE**
- **EXPERIMENTS**

5

Impact-Avalanche Transit-Time Diodes (IMPATT Diodes)

I INTRODUCTION

IMPATT stands for *IMP*act ionization *A*valanche *T*ransit *T*ime. IMPATT diodes employ impact-ionization and transit-time properties of semiconductor structures to produce negative resistance at microwave frequencies. A *p-n* junction can be operated in its IMPATT mode when it is biased into reverse avalanche breakdown and mounted in a microwave cavity. At the present time, the IMPATT diode is one of the most powerful solid-state sources of microwave power.

The negative resistance arising from transit time in semiconductor diodes was first considered by Shockley.[1] In 1958 Read[2] proposed a high-frequency semiconductor diode consisting of an avalanche region at one end of a relatively high resistance region serving the transit-time drift space for the generated charge carriers (i.e., p^+nin^+ or n^+pip^+). The first IMPATT operation[3] as reported by Johnston, DeLoach, and Cohen in 1965, however, was obtained from a simple *p-n* junction. Three weeks later the first Read IMPATT diode was reported by Lee et al.[4] This indicates that besides the particular doping profile proposed by Read there are broad classes of structures which also possess negative resistance due to their IMPATT properties. From the small-signal theory developed by Misawa[5] it is confirmed that a negative resistance of the IMPATT nature can be possessed by a junction diode with any doping profile.

As the name implies, the operation of IMPATT diodes involves interaction between two physical phenomena—impact ionization and the transit time of charge carriers. From the small-signal point of view, when a differential voltage is applied to an avalanching diode, an increase in differential current results. The current is out of phase with the voltage by two effects:

(1) After the voltage increment is applied, the carrier population builds up toward a new level with a delay time τ_A characteristic of the avalanche.

(2) The effect on the external terminals, i.e., the terminal current, is further delayed because of the "transit time" τ_t during which the carriers are collected by the electrodes. When the total delay in the current exceeds a quarter cycle, the in-phase component of the current becomes negative; that is, a negative conductance exists, and in an appropriate circuit the device may therefore oscillate spontaneously.

Misawa[5] has considered the case of the "uniformly avalanching" diode with thickness W and area A, in which impact ionization produces carriers at a uniform rate throughout the active region. Such a diode can be represented by an equivalent circuit consisting of its passive capacitance $C = \varepsilon_s A/W$ in shunt with an electronic admittance $Y = G + jB$. Both G and B are negative and proportional to the current density J_0. Therefore the electronic inductive susceptance B resonates with the capacitance C at a frequency which increases proportionally to $\sqrt{J_0}$. Thus the characteristic frequency of the avalanche increases as $\sqrt{J_0}$. A similar expression has been obtained by Gilden and Hines.[10]

Many IMPATT diodes consist of an avalanching region appended to a drift region where the field is low enough that the carriers pass through it without avalanching. The drift region then produces a current delay equal to some fraction, frequently about one-half, of the transit-time across it. The interaction of the transit-time delay with the avalanche delay produces rather complicated behavior which will be described in detail below.

The small-signal point of view therefore explains qualitatively that IMPATT diodes oscillate because of the avalanche and transit-time delays. Small-signal analysis is a good qualitative guide to the impedance characteristics of the device; but to calculate large-signal properties, such as power output and efficiency, small-signal analysis is inadequate and should be supplemented by large-signal studies.

We shall first consider in Section 2 the static characteristics of the IMPATT diode family including simple p-n junctions, the p-i-n diode, and the Read diode. Section 3 presents the basic dynamic characteristics of an IMPATT diode. The generalized small-signal analysis is considered in Section 4. The large-signal operations such as the power-frequency limitation and efficiency are discussed in Section 5. Section 6 considers the noise performance. Finally, the experiments and typical results of the IMPATT diodes are presented in Section 7.

2 STATIC CHARACTERISTICS

The basic members of the IMPATT diode family are the Read diode, the one-sided abrupt p-n junction, the linearly graded p-n junction, and the p-i-n diode.

We shall now consider their static characteristics such as the field distribution, the breakdown voltage, and the space-charge effect. Figure 1 shows the doping profile, the electric field distribution, and the ionization integrand at breakdown condition for an idealized Read diode (p^+nin^+ or its dual n^+pip^+). The ionization integrand is given by

$$\langle\alpha\rangle \equiv \alpha_n \exp\left[-\int_x^W (\alpha_n - \alpha_p)\, dx'\right] \tag{1}$$

where α_n and α_p are the ionization rates of electrons and holes, respectively, and W is the depletion width.

The avalanche breakdown condition as discussed in Chapter 3 is given by

$$\int_0^W \langle\alpha\rangle\, dx = 1. \tag{2}$$

Because of the strong dependence of α's on an electric field, we note that the "avalanche region" is highly localized, i.e., most of the multiplication processes occur in a narrow region near the highest field between $0 \leq x \leq x_A$ where x_A is defined as the width of the avalanche region (to be discussed later).

The voltage drop across the avalanche region is called V_A. It will be shown that both x_A and V_A have profound effect on the optimum current density and the maximum efficiency of an IMPATT diode. The layer outside the avalanche region ($x_A \leq x \leq W$) is called the drift region.

The corresponding results for a typical one-sided abrupt p-n junction are shown in Fig. 2. Again we notice the highly localized avalanche region. Figure 3 shows the results for a linearly graded silicon p-n junction. The avalanche region is located near the center of the depletion layer. The slight asymmetry of the integrand, $\langle\alpha\rangle$, with respect to the location of the maximum field is because of the large difference between α_n and α_p in Si. If $\alpha_n \simeq \alpha_p$ as in the case of Ge and GaAs, $\langle\alpha\rangle$ reduces to

$$\langle\alpha\rangle = \alpha_n = \alpha_p, \tag{3}$$

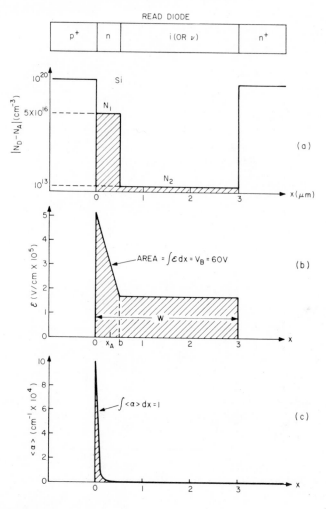

Fig. 1 Read diode
(a) doping profile (p^+nin^+),
(b) electric field distribution, and
(c) ionization integrand at avalanche breakdown.

and the avalanche region is symmetrical with respect to $x = 0$. As shown in Fig. 3 the avalanche region of a linearly graded junction, however, is wider than that of a one-sided abrupt junction having the same breakdown voltage. Figure 4 shows the results for a *p-i-n* diode which has a uniform field across

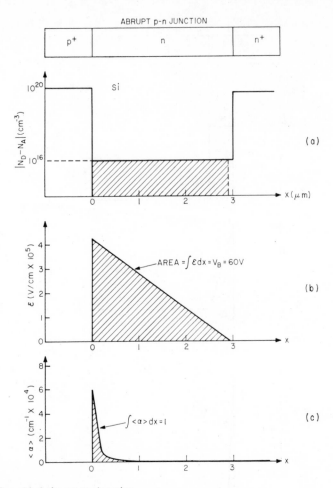

Fig. 2 One-sided abrupt *p-n* junction
(a) doping profile (p^+n),
(b) electric field, and
(c) ionization integrand at breakdown.

the intrinsic layer. The avalanche region in this case corresponds to the full intrinsic layer width.

(1) Breakdown Voltages

The breakdown voltages of the abrupt and the linearly graded *p-n* junctions[6] have been considered in Chapter 3. We can use the same method as

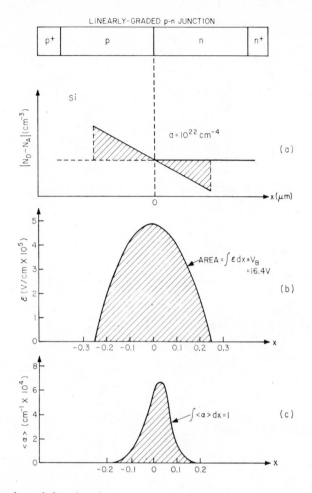

Fig. 3 Linearly graded p-n junction
(a) doping profile (with gradient a),
(b) electric field, and
(c) ionization integrand at breakdown.

outlined in that chapter to calculate the breakdown voltages of the Read diode and the *p-i-n* diode. The most important parameters for a Read diode, Fig. 1(a), are (1) N_1, the impurity concentration in the *n*-region, (2) *b*, the width of the *n* region, and (3) $(W - b)$, the width of the intrinsic region. This impurity distribution yields two special cases for large and small values of *b* as follows: (1) as $b \to 0$, it represents a *p-i-n* diode with an intrinsic width of W;

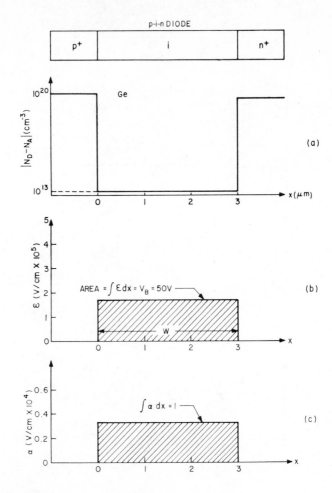

Fig. 4 *p-i-n* diode
(a) doping profile (*p-i-n*),
(b) electric field, and
(c) ionization integrand at breakdown.

(2) for $b > W(N_1)$, where $W(N_1)$ is the depletion layer width at breakdown for a p^+n abrupt junction with impurity concentration N_1, it represents an abrupt junction with uniform background impurity density. For the intermediate values of b (the Read diode), the field profile as shown in Fig. 1(b) is given by

$$\mathscr{E}(x) = \mathscr{E}_m - \frac{qN_1x}{\varepsilon_s} \qquad 0 \leq x \leq b \tag{4a}$$

$$\mathscr{E}(x) = \mathscr{E}_m - \frac{qN_1b}{\varepsilon_s} - \frac{qN_2(x-b)}{\varepsilon_s}, \qquad b \le x \le W \tag{4b}$$

where $\mathscr{E}_m$ is the maximum field which occurs at $x = 0$.

The calculated breakdown voltages for Ge and Si p^+nin^+ structures as a function of b (the width of the n-type region) are shown in Figs. 5 and 6 respectively.[7] At the far left where $b \to 0$, we show the depletion width (W) and the corresponding breakdown voltage of the p-i-n diode. For example, V is 150 V for a Ge p-i-n diode with 10 μm intrinsic width and is about 20 volts for 1 μm. For intermediate values of b the distribution is of the type proposed by Read, and the breakdown voltage for a given set of W and N_1 decreases as b increases. For example, with $W = 10$ μm and $N_1 = 5 \times 10^{15}$ cm^{-3}, V_B is 100 volts for a Ge p^+nin^+ diode with $b = 2.5$ μm. The dotted line indicates the limit at which the values of b are exactly equal to the depletion layer width $W(N_1)$. For larger values of b the breakdown voltage is independent of b and

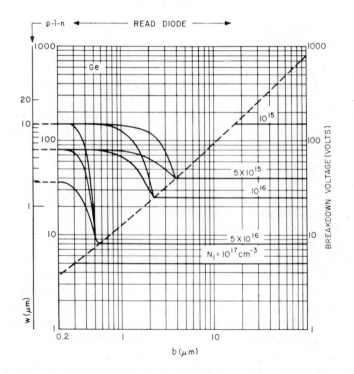

Fig. 5 Breakdown voltage versus b (the width of the n-region) for Ge Read and p-i-n diodes. The total depletion width W is a parameter. (After Gibbons and Sze, Ref. 7.)

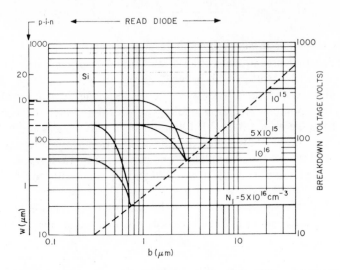

Fig. 6 Breakdown voltage versus b for Si Read and p-i-n diodes. The total depletion width W is a parameter (Ref. 7.)

is equal to that of the corresponding p^+n diode with the same doping N_1 (see Chapter 3).

The maximum field of a Read diode (with a given N_1) is found to be essentially the same (within 1 percent) as the value of the one-sided abrupt junction with the same N_1, provided that the avalanche region x_A is smaller than b.

The idealized multiple layer structure with impurity discontinuities between layers can be approached by the present fabrication techniques such as epitaxial growth or alloying. For diffused diodes, however, the impurity gradient may not be negligible. We shall consider two cases, as shown in the insert of Fig. 7, where, in case (1), the impurity concentration in the n region is linearly graded from the metallurgical junction to the intrinsic layer and, in case (2), it is uniformly close to the metallurgical junction but tails off in a linear fashion to the intrinsic layer. Also shown in this figure for comparison purposes is the idealized case already discussed. The electric field profile[33] is shown in Fig. 7 for each case. For case (1) we first calculate numerically the breakdown field for a one-sided linearly graded junction as a function of the impurity gradient, a, for both Ge and Si. The results are shown in Fig. 8. As in the idealized Read structure, the maximum field at breakdown for the one-sided linearly graded junction can be used for the structure with the doping profile of case (1) with an error of less than 1 percent. Since we know the maximum field (at the metallurgical junction when the diode is biased into

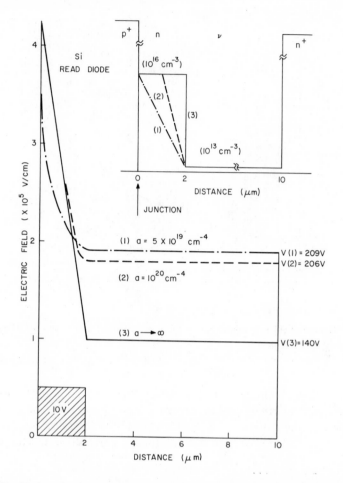

Fig. 7 Graphical approach to obtain breakdown voltage for diodes with impurity gradient. (Ref. 7.)

breakdown), the field profile through the structure is given by

$$\mathscr{E}(x) = \mathscr{E}_m - \frac{qN_1 x}{\varepsilon_s} + \frac{qax^2}{2\varepsilon_s} \qquad 0 \leq x \leq b \tag{5a}$$

$$\mathscr{E}(x) = \mathscr{E}_m - \frac{qN_1 x}{\varepsilon_s} + \frac{qab^2}{2\varepsilon_s} - \frac{qN_2(x-b)}{\varepsilon_s} \qquad b \leq x \leq W. \tag{5b}$$

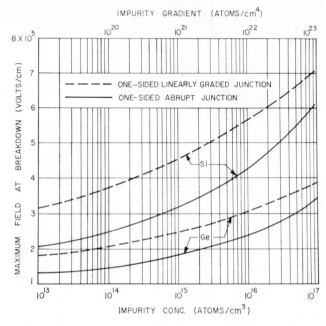

Fig. 8 Maximum field at breakdown condition for one-sided abrupt and one-sided linearly graded Ge and Si p-n junctions.
(Ref. 7.)

The breakdown voltage can then be calculated graphically. Similarly a graphical method can be used to determine V_B in case (2). The values of V_B for each of the three structures are indicated in Fig. 7. Clearly cases (1) and (3) produce upper and lower bounds for V_B for all intermediate structures such as case (2).

(2) Avalanche Region and Drift Region

The avalanche region of an ideal p-i-n diode is the full intrinsic layer width. For the Read diode and p-n junctions, however, the region of carrier multiplication is restricted to a narrow region close to the metallurgical junction. The contribution to the integral in Eq. (2) decreases rapidly as x departs from the metallurgical junction. Thus a reasonable definition of the avalanche region width, x_A, is obtained by taking the distance over which 95 percent of the contribution to the integral is obtained, i.e.,

$$\int_0^{x_A} \langle \alpha \rangle \, dx = 0.95 \qquad (6)$$

for the Read diodes and abrupt p-n junctions, and

$$\int_{-x_A/2}^{x_A/2} \langle \alpha \rangle \, dx = 0.95 \tag{7}$$

for the linearly graded junctions when the diodes are biased into breakdown.

Figure 9 shows a plot of x_A versus b for Si diodes with $N_1 = 10^{16}$ cm^{-3} and $W = 2$, 5, and 10 μm. For abrupt junctions ($b = 2.8$ μm) the width of the avalanche region x_A is ~ 0.9 μm while for small values of b, $x_A \sim W$ (corresponding to the p-i-n diode). For intermediate values of b, x_A is essentially constant and equal to the value of an abrupt junction. Corresponding values for the electric field ($\mathscr{E}_D$) in the intrinsic region are shown in Fig 9: since $\mathscr{E}_D \simeq \mathscr{E}_m - qN_1 b/\varepsilon_s$, $\mathscr{E}_D$ increases as b decreases. It is seen from Fig. 9 that

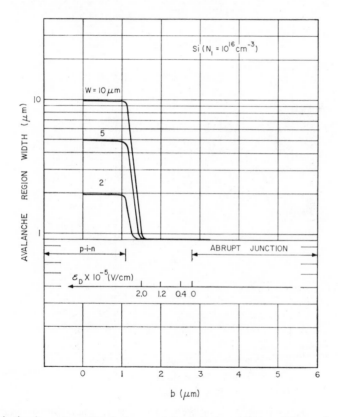

Fig. 9 Avalanche region width (x_A) versus b for Si diodes with $N_1 = 10^{16}$ cm^{-3} and $W = 2$, 5, and 10 μm. (Ref. 7.)

x_A for the Read diode does not vary by more than a few percent from the abrupt junction value up to fields of $\sim 2 \times 10^5$ V/cm. This essentially covers the design range for Si Read diodes, and we can thus conclude that the width of the avalanche region in a Read diode is determined only by the value of N_1 and is equal to the value obtained for an abrupt junction with doping density N_1. A similar result is obtained for the voltage drop, V_A, across the avalanche region. Figure 10 shows plots of avalanche width and voltage drop across the avalanche region for Ge and Si as a function of N_1. These curves apply to both the abrupt junction and the Read structure. For the linearly graded junction the avalanche region is slightly larger ($\sim 10\%$) than for the one-sided abrupt

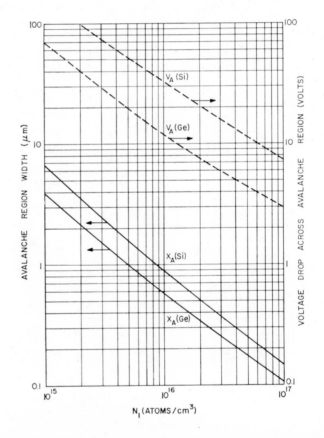

Fig. 10 Avalanche width (x_A) and voltage drop (V_A) across the avalanche region for Ge and Si diodes as a function of the n-type doping density (N_1). These curves apply to both abrupt and Read diodes.
(Ref. 7.)

junction with the same breakdown voltage V_B. This is because, for a given V_B, the maximum field of a linearly graded junction is smaller. This in turn gives a lower ionization rate. To accommodate an equal amount of multiplication, a wider avalanche region is thus required.

The drift region is the depletion layer excluding the avalanche region, or $x_A \leq x \leq W$. The most important parameter in the drift region is the carrier drift velocity. To obtain minimum carrier transit time across the drift region, the electric field in this region should be high enough that the generated carriers can travel at their scattering-limited velocities, v_{sl}. For silicon the electric field should be larger than 10^4 V/cm (see Chapter 2).

For p-i-n diodes this requirement is fulfilled automatically, since at breakdown the field (which is approximately constant over the full intrinsic width) is much larger than the required field for velocity saturation. For a Read diode the minimum field in the drift region is given by $\mathcal{E}_{min} = \mathcal{E}_m - q[N_1 b + N_2(W - b)]/\varepsilon_s$ from Eq. (4b). From the previous discussion it is clear that one can so design a diode that $\mathcal{E}_{min} > 10^4$ V/cm. For abrupt and linearly graded p-n junctions there are always some regions with fields smaller than the minimum required field. The low-field region, however, constitutes only a small percent of the total depletion region. For example, for a Si p^+n junction with 10^{16} cm^{-3} background doping, the maximum field at breakdown is 4×10^5 V/cm. The ratio of the low-field region (for a field less than 10^4 V/cm) to the total depletion layer is $10^4/(4 \times 10^5) = 2.5$ percent. Thus the low-field region has negligible effect on the reduction of the carrier transit time across the depletion layer.

(3) Temperature and Space-Charge Effects

The breakdown voltages shown in Figs. 5 and 6 (and also in Chapter 3 for p-n junctions) are the results for room temperature under isothermal conditions and free from space-charge effects. Under operating condition, however, the IMPATT diode is biased well into avalanche breakdown and the current density is usually very high. This results in a considerable rise in the temperature of the junction and a large space-charge effect.

The ionization rates of electrons and holes decrease with increasing temperature;[8] thus for an IMPATT diode with a given doping profile the breakdown voltage will increase with increasing temperature. As the dc power (product of reverse voltage and reverse current) increases, both the junction temperature and the breakdown voltage increase. Eventually the diode fails to operate, mainly because of permanent damage that results from excessive heating in localized spots. Thus the rising temperature of the junction imposes a severe limit on device operation. To prevent the rise in temperature, one must use a suitable heat sink. This will be considered in the final section of this chapter.

The space-charge effect is the variation of electric field in the depletion region due to generated carrier space charge. This effect gives rise to a positive dc incremental resistance for abrupt junctions and a negative dc incremental resistance for p-i-n diodes.

Consider first a one-sided p^+nn^+ abrupt junction. The depletion layer extends through the n region with a doping of N_D, and is bounded by the planes at $x = 0$ and $x = W$. When the applied voltage V is equal to the breakdown voltage V_B, the electric field $\mathscr{E}(x)$ has its maximum absolute value, $\mathscr{E}_m$, at $x = 0$. If we assume that the electrons travel at their scattering-limited velocity v_{sl} all across W, the space-charge current, I, is given by

$$I = v_{sl}\rho A \tag{8}$$

where ρ is the carrier-charge density and A the area. The disturbance $\Delta\mathscr{E}(x)$ in the electric field due to the space charge is obtained from Eq. (8) and Poisson's equation:[9]

$$\Delta\mathscr{E}(x) \simeq \frac{Ix}{A\varepsilon_s v_{sl}}. \tag{9}$$

If we assume that all the carriers are generated within the avalanche width x_A, the disturbance in voltage caused by the carriers in the drift region $(W - x_A)$ is obtained by integrating $\Delta\mathscr{E}(x)$:

$$\Delta V_B \simeq \int_0^{W-x_A} \frac{Ix}{A\varepsilon_s v_{sl}}\, dx = I\frac{(W - x_A)^2}{2A\varepsilon_s v_{sl}}. \tag{10}$$

The total applied voltage is thus

$$V = V_B + \Delta V_B = V_B + IR_{SC} \tag{11}$$

where R_{SC} is defined as the space-charge resistance and is obtained from Eqs. (10) and (11):

$$R_{SC} \equiv \frac{\Delta V_B}{I} \simeq \frac{(W - x_A)^2}{2A\varepsilon_s v_{sl}}. \tag{12}$$

For wide depletion width and high current density the space-charge effect may give rise to large ΔV_B. For example, in a Si p^+n diode with $N_D = 10^{15}$ cm^{-3} and $A = 5 \times 10^{-4}$ cm^2, the depletion-layer width at breakdown is 18 μm and x_A is about 6 μm (Fig. 10). The space-charge resistance is determined from Eq. (12) to be about 140 ohms. The value of ΔV_B for a current density of 1000 amp/cm^2 ($I = 0.5$ amp) is 70 V which amounts to about a 20-percent increase over the breakdown voltage.

For a p-i-n diode the situation is different from that of a p^+n junction (refer to Section 7 of Chapter 3); when the applied reverse voltage is just large enough to cause avalanche breakdown, the reverse current is small. The space-charge effect can be neglected and the electric field is essentially uniform across the depletion layer (assuming $\alpha_n = \alpha_p$ as in Ge diodes). As the current increases, more electrons are injected from the p-i boundary and more holes are injected from the n-i boundary. These space charges will cause a reduction of the field $\mathscr{E}_i$ in the center of the i region. At breakdown the maximum fields which occur at the boundaries are essentially fixed. Thus as the current increases, $\mathscr{E}_i$ decreases, and the voltage which equals $\int_0^W \mathscr{E}\, dx$ is thus reduced. This results in a negative incremental resistance for the p-i-n diode.

3 BASIC DYNAMIC CHARACTERISTICS

In this section we shall present the basic small-signal analysis of the Read diode to illustrate the physical significance of the avalanche-multiplication and the transit-time effects. This analysis was first considered by Read[2] and developed further by Gilden and Hines.[10] For simplicity it is assumed that $\alpha_n = \alpha_p = \alpha$, and that the scattering-limited drift velocities of holes and electrons are equal. Figure 11(a) shows the model of a Read diode. According to the discussion in Section 2, we have divided the diode into three regions: (1) the avalanche region which is assumed to be thin so that space-charge and signal delay can be neglected; (2) the drift region where no carriers are generated, and all carriers entering from the avalanche region travel at their scattering-limited velocities; and (3) an inactive region which adds undesirable parasitic resistance.

The two active regions interact with one another, since the ac electric field is continuous across the boundary between them. We shall use a zero subscript to indicate dc quantities, and tilda ($\sim$) to indicate small-signal ac quantities. For quantities including both dc and ac components, no zero subscript or tilda will be added. We first define $\tilde{J}_A$ as the avalanche current density which is the alternating conduction (particle) current density in the avalanche region, and $\tilde{J}$ as the total alternating current density. With our assumption of a thin avalanche region, $\tilde{J}_A$ is presumed to enter the drift region without delay. With the assumption of a saturated drift velocity v_{sl}, the alternating conduction current density $\tilde{J}_c(x)$ in the drift region propagates as an unattenuated wave (with only phase change) at this drift velocity,

$$\tilde{J}_c(x) = \tilde{J}_A e^{-j\omega x/v_{sl}}$$

$$\equiv \gamma \tilde{J} e^{-j\omega x/v_{sl}} \tag{13}$$

where $\gamma \equiv \tilde{J}_A / \tilde{J}$ is the complex fraction relating the avalanche current density to the total alternating current density.

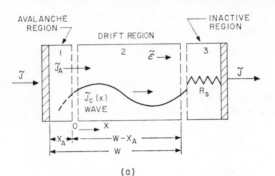

(a)

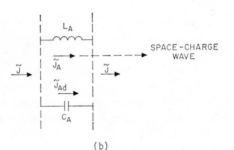

(b)

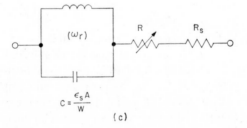

(c)

Fig. 11

(a) Model of Read diode with avalanche region, drift region, and inactive region.

(b) Equivalent circuit of the avalanche region.

(c) Equivalent circuit of Read diode for small transit angle.

At any cross section, the total alternating current density $\tilde{J}$ is equal to the sum of the conduction current density $\tilde{J}_c$ and the displacement current density $\tilde{J}_d$, and this sum is a constant, independent of position x:

$$\tilde{J} = \tilde{J}_c(x) + \tilde{J}_d(x) \neq f(x). \tag{14}$$

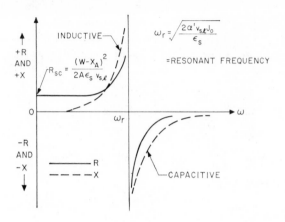

(d) Real and imaginary parts of the impedance versus frequency. ω_r is the resonant angular frequency.
(After Gilden and Hines, Ref. 10.)

The displacement current density is related to the ac field $\tilde{\mathscr{E}}(x)$ by

$$\tilde{J}_d = j\omega\varepsilon_s\,\tilde{\mathscr{E}}(x). \tag{15}$$

Combination of Eqs. (13) through (15) yields an expression for the ac electric field in the drift region as a function of x and $\tilde{J}$,

$$\tilde{\mathscr{E}}(x) = \tilde{J}\frac{(1 - \gamma e^{-j\omega x/v_{sl}})}{j\omega\varepsilon_s}. \tag{16}$$

Integration of $\tilde{\mathscr{E}}(x)$ gives the voltage across the drift region in terms of $\tilde{J}$. The coefficient γ is derived in the analysis below.

(1) Avalanche Region

Consider the avalanche region first.[10] Under the dc condition, the direct current density $J_0(= J_{po} + J_{no})$ is related to the thermally generated reverse saturation current density $J_s(= J_{ns} + J_{ps})$ by

$$\frac{J_s}{J_0} = 1 - \int_0^W \langle\alpha\rangle\,dx. \tag{17}$$

At breakdown, J_0 approaches infinity and $\int_0^W \langle\alpha\rangle\,dx = 1$. In the dc case the ionization integral cannot be greater than unity. This is not necessarily so for a rapidly varying field. The differential equation for the current as a function of time will now be derived. Under the conditions that (1) electrons and holes have equal ionization rates and equal scattering-limited velocities, and (2) the drift current components are much larger than the diffusion component, the basic device equations, in the one-dimensional case, can be given as follows:

$$\frac{\partial \mathscr{E}}{\partial x} = \frac{q}{\varepsilon_s}(N_D^+ - N_A^- + p - n) \qquad \text{Poisson's Equation} \qquad (18)$$

$$\left. \begin{array}{l} J_n = qv_{sl}n \\[2mm] J_p = qv_{sl}p \\[2mm] J = J_n + J_p \end{array} \right\} \quad \text{Current Density Equations,} \qquad (19)$$

$$\left. \begin{array}{l} \dfrac{\partial n}{\partial t} = \dfrac{1}{q}\dfrac{\partial J_n}{\partial x} + \alpha v_{sl}(n + p) \qquad\qquad\qquad\quad (20\text{a}) \\[4mm] \dfrac{\partial p}{\partial t} = -\dfrac{1}{q}\dfrac{\partial Jp}{\partial x} + \alpha v_{sl}(n + p) \qquad\qquad\qquad\quad (20\text{b}) \end{array} \right\} \text{Continuity Equations.}$$

The second terms on the right-hand side of Eq. (20) correspond to the generation rate of the electron-hole pairs by avalanche multiplication. This is so large compared to the rate of thermal generation that the latter can be neglected. Addition of Eqs. (20a) and (20b), using Eq. (19), and integration from $x = 0$ to $x = x_A$ gives

$$\tau_A \frac{dJ}{dt} = -(J_p - J_n)_0^{x_A} + 2J \int_0^{x_A} \alpha \, dx \qquad (21)$$

where $\tau_A = x_A/v_{sl}$ is the transit time across the multiplication region. The boundary conditions are that the electron current at $x = 0$ consists entirely of the reverse saturation current J_{ns}. Thus at $x = 0$, $J_p - J_n = -2J_n + J = -2J_{ns} + J$. At $x = x_A$ the hole current consists of the reverse saturation current J_{ps} generated in the space-charge region, so $J_p - J_n = 2J_p - J = 2J_{ps} - J$. With these boundary conditions Eq. (21) becomes

$$\frac{dJ}{dt} = \frac{2J}{\tau_A}\left[\int_0^{x_A} \alpha \, dx - 1\right] + \frac{2J_s}{\tau_A}. \qquad (22)$$

In the dc case J is the direct current J_o, so that Eq. (22) reduces to Eq. (17).

We will simplify Eq. (22) by substituting $\bar{\alpha}$ in place of α, where $\bar{\alpha}$ is an average value of α obtained by evaluating the integral over the extent of the avalanche region. We obtain (by neglecting the term J_s)

$$\frac{dJ}{dt} = \frac{2J}{\tau_A}(\bar{\alpha}x_A - 1). \qquad (23)$$

The small-signal assumptions are now made:

$$\bar{\alpha} = \bar{\alpha}_0 + \tilde{\alpha}e^{j\omega t} \simeq \bar{\alpha}_0 + \alpha'\tilde{\mathscr{E}}_A e^{j\omega t}$$

$$\bar{\alpha}x_A = 1 + x_A \alpha'\tilde{\mathscr{E}}_A e^{j\omega t}$$

$$J = J_0 + \tilde{J}_A e^{j\omega t}$$

$$\mathscr{E} = \mathscr{E}_0 + \tilde{\mathscr{E}}_A e^{j\omega t} \tag{24}$$

where $\alpha' \equiv \partial\alpha/\partial\mathscr{E}$ and the substitution $\tilde{\alpha} = \alpha'\tilde{\mathscr{E}}_A$ has been employed. Substitution of the above expressions into Eq. (23), neglecting products of higher-order terms, leads to the expression for the ac component of the avalanche conduction current density

$$\tilde{J}_A = \frac{2\alpha'x_A J_0 \tilde{\mathscr{E}}_A}{j\omega\tau_A}. \tag{25}$$

The displacement current in the avalanche region is given by

$$\tilde{J}_{Ad} = j\omega\varepsilon_s \tilde{\mathscr{E}}_A. \tag{26}$$

These are the two components of the total circuit current in the avalanche region. For a given field the avalanche current $\tilde{J}_A$ is reactive and varies inversely with ω as in an inductor. The other is also reactive and varies directly with ω as in a capacitor. Thus the avalanche region behaves as an LC parallel circuit. The equivalent circuit is shown in Fig. 11(b) where the inductance and capacitance are given as (where A is the diode area):

$$L_A = \tau_A/2J_0\alpha'A$$

$$C_A = \varepsilon_s A/x_A. \tag{27}$$

The resonant frequency of this combination is given by

$$f_r \equiv \frac{\omega_r}{2\pi} = \frac{1}{2\pi}\sqrt{\frac{2\alpha'v_{sl}J_0}{\varepsilon_s}}. \tag{28}$$

The impedance of the avalanche region has the simple form

$$Z_A = \frac{x_A}{j\omega\varepsilon_s A}\left[\frac{1}{1 - \dfrac{\omega_r^2}{\omega^2}}\right] = \frac{1}{j\omega C_A}\left[\frac{1}{1 - \dfrac{\omega_r^2}{\omega^2}}\right]. \tag{29}$$

The factor γ has the form

$$\gamma \equiv \frac{\tilde{J}_A}{\tilde{J}} = \frac{1}{1 - \dfrac{\omega^2}{\omega_r^2}}. \tag{30}$$

A thin avalanche region, therefore, behaves as an antiresonant circuit with a resonant frequency proportional to the square root of the direct current density J_0, Eq. (28).

(2)   Drift Region and Total Impedance

Substitution of γ in Eq. (30) into Eq. (16) and integration over the drift length $(W - x_A)$ give an expression for the ac voltage across this region,

$$\tilde{V}_d = \frac{(W - x_A)\tilde{J}}{j\omega\varepsilon_s}\left[1 - \frac{1}{1 - \dfrac{\omega^2}{\omega_r^2}}\left(\frac{1 - e^{-j\theta_d}}{j\theta_d}\right)\right] \tag{31}$$

where θ_d is the transit angle of the drift space

$$\theta_d \equiv \frac{\omega(W - x_A)}{v_{sl}} \equiv \omega\tau_d \tag{32}$$

and

$$\tau_d = \frac{(W - x_A)}{v_{sl}}. \tag{32a}$$

We may also define $C_d \equiv A\varepsilon_s/(W - x_A)$ as the capacitance of the drift region. From Eq. (31) we obtain the impedance for the drift region,

$$Z_d \equiv \frac{\tilde{V}_d}{\tilde{J}A} = \frac{1}{\omega C_d}\left[\frac{1}{1 - \dfrac{\omega^2}{\omega_r^2}}\left(\frac{1 - \cos\theta_d}{\theta_d}\right)\right] + \frac{j}{\omega C_d}\left[-1 + \frac{1}{1 - \dfrac{\omega^2}{\omega_r^2}}\left(\frac{\sin\theta_d}{\theta_d}\right)\right]$$

$$= R + jX \quad \text{(ohms)} \tag{33}$$

where R and X are the resistance and reactance respectively. It is obvious that the real part (resistance) will be negative for all frequencies above ω_r except for nulls at $\theta_d = 2\pi \times$ integer. The resistance is positive for frequencies below ω_r and approaches a finite value at zero frequency:

$$R(\omega \to 0) = \frac{\tau_d}{2C_d} = \frac{(W - x_A)^2}{2A\varepsilon_s v_{sl}} \quad \text{(ohms)}.$$

The low-frequency small-signal resistance is a consequence of the space-charge in the finite thickness of the drift region, and the above expression is identical to Eq. (12) derived previously.

The total impedance is the sum of the impedances of the avalanche region, the drift region, and the passive resistance R_s of the inactive region:

$$Z = \frac{(W - x_A)^2}{2A\varepsilon_s v_{sl}} \left(\frac{1}{1 - \frac{\omega^2}{\omega_r^2}}\right) \frac{1 - \cos \theta_d}{\frac{\theta_d^2}{2}} + R_s$$

$$+ \frac{j}{\omega C_d} \left[\left(\frac{\sin \theta_d}{\theta_d} - 1\right) - \left(\frac{\frac{\sin \theta_d}{\theta_d} + \frac{x_A}{W - x_A}}{1 - \frac{\omega_r^2}{\omega^2}}\right)\right] \quad \text{(ohms)}. \tag{35}$$

Equation (35), which is given by Gilden and Hines,[10] has been cast in a form which can be simplified directly for the case of small transit angle θ_d. For $\theta_d < \pi/4$, Eq. (35) reduces to

$$Z = \frac{(W - x_A)^2}{2Av_{sl}\varepsilon_s\left(1 - \frac{\omega^2}{\omega_r^2}\right)} + R_s + \frac{j}{\omega C}\left(\frac{1}{\frac{\omega_r^2}{\omega^2} - 1}\right) \tag{36}$$

where $C \equiv \varepsilon_s A/W$ corresponding to the total depletion capacitance. From the above equation we note that the first term is the active resistance which becomes negative for $\omega > \omega_r$. The third term is reactive and corresponds to a parallel resonant circuit which includes the diode capacitance and a shunt inductor. The reactance is inductive for $\omega < \omega_r$ and capacitive for $\omega > \omega_r$. In other words, the resistance becomes negative at the frequency where the reactive component changes sign. It is interesting to note that large negative resistance can be obtained when the transit angle θ_d is substantially less than π. The equivalent circuit and the frequency dependence of the real and imaginary parts of the impedance are shown in Figs. 11(c) and (d) respectively.

4 GENERALIZED SMALL-SIGNAL ANALYSIS

In this section we shall first define some quantities pertinent to the dynamic operation of a general IMPATT diode. The elementary small-signal solutions of a *p-i-n* diode will then be considered. The characteristics of a general IMPATT diode are analyzed with the help of the elementary solutions.

(1) Dynamic Quantities

(A) Transit Time (τ) and Transit Angle (θ):

$$\tau \equiv W/v_{sl} \qquad (37)$$

$$\theta \equiv \omega W/v_{sl} = \omega\tau \qquad (38)$$

where W is the total depletion layer width, and v_{sl} the scattering-limited drift velocity. For a given angular frequency ω, the transit angle increases with increasing width. For a given width the transit angle will also increase with increasing frequency.

(B) AC Impedance (Z) and Admittance (Y):

$$Z = R + jX = \frac{1}{G + jB} = \frac{1}{Y} \equiv \frac{\tilde{V}}{\tilde{J}A} \qquad (39)$$

where $\tilde{V}$ and $\tilde{J}$ are the total ac voltage and total current density (particle current and displacement current) respectively, and A is the device area.

(C) Resonant Frequency ($f_r \equiv \omega_r/2\pi$): f_r is the frequency at which the imaginary part (B) of the admittance changes from inductive to capacitive. For a Read diode, f_r is given previously in Eq. (28) as $(2\alpha'v_{sl}J_o/\varepsilon_s)^{1/2}/2\pi$. This relationship will be shown to be valid even for a general IMPATT diode.

(D) Cutoff Frequency ($f_c \equiv \omega_c/2\pi$): f_c is the minimum frequency at which the real part (G) of the admittance changes from positive to negative. For a Read diode, f_c is exactly equal to f_r. For a general IMPATT diode it will be shown, however, that f_c is lower than f_r and that, as the avalanche width increases, the difference becomes larger.

(E) Quality Factor Q and Growth Factor g: The small-signal quality factor Q is defined[5] as the time average of the ac field energy, $\langle E \rangle$, divided by the ac power dissipation, $\langle dE/dt \rangle$, per cycle, or

$$Q \equiv \frac{\omega \langle E \rangle}{-\left\langle \dfrac{dE}{dt} \right\rangle} = \frac{B}{G}. \qquad (40)$$

If the conductance is negative, then Q is negative. The factor Q gives information about threshold and the buildup rate of oscillation when the negative resistance is used as an oscillator. A smaller magnitude of negative Q is preferable. The growth factor g is related to Q and is defined as

$$g \equiv -\frac{1}{2Q}. \qquad (41)$$

(F) Optimum Frequency (f_{opt}): For a given bias current the optimum frequency is defined as the frequency at which the quality factor $|Q|$ takes the smallest magnitude.

(2) Elementary Solutions

The basic device equations (18) through (20) can be written as

$$\frac{\partial \mathscr{E}}{\partial x} = \frac{q}{\varepsilon_s}(N_D^+ - N_A^-) + \frac{1}{v_{sl}\varepsilon_s}(J_p - J_n), \tag{42}$$

$$\frac{1}{v_{sl}}\frac{\partial J_n}{\partial t} = \frac{\partial J_n}{\partial x} + \alpha(J_n + J_p), \tag{43}$$

$$\frac{1}{v_{sl}}\frac{\partial J_p}{\partial t} = -\frac{\partial J_p}{\partial x} + \alpha(J_n + J_p). \tag{44}$$

With small-signal assumptions similar to those made in Eq. (24) such that the zero subscripts indicate time-independent steady-state (dc) solutions, and the tilda indicates small ac signals, the time-dependent parts of the system of equations, Eqs. (42) through (44), are

$$\frac{\partial \tilde{\mathscr{E}}}{\partial x} = \frac{-1}{\varepsilon_s v_{sl}}(\tilde{J}_p - \tilde{J}_n) \tag{45}$$

$$\frac{\partial \tilde{J}_n}{\partial x} = +\bar{\alpha}_0(\tilde{J}_n + \tilde{J}_p) + \alpha'\tilde{\mathscr{E}}(J_{no} + J_{po}) - \frac{j\omega}{v_{sl}}\tilde{J}_n \tag{46}$$

$$\frac{\partial \tilde{J}_p}{\partial x} = -\bar{\alpha}_0(\tilde{J}_n + \tilde{J}_p) - \alpha'\tilde{\mathscr{E}}(J_{no} + J_{po}) + \frac{j\omega}{v_{sl}}\tilde{J}_p \tag{47}$$

where $\alpha' \equiv \partial\alpha/\partial\mathscr{E}$.

Substitution of Eq. (47) into Eq. (46) and combination of the resultant equation with Eq. (45) yield one solution of the above equations:

$$\tilde{J}_n + \tilde{J}_p + j\omega\varepsilon_s\tilde{\mathscr{E}} = \tilde{J} = \text{constant}, \tag{48}$$

which is actually the required condition that the total alternating current is equal to the sum of the conduction current and the displacement current. Differentiation of Eq. (45) with respect to x and substitution of Eqs. (46) and (47) yield the second-order differential equation:

$$\frac{\partial^2 \tilde{\mathscr{E}}}{\partial x^2} + \left[\frac{\omega^2}{v_{sl}^2} - \frac{2\alpha'J_0}{v_{sl}\varepsilon_s} + \frac{j2\bar{\alpha}_0\omega}{v_{sl}}\right]\tilde{\mathscr{E}} = \tilde{J}\left(\frac{2\bar{\alpha}_0}{v_{sl}\varepsilon_s} - \frac{j\omega}{\varepsilon_s v_{sl}^2}\right). \tag{49}$$

When the ac electric field is constant over the whole depletion region (as for a p-i-n diode), $\bar{\alpha}_o$ is independent of x, and the above equation has constant coefficients. The solution is simply

$$\tilde{\mathscr{E}} = C_1 e^{jkx} + C_2 e^{-jkx} + \frac{\left(2\dfrac{\bar{\alpha}_0}{\varepsilon_s v_{sl}} - \dfrac{j\omega}{\varepsilon_s v_{sl}^2}\right)}{k^2}\tilde{J} \tag{50}$$

where C_1 and C_2 are integration constants to be determined by the boundary conditions and the complex wave number k is given by

$$k = \left(\frac{\omega^2}{v_{sl}^2} - \frac{2\alpha' J_0}{v_{sl} \varepsilon_s} + \frac{j 2 \bar{\alpha}_0 \omega}{v_{sl}}\right)^{1/2}. \tag{51}$$

The voltage is obtained by integrating the field

$$\tilde{V}(x) = \int_0^x \tilde{\mathscr{E}} \, dx = \frac{C_1}{jk}(e^{jkx} - 1) - \frac{C_2}{jk}(e^{-jkx} - 1) + \frac{\left(\dfrac{2\bar{\alpha}_0}{\varepsilon_s v_{sl}} - \dfrac{j\omega}{\varepsilon_s v_{sl}^2}\right)}{k^2}\tilde{J}x. \tag{52}$$

Since in breakdown $\int \bar{\alpha}_o \, dx = 1$, it follows that $\bar{\alpha}_o = 1/W$. From Eqs. (50) and (45) through (47) the alternating current densities are given as[5]

$$\tilde{J}_n = \frac{j\varepsilon_s v_{sl}}{2}\left(k - \frac{\omega}{v_{sl}}\right)C_1 e^{jkx} - \frac{j\varepsilon_s v_{sl}}{2}\left(k + \frac{\omega}{v_{sl}}\right)C_2 e^{-jkx} - \frac{\alpha' J_0}{k^2 \varepsilon_s v_{sl}}\tilde{J}, \tag{53}$$

$$\tilde{J}_p = -\frac{j\varepsilon_s v_{sl}}{2}\left(k + \frac{\omega}{v_{sl}}\right)C_1 e^{jkx} + \frac{j\varepsilon_s v_{sl}}{2}\left(k - \frac{\omega}{v_{sl}}\right)C_2 e^{-jkx} - \frac{\alpha' J_0}{k^2 \varepsilon_s v_{sl}}\tilde{J}. \tag{54}$$

The currents and field consist of three components; two propagate in the positive and negative x directions with increasing amplitude in the direction of propagation, the third component is a constant. When C_1 and C_2 are determined from the boundary conditions, the small-signal impedance is then given by

$$Z \equiv R + jX = \frac{1}{G + jB} = \frac{\tilde{V}(W)}{\tilde{J}A}. \tag{55}$$

The calculated values of $\alpha' \equiv \partial\alpha/\partial\mathscr{E}$ are plotted in Fig. 12 as a function of α. We note that α' increases as α increases; and at any given field, larger α corresponds to larger α'. Figure 13 shows a typical result[5] of the impedance as a function of frequency for a p-i-n diode with $W = 5$ μm, $\varepsilon_s/\varepsilon_o = 12$, and $v_{sl} = 8.5 \times 10^6$ cm/sec. The impedances are normalized to $W^2 v_{sl}/\varepsilon_s = 2.77 \times 10^{-2}$ ohm-cm^2, and the frequency is normalized to $v_{sl}/2\pi W = 2.71$ GHz. The three sets of curves correspond to three different direct bias currents. The real part shown is all negative. Its magnitude, increased rapidly from low frequency, has a maximum value at the resonant frequency f_r, and

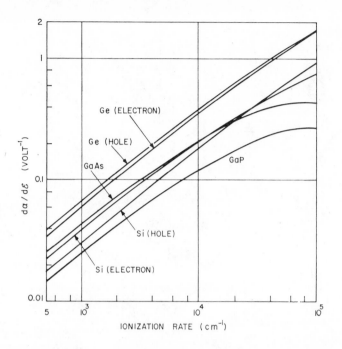

Fig. 12 $\alpha' \equiv \partial\alpha/\partial\mathscr{E}$ versus the ionization coefficient α for Ge, Si, GaAs, and GaP. (After Misawa, Ref. 11.)

then decreases. The reactance is inductive at lower frequencies and changes to capacitive after the resonance. We note that the resonant frequency increases approximately proportionally to the square root of the direct bias current. This is in agreement with the results discussed previously in Section 3 for the Read diode. The resistance of a *p-i-n* diode, however, is always negative even at zero frequency. This is in contrast to the Read diode whose resistance becomes negative right at the frequency (f_r) where the reactance changes sign.

(3) General Solutions

To analyze a general IMPATT diode, we consider the multiple-uniform-layer model proposed by Misawa.[11] This model consists in dividing the entire depletion region into several successive layers having constant avalanche multiplication (including zero multiplication) in each layer, and in connecting the analytical solutions for the individual layers having proper boundary conditions. In the previous studies we have considered two limiting cases: the Read diode and the *p-i-n* diode. In the Read diode the avalanche region is

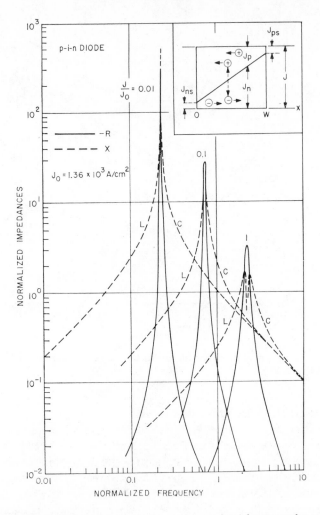

Fig. 13 Normalized small-signal impedance versus normalized frequency for a *p-i-n* diode with $W = 5$ μm. $L =$ inductive, $C =$ capacitive. (After Misawa, Ref. 5.)

so narrow that the phase shift of the signal within the region is neglected; in the *p-i-n* diode there is no avalanche-free drift region. By the use of the multiple-uniform-layer model we can investigate the intermediate case between the two limiting cases.

An intermediate case is shown in Fig. 14 where region 1 having width $W_1(\simeq x_A)$ is approximated by a constant ionization coefficient $\langle \bar{\alpha} \rangle$, and region

2 with width W_2 has zero multiplication. In region 1 the solutions are the same as given by Eqs. (50) through (54). In region, 2, where $\alpha = \alpha' = 0$, the wave number k is a real number and equal to ω/v_{sl}. Equations (50) through (54) reduce to

$$\tilde{\mathscr{E}} = C_1 e^{j\omega x/v_{sl}} + C_2 e^{-j\omega x/v_{sl}} + \frac{\tilde{J}}{j\omega\varepsilon_s} \tag{56}$$

$$\tilde{V} = \frac{C_1 v_{sl}}{j\omega} (e^{j\omega x/v_{sl}} - 1) - \frac{2C v_{sl}}{j\omega} (e^{-j\omega x/v_{sl}} - 1) + \frac{\tilde{J}x}{j\omega\varepsilon_s} \tag{57}$$

$$\tilde{J}_n = -j\omega\varepsilon_s C_2 e^{-j\omega x/v_{sl}}, \tag{58}$$

$$\tilde{J}_p = -j\omega\varepsilon_s C_1 e^{j\omega x/v_{sl}}. \tag{59}$$

The waves now have constant amplitude.

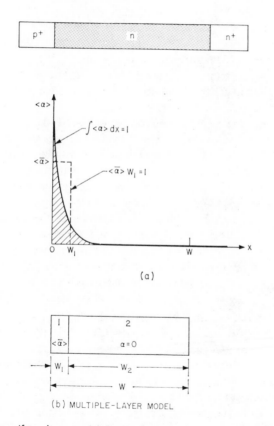

(a)

(b) MULTIPLE-LAYER MODEL

Fig. 14 Multiple-uniform-layer model. Region 1 is the avalanche region. Region 2 is the drift region.

The above elemental solutions can be matched at the boundaries. For the case shown in Fig. 14, at $x = 0$, $\tilde{J}_n(1) = 0$; at $x = W_1$, $\tilde{J}_n(1) = \tilde{J}_n(2)$, $\tilde{J}_p(1) = \tilde{J}_p(2) = 0$, since there is no hole current in region 2. From these conditions the constants C_1's and C_2's can be determined, and the small-signal admittance can be obtained from Eq. (55).

Figures 15(a) and (b) show the real and imaginary parts, respectively, of the admittance of a diode with 10% avalanche region (approximately a Read diode) for five bias currents. The admittance is normalized to $v_{sl}\varepsilon_s/W^2$ which is equal to 10.76 mho/cm^2 for a diode with $W = 10$ μm, $v_{sl} = 10^7$ cm/sec, and $\varepsilon_s/\varepsilon_o = 12$ (corresponding to silicon). The normalizing current J_1 is given by $(\pi^2 \varepsilon_s v_{sl}/2\alpha' W^2)$ so that for $J/J_1 = 1$, the resonant frequency $\omega_r = (2\alpha' v_{sl} J_1/\varepsilon_s)^{1/2} = v_{sl}\pi/W$, and $\theta = \pi$. We note that, as expected, both the real and imaginary parts of the admittance for $J = J_1$ change their sign near $\theta = \pi$. The slightly larger transit angle at $G = 0$ is due to the effect of phase shift in the avalanche region. At $J = 4J_1$, the transit angle $\theta = 2\pi$. From Eq. (35) the real part is zero (assuming $R_s = 0$) and the imaginary part can be approximated by a capacitance with width $(W - x_A)$. These are in good agreement with the results shown in Fig. 15. Figure 16 shows the admittance presented in a different fashion[12] where more realistic values $(\alpha_n \neq a_p)$ are used in the calculation. We note that the general behavior of the admittance is similar to that of Fig. 15.

Figure 17 shows the admittance[11] of six IMPATT diodes with identical total depletion-layer widths but different avalanche region widths $(W_1/W = 1/10, 1/3, 1/2, 2/3, 9/10, 1)$. The current density is J_1 as given previously. As the avalanche region widens, the negative resistance band also widens. This happens because the avalanche region, which possesses a negative resistance over a wide frequency band, occupies a larger and larger fraction of the depletion layer. The cutoff frequency f_c decreases as the avalanche region widens. In Fig. 18(a) the cutoff frequency and the resonant frequency are plotted as a function of the bias current for the above six diodes. Both frequencies vary approximately as the square root of the bias current. These frequencies are replotted in Fig. 18(b) as a function of the avalanche region width (W_1/W). For a given current the resonant frequency decreases with W_1. This is because $f_r \sim \sqrt{\alpha'}$, and for larger avalanche region W_1, the average ionization coefficient is smaller $(\langle \bar{\alpha} \rangle = 1/W_1)$, which in turn gives smaller α'. The cutoff frequency also decreases with increasing W_1 and becomes smaller than the resonant frequency. This is because the power dissipation in the avalanche region is still negative even at very low frequencies.

The quality factor Q for the two limiting cases $(W/W_1 = 1/10, 1)$ is shown in Fig. 18(c). At the smallest bias current, the Q of the 10% unit has the optimum (i.e., minimum negative) value at about $\theta = \pi$. As the bias current increases, the optimum frequency increases, and the best Q degrades.

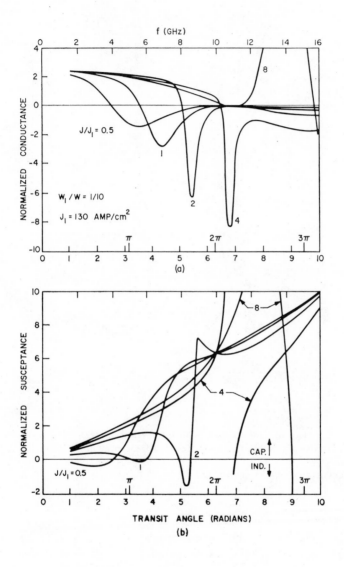

Fig. 15
(a) Real part of the admittance of a diode with 10% avalanche region for five bias currents.
(b) Imaginary part of the admittance.
(After Misawa, Ref. 11.)

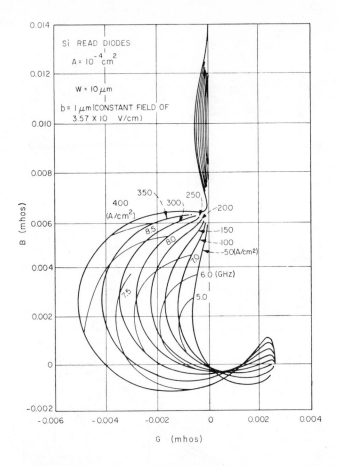

Fig. 16 Admittance of a silicon diode with 10% avalanche region. The admittance is calculated using realistic values ($\alpha_n \neq \alpha_p$). Selected frequencies in GHz are marked off on each curve.
(After Gummel and Scharfetter, Ref. 12.)

Although not shown, as the bias current decreases below $J_1/2$, the optimum frequency decreases and the best Q degrades. Thus the overall best Q for the 10% unit is obtained at about $\theta \approx \pi$. For the 100% unit ($p\text{-}i\text{-}n$ diode) the quality factor Q can approach very small negative values at high bias currents and small transit angles. The above results can be readily used for other structures with different sizes and parameters. For a set of depletion width

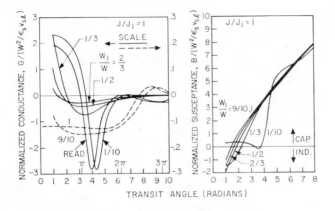

TRANSIT ANGLE (RADIANS)

Fig. 17 Admittance of six IMPATT diodes with identical total depletion layer width but different avalanche region width.
(After Misawa, Ref. 11.)

W_i, scattering-limited velocity v_{di}, and α_i', we can obtain pertinent quantities by some simple multiplication factors. For example, the factor is

$$\left(\frac{W_i}{W_f}\frac{v_{df}}{v_{di}}\right)$$

for frequency,

$$\left(\frac{W_f}{W_i}\right)^2\left(\frac{v_{di}}{v_{df}}\right)$$

for impedance, and

$$\frac{\alpha'_i}{\alpha'_f}\left(\frac{W_i}{W_f}\right)^2\left(\frac{v_{df}}{v_{di}}\right)$$

for current where the subscript f indicates the new (or final) parameter.

From the above discussions we can summarize the basic small-signal properties of the IMPATT diodes:

(1) under proper biasing conditions junction diodes with various doping and field profiles can give rise to incremental negative resistance,

(2) both the cutoff frequency f_c and the resonant frequency f_r are proportional to the square root of the direct current density,

(3) at a given average current density, both f_c and f_r decrease with increasing avalanche region,

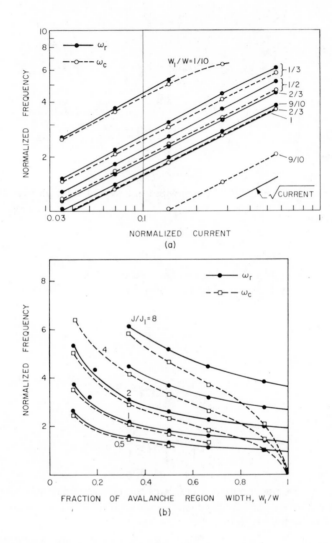

Fig. 18

(a) Cutoff frequency and resonant frequency for the six diodes shown in Fig. 17 versus direct current density.

(b) Cutoff frequency and resonance frequency versus avalanche region width. Note that when $W_1/W = 1$ (p-i-n), the cutoff frequency is zero.

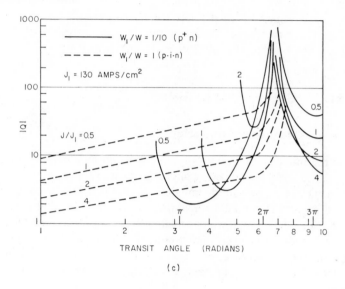

TRANSIT ANGLE (RADIANS)

(c)

(c) Quality factor Q for two limiting cases ($W_1/W = 1/10, 1$).
(Ref. 11.)

(4) for a given depletion width, as the avalanche region widens, the negative conductance band widens and the values of negative conductance decrease,

(5) for a *p-i-n* diode with uniform avalanche multiplication ($W_1/W = 1$) there is no cutoff frequency due to the intrinsic instability of the electron-hole plasma over the entire depletion region,

(6) for a Read diode ($W_1/W \simeq 1/10$) the optimum quality factor Q is obtained at a current density such that the frequency is approximately equal to one-half the reciprocal transit time ($f_{opt} \simeq 1/2\tau = v_{sl}/2W$, or $\theta = \pi$),

(7) for an intermediate case, as when the ratio W_1/W increases, the optimum frequency increases to a maximum value of $f = 1/\tau = v_{sl}/W$ (or $\theta = 2\pi$) at $W_1/W = \frac{1}{2}$ and then decreases to $f = 1/2\tau$ at $W_1/W = 9/10$; the corresponding average current density at the optimum frequency increases with increasing W_1/W,

(8) it is required that, for higher frequency oscillation, a narrower depletion layer should be used,

(9) from the behavior of the small-signal negative Q, it is expected that for low bias currents the oscillation performance improves when the avalanche region becomes relatively narrow, and

(10) in materials with larger ionization rate, a negative resistance of a given Q can be obtained at lower direct current density.

5 LARGE-SIGNAL ANALYSIS

(1) Power-Frequency Limitation

Due to the inherent limitations of semiconductor materials and the attainable impedance levels in microwave circuitry, the maximum output power at a given frequency of a single solid-state device is limited. The two most important limitations of semiconductor materials are (1) the critical electric field at which the avalanche breakdown occurs and (2) the scattering-limited velocity which corresponds to the maximum attainable velocity in the semiconductor.

If we make the simplified assumption that the ionization coefficient is proportional to $\mathscr{E}^m$ where m is a constant, $\alpha = \alpha_0(\mathscr{E}/\mathscr{E}_0)^m$, the breakdown condition is given by

$$\int_0^W \alpha_0 \left[\frac{\mathscr{E}(x)}{\mathscr{E}_0}\right]^m dx = 1 \tag{60}$$

where $\mathscr{E}(x)$ is in general a function of x, and the breakdown voltage is given by $V_B = \int_0^W \mathscr{E}(x)\, dx$. Under the constraint given by Eq. (60) the maximum value of V_B is obtained when $\mathscr{E}(x) = \text{constant} \equiv \mathscr{E}_c$. Thus under avalanche limitation the maximum voltage that can be applied across a semiconductor sample is given by

$$V_m = \mathscr{E}_c W = \alpha_0^{-1/m}\mathscr{E}_0 W^{(1-1/m)}, \tag{61}$$

and

$$\mathscr{E}_c = \alpha_0^{-1/m}\mathscr{E}_0 W^{-1/m}. \tag{61a}$$

We next consider the maximum current that can be carried by the semiconductor sample. The basic limitation comes from the fact that the current in the space-charge region causes an increase of the electric field (from Poisson's equation), and the maximum increase in the field is again limited by the avalanche breakdown process. From Eq. (9), assuming $\Delta\mathscr{E}(W) \approx \mathscr{E}_c$, we obtain

$$I_m \simeq \frac{\mathscr{E}_c \varepsilon_s v_{s1} A}{W}. \tag{62}$$

The optimistic upper limit on the ac power deliverable to the carriers within the length of semiconductor W is given by the product of V_m and I_m from Eqs. (61) and (62):

$$P_m = V_m I_m = \mathscr{E}_c^{\,2} \varepsilon_s v_{sl} A. \tag{63}$$

If the semiconductor is swept free of mobile charges and has ohmic contacts at both ends, it has a capacity given by $C = \varepsilon_s A / W$. We define a transit time τ of carriers across W as $\tau \equiv W/v_{sl}$, and further label $f \equiv 1/2\pi\tau$, as a characteristic cutoff frequency of the device. Equation (63) can be rewritten as

$$P_m X_c f^2 = \frac{\mathscr{E}_c^{\,2} v_{sl}^{\,2}}{(2\pi)^2} = \frac{(\alpha_0^{-2/m} \mathscr{E}_0^{\,2} W^{-2/m}) v_{sl}^{\,2}}{(2\pi)^2} \tag{64}$$

where X_c is the reactance $1/2\pi f C$. Equation (64) is obtained by DeLoach[13] based on a generalization of the Early-Johnson approach.[14,15]

If we assume from skin-effect considerations that the impedance level varies as the square root of frequency in a practical microwave circuit, from Eq. (64) we observe that the maximum power attainable is limited by the material parameters of the semiconductor and varies approximately as f^{-2}.

(2) Fundamental IMPATT Mode

For efficient operation of a Read diode, as carriers move through the drift region, we require the generation of as large a charge pulse, Q_m, as possible in the avalanche region without a reduction of the electric field in the drift region below that required for velocity saturation. The motion of Q_m through the drift region results in an ac voltage amplitude about one-half the average voltage, V_D, developed across the drift region. At the optimum frequency $(v_{sl}/2W)$ the motion of Q_m also results in an alternating particle current which is 180° out of phase with the ac voltage across the diode. The average of the particle current is the average current J_0. The particle current swing is therefore at most from zero to $2J_0$. For a square wave of particle current, and a sinusoidal variation of drift voltage, both with magnitude and phase as described above, the microwave power generating efficiency, η, is[16]

$$\eta \equiv \frac{\text{ac power output}}{\text{dc power input}} = \frac{\left(\dfrac{2J_0}{\pi}\right)\left(\dfrac{V_D}{2}\right)}{J_0(V_A + V_D)}$$

$$= \frac{1}{\pi} \frac{V_D}{(V_A + V_D)} \tag{65}$$

where V_A is the dc voltage drop across the avalanche region, and the sum of V_A and V_D is the total applied dc voltage. The ac power contribution from the avalanche region is neglected. This is because the avalanche region voltage is inductively reactive relative to the particle current. The displacement current is capacitively reactive relative to the diode voltage and therefore contributes no average ac power.

It is clear from Eq. (65) that to improve the efficiency one must reduce V_A. At the limit that $V_A \ll V_D$, the efficiency is $1/\pi$ or 30%. For a practical Read diode under the condition that the operating frequency is approximately equal to the resonant frequency (this is the optimum condition from small-signal analysis), it is found that the ratio V_A to V_D is given by[16]

$$\frac{V_A}{V_D} \approx \bar{\alpha}_n x_A / 3 \tag{66}$$

with the breakdown condition in the avalanche region

$$\int_0^{x_A} \bar{\alpha}_n e^{-(\bar{\alpha}_n - \bar{\alpha}_p)x} \, dx = 1 \tag{67}$$

where $\bar{\alpha}_n$ is the average electron ionization rates in the avalanche region. For Ge and GaAs or other semiconductors with nearly equal ionization coefficients $\bar{\alpha}_n x_A = 1$, $V_A/V_D = 1/3$, and the efficiency from Eq. (65) is about 23%. For Si, however, the ionization rate of electrons is about 10 times larger than that of holes, the product $\bar{\alpha}_n x_A$ is about 3, thus $V_A/V_D \approx 1$, and the efficiency is about 15% which is a factor 2 lower than the efficiency of an idealized Read diode.

Because of the complication of the system of equations, Eqs. (18) through (20), under large-signal operation with realistic semiconductor parameters, the numerical approach is generally used for large-signal analysis. Basically, the approach is to obtain self-consistent numerical solutions for the equations describing carrier transport, carrier generation, and space charge balance in a given semiconductor device structure. The solutions describe the evolution in time of the diode and its associated resonant circuit. For a silicon Read diode of p^+nvn^+ with $N_1 \simeq 10^{16}$ cm^{-3}, $b = 1$ μm, $N_2 = 10^{15}$ cm^{-3}, and $W \simeq 6$ μm (see Fig. 1), the computed results[16] by Scharfetter and Gummel are shown in Fig. 19 for four cases at approximately 1/4-cycle intervals. The electric field, the hole density, and the electron density are shown as functions of the distance over the depletion region. A phase plot of the terminal current and voltage of the oscillation is included. We note that (1) the generation of pulses of holes and electrons begins when the voltage is maximum and a quarter of a cycle later the charge pulses are fully formed and begin drifting into their respective drift spaces, (2) the holes disappear quickly from the active region while the electrons drift for approximately a half cycle and constitute positive particle current while the ac voltage is negative, (3) for the

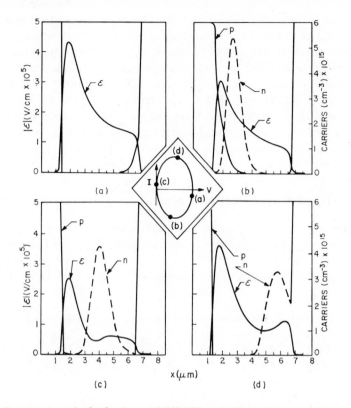

Fig. 19 Computed results for fundamental IMPATT mode. Four cases are shown at $\frac{1}{4}$-cycle intervals. A phase plot of the terminal current and voltage of the oscillation is included in the center insert.
(After Scharfetter and Gummel, Ref. 16.)

remaining quarter cycle the remnants of the electron charge pulse are swept out of the picture as the voltage again approaches its maximum value, (4) the displacement current is quite large and has an appreciable swing into the forward direction (positive on p^+), while the terminal voltage always remains in the reverse polarity, and (5) for sufficient modulation of diode voltage and particle current to give efficient oscillation, an extended avalanche region is required.

In Fig. 19 the direct current density is 1000 amp/cm². For this current density a maximum efficiency of 9% is obtained at an operating frequency of 13.4 GHz. For a lower current density of 200 amp/cm², a maximum efficiency of 18% is obtained at an operating frequency of 9.6 GHz. The nearly ideal phase relations obtained between diode voltage and particle current are illustrated in Fig. 20. In this figure we plot the waveforms in time of the particle current and terminal voltage for an ac steady-state solution at 9.6 GHz and direct current density of 200 amp/cm². The ac voltage amplitude is 17 volts. The waveforms for larger amplitudes are found to be similar. The above results thus are in reasonable agreement with the relatively simple design theory of Eqs. (65) through (67). It is interesting to point out that (1) Eq. (66) is obtained from a combination of small-signal results for the avalanche region and large-signal constraints on the drift region, and (2) a higher efficiency is obtained at lower bias current in agreement with the expected results from small-signal analysis.

(3) High-Efficiency IMPATT Mode

Under large-signal oscillation conditions the dynamic space-charge effect of generated carriers can give rise to multifrequency large-amplitude terminal waveforms. It is possible that high carrier concentrations of both electrons and holes are involved in the central region of the depletion layer. This is in

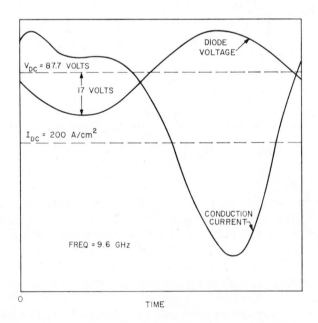

Fig. 20 Terminal voltage and particle current for the diode shown in Fig. 19 operated at 9.6 GHz and direct current density of 200 amp/cm². The ac voltage amplitude is 17 volts. (Ref. 16.)

contrast to the fundamental IMPATT mode from Read's original theory in which only one type of carrier (e.g., electrons as shown in Fig. 19) is involved in the transit through the depletion layer.

It has been shown theoretically that this mode of oscillation can give rise to considerably higher efficiency (of the order of 50 percent) and is called the high-efficiency mode. The result computed by Johnston et al.[16a] for an IMPATT diode with a doping profile similar to that for Fig. 19 is shown in Fig. 21. The insert shows the current-voltage phase loop. A full cycle of the phase loop involves a normal IMPATT period represented by (a), the beginning of a second IMPATT cycle at (b), the beginning of an avalanche generation of magnitude exceeding the Read limitation at (c), and a low-voltage high-carrier concentration (trapped plasma) state starting at (d). The cycle then repeats. The very large carrier concentrations following point (c) result from multiplication over an extensive portion of what is normally considered the drift region in Read structure. The subsequent separation of carriers results in

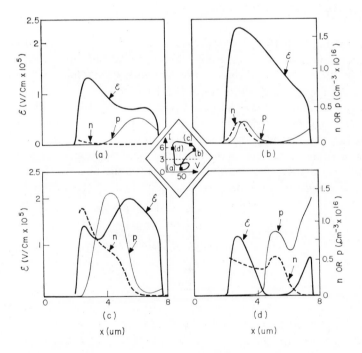

Fig. 21 Computed results for high-efficiency IMPATT mode. Four cases are shown at $\frac{1}{4}$-cycle intervals. A phase plot of the terminal current and voltage is show in the insert. (After Johnston et al., Ref. 16a.)

a reduced field region and resultant trapping of a portion of the carriers. The period of gradual escape of these carriers is determined by the external circuit conditions.

Besides large voltage swing and thus high efficiency, a marked benefit is realized from this mode in that the ratio of susceptance to negative conductance, i.e., diode Q, can become quite small, so that contact or other series resistance will then have reduced detrimental effect. This internal tuning of the normal depletion-layer capacitance is obtained by the inductive behavior of the avalanche; witness the abrupt reversal of the direction of rotation of the phase loop at point (c) in Fig. 21. A similar improvement of diode Q can be realized in these diodes without the trapped plasma. In this case a large avalanche occurs, as in Fig. 21(c), but it is not large enough to collapse the field below the velocity saturation knee. The phase loop will then involve only a single avalanche per cycle, but a multifrequency waveform will be necessary. The above result has been used to explain the observed high-efficiency IMPATT mode[27] to be considered in Section 7.

6 NOISE

The noise in an IMPATT diode arises mainly from the statistical nature of the generation rates of electron-hole pairs in the avalanche region. Since the noise sets a lower limit to the microwave signals to be amplified, it is important to consider the noise theory of the IMPATT diode.

For amplification the IMPATT diode can be inserted into a resonator which is coupled to a transmission line.[17] The line is coupled to separate input and output lines by means of a circulator as shown in Fig. 22(a). Figure 22(b) shows the equivalent circuit upon which the small-signal analysis is based. We shall now introduce two useful expressions for the noise performance: the noise figure and the noise measure. The noise figure, NF, is defined as

$$NF \equiv 1 + \frac{\left(\begin{array}{c}\text{output Noise Power}\\ \text{arising in the Amplifier}\end{array}\right)}{(\text{Power Gain})(kT_0 B_1)}$$

$$= 1 + \frac{\bar{I}_n^2 R_L}{P_G kT_0 B_1} \tag{68}$$

where P_G is the amplifier power gain, R_L the load resistance, k Boltzmann's constant, $T_0 = 290°K$, B_1 the noise bandwidth, and $\bar{I}_n^2$ the mean-square noise current caused by the diode and induced in the loop of Fig. 22(b). The noise measure M is defined as

$$M \equiv \frac{\bar{I}_n^2}{4kT_0 GB_1} \tag{69a}$$

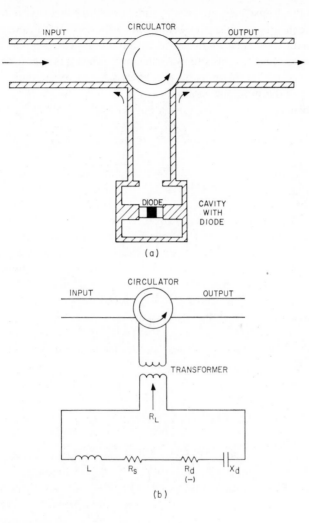

Fig. 22
(a) IMPATT diode inserted into a resonator.
(b) Equivalent circuit.
(After Hines, Ref. 17.)

or

$$M \equiv \frac{\overline{V}_n^2}{4kT_0(-Z_{\text{real}})B_1} \tag{69b}$$

where G is the negative conductance, $-Z_{\text{real}}$ the real part of the diode impedance, and $\overline{V}_n^2$ the mean-square noise voltage. We note that both the noise figure and the noise measure are dependent on the mean-square noise current (or the mean-square noise voltage). It will be shown that for frequencies above the resonant frequency f_r, the noise in the diode decreases, but so does the negative resistance. In this situation the appropriate quantity for assessing the performance of the diode as an amplifier is the noise measure, and we are interested in the minimum noise measure.

The noise figure for a high-gain amplifier is given by[17]

$$NF = 1 + \frac{1}{4m\omega^2\tau_A^2} \frac{\dfrac{qV_A}{kT_0}}{1 - \dfrac{\omega_r^2}{\omega^2}} \tag{70}$$

where m is the factor associated with the expression $\alpha = \alpha_0(\mathscr{E}/\mathscr{E}_0)^m$, τ_A and V_A are respectively the time and voltage drop across the avalanche region, and f_r is the resonant frequency given in Section 3. The above expression is obtained under the simplified assumptions that the avalanche region is narrow and that the ionization coefficients of holes and electrons are equal. For $m = 6$ (for Si), $\omega = 2\omega_r$, and $V_A = 3$ volts, the noise figure at $f = 10$ GHz is predicted to be 11,000 or 40.5 dB.

With realistic ionization coefficients ($\alpha_n \neq \alpha_p$ for Si) and an arbitrary doping profile, the low-frequency expression for the mean-square noise voltage is given by[18]

$$\overline{V}_n^2 = \frac{2qB_1}{J_0 A} \left[\frac{1 + \dfrac{W}{x_A}}{\alpha'} \right]^2 \sim \frac{1}{J_0} \tag{71}$$

where $\alpha' = \partial\alpha/\partial\mathscr{E}$. Figure 23 shows $\overline{V}_n^2/B_1$ as a function of frequency for a silicon IMPATT diode with $A = 10^{-4}$ cm^2, $W = 5$ μm, and $x_A = 1$ μm. At low frequencies we note that the noise voltage $\overline{V}_n^2$ is inversely proportional to the direct current density, Eq. (71). Near the resonant frequency (which varies as $\sqrt{J_0}$) $\overline{V}_n^2$ reaches a maximum and then decreases roughly as the fourth power of frequency, with a superimposed structure that is related to the width of the drift region. Figure 24 shows the noise measures as a function of frequency for Si and Ge diodes at current densities of 100 and 1000 amp/cm^2. As is seen, both for the hypothetical case of zero series resistance and for the more realistic assumption of a 1-ohm series resistance, the Ge structure has a lower noise measure. This can be understood from the fact that in Eq. (71) the factor α' for Ge is larger than that for Si. We also note that for the case of 1-ohm resistance, the minimum noise measure decreases with current density for both the Ge and Si diodes.

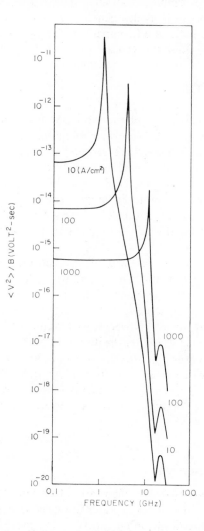

Fig. 23 Mean-square noise voltage per bandwidth versus frequency for a Si IMPATT diode. (After Gummel and Blue, Ref. 18.)

We can summarize the noise characteristics as follows: (1) the mean-square noise voltage (or current) varies inversely as the direct current density, (2) a wider avalanche region gives rise to lower $\overline{V}_n^2$, and (3) a lower noise measure is obtained in a semiconductor with higher ionization coefficients. A noise figure of 30 dB has been reported for an n^+p Ge IMPATT diode.[19] Measurements of silicon IMPATT diodes generally give a 6- to 8-dB-higher noise figure than germanium diodes of comparable doping profile. This is indeed to be expected from the noise theory.

7   EXPERIMENTS

The IMPATT diode is one of the most powerful solid-state sources for generation of microwaves. A microwave oscillator or amplifier can be made using an IMPATT diode biased in reverse and mounted in a microwave cavity. The impedance of the cavity is mainly inductive and can be matched to the mainly capacitive impedance of the diode so as to form a resonant system. From our small-signal and large-signal analyses, we show that an IMPATT diode can have a negative ac resistance under proper biasing conditions so that it can deliver power from the dc bias to the oscillation with a small negative Q and high efficiency.

In this section we shall discuss the diode geometries, the basic measurement setup, and typical experimental results; and we shall consider the effects due to series resistance and temperature limitation. To demonstrate the rapid progress in the experimental study of the IMPATT diode, we first show the measured output microwave power versus frequency[13,34] in Fig. 25. In November 1965 (about a year after the discovery of the first IMPATT diode)[3] the maximum CW power[13] at 10 GHz was only 50 mW. The maximum CW power has now increased by two orders of magnitude. The power-frequency relationship $p \sim f^2$ (as given by Eq. (64)) is followed at higher frequencies. At lower frequencies, the output power is proportional to l/f and is due to thermal limitations.[34]

(1)   Diode Geometries

Figure 26 shows the schematic diagrams of four different IMPATT diodes. The first diode[3] that gave microwave oscillations is shown in Fig. 26(a) which is a simple boron-diffused silicon p-n junction. The pulsed output power is 80 mW at 12 GHz with 0.5 percent efficiency. The first Read diode ($n^+p\pi p^+$) reported[4] is shown in Fig. 26(b) which was operated at 180 MHz with CW output of 1 μW with an efficiency of the order of 10^{-6}. The improved versions of the IMPATT diodes are shown in Fig. 26(c) and (d) where accurate controls of epitaxial layer thicknesses are employed.[21,22] Two typical microwave packages are shown in Fig. 27. In both cases the diode is mounted with its diffused side in contact with a copper heat sink so that the heat generated at the junction can be readily conducted away.

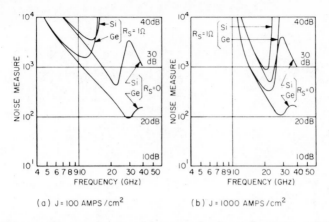

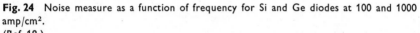

(a) J = 100 AMPS/cm²   (b) J = 1000 AMPS/cm²

Fig. 24 Noise measure as a function of frequency for Si and Ge diodes at 100 and 1000 amp/cm².
(Ref. 18.)

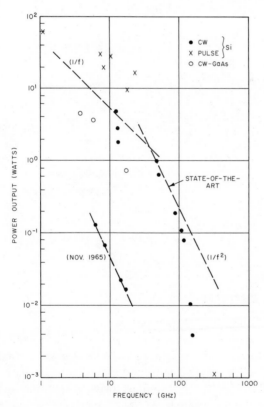

Fig. 25 Output microwave power versus frequency.
(Nov. 1965 line after DeLoach, Ref. 13; state-of-the-art line after Sze and Ryder, Ref. 34.)

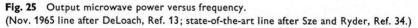

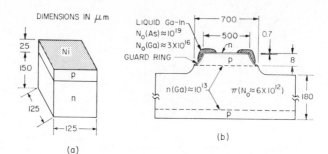

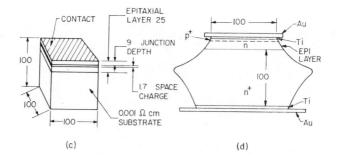

Fig. 26
(a) Simple boron-diffused Si p-n junction.
(After Johnston et al., Ref. 3.)
(b) The first Read diode — $n^+p\pi p^+$.
(After Lee et al., Ref. 4.)
(c) Improved version.
(After DeLoach and Johnston, Ref. 21.)
(d) Improved version.
(After Misawa and Marinaccio, Ref. 22.)

(2) Microwave Measurements

A wide-range measurement setup[23] for IMPATT diodes is shown in Fig. 28. The most important features are (1) the diode mount which provides adequate heat conduction and has movable tuning sleeves to optimize the resonant circuit, (2) the power meter to give the microwave output power, (3) the sweeper to detect the microwave frequencies, and (4) the oscilloscope to display the output power signal. A typical spectrum of oscillation as obtained from an oscilloscope display is shown in Fig. 29 for a Si p-v-n

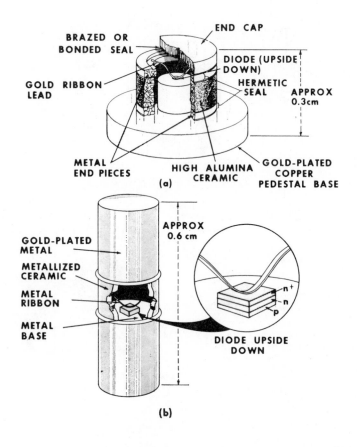

Fig. 27 Two microwave packages with IMPATT diodes mounted.

diode.[22] A 250-mW CW power output is obtained from this diode at X band with a spectrum width less than 1 kHz. The bands and the corresponding frequency ranges are listed in Table 5.1.

A typical result of the small-signal admittance measurements of an IMPATT diode is shown in Fig. 30. The diode can be approximated by a p-v-n diode with $W = 1.8$ μm. The variation of diode admittance as a function of frequency at a bias current of 15 ma is shown in Fig. 30(a) where the experimental points are plotted in extended Smith coordinates normalized to 20

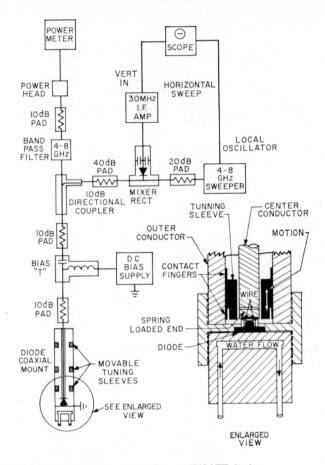

Fig. 28 Wide-range test circuit for high-efficiency IMPATT diodes. (After Iglesias, Ref. 23.)

millimhos.[24] We note that the negative conductance extends over the entire frequency range (4 to 9.8 GHz). Figure 30(b) shows the nonlinear behavior of the conductance as a function of direct bias current at five frequencies. At a given frequency, as the current increases the conductance reaches a maximum negative value and then becomes positive at sufficiently large current. These results are in good agreement with the small-signal analysis considered in Section 4.

Microwave oscillations well into the millimeter wave range (up to 300 GHz) have been obtained from diffused *p-n* junctions under pulse conditions.[25]

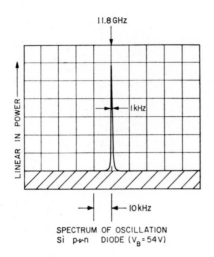

Fig. 29 Spectrum of oscillation of a Si *p-i-n* diode.
(After Misawa and Marinaccio, Ref. 22.)

TABLE 5.1

BANDS AND FREQUENCY RANGE[32]

Band	Waveguide Size (cm)	Frequency Range (GHz)
L	—	1.0 to 2.6
S	7.6 × 3.8	2.60 to 3.95
G	5 × 2.5	3.95 to 5.85
C	4.4 × 2.3	4.90 to 7.05
J	3.8 × 1.9	5.30 to 8.20
H	3.2 × 1.3	7.05 × 10.00
X	2.5 × 1.25	8.20 to 12.40
M	2.1 × 1.2	10.00 to 15.00
P	1.8 × 1.0	12.40 to 18.00
N	1.5 × 0.85	15.00 to 22.00
Ku	—	15.3 to 18.0
K	1.2 × 0.65	18.00 to 26.50
R	0.9 × 0.56	26.50 to 40.00
Millimeter		above 30 to 300
Submillimeter		above 300

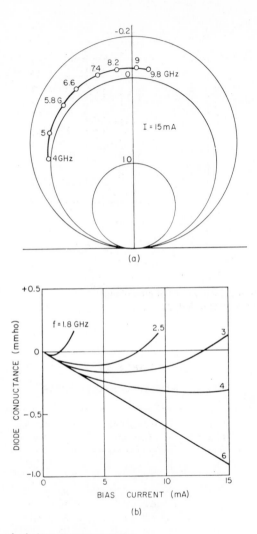

Fig. 30 Small-signal admittance measurements.
(After Josenhaus and Misawa, Ref. 24.)

The IMPATT diode is mounted in a cavity shown in Fig. 31(a). The measured threshold frequency (i.e., the lowest frequency of operation) and the corresponding breakdown voltage are shown in Fig. 31(b). The solid line is calculated based on the facts that the frequency is inversely proportional to the depletion layer width due to the transit-time effect and that the breakdown

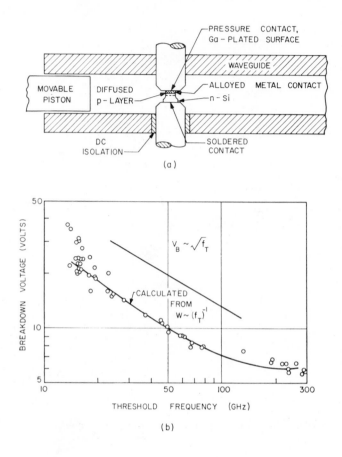

Fig. 31
(a) Microwave cavity.
(b) Measured threshold frequency versus breakdown voltage.
(After Bowman and Burrus, Ref. 25.)

voltage decreases with decreasing depletion width. Thus for lower breakdown voltages, the depletion width is narrower resulting in higher frequency.

Figure 32 shows the dependence of the threshold frequency on bias current. We note that the frequency increases approximately as the square root of the direct current density, in agreement with the general behavior of the cutoff frequency and the resonant frequency. It is interesting to point out that at 50 GHz, the depletion layer width is about 2000 Å. This very narrow width gives some indication of the difficulty inherent in fabricating a Read diode structure at this frequency. At 300 GHz with a total depletion width of only

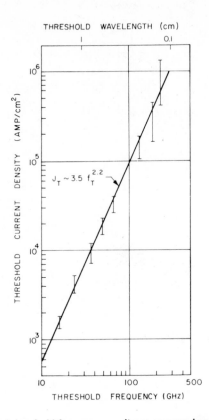

Fig. 32 Dependence of threshold frequency on direct current density. (Ref. 25.)

300 Å, a Read structure which requires a very narrow avalanche region and a separate drift region is beyond the present state of semiconductor technology.

We now consider some experimental results of the subtransit-time high-efficiency IMPATT diodes. Germanium diodes with a fundamental IMPATT frequency of 6 GHz have been operated under pulse bias in the circuit as shown in Fig. 28 and have produced efficiencies as high as 40% at 3 GHz.[16a] *CW* operation at room temperature with efficiency up to 43% has been reported for Ge IMPATT diodes.[26] Silicon diodes have also been reported[27] to give high efficiency operation at somewhat lower frequencies (0.1 to 1.1 GHz).

Figure 33(a) shows the measured output power and efficiency of a Ge IMPATT diode. The result shows an efficiency of 43% with 5.3 watts power

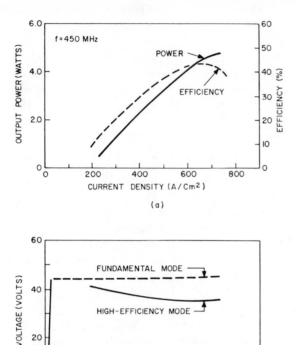

Fig. 33 Experimental measurement of Ge IMPATT diode operated in high-efficiency mode,
(a) output power and efficiency versus current density and
(b) voltage versus bias current density.
(After Iglesias and Evans, Ref. 26.)

output at 450 MHz. The diode was made from epitaxial Ge material and
designed to operate as a 5-GHz IMPATT oscillator in the fundamental mode.
Typical device parameters are: 1-pf junction capacitance at -20 V, junction
area of 4×10^{-4} cm², and breakdown voltage of 40 V. To operate the device
in high-efficiency mode, the circuit setup as shown in Fig. 28 is modified by
first providing a local cavity for the microwave (~ 5 GHz) oscillation. This is
obtained by positioning a single tuning sleeve close to the diode. Then a
second tuning element is positioned a half wavelength (at UHF) from the
first tuning element. Probe measurements in the cavity that is formed by

the tuning elements have shown that the diode is also oscillatory at a microwave frequency (~ 5GHz) which is harmonically related to the UHF oscillations.

It is interesting to note that the results shown in Fig. 33(a) are obtained at current densities comparable to the densities required to operate the diode as a 5-GHz oscillator. Figure 33(b) shows the current-voltage characteristic for the oscillating diode. There is a significant drop in dc voltage when the diode is operated in the high-efficiency mode.

(3) Effects of Unswept Layer and Heat Dissipation

The importance of the epitaxial layer and the heat sink on microwave performance will now be considered. Figure 34(a) shows the impurity distribution and electric field profile at breakdown of a Ge IMPATT diode[28] which is designed as a microwave power source at 6 GHz. It is evident from the figure that about 7 μm of epitaxial material is required to support the diffused layer and the depletion layer at breakdown. Any epitaxial material in excess of this 7 μm will remain as an unswept neutral region at breakdown and will therefore contribute to the positive series resistance of the diode. The effect of this unswept epitaxial layer on the microwave performance is illustrated in Fig. 34(b) which shows that the efficiency decreases as the unswept epitaxial thickness increases. This is because the negative resistance due to the impact-avalanche transit-time effect is partly cancelled by the parasitic series resistance. It is clear from Fig. 34(b) that to produce high-efficiency IMPATT diodes, the epitaxial layer thickness should be tailored so that no unswept layer remains at breakdown. The elimination of the series resistance is also important in the noise consideration as shown in Fig. 24.

An IMPATT diode generally dissipates several times as much power in the form of heat as is converted to microwave power. In addition, high current densities are required to obtain good efficiency. It is thus important to have an adequate heat sink to dissipate the associated high power densities without excessive diode heating. Figure 35(a) shows an example of a diode mounted on copper with metallizations used for a gold-to-gold thermal compression bond.[29] A simplified diode and heat-sink structure is shown in Fig. 35(b). The total thermal resistance for a circular heat source of radius r at a depth d_s in the silicon is given by[29]

$$R_T = R_s + R_t + R_g + R_n + R_c$$
$$= \frac{1}{A}\left(\frac{d_s}{\kappa_s} + \frac{d_t}{\kappa_t} + \frac{d_g}{\kappa_g} + \frac{d_n}{\kappa_n}\right) + \frac{1}{4\pi\kappa_c}. \tag{72}$$

The symbols are defined in Fig. 35(b). The last term gives the thermal spreading resistance for the infinite half-space heat sink. The various components of

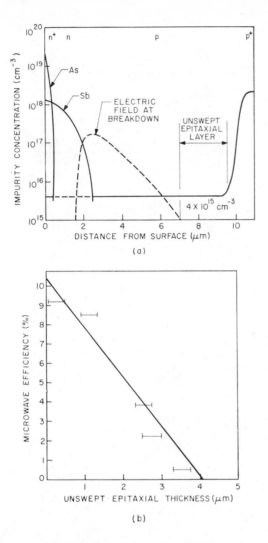

Fig. 34
(a) Impurity distribution and electric field profile at breakdown of a Ge IMPATT diode.
(b) Efficiency versus unswept epitaxial layer thickness.
(After Kovel and Gibbons, Ref. 28.)

the thermal resistance are shown in Fig. 35(c). The dashed curves show the thermal spreading resistance for a diamond (type II) heat sink and the corresponding total thermal resistance R_T. The thermal conductivity at 300°K of the diamond is assumed to be three times that of copper, and the thermal conductivity for silicon is for a temperature of 500°K which is assumed as a

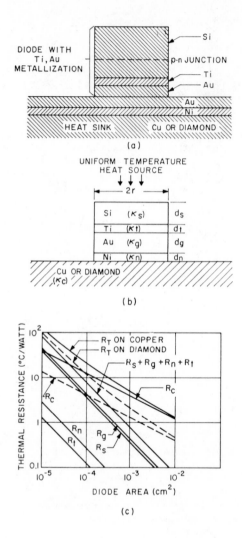

Fig. 35
(a) Diode mounted on copper with metallization used for gold-to-gold thermal compression bond.
(b) Simplified diode and heat-sink structure.
(c) Various components and the total thermal resistance versus diode area.
(After Swan et al., Ref. 29.)

maximum operating temperature (see Table 5.2). We see clearly that a diamond heat sink reduces the thermal resistance R_T by a factor of about two, and that R_T decreases as the diode area increases.

Figure 36(a) compares the power densities and the total dissipation power levels that can be achieved on copper and on diamond heat sinks for a diode

TABLE 5.2

THERMAL CONDUCTIVITY AND TYPICAL LAYER THICKNESS FOR
MATERIALS IN A Ku BAND DIODE (300°K)

Material	Thermal Conductivity* κ (Watts/cm°C)	Thickness d (μm)	d/κ ($10^{-4} \times$ cm² °C/Watt)
Silicon	0.80	3	3.8
Titanium	0.16	0.02	0.13
Gold	3.0	12.5	4.2
Nickel	0.71	0.2	0.28
Copper	3.9	—	—
Diamond	20	—	—

* κ for Si is for a temperature of 500°K which is assumed as a maximum operating temperature.

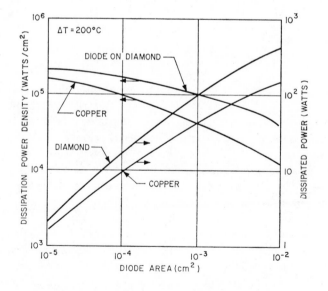

Fig. 36

(a) Power density and total dissipation power versus diode area.

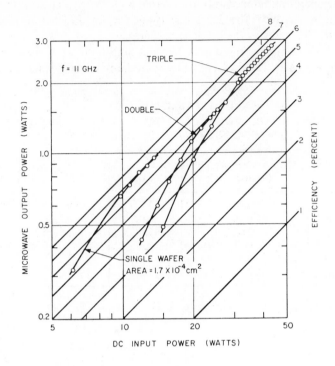

(b) Output power versus input power for parallel connections.
(After Swan et al., Ref. 29.)

temperature rise of 200°C. It is seen that a significantly higher power can be
handled with diamond. Because of the temperature rise associated with the
power density requirements (i.e., high current density), for efficient oscillation
a single-diode oscillator at any specified efficiency is limited by the maximum
area that can be used. To increase the output power we can use parallel,
series, and hybrid connections. An example of parallel connection is shown in
Fig. 36(b) in which data are given for a single, a double, and a triplet unit.[29]
The output power of the triple circuit is increased by a factor of about 3
without significant reduction in efficiency. For series connection[30] of IMPATT
diodes it is shown that the power output is the sum of the individual diodes.
Since the impedance level is not lowered by the series connection, the f^{-2}
limitation of the power-impedance product for IMPATT diodes is no longer
important. The hybrid connection[31] (combination of series and parallel connections) can also give considerable increase in the output power.

REFERENCES

1. W. Shockley, "Negative Resistance Arising From Transit Time in Semiconductor Diodes," Bell Syst. Tech. J., *33*, 799 (1954).

2. W. T. Read, "A Proposed High-Frequency Negative Resistance Diode," Bell Syst. Tech. J., *37*, 401 (1958).

3. R. L. Johnston, B. C. DeLoach, Jr., and B. G. Cohen, "A Silicon Diode Microwave Oscillator," Bell Syst. Tech. J., *47*, 369 (1964).

4. C. A. Lee, R. L. Batdorf, W. Wiegman, and G. Kaminsky, "The Read Diode an Avalanche, Transit-Time, Negative-Resistance Oscillator," Appl. Phys. Letters, *6*, 89 (1965).

5. T. Misawa, "Negative Resistance on *p-n* Junction Under Avalanche Breakdown Conditions, Part I and II," IEEE Trans. Electron Devices, *ED-13*, 137–151 (1966).

6. S. M. Sze and G. Gibbons, "Avalanche Breakdown Voltages of Abrupt and Linearly Graded *p-n* Junctions in Ge, Si, GaAs, and GaP," Appl. Phys. Letters, *8*, 111 (1966).

7. G. Gibbons and S. M. Sze, "Avalanche Breakdown in Read and *p-i-n* Diodes," Solid State Electron., *11*, 225 (1968).

8. C. R. Crowell and S. M. Sze, "Temperature Dependence of Avalanche Multiplication in Semiconductors," Appl. Phys. Letters, *9*, 242 (1966).

9. S. M. Sze and W. Shockley, "Unit-Cube Expression for Space-Charge Resistance," Bell Syst. Tech. J., *46*, 837 (1967).

10. M. Gilden and M. F. Hines, "Electronic Tuning Effects in the Read Microwave Avalanche Diode," IEEE Trans. Electron Devices, *ED-13*, 169 (1966).

11. T. Misawa, "Multiple Uniform Layer Approximation in Analysis of Negative Resistance in *p-n* Junctions in Breakdown," IEEE Trans. Electron Devices, *ED-14*, 795 (1967).

12. H. K. Gummel and D. L. Scharfetter, "Avalanche Region of IMPATT Diodes," Bell Syst. Tech. J., *45*, 1797 (1966).

13. B. C. DeLoach, Jr., "Recent Advances in Solid State Microwave Generators," a chapter in *Advances in Microwaves*, Vol. 2, Academic Press, New York, pp. 43–88 (1967).

14. J. M. Early, "Maximum Rapidly Switchable Power Density in Junction Triodes," IRE Trans. Electron Devices, *ED-6*, 322 (1959).

15. E. O. Johnson, "Physical Limitation on Frequency and Power Parameters of Transistors," IEEE Intern. Conv. Record, pt. 5, p. 27 (1965).

16. D. L. Scharfetter and H. K. Gummel, "Large-Signal Analysis of a Silicon Read Diode Oscillator," Solid-State Device Research Conference, Evanston, June 1966.

16a. R. L. Johnston, D. L. Scharfetter, and D. J. Bartelink, "High-Efficiency Subtransit Time Oscillations in Germanium Avalanche Diode," Proc. IEEE, *56* (1968).

17. M. F. Hines, "Noise Theory for Read Type Avalanche Diode," IEEE Trans. Electron Devices, *ED-13*, 158 (1966).

18. H. K. Gummel and J. L. Blue, "A Small-Signal Theory of Avalanche Noise on IMPATT Diodes," IEEE Trans. Electron Devices, *ED-14*, 569 (1967).

19. R. L. Rulison, G. Gibbons, and J. G. Josenhaus, "Improved Performance of IMPATT Diodes Fabricated From Ge," Proc. IEEE, *55*, 223 (1967).

20. C. B. Swan, T. Misawa, and C. H. Bricker, "Continuous Oscillations at Millimeter Wavelengths With Silicon Avalanche Diodes," Proc. IEEE, *55*, 1747 (1967).

21. B. C. DeLoach and R. L. Johnston, "Avalanche Transit-Time Microwave Oscillators and Amplifiers," IEEE Trans. Electron Devices, *ED-13*, 18 (1966).

22. T. Misawa and L. P. Marinaccio, "A 1/4 Watt Si *p-v-n* X-Band IMPATT Diode," Inter. Electron Device Meeting, Washington, D.C. (Oct. 1966).

23. D. E. Iglesias, "Circuit for Testing High Efficiency IMPATT Diodes," Proc. IEEE, *55*, 2065 (1967).

24. J. G. Josenhaus and T. Misawa, "Experimental Characterization of a Negative-Resistance Avalanche Diode," IEEE Trans. Electron Devices, *ED-13*, 206 (1966).

25. L. S. Bowman and C. A. Burrus, Jr., "Pulse-Driven Silicon *p-n* Junction Avalanche Oscillators for the 0.9 to 20 mm Band," IEEE Trans. Electron Devices, *ED-14*, 411 (1967).

26. D. E. Iglesias and W. J. Evans, "High Efficiency CW IMPATT Operation," Proc. IEEE (1968).

27. H. J. Prager, K. K. N. Chang, and S. Weisbrod, "High Power, High Efficiency Silicon Avalanche Diodes at Ultrahigh Frequencies," Proc. IEEE, *55*, 586 (1967).

28. S. R. Kovel and G. Gibbons, "The Effect of Unswept Epitaxial Material on the Microwave Efficiency of IMPATT Diodes," Proc. IEEE, *55*, 2066 (1967).

29. C. B. Swan, T. Misawa, and L. Marinaccio, "Composite Avalanche Diode Structures for Increased Power Capability," IEEE Trans. Electron Devices, *ED-14*, 684 (1967).

30. F. M. Magalhaes and W. O. Schlosser, "A Series Connection of IMPATT Diodes," Inter. Solid. State Circuit Conf., Philadelphia (Feb. 1968).

31. H. Fukui, "Frequency Locking and Modulation of Microwave Silicon Avalanche Diode Oscillators," Proc. IEEE, *54*, 1475 (1966).

32. *Hewlett-Packard Electronic Test Instruments*, Hewlett-Packard Co., Palo Alto, California, (1961).

33. R. Svensson, private communication.

34. S. M. Sze and R. M. Ryder, "Microwave Avalanche Diodes," Proc IEEE, Special Issue on Microwave Semiconductor Devices (August 1971).

■ INTRODUCTION

■ STATIC CHARACTERISTICS

■ MICROWAVE TRANSISTOR

■ POWER TRANSISTOR

■ SWITCHING TRANSISTOR

■ UNIJUNCTION TRANSISTOR

6

Junction Transistors

I INTRODUCTION

Of all semiconductor devices the junction transistor is the most important. The invention of the transistor (contraction for *tran*sfer re*sistor*) by a research team of the Bell Laboratories in 1948 has had an unprecedented impact on the electronic industry in general and on solid-state research in particular. Prior to 1948 semiconductors had found application only as thermistors, photodiodes, and rectifiers. In 1948 the development of the point-contact transistor by John Bardeen and Walter Brattain was announced.[1] In the following year William Shockley's classic paper on junction diodes and transistors was published.[2]

Since then the transistor theory has been extended to include high-frequency high-power, and switching behaviors. Transistor technology has enjoyed many breakthroughs, particularly in the alloy-junction[3] and grown-junction techniques[3a] and in zone-refining,[3b] diffusion,[3c, d, e] epitaxial,[4] planar,[5] beam-lead,[6] and ion implantation[7] technologies. These breakthroughs have helped to increase the power and frequency capabilities of transistors, as well as their reliability, by millions of times. In addition, application of semi-conductor physics, transistor theory, and transistor technology has broadened our knowledge and improved other semiconductor devices as well.

It is important to point out that transistors and related semiconductor devices are not mere replacements for vacuum tubes. The true essence of these devices lies in their overwhelming potential to create for science and industry novel developments that could never have been derived from tubes alone. Transistors are now key elements, for example, in high-speed computers, in space vehicles and satellites, and in all modern communication and power systems.

Because of the importance of transistors, over 250 books have been written to date on the subjects of transistor physics, design, and application. Among them are standard texts by such editors and authors as Biondi,[8] Shive,[9] Philips,[10] Gartner,[11] and Pritchard,[11a] and a series of books[12] (Vol. 1 through 4) by the Semiconductor Electronics Education Committee (SEEC) which gives a lucid and penetrating presentation of transistor theory and circuits.

In Section 2 of this chapter we discuss briefly the static characteristics of a transistor. The high-frequency and microwave transistor are considered in Section 3 in which emphasis has been placed on device geometry, cutoff frequency, power gain, and noise figures. In Section 4 we consider the power transistor where the main concern is with the absolute values of power and the limitation of operation imposed by second breakdown. The switching transistor is considered in Section 5 in which the important parameters are current gain and switching speed. The classification of junction transistors in the above categories, however, is quite arbitrary. For example, since the power-frequency product is mainly limited by material parameters,[13] there is no clear-cut boundary between power and microwave transistors. In Section 6 we discuss the unijunction transistor which is a three-terminal device having one emitter junction and two base contacts. It can exhibit negative-resistance characteristics and is used in timing and switching circuits.

2 STATIC CHARACTERISTICS

(1) Basic Current-Voltage Relationship

In this section we shall consider the basic dc characteristics of *p-n-p* and *n-p-n* junction transistors. Figure 1 shows the symbols and nomenclatures for *p-n-p* and *n-p-n* transistors. The arrow indicates the direction of current flow under normal operating conditions, i.e., forward-biased emitter junction and reverse-biased collector junction. A transistor can be connected in three circuit configurations depending on which lead is common to the input and output circuits. Figure 2 shows the common-base, common emitter, and common-collector configurations for a *p-n-p* transistor. The current and voltage conventions are given for normal operations. All the signs and polarities should be inverted for an *n-p-n* transistor. In the following discussion we shall consider *p-n-p* transistors; the results are applicable to the *n-p-n* transistor with an appropriate change of polarities.

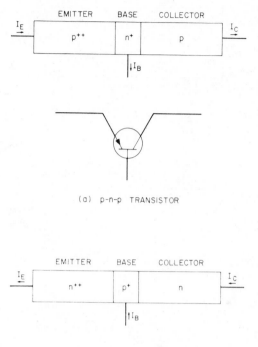

(a) p-n-p TRANSISTOR

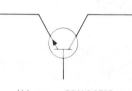

(b) n-p-n TRANSISTOR

Fig. 1 Symbols and nomenclatures of *n-p-n* and *p-n-p* transistors.

Figure 3(a) is a schematic of a *p-n-p* transistor connected as an amplifier with common-base configuration. A schematic doping profile for the transistor with regions of uniform impurity density is shown in Fig. 3(b), and the corresponding band diagram under normal operating conditions is shown in Fig. 3(c).

The static characteristics can be readily derived from the *p-n* junction theory discussed in Chapter 3. To illustrate the major properties of a transistor, we shall assume that the current-voltage relationship of the emitter and collector junctions is that given by the Shockley equation,[2] i.e., the

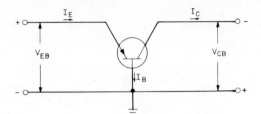

(a) COMMON-BASE CONFIGURATION

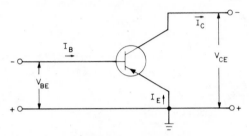

(b) COMMON-EMITTER CONFIGURATION

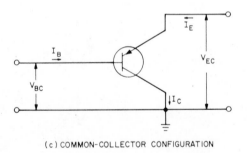

(c) COMMON-COLLECTOR CONFIGURATION

Fig. 2 Three configurations of a *p-n-p* transistor.

effects due to recombination-generation, series resistance, and high-level injection are neglected. These effects will be considered later.

As in Fig. 3(b) where all the potential drops occur across the junction depletion region, the equations that govern the steady-state characteristics are those of continuity and current density. For the neutral base region these equations are given by

$$0 = -\frac{p - p_B}{\tau_B} + D_B \frac{\partial^2 p}{\partial x^2} \qquad (1)$$

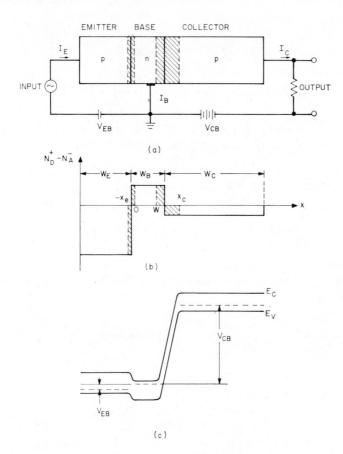

Fig. 3
(a) A *p-n-p* transistor connected in common-base configuration for amplifier application.
(b) Doping profiles of the transistor with abrupt impurity distributions.
(c) Energy band diagram under normal operating conditions.

$$J_p = -qD_B \frac{\partial p}{\partial x} \tag{2a}$$

$$J_n = J_{tot} + qD_B \frac{\partial p}{\partial x} \tag{2b}$$

where p_B is the equilibrium minority carrier density in the base, J_{tot} is the total conduction current density, and τ_B and D_B are the minority-carrier lifetime and diffusion coefficients respectively. The conditions at the emitter depletion-layer edges for the excess carrier concentrations are

$$p'(0) \equiv p(0) - p_B = p_B \left[\exp\left(\frac{qV_{EB}}{kT}\right) - 1 \right]$$

$$n'(-x_E) = n(-x_E) - n_E = n_E \left[\exp\left(\frac{qV_{EB}}{kT}\right) - 1 \right] \tag{3}$$

where n_E is the equilibrium minority carrier density (electrons) in the emitter. A similar set of equations can be written for the collector junction, i.e.,

$$p'(W) = p(W) - p_B = p_B \left[\exp\left(\frac{qV_{CB}}{kT}\right) - 1 \right]$$

$$n'(x_C) = n(x_C) - n_C = n_C \left[\exp\left(\frac{qV_{CB}}{kT}\right) - 1 \right]. \tag{4}$$

The solutions for the minority carrier distributions, i.e., the hole distribution in the base from Eq. (1) and electron distributions in the emitter and collector, are given by[11]

$$p(x) = p_B + \left[\frac{p'(W) - p'(0)e^{-W/L_B}}{2 \sinh(W/L_B)} \right] e^{x/L_B} - \left[\frac{p'(W) - p'(0)e^{W/L_B}}{2 \sinh(W/L_B)} \right] e^{-x/L_B} \tag{5}$$

$$n(x) = n_E + n'(-x_E) \exp\left[\frac{(x + x_E)}{L_E} \right], \qquad x < -x_E \tag{6}$$

$$n(x) = n_C + n'(x_C) \exp\left[-\frac{(x - x_C)}{L_C} \right], \qquad x > x_C \tag{7}$$

where $L_B = \sqrt{\tau_B D_B}$ is the diffusion length of holes in the base, and L_E and L_C are the diffusion lengths in the emitter and collector respectively. Equation (5) is important because it correlates the base width W to the minority carrier distribution. If $W \to \infty$ or $W/L_B \gg 1$, Eq. (5) reduces to

$$p(x) = p_B + p(0)e^{-x/L_B} \tag{8}$$

which is identical to the case of a p-n junction. In this case there is no communication between the emitter and collector currents which are determined by the density gradient at $x = 0$ and $x = W$ respectively. The "transistor" action is thus lost. From Eqs. (2) and (3) we can obtain the total dc emitter current as a function of the applied voltages:

$$I_E = AJ_p(x = 0) + AJ_n(x = -x_E)$$

$$= A\left(-qD_B \frac{\partial p}{\partial x}\bigg|_{x=0}\right) + A\left(-qD_E \frac{\partial n}{\partial x}\bigg|_{x=-x_E}\right)$$

$$= Aq \frac{D_B p_B}{L_B} \coth\left(\frac{W}{L_B}\right)\left[\left(e^{qV_{EB}/kT} - 1\right) - \frac{1}{\cosh\left(\frac{W}{L_B}\right)}\left(e^{qV_{CB}/kT} - 1\right)\right]$$

$$+ Aq \frac{D_E n_E}{L_E}\left(e^{qV_{EB}/kT} - 1\right), \tag{9}$$

and for the total dc collector current

$$I_C = AJ_p(x = W) + AJ_n(x = x_C)$$

$$= A\left(-qD_B \frac{\partial p}{\partial x}\bigg|_{x=W}\right) + A\left(-qD_C \frac{\partial n}{\partial x}\bigg|_{x=x_C}\right)$$

$$= Aq \frac{D_B p_B}{L_B} \frac{1}{\sinh\left(\frac{W}{L_B}\right)}\left[\left(e^{qV_{EB}/kT} - 1\right) - \coth\left(\frac{W}{L_B}\right)\left(e^{qV_{CB}/kT} - 1\right)\right]$$

$$- Aq \frac{D_C n_C}{L_C}\left(e^{qV_{CB}/kT} - 1\right) \tag{10}$$

where A is the cross-sectional area of the transistor. The difference between these two currents is small and appears as the base current:

$$I_B = I_E - I_C. \tag{11}$$

We shall now modify the doping distribution in the base layer of Fig. 3(b) and consider a more general base impurity distribution[14] as shown in Fig. 4. A transistor with such doping distribution is called a drift transistor, since there is a built-in electric field to enhance the hole drift in the base. The donor density N and the electron density in the base for $N \gg n_i$ are given by

$$n \approx N = n_i \exp\left[\frac{q(\psi - \phi)}{kT}\right] \tag{12}$$

where n_i is the intrinsic carrier concentration, ϕ the Fermi potential, and ψ is the intrinsic Fermi potential. From Eq. (12) we obtain for the built-in field

$$\mathscr{E} \equiv -\frac{d\psi}{dx} = -\frac{kT}{q}\frac{1}{N}\frac{dN}{dx}. \tag{13}$$

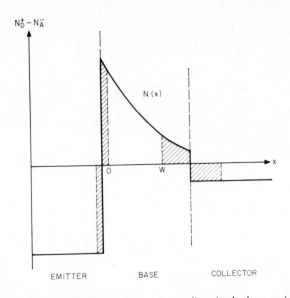

Fig. 4 Transistor doping profile with an impurity gradient in the base region. (After Moll and Ross, Ref. 14.)

The hole current density is given by

$$J_p = q\mu_B p\mathscr{E} - qD_B \frac{dp}{dx}. \tag{14}$$

Substitution of Eq. (13) into (14) yields

$$J_p = -qD_B \left[\frac{p}{N}\frac{dN}{dx} + \frac{dp}{dx} \right]. \tag{15}$$

The steady-state solution to Eq. (15) with the boundary condition that $p = 0$ at $x = W$ is

$$p = \frac{J_p}{qD_B}\frac{1}{N(x)} \int_x^W N(x)\,dx. \tag{16}$$

The hole concentration at $x = 0$ is given by

$$p(x = 0) = \frac{J_p}{qD_B}\frac{1}{n_{BO}} \int_0^W N(x)\,dx \simeq p_{BO}\exp\!\left(\frac{qV_{EB}}{kT}\right) \tag{17}$$

where n_{BO} is defined as the donor concentration at $x = 0$, and p_{BO} is the equilibrium hole concentration at $x = 0$ (so that $n_{BO}p_{BO} = n_i^2$). The current $I_p = AJ_p$, where A is the area, is given by

$$I_p = \frac{qAD_B n_i^2}{\displaystyle\int_0^W N(x)\,dx} \exp\!\left(\frac{qV_{EB}}{kT}\right) = I_1 \exp\!\left(\frac{qV_{EB}}{kT}\right). \qquad (18)$$

The total collector current is given by

$$I_C = I_1 \exp\!\left(\frac{qV_{EB}}{kT}\right) + I_2 \qquad (19)$$

where I_2 is the saturation current. A typical experimental result[15] is shown in Fig. 5. We note that the exponential law of Eq. (19) is very closely obeyed at currents low enough that the voltage drop of the base current flowing through the base resistance is negligible. The constant I_1 can be obtained

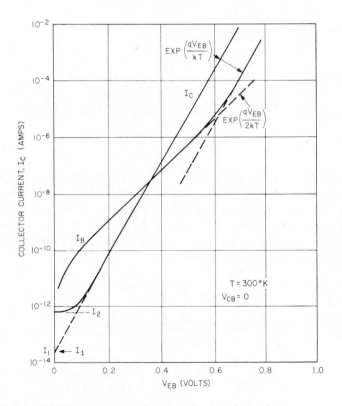

Fig. 5 Collector and base current as a function of emitter-base voltage. (After Iwersen et al., Ref. 15.)

by subtracting the saturation current I_2 from the measured collector current and by plotting the difference in a semilog plot as indicated in Fig. 5. The number of impurities per unit area (the so-called Gummel number)[16] can be obtained from Eqs. (18) and (19):

$$N_B = \int_0^W N(x)\, dx = \frac{q}{I_1}\, A D_B\, n_i^2. \tag{20}$$

(2) Current Gain

The static common-base current gain α_0, also referred to as h_{FB} from the four-terminal hybrid parameters (where the subscript F and B refer to *forward* and common-*base* respectively), is defined as

$$\alpha_0 \equiv h_{FB} = \frac{\Delta I_C}{\Delta I_E} \quad (\text{for } f \to 0). \tag{21}$$

The static common-emitter current gain β_0, also referred to as h_{FE}, is defined as

$$\beta_0 \equiv h_{FE} = \frac{\Delta I_C}{\Delta I_B} \quad (\text{for } f \to 0). \tag{22}$$

From Eq. (11) we note that α_0 and β_0 are related to each other by

$$\beta_0 = \frac{\alpha_0}{1 - \alpha_0}. \tag{23}$$

A plot of β_0 versus α_0 is shown in Fig. 6. We note that as α_0 approaches unity, the value of β_0 increases extremely rapidly. For a 1% change in α_0, from 0.98 to 0.99, the value of β_0 increases by about 100%, from 48 to 99.

Under normal operation $V_{EB} > 0$ and $V_{CB} \ll 0$, so that the terms in Eqs. (9) and (10) associated with V_{CB} can be neglected. The current gain α_0 can be obtained from Eqs. (9) and (10) as

$$\alpha_0 = \gamma \alpha_T \tag{24}$$

where

$$\gamma(\text{emitter efficiency}) = \frac{\text{hole current from the emitter}}{\text{total emitter current}}$$

$$= \frac{J_p(x = 0)}{J_p(x = 0) + J_n(x = -x_E)} \approx \frac{1}{1 + \left(\dfrac{n_E}{p_B}\dfrac{D_E}{D_B}\right)\dfrac{L_B}{L_E}\tanh\left(\dfrac{W}{L_B}\right)} \tag{25}$$

and

$$\alpha_T(\text{transport factor}) = \frac{\text{hole current reaching collector}}{\text{total emitted hole current into the base}}$$

$$= \frac{\left(\dfrac{\partial p}{\partial x}\right)_{x=W}}{\left(\dfrac{\partial p}{\partial x}\right)_{x=0}} \approx \frac{1}{\cosh\left(\dfrac{W}{L_B}\right)}. \tag{26}$$

Both γ and α_T are less than unity, and the extent to which they depart from unity represents an electron current which must be supplied from the base contact.

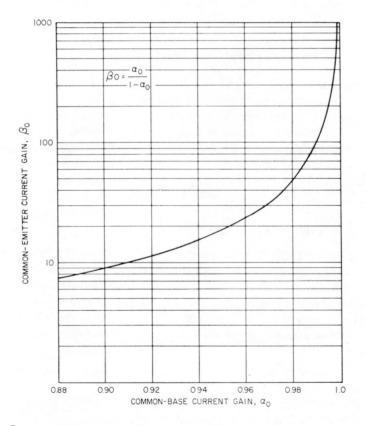

Fig. 6 Common-emitter current gain vs. common-base current gain.

In Eq. (25) the dominant factor is the ratio $n_E/p_B \simeq N_B/N_E$ where N_B and N_E are the impurity doping concentrations in the base and emitter respectively. For $\gamma \to 1$, it is required that $N_B/N_E \ll 1$, or that the emitter should be much more heavily doped than the base. The transport factor α_T, Eq. (26), is determined by the ratio of the base width to the diffusion length in the base. To improve α_T, one must reduce this ratio. The current gain in Eq. (24) is independent of emitter current density. This is true only under the assumptions of ideal junctions and low-injection level.

For a practical transistor, however, α_0 varies with the emitter current. At very low emitter currents the contribution of the recombination-generation current (the so-called Sah-Noyce-Shockley current)[17] in the emitter depletion region as well as of the surface leakage current may be large compared with the useful diffusion current of minority carriers across the base, so that the emitter efficiency is low. With increasing total emitter current the diffusion current begins to dominate over the recombination-generation current, and the emitter efficiency increases. For still higher emitter current (the high-injection condition where the injected minority carrier density approaches the majority carrier density in the base), the injected carriers effectively increase the base doping, which, in turn, causes the emitter efficiency to decrease. The detailed analysis can be obtained by solving the continuity equation and current equations with both diffusion and drift components. The decrease of current gain with increasing I_E is referred to as the Webster effect.[18] Thus the current gain goes through a maximum and continues to decrease for higher current. A typical result is shown in Fig. 7. It is of interest to point out that while the value of β_0 varies from about 60 to a maximum of 65 then to 20 at $I_E = 25$ ma, the value of α_0 varies only from 0.983 to 0.984 then to 0.954.

(3)   Common-Base Configuration

In the previous section we have seen that the currents in the three terminals of a transistor are related by the minority carrier distribution in the base region. For a transistor with high emitter efficiency, the expressions for the dc emitter and collector currents, Eqs. (9) and (10), reduce to terms proportional to the minority-carrier gradient, $(\partial p/\partial x)$, at $x = 0$ and $x = W$ respectively. We can thus summarize the fundamental relationships of a transistor as follows:

(a) the applied voltages control the boundary densities through the terms $\exp(qV/kT)$,

(b) the emitter and collector currents are given by the minority (hole) density gradients at the junction boundaries, i.e., $x = 0$ and $x = W$, and

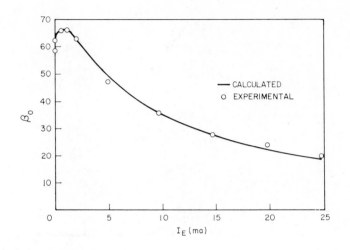

Fig. 7 Variation of dc common-emitter current gain with emitter current. (After Webster, Ref. 18.)

   (c) the base current is the difference between the emitter and collector currents.

Figure 8 shows the hole distribution in the base region of a *p-n-p* transistor for various applied voltages;[19] all dc characteristics can be interpreted by means of these diagrams.

   For a given transistor the emitter current I_E and collector current I_C are functions of the applied voltages V_{EB} and V_{CB}, i.e., from Eqs. (9) and (10) $I_E = f_1(V_{EB}, V_{CB})$ and $I_C = f_2(V_{EB}, V_{CB})$. To give all the relationships between the terminal voltages and currents, two sets of curves are thus required. Of the many possible characteristics the input and output current-voltage curves are of greatest practical importance.

   A typical set of common-base characteristics is shown in Fig. 9. The input characteristics I_E versus V_{EB} with V_{CB} as a parameter, as shown in Fig. 9(a), are similar to those of a forward-biased *p-n* junction. The emitter characteristic depends slightly on the collector voltage V_{CB} because of the variation of the collector depletion width with V_{CB}. This is referred to as the Early effect.[20] In Fig. 8(b), as V_{CB} increases, the depletion layer edge moves from W to W'. The gradient at $x = 0$ is slightly increased; this, in turn, causes a slight increase of the emitter current for a given emitter voltage.

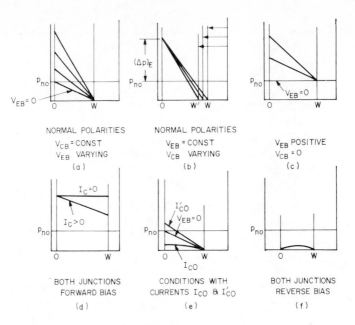

Fig. 8 Hole density in the base region of a *p-n-p* transistor for various applied voltages. (After Morant, Ref. 19.)

The common-base output characteristics [I_C versus V_{CB} with I_E as a parameter as shown in Fig. 9(b)] show that the collector current is practically equal to the emitter current ($\alpha_0 \approx 1$) and virtually independent of V_{CB}. The collector current remains practically constant, even down to zero voltage where the excess holes are still extracted by the collector as indicated by the hole profile shown in Fig. 8(c). To reduce the collector current to zero it is necessary to apply a small forward voltage (~ 1 volt for Si) to the collector. This sufficiently increases the hole density at W to make it equal to that of the emitter at $x = 0$, as shown in Fig. 8(d).

The collector saturation current I_{CO} (also denoted by I_{CBO}) is measured with the emitter open circuit. This current is considerably smaller than the ordinary reverse current of a *p-n* junction because the presence of the emitter junction with a zero hole gradient at $x = 0$ (corresponding to zero emitter current) reduces the hole gradient at $x = W$ as shown in Fig. 8(e). The current I_{CO} is therefore smaller than when the emitter junction is short-circuited ($V_{EB} = 0$). The current I'_{CO} is associated with the common-emitter configuration and will be considered later.

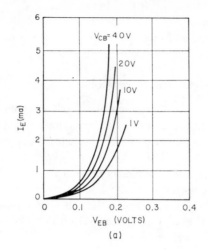

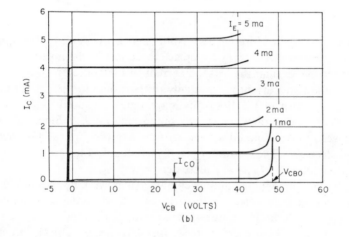

Fig. 9
(a) Input characteristic and
(b) output characteristic of a typical *p-n-p* transistor in common-base configuration.
(Ref. 19.)

As V_{CB} increases to the value V_{CBO} the collector current starts to increase rapidly. Generally this is due to the avalanche breakdown of the collector-base junction, and the breakdown voltage is similar to that considered in Chapter 3 for *p-n* junctions. For a very narrow base width or a base with relatively low doping, the breakdown may also be due to the punch-through effect, i.e., the neutral base width is reduced to zero at a sufficient V_{CB} and the collector depletion region is in direct contact with the emitter depletion region. At this point the collector is effectively short-circuited to the emitter, and a large current can flow.

(4) Common-Emitter Configuration

We now consider the $I\text{-}V$ characteristics of the common-emitter configuration. The only difference between the common-emitter and common-base configurations is that in one case the voltages are referred to the emitter and in the other to the base. Therefore, the $I\text{-}V$ characteristics of a common-emitter configuration can be completely derived from those of the common-base configuration discussed previously. The common-emitter input (I_B versus V_{BE}) and output (I_C versus V_{CE}) characteristics of a typical $p\text{-}n\text{-}p$ transistor are shown in Fig. 10(a) and 10(b) respectively. From the sign convention of Fig. 2 the base current is given by

$$I_B = I_E - I_C = I_E - (I_{CO} + \alpha_0 I_E)$$

$$= (1 - \alpha_0)I_E - I_{CO}. \tag{27}$$

The base current therefore has two components: the current $(1 - \alpha_0)I_E$ due to an inward flow of electrons to replace those lost in the hole injection and diffusion processes, and the leakage current of the collector I_{CO}. Thus for a given V_{CE} the base current is essentially proportional to the emitter current. However as shown in Fig. 10(a), its magnitude is reduced by $(1 - \alpha_0)$ and it is displaced along the current axis by an amount I_{CO}. Since the sum of the voltages around a closed loop is zero, for a given V_{BE}, as V_{CE} increases, V_{CB} will decrease. The current I_E is thus reduced, Fig. 8(b); this in turn causes a reduction of I_B as V_{CE} increases.

For the common-emitter output characteristics we note that in Fig. 10(b) there is considerable current gain since, for $\alpha_0 \approx 1$, the common-emitter current gain $\beta_0 \gg 1$. The saturation current I_{CO}', which is the collector current with zero base current (base open-circuited), is much larger than I_{CO}. This is because, from Eq. (27), the emitter current for zero base current is given by

$$I_E'(I_B = 0) = \frac{I_{CO}}{1 - \alpha_0}. \tag{28}$$

Since the emitter and collector currents are equal in this condition, Fig. 8(e), $I_{CO}' = I_E'$ and therefore

$$I_{CEO} \equiv I_{CO}' = \frac{I_{CO}}{(1 - \alpha_0)} \approx \beta_0 I_{CO} = \beta_0 I_{CBO} \tag{29}$$

for $\alpha_0 \approx 1$ (I_{CO}' may also be denoted as I_{CEO}).

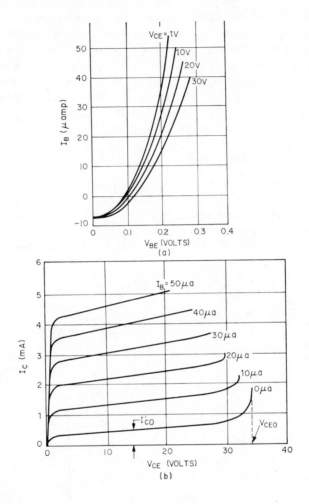

Fig. 10
(a) Input characteristic and
(b) output characteristic of a typical p-n-p transistor in common-emitter configuration.
(Ref. 19).

Because of the variation of base width with collector voltage, β_0 also increases with V_{CE}. The lack of saturation in the common-emitter output characteristic is due to the large increase of β_0 with V_{CE}. For small collector-emitter voltages the collector current falls rapidly to zero. The voltage V_{CE} is divided between the two junctions to give the emitter a small forward bias and the collector a larger reverse bias. To maintain a constant base current, the potential across the emitter junction must remain essentially constant. Thus when V_{CB} is reduced below a certain value (~ 1 volt for the Si transistor), the collector junction will reach zero bias as shown in Fig. 8(c). With further reduction in V_{CE} the collector is actually forward-biased as shown in Fig. 8(d), and the collector current falls rapidly because of the rapid decrease of the hole gradient at $x = W$. The breakdown voltage under the open-base condition can be obtained as follows. Let M be the multiplication factor at the collector junction and be approximated by

$$M = \frac{1}{1 - \left(\dfrac{V}{V_{CBO}}\right)^n} \tag{30}$$

where V_{CBO} is the common-base breakdown voltage, and n is a constant. When the base is open-circuited, we have $I_E = I_C = I$. The currents I_{CO} and $\alpha_0 I_E$ as shown in Fig. 11 are multiplied by M when they flow across the collector junction. We have

$$M (I_{CO} + \alpha_0 I) = I$$

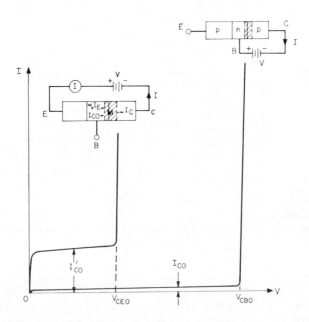

Fig. 11 Breakdown voltage V_{CBO} and saturation current I_{CO} for common-base configuration, and corresponding qualities V_{CEO} and I_{CO}' for common-emitter configuration. (After Gartner, Ref. 11.)

or

$$I = \frac{MI_{CO}}{1 - \alpha_0 M}.$$ (31)

Current I will be limited only by external resistances when $\alpha_0 M = 1$. From the condition $\alpha_0 M = 1$ and Eq. (30), the breakdown voltage V_{CEO} for the common-emitter configuration is given by

$$V_{CEO} = V_{CBO}(1 - \alpha_0)^{1/n}.$$ (32)

For $\alpha_0 \approx 1$, the value of V_{CEO} is much smaller than V_{CBO}.

3 MICROWAVE TRANSISTOR

(1) Device Geometry

In this section we shall consider transistors operated in the high-frequency region where the range goes up to a few GHz. Figure 12 shows two typical geometries for high-frequency transistors. Both are made using planar technology. An epitaxial wafer of n on n^+ is used as the substrate. An insulating layer is formed on the surface (such as SiO_2 thermally grown on Si). The

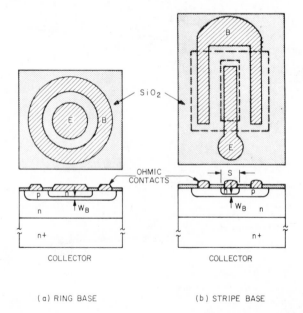

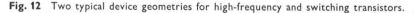

Fig. 12 Two typical device geometries for high-frequency and switching transistors.

base-layer diffusion pattern is then cut into the insulating layer employing the photoresist method. After the *p*-type base diffusion the emitter-layer pattern is formed using similar methods. The emitter junction can be formed by diffusion (in which case the device is called a double-diffused transistor) or by alloy (in which case the device is called a diffused-base transistor).

A final metallization process is used to make ohmic contacts to the emitter, base, and collector. A particularly important metallization process is the use of beam-lead technology.[6] Figure 13 shows a silicon high-frequency beam-lead transistor with stripe-base geometry as shown in Fig. 12(b). Metal leads which are about 10 μm thick are used for structural support of the silicon chip as well as for electrical contacts. This technology also permits making integrated circuits with excellent reliability and electrical performance.

Variations in horizontal geometry (such as employing various numbers of

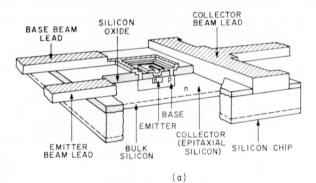

(a)

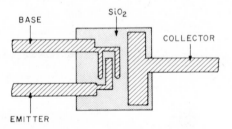

Fig. 13 Beam-lead transistor structure
(a) schematic view
(b) top view of an actual transistor.
(After Lepselter, Ref. 6.)

emitter and base stripes) yield the different current capabilities of the transistor while changes in doping profiles result in frequency and breakdown voltage differences. For high-frequency applications it is essential that an epitaxial substrate be used in order to reduce the collector series resistance. Processes similar to those described above can be used to fabricate *p-n-p* transistors. The choice of an *n-p-n* or *p-n-p* transistor, as well as the selection of semiconductor material, depends on the circuit application and device technology. We shall compare the performance of *n-p-n* and *p-n-p* transistors in Ge, Si, and GaAs in later sections.

The differences between a low-frequency transistor and a microwave transistor are in the dimensions of the active areas, and in the control of the wafer and package parasitics. To achieve microwave capability, the dimensions of the active areas and the parasitics should be considerably reduced. For microwave applications the stripe-base geometry as shown in Fig. 12(b) is preferred, and the two critical dimensions are the emitter stripe width (S) and the base thickness W_B. The reduction of these dimensions over the period from 1952 to 1968 is shown in Fig. 14. Also indicated are the major events

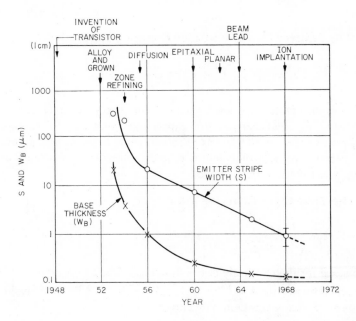

Fig. 14 Reduction of the two critical dimensions, emitter-stripe width and base-layer thickness, over the period 1952 to 1968. Also indicated are the major events associated with transistor development.
(After Edwards, Ref. 29.)

associated with transistor development which have been discussed in Section 1. Of particular interest is the development of the diffusion process which is mainly responsible for the reduction of these critical dimensions. At the present time the emitter stripe width can be reduced to less than 1 μm, and the base thickness, to about 0.1 μm.

(2) Cutoff Frequency

The cutoff frequency f_T is an important figure-of-merit for high-frequency transistors and is defined[21, 22] as the frequency at which the common emitter, short-circuit current gain h_{fe} ($\equiv \partial I_C / \partial I_B$) is unity. The cutoff frequency is related to the physical structure of the transistor through the emitter-to-collector delay time, τ_{ec}, by

$$f_T = \frac{1}{2\pi\tau_{ec}}. \tag{33}$$

Delay time τ_{ec} represents the sum of four delays encountered sequentially by the minority carriers as they flow from the emitter to the collector, as follows.

A. The Emitter Depletion-Layer Charging time

$$\tau_E = r_e[C_e + C_c + C_p] \approx \frac{kT}{qI_E}[C_e + C_c + C_p] \tag{34}$$

where r_e is the emitter resistance, C_e the emitter capacitance, C_c the collector capacitance, C_p any other parasitic capacitance connected to the base lead, and I_E the emitter current which is essentially equal to the collector current I_C. The expression for r_e is obtained by differentiation of the Shockley equation with respect to the emitter voltage.

B. The Base-Layer Charging Time

$$\tau_B = \frac{W^2}{\eta D_B} \tag{35}$$

where $\eta = 2$ for the uniformly doped base layer. The expression Eq. (35) can be obtained by substituting $L_B = \sqrt{D_B \tau_B/(1 + j\omega\tau_B)}$ in Eqs. (25) and (26) to give the small-signal common-base current gain:[23]

$$\alpha \approx \frac{1}{\cosh\left(W\sqrt{\dfrac{1 + j\omega\tau_B}{D_B \tau_B}}\right)} \approx \frac{1}{1 + \dfrac{jW^2\omega}{2D_B \tau_B}}. \tag{36}$$

The charging time τ_B is defined as $1/2\pi f_\alpha$ where f_α is generally called the alpha cutoff frequency at which the gain has fallen to $1/\sqrt{2}$ of its low-frequency value. In Eq. (36) the contribution of the emitter efficiency γ to the charging time is small and is neglected. For a nonuniformly doped base, e.g., the drift transistor shown in Fig. 4, the factor η in Eq. (35) should be replaced by a larger number. If the built-in field $\mathscr{E}_{bi}$ is a constant the factor η is given by[24, 25]

$$\eta \approx 2\left[1 + \left(\frac{\mathscr{E}_{bi}}{\mathscr{E}_0}\right)^{3/2}\right] \tag{37}$$

where $\mathscr{E}_0 = 2D_B/\mu_B W$. For $\mathscr{E}_{bi}/\mathscr{E}_0 = 10$, η is about 60; thus considerable reduction in τ_B can be achieved by a large built-in field. This built-in field can be obtained automatically in a practical transistor using the base-diffusion process. A typical example is shown in Fig. 15 for a high-frequency double-diffused epitaxial n-p-n transistor. Doping in the base varies from about 10^{17} to less than 10^{15} cm^{-3} within 2 μm; this gives rise to a built-in field of about 150 V/cm [Eq. (13)].

C. Collector Depletion-Layer Transit Time (Fig. 3)

$$\tau_C = \frac{(x_c - W)}{2v_{sl}} \tag{38}$$

where v_{sl} is the scattering-limited drift velocity in the collector.

D. Collector Charging Time

$$\tau_C' = r_c C_c \tag{39}$$

where r_c is the collector series resistance and C_c the collector capacitance. For epitaxial transistors, r_c can be substantially reduced and the charging time τ_C' can be neglected in comparison with other delay times.

The time constant $r_b C_c$ does not appear in f_T but does affect high-frequency gain. [See Eqs. (53)–(55).] The alpha cutoff frequency f_α defined in Eq. (36) is meaningful for noise calculations, Eq. (56), in which phase relations between emitter and collector current are important. For calculations of cutoff frequency f_T, however, we should consider the sum of the delay times, Eqs. (34), (35), and (38):

$$\tau_{ec} = \tau_E + \tau_B + \tau_c. \tag{40}$$

For microwave transistors, τ_B can be comparable or even smaller than other delay times. The cutoff frequency f_T is given by:

$$f_T = \frac{1}{2\pi\tau_{ec}} = \left\{2\pi\left[\frac{kT(C_e + C_c + C_p)}{qI_c} + \frac{W^2}{\eta D_B} + \frac{(x_c - W)}{2v_{sl}}\right]\right\}^{-1}. \tag{41}$$

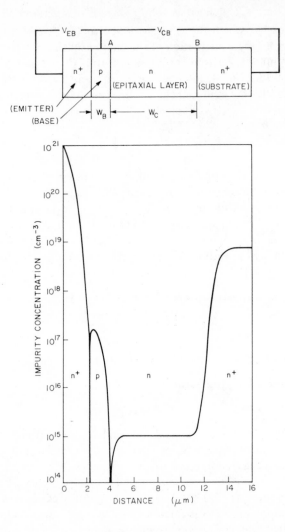

Fig. 15 One-dimensional double-diffused *n-p-n* transistor for high-frequency operation.

It is clear from the above equation that, to increase the cutoff frequency, the transistor should have a very narrow base thickness, which is one of the critical dimensions shown in Fig. 14, and a narrow collector region, and should be operated at a high-current level. As the collector width decreases, however, there is a corresponding decrease in breakdown voltage. Therefore compromises must be made for high-frequency and high-voltage operation.

As the operating current increases, the cutoff frequency decreases because the emitter charging time τ_E is inversely proportional to the current. However, as the current becomes sufficiently high that the injected minority carrier density is comparable to or larger than the base doping concentration (the so-called high-injection level), the high-field region originally located at the transition region between the base and the epitaxial layer is relocated to the interface between the epitaxial layer and the substrate, i.e., the effective base thickness increases from W_B to $W_B + W_C$ (see Fig. 15). This high-field-relocation phenomenon is referred to as the Kirk effect[26] which may considerably increase the delay time τ_{ec}. It is important to point out that under a high-injection condition where the currents are large enough to produce substantial fields in the collector region, the classic concept of well-defined transition regions at emitter-base and base-collector junctions is no longer valid. One must solve the basic differential equations (current density, continuity, and Poisson's equations) numerically with boundary conditions applied only at the electric terminals. The computed results[27] of the electric field distributions for $|V_{CB}| = 2$ volts and various collector current densities are shown in Fig. 16 for the doping profile of Fig. 15. We note that, as the current increases, the peak electric field moves from point A to point B.

Figure 17 shows the corresponding hole and electron distributions for $|V_{CB}| = 2$ volts. We note that, as the current increases, the boundary between the base and collector junction becomes less well-defined. Figure 18 shows the calculated emitter-to-collector delay time τ_{ec} which is defined in its most rigorous sense as $\partial Q / \partial J_C$ where Q is the total charge per unit area of all the holes in the device (for an *n-p-n* transistor), i.e., of those carriers that communicate with the base terminal. At low current densities, τ_{ec} decreases with J_c as predicted by Eq. (41), and the collector current J_C is carried mainly by the drift component such that

$$J_C \approx q\mu_C N_C \mathscr{E}_C \tag{42}$$

where μ_C, N_C, and $\mathscr{E}_C$ are the mobility, impurity doping, and electric field respectively in the collector epitaxial layer. As the current increases, delay time τ_{ec} reaches a minimum and increases rapidly around J_1 where J_1 is the current at which the largest uniform electric field $\mathscr{E}_C = (V_{CO} + |V_{CB}|)/W_C$ can exist where V_{CO} is the collector built-in potential and V_{CB} is the applied collector-base voltage. Beyond this point the current cannot be carried totally by the drift component throughout the collector epitaxial region. The current J_1 is given from Eq. (42) as

$$J_1 = q\mu_C N_C(V_{CO} + |V_{CB}|)/W_C. \tag{43}$$

Because of the above-mentioned Kirk effect, there is an optimum collector

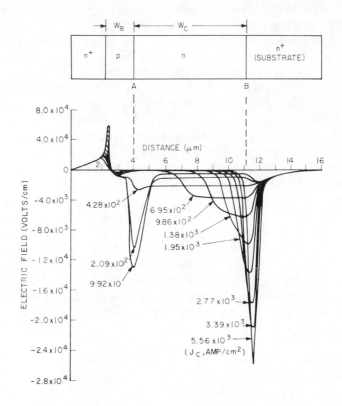

Fig. 16 Electric field vs. distance for $|V_{CB}| = 2$ volts and various collector current densities. The doping profile is shown in Fig. 15. (After Poon et al., Ref. 27.)

current which gives the maximum cutoff frequency. It should be pointed out that as $|V_{CB}|$ increases the corresponding value of J_1 will also increase.

We shall now compare the theoretical and experimental cutoff frequencies of various transistors on the basis of unique device geometry and typical material parameters.[28, 29] The device geometry that we shall use is shown in Fig. 12(b) with state-of-the-art dimensions of $W_B = 0.15$ μm, $W_C = 1 \mu$m, and $S = 1$ μm. The typical material parameters to be used for the calculation are listed in Table 6.1. The theoretical cutoff frequencies [as calculated from Eq. (41) with $J_C = 1000$ amp/cm^2] and the experimental results are listed in Table 6.2. We note that, for a given semiconductor, the n-p-n transistor has a higher cutoff frequency than does the p-n-p transistor. This is mainly because

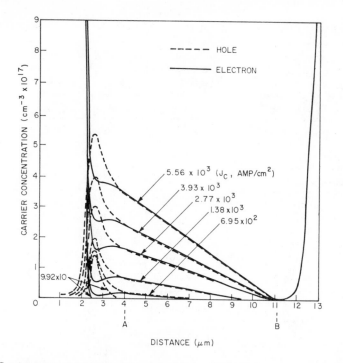

Fig. 17 Carrier concentrations vs. distance for $|V_{CB}| = 2$ volts and various collector current densities. The doping profile is shown in Fig. 15. (Ref. 27.)

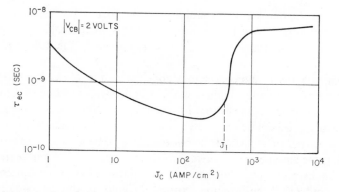

Fig. 18 Emitter-collector delay time τ_{ec} as a function of collector current density for the device shown in Fig. 15. (Ref. 27.)

TABLE 6.1

PARAMETERS FOR CALCULATION OF TRANSISTOR PERFORMANCE

Parameters	Unit	Ge	Si	GaAs
μ_n (for $N_B = 4 \times 10^{17}$ cm^{-3})	cm²/V-sec	2300	480	2800
μ_p (for $N_B = 4 \times 10^{17}$ cm^{-3})	cm²/V-sec	540	270	200
v_{sl} (electrons)	cm/sec	6×10^6	10^7	10^7
v_{sl} (holes)	cm/sec	6×10^6	6×10^6	10^7
Dielectric constant		16	12	12
Breakdown field $\mathscr{E}_m$ (for $N_C = 3 \times 10^{15}$ cm^{-3})	V/cm	2×10^5	3.4×10^5	3.8×10^5

TABLE 6.2

CUTOFF FREQUENCY f_T (GHz)

	Transistor	Ge	Si	GaAs
Theoretical	n-p-n	10.4	8.6	18.5
	p-n-p	6.7	5.2	5.0
Experimental	n-p-n	>7	>8	1.4
	p-n-p	>6	>5	0.7

the minority carrier (electron) mobility in the base region and the majority carrier (also electron) scattering-limited velocity in the collector region are larger for an *n-p-n* transistor than the corresponding quantities for a *p-n-p* transistor.

Among the three semiconductors considered, the GaAs *n-p-n* transistor can theoretically give the highest cutoff frequency, and Ge transistors have some margins over Si transistors. In practice, however, the Si transistor with its more advanced device and material technology, gives a performance comparable or superior to that of the Ge transistor. The GaAs transistors suffer from many technological problems which include poor starting material (an epitaxial substrate with a high density of defect and trapping centers), difficulty in control of the impurity profile, difficulty in the insulator masking process, thermal conversion (e.g., it converts from *p*-type to *n*-type under high-temperature treatment), and copper contamination.[29a] In addition the

GaAs transistors also suffer from fundamental physical limitations such as low hole mobility, low thermal conductivity, and low minority-carrier lifetime (as a result of its direct band gap). Nevertheless, it should be pointed out that if the technological problems of GaAs are solved eventually, one will be able to take advantage of many interesting properties of GaAs to extend the Ge and Si attainable performance (a) to higher frequencies because of the high μ_n and v_{sl} in GaAs, (b) to higher power because of its high breakdown field and wide band gap, and (c) to cryogenic operation because of the shallow impurity ionization energies in GaAs (refer to Fig. 12 in Chapter 2).

(3) High-Frequency Characteristics

There are various approaches to characterizing the performance of high-frequency transistors. The most common approach is a combination of internal-parameter and two-port (four-terminal) analyses. We shall use the simplified equivalent circuits as shown in Fig. 19 with the following internal parameters: emitter resistance r_e, base resistance r_b, emitter depletion-layer capacitances C_e, collector depletion-layer capacitance C_c, and small-signal common-base current gain α. We shall consider the stripe-base geometry shown in Fig. 12(b) with emitter stripe width S, length L, and spaced S from the base stripes on either side. The collector capacitance can be approximated by $C_c = C_0 SL$ where C_0 is the collector capacitance per unit area. The base resistance for this geometry is approximately $r_b = r_0 S/L$ with $r_0 \simeq \rho_B/W$ where ρ_B is the average resistivity of the base layer.

The small-signal common-base current gain α is defined as

$$\alpha \equiv h_{fb} = \frac{\partial I_C}{\partial I_E}. \tag{44}$$

Similarly the small-signal common-emitter current gain β is defined as

$$\beta = h_{fe} = \frac{\partial I_C}{\partial I_B}. \tag{45}$$

From Eqs. (21), (22), (44), and (45) we obtain

$$\alpha = \alpha_0 + I_E \frac{\partial \alpha_0}{\partial I_E}$$

$$\beta = \beta_0 + I_B \frac{\partial \beta_0}{\partial I_B} \tag{46}$$

and

$$\beta = \frac{\alpha}{(1 - \alpha)}.$$

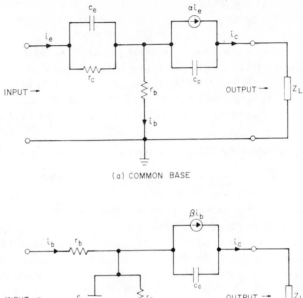

(a) COMMON BASE

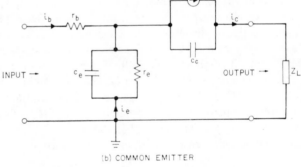

(b) COMMON EMITTER

Fig. 19 Simplified high-frequency equivalent current for
(a) common-base and
(b) common-emitter configurations.

At low-current levels both α_0 and β_0 increase with current as shown in Fig. 7, and α and β are larger than their corresponding static values. At high-current levels, however, the opposite is true.

We shall now define several figures of merit for a high-frequency transistor:

A. Power Gain: The ratio of power into load impedance Z_L to the input power into the two-port network. In terms of hybrid two-port parameters, i.e.,

$$\begin{bmatrix} V_1 \\ I_2 \end{bmatrix} = \begin{bmatrix} h_{11} & h_{12} \\ h_{21} & h_{22} \end{bmatrix} \begin{bmatrix} I_1 \\ V_2 \end{bmatrix},$$ (47)

we have

$$G_p(\text{Power Gain}) \equiv \frac{\mathscr{R}e(Z_L)|h_{21}|^2}{\mathscr{R}e[(h_{11} + \Delta^h Z_L)(1 + h_{22} Z_L)^*]} \tag{48}$$

where $\mathscr{R}e$ means the real part, $\Delta^h \equiv (h_{11}h_{22} - h_{12}h_{21})$, Z_L is the load impedance, and the asterisk denotes the complex conjugate.

B. Stability Factor: The value of the stability factor K which is indicative of whether or not a transistor will oscillate upon application of a combination of passive load and source admittance with no external feedback. The factor is given by

$$K \equiv \frac{[2\mathscr{R}e(h_{11})\mathscr{R}e(h_{22}) - \mathscr{R}e(h_{12}h_{21})]}{|h_{12}h_{21}|} \tag{49}$$

If $K > 1$, the device is unconditionally stable, i.e., in the absence of external feedback, a passive load or source impedance will not cause oscillation. If $K < 1$, the device is potentially unstable, i.e., the application of certain combinations of passive load and source impedance could induce oscillation.

C. Maximum Available Gain (MAG): The maximum power gain that can be realized by a particular transistor without external feedback. It is given by the forward power gain of the transistor when the input and output are simultaneously and conjugately matched. MAG is defined only for an unconditionally stable transistor ($K > 1$), and is stated:

$$MAG = \frac{|h_{21}|^2/|h_{12}h_{21}|}{K + \sqrt{K - 1}}. \tag{50}$$

It is obvious from Eq. (50) that, when $K < 1$, the denominator becomes a complex number and MAG is not defined. The measured MAG's for state-of-the-art transistors[29] are shown in Fig. 20. At 4 GHz, about 5-dB MAG has been realized in both Si and Ge transistors. The GaAs device is a junction field-effect transistor (JFET) with metal-semiconductor gate contact. This device will be discussed in Chapter 8. Also shown are the noise figures to be discussed later.

D. Unilateral Gain: The forward power gain in a feedback amplifier having its reverse power gain set to zero by adjustment of a lossless reciprocal feedback network around the transistor. Unilateral gain is independent of header reactances and common-lead configuration. It is defined as

$$U \equiv \frac{|h_{21} + h_{12}|^2}{4[\mathscr{R}e(h_{11})\mathscr{R}e(h_{22}) + \mathscr{I}m(h_{12})\mathscr{I}m(h_{21})]}. \tag{51}$$

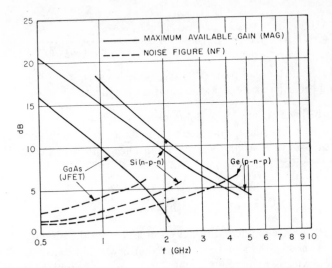

Fig. 20 Maximum available gain (MAG) and noise figure (NF) vs. operating frequency for Ge, Si junction transistors and GaAs junction field-effect transistor. (After Edwards, Ref. 29.)

For the equivalent circuit shown in Fig. 19(a) the gain is given by [30]

$$U \equiv \frac{|\alpha(f)|^2}{8\pi f r_b C_c \left\{ -\mathscr{I}m[\alpha(f)] + \frac{2\pi f r_e C_c}{1 + 4\pi^2 f^2 r_e^2 C_e^2} \right\}} \tag{52}$$

where $\mathscr{I}m[\alpha(f)]$ is the imaginary part of the common-base current gain. If $\alpha(f)$ can be expressed as $\alpha_0/(1 + jf/f_T)$, and if $f < f_T$, $\mathscr{I}m[\alpha(f)]$ can be approximated by $-\alpha_0 f/f_T$ or $-\alpha_0 \omega\tau_{ec}$. The gain is then given by

$$U \simeq \frac{\alpha_0}{16\pi^2 r_b C_c f^2 \left(\tau_{ec} + \frac{r_e C_c}{\alpha_0} \right)}$$

$$= \frac{\alpha_0/f^2}{16\pi^2 S^2 r_0 C_0 \tau_{ec}^*} \tag{53}$$

where the relationships $r_b = r_0 S/L$ and $C_c = C_0 SL$ have been used, and τ_{ec}^* is the sum of τ_{ec} and $r_e C_c/\alpha_0$. If $\alpha_0 \approx 1$ and $\tau_{ec} > r_e C_c$, Eq. (53) reduces to the simplified form

$$U = \frac{f_T}{8\pi f^2 r_b C_c} = \frac{f_T/f^2}{8\pi S^2 r_0 C_0}. \tag{54}$$

E. Maximum Oscillation Frequency: The frequency at which uni-lateral gain becomes unity. From Eqs. (53) and (54) the extrapolated value of f_{max} is given by

$$f_{max} \simeq \frac{1}{4\pi S} \left[\frac{\alpha_0}{r_0 C_0 \tau_{ec}^*} \right]^{1/2} \tag{55a}$$

or

$$f_{max} \simeq \frac{1}{2S} \left[\frac{f_T}{2\pi r_0 C_0} \right]^{1/2}. \tag{55b}$$

We note that both unilateral gain and maximum oscillation frequency will increase with decreasing S. This is the reason that the emitter stripe width S is one of the critical dimensions for microwave application.

F. Noise Figure: The ratio of total mean square noise voltage at the output of the transistor to mean square noise voltage at the output resulting from thermal noise in source resistance R_g. At lower frequencies the dominant noise source in a transistor is due to the surface effect which gives rise to the $1/f$ noise spectrum. At medium and high frequencies the noise figure is given by[30a]

$$NF = 1 + \frac{r_b}{R_g} + \frac{r_e}{2R_g} + \frac{(1 - \alpha_0)[1 + (1 - \alpha_0)^{-1}(f/f_\alpha)^2](R_g + r_b + r_e)^2}{2\alpha_0 r_e R_g} \tag{56}$$

where R_g is the generator resistance. From Eq. (56) it can be shown that at medium frequencies where $f \ll f_\alpha$, the noise figure is essentially a constant determined by r_b, r_e, $(1 - \alpha_0)$, and R_g. There is an optimum termination R_g which can be calculated from the condition $d(NF)/dR_g = 0$. For a low-noise design, a low value of $(1 - \alpha_0)$, that is, a high β_0, is very important. At high frequencies beyond the "corner" frequency $f = \sqrt{1 - \alpha_0} f_\alpha$ the noise figure will increase approximately as f^2.

The calculated unilateral gain from Eq. (54) and the optimum noise figure at 4 GHz are listed in Table 6.3. We have used the device geometry and material parameters considered in Section 3(2) and Table 6.1. We note that for a given semiconductor the p-n-p transistor has a slightly higher unilateral gain. This is mainly due to the smaller resistance r_b in an n-type base. The same reason is also responsible for the lower noise figure in p-n-p transistors. The noise figure as a function of frequency in the microwave frequency region is shown in Fig. 20. The Ge p-n-p transistor with its small base resistance shows the lowest noise figure. Figure 21 shows the optimum noise figure versus frequency for a state-of-the-art Ge transistor. We note that the curve

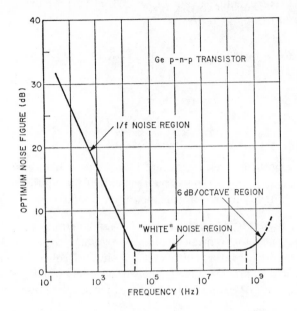

Fig 21 Optimum noise figure vs. frequency for a Ge microwave transistor. (After Sevick, Ref. 45.)

TABLE 6.3

UNILATERAL GAIN AND NOISE FIGURE
OF MICROWAVE TRANSISTORS AT 4 GHz (dB)

U and NF	Type	Ge	Si	GaAs
Unilateral Gain	*n-p-n*	18.1	12.4	16.1
	p-n-p	22.1	12.6	19.8
Optimum Noise Figure	*n-p-n*	3.6	5.0	3.3
	p-n-p	2.5	4.8	2.3

can be roughly divided into three regions: the $1/f$ noise region, the "white" noise region in which the noise figure is independent of frequency, and the 6-dB/octave noise region in which the term $(f/f_a)^2$ dominates.

4 POWER TRANSISTOR

(1) General Consideration

Power transistors are designed for power amplification and for handling large amounts of power. Although there is no well-defined boundary between power transistors and microwave transistors, usually power gain and efficiency are the prime considerations for a power transistor while cutoff frequency and noise figure are the prime considerations for a microwave transistor. Power output versus frequency for Si transistors[31] is shown in Fig. 22. At higher frequencies the power output varies approximately as $1/f^2$ as a result of the limitations[13] of avalanche breakdown field and carrier scattering-limited velocity as discussed in Chapter 5. At lower frequencies, however, the limitation on power output is mainly due to the thermal effect. As the power increases, the junction temperature T_j increases The maximum T_j is limited by the temperature at which the base region becomes intrinsic. Above T_j the transistor action ceases, since by then the collector is effectively short-circuited to the emitter. To improve transistor performance, one must improve the encapsulation sufficiently to provide an adequate heat sink for efficient thermal dissipation, and must use materials with large band gaps which will allow higher-temperature operation.

Because of these requirements Ge transistors have only limited application, and most power transistors are made from Si. GaAs transistors, despite their technological problems, have already shown excellent temperature performance and potentials as high-power transistors.[32-35] Figure 23 illustrates the temperature dependence of the dc common-emitter current gain and the 50-MHz small-signal power gain of a double-diffused (zinc for base dopant and tin for emitter dopant) n-p-n GaAs transistor.[32] The power gain decreases only about 3 dB from room temperature to 350°C. The current gain remains essentially constant over the whole temperature range. This excellent performance at high temperatures can be expected for GaAs because of its wide band gap. It has also been established that the above transistor can be fully operated at liquid helium temperature (4°K); this is mainly because the impurity levels are shallow enough that no substantial carrier "freeze-out" occurs at 4°K.

We shall now compare the performances of Ge, Si, and GaAs power transistors using the device geometry shown in Fig. 12(b), and the material parameters listed in Table 6.1. To handle a large amount of power, the stripe width (S) and the base thickness (W_B) should be appropriately adjusted. In addition, more stripes for emitter and base contacts or overlay structures[31] should be used to handle the large input current and to distri-

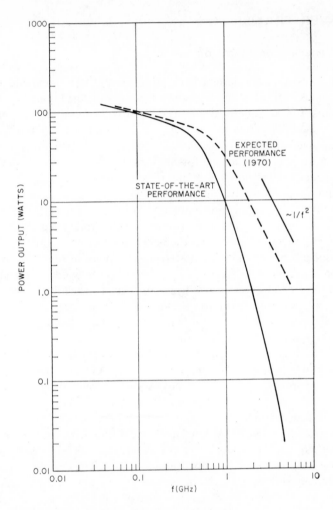

Fig. 22 Power output vs. frequency for silicon transistors.
(After Carley, Ref. 31.)

bute the current more uniformly. The overlay structure consists of (a) many small separate emitter elements, instead of a continuous emitter stripe, to increase the overall emitter periphery, (b) an extra diffused base region to distribute base current uniformly over all the separate emitter segments, and (c) an emitter metallization which overlies the base region and connects all the separate emitter elements in parallel. The basic comparison is shown in

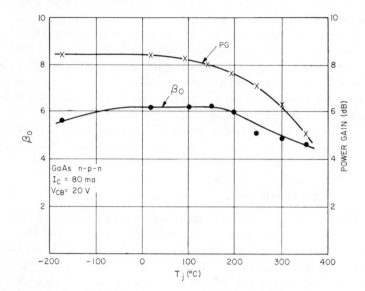

Fig. 23 Common-emitter current gain and power gain vs. temperature for a GaAs n-p-n transistor.
(After Becke et al., Ref. 32.)

TABLE 6.4

PERFORMANCE OF POWER TRANSISTORS

Quality	Ge (p-n-p)	Si (n-p-n)	GaAs (p-n-p)
Band Gap (eV) at 300°K	0.66	1.12	1.43
Operating Voltage (volts)	20	50	55
Max. Junction Temp. (T_j)	100°C	200°C	450°C
Thermal Cond. (κ)	0.5 :	1 :	0.3
$\sqrt{\text{Power} \times \text{Impedance}}\ f \cong \dfrac{\mathscr{E}_m v_{sl}}{2\pi}$ (V/sec)	2×10^{11}	4×10^{11}	4.6×10^{11}

Table 6.4. It is apparent from this table that, for a given collector doping profile, the operating voltage increases with the semiconductor band gap, E_g, since the breakdown voltage[36] increases with E_g. The maximum junction temperature is the temperature at which the base region becomes intrinsic.

In this respect GaAs is superior by far in comparison with Ge and Si. For heat dissipation, an important parameter is thermal conductivity; among the three best-known semiconductors, Si has the largest value. The ultimate limitation[13] on power-frequency performance is given by the expression $\mathscr{E}_m v_{sl}/2\pi$ where $\mathscr{E}_m$ is the breakdown field and v_{sl} is the scattering-limited velocity. Of the three semiconductors considered, Ge has the lowest value, and values for Si and GaAs are comparable. With the improvement of device technology it is expected that in the near future a substantial increase in output power can be achieved in the whole frequency spectrum and in particular in the microwave region as shown in Fig. 22 in which the power transistor and microwave transistor emerge.

(2) Second Breakdown

The use of power transistors and other semiconductor devices is often limited by a phenomenon called "second breakdown" whose initiation is manifested by an abrupt decrease in device voltage with a simultaneous internal constriction of current. The second breakdown phenomenon was first reported by Thornton and Simmons,[37] and has since been under extensive study in high-power semiconductor devices.[38,39] For high-power transistors it is important to operate the device within a certain safe region so that one can avoid the permanent damage caused by the second breakdown.

The general features of the I_C versus V_{CE} characteristics of a transistor under second breakdown conditions are shown[38] in Fig. 24. The symbols F, 0, and R stand respectively for constant forward-, zero-, and reverse-base current drive. The initiation of second breakdown for each of the three base drive conditions is indicated by the abrupt drop in V_{CE} at the instability points $I(F)$, $I(0)$, and $I(R)$. The experimental results can generally be treated as consisting of four stages: the first stage leads to instablilty, I, at the breakdown or breakover voltage, the second, to switching from the high- to the low-voltage region, the third to the low-voltage high-current range; the fourth stage to destruction as marked by D in Fig. 24.

The initiation of instability is mainly due to the temperature effect. When a pulse with given power $P = I_C \cdot V_{CEO}$ is applied to a transistor, there is a time delay before the device is triggered into the second breakdown condition. This time is called the triggering time. A typical plot[40] of the triggering time versus applied pulse power for various ambient temperatures is shown in Fig. 25. For the same triggering time τ, the triggering temperature T_{tr}, which is the temperature at the "hot" spot prior to second breakdown, is found to be approximately related to the pulse power P at different ambient temperatures T_0 by the thermal relation:

$$T_{tr} - T_0 = C_1 P \tag{57}$$

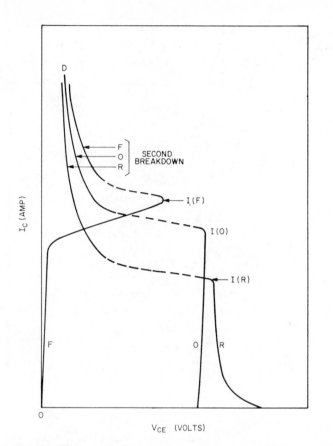

Fig. 24 Collector current vs. collector-emitter voltage under second breakdown condition. F, 0, or R indicates forward-, zero-, or reverse-base drive, respectively. I and D indicate the initiation of instability and destruction, respectively. (After Schafft, Ref. 38.)

where C_1 is a constant. From Fig. 25 we note that for a given ambient temperature the relationship between the pulse power and the triggering time is given approximately as

$$\tau \sim \exp[-C_2 P] \tag{58}$$

where C_2 is a constant.

Substitution of Eq. (57) into the above equation yields

$$\tau \sim \exp\left[-\frac{C_2}{C_1}(T_{tr} - T_0)\right]. \tag{59}$$

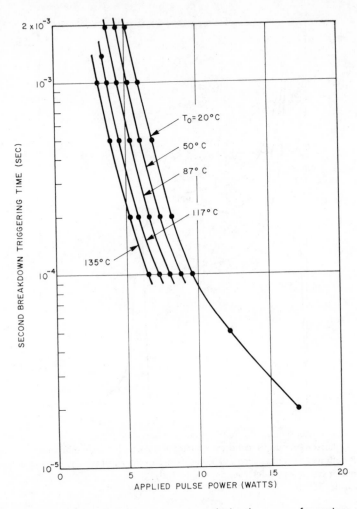

Fig. 25 Second breakdown triggering time vs. applied pulse power for various ambient temperatures.
(After Melchior and Strutt, Ref. 40.)

The triggering temperature T_{tr} depends on various device parameters and geometry. For most silicon diodes and transistors it is found that T_{tr} is the temperature at which the intrinsic concentration n_i equals the collector doping concentration. The location of the "hot" spot is usually near the center of the device. For different doping concentrations the values of T_{tr} will vary, and for different device geometries the value of C_2/C_1 will vary. This results in a large variation of the triggering time as a result of its exponential dependence on the above parameters in Eq. (59).

After instability the voltage collapses across the junction, and this is the second stage of the breakdown process. During this stage the resistance of the breakdown spot becomes drastically reduced. In the third low-voltage stage the semiconductor is at high temperature and is intrinsic in the vicinity of the breakdown spot. As the current continues to increase, the breakdown spot will melt, resulting in the fourth stage of destruction.

To safeguard a transistor from permanent damage it is necessary to specify the safe operating area. A typical example is shown in Fig. 26 for a silicon power transistor operated in common-emitter configuration.[41] The collector load lines for specific circuits must fall below the limits indicated by the applicable curve. The data are based upon a peak junction temperature T_j of 150°C. The solid curves are the upper limits imposed by the second break-for dc and various pulse operating conditions with a 10% duty cycle and with indicated pulse durations. For example, for $V_{CE} = 50$ volts and a pulse width of 1 ms, the maximum collector current for reliable operation should be about 1 amp. There are two other limitations which are also indicated in Fig. 26. One is the first breakdown voltage V_{CEO} as indicated by the vertical dotted line. The other is the thermal limitation as indicated by the horizontal

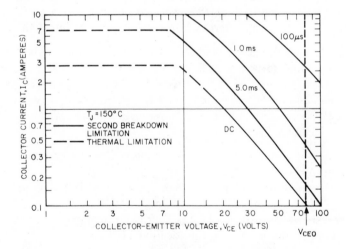

Fig. 26 Limitations on transistor operation. The vertical dotted lines indicate the maximum voltage due to avalanche breakdown. The horizontal dotted lines indicate the maximum currents due to thermal effect. The solid lines indicate the limitations imposed by second breakdown for various pulse widths of 10% duty cycle. (After Lehner, Ref. 41.)

dotted lines. At high ambient temperatures the thermal limitation reduces the power which can be handled to values less than the limitation imposed by the second breakdown.

5 · SWITCHING TRANSISTOR

A switching transistor is a transistor, designed to function as a switch, which can change its state, say from the high-voltage low-current (off) condition to the low-voltage high-current (on) condition, in a very short time. The basic operating conditions of switching transistors are different from that of microwave transistors. This is because switching is a large-signal transient process, while microwave transistors are generally concerned with small-signal amplification. The basic device geometries, however, are similar to those for microwave transistors as shown in Fig. 12(a) and (b).

The most important parameters for a switching transistor are current gain and switching time. To improve the current gain, usually lower doping in the base region is used. The transistor can be doped with gold to introduce midgap recombination centers and thus reduce the switching time.

A switching transistor can be operated in various switching modes. The three basic modes and their corresponding load lines are shown in Fig. 27(a) and are classified as saturated, current, and avalanche modes which are determined by the portion of the transistor output characteristic curve utilized.[42] The output characteristic curve can be divided into four regions:

Region I: cutoff region, collector current off, emitter and collector junctions reverse-biased,

Region II: active region, emitter forward and collector reverse-biased,

Region III: saturation region, emitter and collector both forward-biased,

Region IV: avalanche region, collector junction under avalanche breakdown.

The corresponding minority carrier distributions[43] in the base for the first three regions are shown in Fig. 27(b).

For all switching modes the switch-off condition is characterized by an excursion of the load line into the cutoff region of the transistor. The operating mode, therefore, is determined primarily by the direct current level in the switch-on condition and by the location of the operating points. The most common mode of operation is the saturated mode, which most nearly duplicates the function of an ideal switch. The transistor is virtually open-circuited between the emitter and collector terminal in the off condition and short-circuited in the on condition. The current-mode operation is useful for high-speed switching, since the storage delay time associated with the excur-

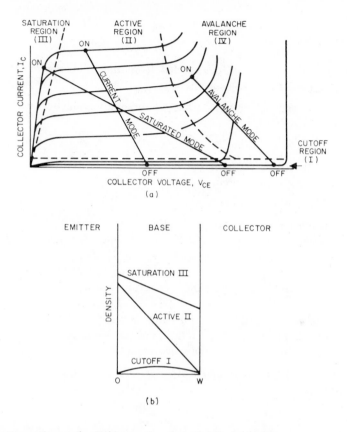

Fig. 27
(a) Operation regions and switching modes of a switching transistor.
(After Roehr and Thorpe, Ref. 42.)
(b) Minority carrier densities in the base for cutoff region (I), active region (II), and satura-
     tion region (III).
(After Moll, Ref. 43.)

sion of the transistor into the saturation region is eliminated. The avalanche
mode operation utilizes the negative-resistance characteristics of transistors
which result from operation in the common-emitter breakdown region. However, because of instability problems associated with the negative resistance
region, the avalanche-mode circuits do not find general use at the present
time.

We shall now consider the switching behavior of a transistor based on the
Ebers-Moll model.[44] Referring to Eqs. (9) and (10) derived previously, one
can write the following general expressions for the total emitter and collector

current:

$$I_E = a_{11}(e^{qV_{EB}/kT} - 1) + a_{12}(e^{qV_{CB}/kT} - 1) \tag{60}$$

$$I_C = a_{21}(e^{qV_{EB}/kT} - 1) + a_{22}(e^{qV_{CB}/kT} - 1). \tag{61}$$

The coefficient a's can be determined from the following four quantities which can be directly measured:

I_{EO}: the reverse saturation current of the emitter junction with collector open-circuited,

$$e^{qV_{EB}/kT} \ll 1, \quad \text{and} \quad I_C = 0.$$

I_{CO}: the reverse saturation current of the collector junction with the emitter open-circuited,

$$e^{qV_{CB}/kT} \ll 1, \quad I_E = 0.$$

α_N: the normal current gain under the normal operating conditions where the emitter is forward-biased, and the collector is reverse-biased. The collector current is given by $I_C = -\alpha_N I_E + I_{CO}$.

α_I: the inverse current gain under inverted operating conditions, i.e., emitter reverse-biased and collector forward-biased. The emitter current is given by $I_E = -\alpha_I I_C + I_{EO}$. For most transistors $\alpha_N > \alpha_I$. Because the emitter area is usually smaller than the collector area, the latter is much more effective in collecting the carriers which diffuse away from the emitter than vice versa.

From the above quantities and from Eqs. (60) and (61), we obtain the coefficients:

$$
\begin{aligned}
a_{11} &= -\frac{I_{EO}}{(1 - \alpha_N \alpha_I)} \\[2mm]
a_{12} &= \frac{\alpha_I I_{CO}}{(1 - \alpha_N \alpha_I)} \\[2mm]
a_{21} &= \frac{\alpha_N I_{EO}}{(1 - \alpha_N \alpha_I)} \\[2mm]
a_{22} &= -\frac{I_{CO}}{(1 - \alpha_N \alpha_I)}.
\end{aligned}
\tag{62}
$$

It can be shown that $a_{12} = a_{21}$, this $\alpha_I I_{CO} = \alpha_N I_{EO}$.

In Regions I and II the collector junction is reverse-biased. Equations (60) through (62) reduce to

$$I_E = -\frac{I_{EO}}{1 - \alpha_N \alpha_I} e^{qV_{EB}/kT} + \frac{(1 - \alpha_N)I_{EO}}{1 - \alpha_N \alpha_I}, \tag{63a}$$

$$I_C = \frac{\alpha_N I_{EO}}{1 - \alpha_N \alpha_I} e^{qV_{EB}/kT} + \frac{(1 - \alpha_I)I_{CO}}{1 - \alpha_N \alpha_I}. \tag{63b}$$

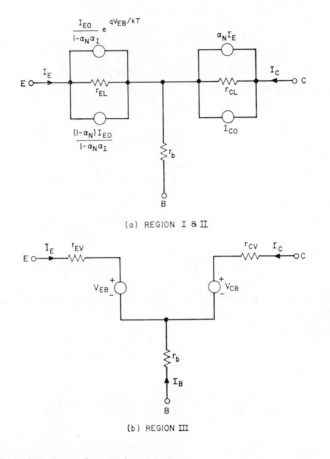

Fig. 28   Equivalent circuits for switching transistor
(a) in Regions I and II,
(b) in Region III.
(After Ebers and Moll, Ref. 44.)

The equivalent circuit corresponding to Eq. (63) is shown in Fig. 28(a). In this circuit, base resistance r_b, emitter leakage resistance r_{EL}, and collector leakage resistance r_{CL} have been added to account for the finite resistivity of the semiconductor and the finite junction conductances. In Region III it is most convenient to consider the currents as independent variables. From Eqs. (60) through (62) we obtain

$$V_{EB} = \frac{kT}{q} \ln \left[-\frac{I_E + \alpha_I I_C}{I_{EO}} + 1 \right]$$ (64a)

$$V_{CB} = \frac{kT}{q} \ln \left[-\frac{I_C + \alpha_N I_E}{I_{CO}} + 1 \right]$$ (64b)

The equivalent circuit for this region is shown in Fig. 28(b) where we have also added the base resistance r_b, the emitter body (volume) resistance r_{EV}, and the collector body resistance r_{CV}. Equations (60) through (62) form the basis on which a nonlinear large-signal switching problem can be analyzed. The above approach is based on a nonlinear two-diode model and is referred to as the Ebers-Moll model.

To characterize a switching transistor, we must consider the following five quantities: current-carrying capability, maximum open-circuit voltage, off impedance, on impedance, and switching time. The current-carrying capability is determined by the allowable power dissipation and is related to the thermal limitation as it would be to a power transistor. The maximum open-circuit voltage is determined by the breakdown or punch-through voltage discussed previously. The impedance at the off or on condition can be obtained from Eqs. (60) through (62) using appropriate boundary conditions. For example, for a common-base configuration the off and on impedances are given by

$$V_C/I_C(\text{off, Region I}) = \frac{V_C(1 - \alpha_N \alpha_I)}{(I_{CO} - \alpha_N I_{EO})}$$ (65)

and

$$V_C/I_C(\text{on, Region III}) = \frac{kT}{qI_C} \ln \left(-\frac{I_C + \alpha_N I_E}{I_{CO}} \right).$$ (66)

It is apparent from Eq. (65) that the off impedance will be high for small reverse saturation currents I_{CO} and I_{EO} of the junctions. The on impedance, Eq. (66), is approximately inversely proportional to the collector current I_C, and is very small when I_C is large. In practice the ohmic resistances as shown in Fig. 28(b) will contribute to the total impedance of the transistor and must be added.

We shall now consider the switching time, which is that required for a transistor to switch from the off to the on condition or vice versa, where in general the turn-on time is different from the turn-off time.[43] Figure 29(a) shows a switching circuit for a transistor connected in the common-base configuration. When a pulse is applied to the emitter terminal as shown in Fig. 29(b), from time $t = 0$ to t_1, Fig. 29(c), the transistor is being "turned on" and the transient is determined by the active region parameters (Region

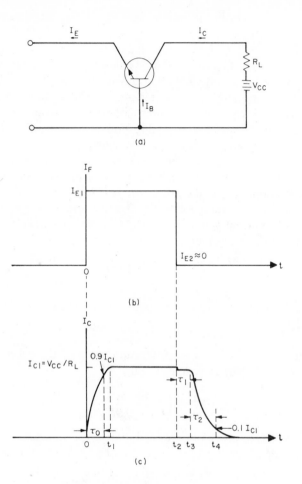

Fig. 29
(a) Switching circuit using an n-p-n switching transistor.
(b) Input emitter current pulse.
(c) Corresponding collector current response. τ_0 is the turn-on time, τ_1 the storage time, and τ_2 the decay time.
(After Moll, Ref. 43.)

II) of the transistor. At time t_1 the operating point enters the current saturation region (Region III). The period of time required for the current to reach
90% of its current saturation value ($= V_{CC}/R_L$) is called the turn-on time τ_0.
At time t_2 the emitter current is reduced to zero, and the turn-off transient
begins. From t_2 to t_3 the minority carrier density in the base layer is large,

corresponding to operation in Region III, Fig. 27(b), but decaying toward zero. During the time τ_1 the collector has a low impedance, and the collector current is determined by the external circuit. At t_3 the carrier density near the collector junction is nearly zero. At this point the collector junction impedance increases rapidly, and the transistor begins to operate in active Region II. The time interval τ_1 is the carrier storage time. After time t_3 transient behavior is calculated from the active region parameters. At time t_4 the collector current has decayed to 10% of its maximum value. The interval of time τ_2, from t_3 to t_4, is called the decay time.

The turn-on time τ_0 can be obtained from the transient response in the active region. For a step input function I_{E1} the Laplace transform is given by I_{E1}/s. If the common-base current gain α is expressed as $\alpha_N/(1 + j\omega/\omega_N)$ where ω_N is the alpha cutoff frequency at which $\alpha/\alpha_N = 1/\sqrt{2}$, the Laplace transform of the current gain is $\alpha_N/(1 + s/\omega_N)$. Thus the collector current in the Laplace transform notation is given by

$$I_C(s) = \frac{\alpha_N}{1 + s/\omega_N} \frac{I_{E1}}{s}. \tag{67}$$

The inverse transform of the above equation is given by

$$I_C = I_{E1}\alpha_N(1 - e^{-\omega_N t}). \tag{68}$$

If we denote $I_{C1} \approx V_{CC}/R_L$ as the saturation value of the collector current, τ_0 is given from Eq. (68) by setting $I_C = 0.9 I_{C1}$:

$$\tau_0 = \frac{1}{\omega_N} \ln\left(\frac{I_{E1}}{I_{E1} - 0.9 I_{C1}/\alpha_N}\right). \tag{69}$$

Based on a similar approach as outlined above, the storage time and decay time for common-base configurations, as obtained by Moll, are given as follows:[43]

$$\tau_1 = \frac{\omega_N + \omega_I}{\omega_N \omega_I (1 - \alpha_N \alpha_I)} \ln\left(\frac{I_{E1} - I_{E2}}{\dfrac{I_{C1}}{\alpha_N} - I_{E2}}\right) \tag{70}$$

$$\tau_2 = \frac{1}{\omega_N} \ln\left(\frac{I_{C1} - \alpha_N I_{E2}}{0.1 I_{C1} - \alpha_N I_{E2}}\right) \tag{71}$$

where ω_I is the inverted alpha cutoff frequency and I_{E1} and I_{E2} are indicated in Fig. 29. From Eq. (70) we note that the storage time τ_1 becomes equal to zero if the transistor does not enter saturation Region III (as in current mode), because in this case $I_{C1} = \alpha_N I_{E1}$. For the common-emitter configuration the above equations can be used with some appropriate changes of quantities: for τ_0 and τ_2, ω_N is replaced by its corresponding beta cutoff frequency, or $\omega_N(1 - \alpha_N)$, I_{E1} and I_{E2} are replaced by I_{B1} and I_{B2} respectively, and α_N by $\alpha_N/(1 - \alpha_N)$; for τ_1 the latter two operations apply: I_{E1} and I_{E2} are replaced by I_{B1} and I_{B2} respectively, and α_N by $\alpha_N/(1 - \alpha_N)$.

It is apparent from the above equations that the switching times, i.e., the turn-on time τ_0 and the turn-off time ($\tau_1 + \tau_2$), are inversely proportional to the cutoff frequencies. To increase the switching speed one must increase the cutoff frequencies. It is important to point out that the cutoff frequencies of most switching transistors (particularly the double-diffused transistors) are limited by the collector storage capacitance which should be reduced in order to increase the cutoff frequency.

The switching times are also functions of the emitter and collector current and increase with the current levels. This is the reason that it is difficult to obtain high speed and high current at the same time. Figure 30 shows a plot of the current versus speed for some state-of-the-art switching transistors.[45] We note that for a switching speed of 1 ns (10^{-9} sec) the maximum current level is limited to about 40 ma. For a speed of 1 μs, however, one can operate the transistor at a current of about 10 amp. With improvement in device and material technology a factor-of-five increase in current handling capability is expected in the near future.

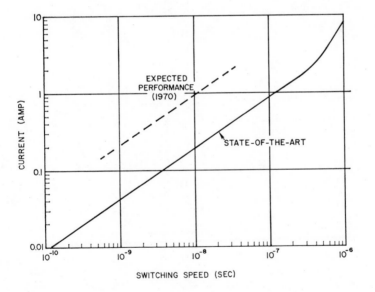

Fig. 30 Collector current vs. switching speed ($\tau_1 + \tau_2$), for the state-of-the-art transistors. Also shown is the expected performance in the near future. (After Sevick, Ref. 45.)

6 UNIJUNCTION TRANSISTOR

In this section we shall briefly consider a different kind of bipolar transistor called the unijunction transistor which is a three-terminal device having one emitter junction and two base contacts. The unijunction transistor (UJT) has evolved from the alloyed germanium bar structure originally discussed by Shockley et al.,[46] At that time the structure was called a filamentary transistor. As the device developed through the cube UJT,[47] the diffused planar structure,[48] and the epitaxial planar structure,[49] the terms double-base diode[50] and finally unijunction transistor [48, 51, 52] were coined for the device.

A schematic diagram of a UJT is shown in Fig. 31(a). The two base ohmic contacts are called base one ($B1$) and base two ($B2$). The p-n junction located

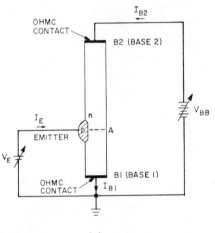

(a)

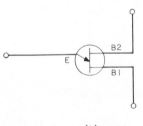

(b)

Fig. 31 Schematic diagram and nomenclature of unijunction transistors.

between $B1$ and $B2$ is called the emitter junction. The symbol and nomenclature of a UJT are shown in Fig. 31(b). Figure 32 shows a diffused and
an epitaxial UJT structure fabricated using the planar technology.[49]

The operation of UJT is mainly dependent upon the conductivity modulation between the emitter and the base one contact. In the normal operating
condition, the base one terminal is grounded and a positive bias voltage,
V_{BB}, is applied at base two as shown in Fig. 31(a). The resistance between
$B1$ and $B2$ is designated by R_{BB}, that between $B2$ and A by R_{B2}, and that
between A and $B1$ by R_{B1} ($R_{BB} = R_{B2} + R_{B1}$). The applied voltage estab-

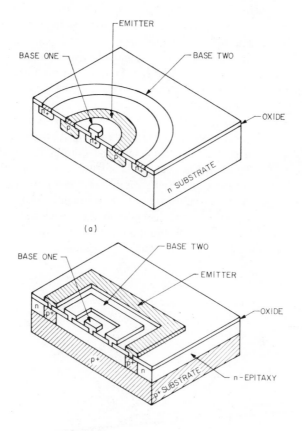

(a)

(b)

Fig. 32
(a) Diffused planar structure and
(b) epitaxial planar structure of unijunction transistor.
(After Senhouse, Ref. 49.)

lishes a current and an electric field along the semiconductor bar and produces a voltage on the n side of the emitter junction which is a fraction η of the applied voltage V_{BB}. The fraction η is called the intrinsic stand-off ratio and is given by

$$\eta \equiv \frac{R_{B1}}{R_{B1} + R_{B2}} = \frac{R_{B1}}{R_{BB}}. \tag{72}$$

When the emitter voltage V_E is less than ηV_{BB}, the emitter junction is reverse-biased and only a small reverse saturation current flows in the emitter circuit. If the voltage V_E exceeds ηV_{BB} by an amount equal to the forward voltage drop of the emitter junction, holes will be injected into the bar. Because of the electric field within the semiconductor bar these holes will move toward base one and increase the conductivity of the bar in the region between the emitter and base one. As I_E is increased, the emitter voltage will decrease because of the increased conductivity and the device will exhibit a negative resistance characteristic.

The emitter characteristic is shown in Fig. 33. The two important points on the curve are the peak point and the valley point. At these two points the slope $dV_E/dI_E = 0$. The region with current less than I_P is called the cutoff region. The region between the peak and valley part is called the negative resistance region; here the conductivity modulation is important. The region with current larger than I_V is called the saturation region. The switching time from the peak to the valley point depends on the device geometry and the biasing condition. It has been found that the time is proportional to the distance between the emitter and the base one contact.[53]

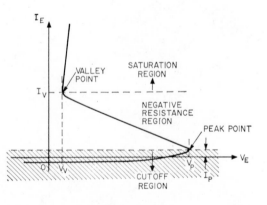

Fig. 33 Emitter current-voltage characteristic of UJT.

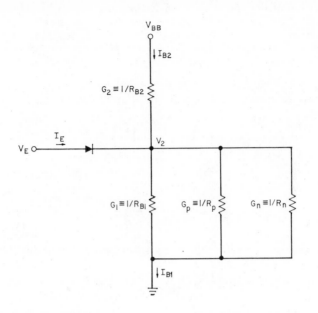

Fig. 34 Equivalent circuit of UJT.
(After Clark, Ref. 52.)

The basic emitter characteristic can be derived using the equivalent circuit shown in Fig. 34 where G_n and G_p are the excess electron and hole conductances between the emitter and base one.[52] We shall denote γ as the emitter efficiency with which holes move from the emitter to the base one contact. We shall neglect the diffusion, recombination, and surface effects. The current node equations for the top point (at voltage V_{BB}) and the central point (at voltage V_2) are

$$I_{B2} = (V_{BB} - V_2)G_2 \tag{73}$$

$$I_E + I_{B2} = V_2(G_1 + G_p + G_n). \tag{74}$$

In addition there are the following relationships

$$\frac{G_n}{G_p} = \frac{\mu_n}{\mu_p}$$

$$\gamma = \frac{V_2 G_p}{I_E}. \tag{75}$$

Combination of the above equations yields

$$V_2 = \eta V_{BB} - \left(\frac{\mu_n}{\mu_p} + 1 - \frac{1}{\gamma}\right)\gamma I_E \frac{R_{B1}R_{B2}}{R_{BB}}. \tag{76}$$

The emitter voltage V_E is given by the sum of V_2 and V_F which is the voltage drop across the diode and is obtainable from the ideal diode equation:

$$V_F = \frac{kT}{q} \ln\left(\frac{I_E}{I_s}\right) \tag{77}$$

where I_s is the diode saturation current. We have from Eqs. (76) and (77)

$$V_E = \eta V_{BB} - \left(\frac{\mu_n}{\mu_p} + 1 - \frac{1}{\gamma}\right)\gamma \frac{R_{B1}R_{B2}}{R_{BB}} I_E + \frac{kT}{q} \ln\left(\frac{I_E}{I_s}\right) \tag{78}$$

At the peak point, $dV_E/dI_E = 0$, we obtain from the above equation:

$$I_P \simeq \frac{kT}{q\gamma} \frac{1}{\left(\dfrac{\mu_n}{\mu_p} + 1 - \dfrac{1}{\gamma}\right)} \left(\frac{1}{R_{B1}} + \frac{1}{R_{B2}}\right) \tag{79}$$

$$V_P \simeq \eta V_{BB} + \frac{kT}{q} \ln\left(\frac{I_P}{I_s}\right). \tag{80}$$

It is clear from Eq. (79) that, to reduce I_P, one must employ semiconductor bars with high resistivity. For a small I_P the peak voltage V_P is approximately given by ηV_{BB} which is essentially independent of temperature. The valley current can be obtained from Eq. (76) under the condition that $V_2 \approx 0$:

$$I_V \approx \frac{1}{\gamma\left(\dfrac{\mu_n}{\mu_p} + 1 - \dfrac{1}{\gamma}\right)} \left(\frac{V_{BB}}{R_{B2}}\right) \tag{81}$$

At the valley point the injection level is high, and the value of the emitter efficiency γ is reduced.

Figure 35 shows some experimental results[54] of a unijunction transistor with $\eta = 0.6$ and $R_{BB} = 8 k\Omega$. When the base two contact is open ($I_{B2} = 0$), the I-V curve shows essentially the forward characteristic of a simple p-n junction. As V_{BB} increases, the peak voltage point V_P and the valley current point also increase. We note that the emitter characteristics are fairly insensitive to temperature variation. This is in agreement with Eqs. (80) and (81).

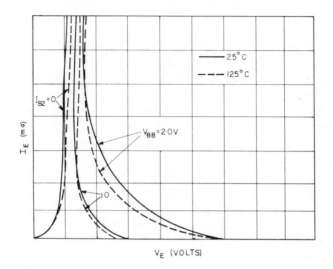

Fig. 35 Measured emitter characteristics of UJT with 0.6 intrinsic stand-off ratio (η), and $R_{BB} = 8k\Omega$.
(After Sylvan, Ref. 54.)

REFERENCES

1. J. Bardeen and W. H. Brattain, "The Transistor, A Semiconductor Triode," Phys. Rev., *74*, 230 (1948).

2. W. Shockley, "The Theory of *p-n* Junctions in Semiconductors and *p-n* Junction Transistors," Bell Syst. Tech. J., *28*, 435 (1949).

3. R. N. Hall and W. C. Dunlap, "*p-n* Junctions Prepared by Impurity Diffusion," Phys. Rev., *80*, 467 (1950).

3a. G. K. Teal, M. Sparks, and E. Buehlor, "Growth of Germanium Single Crystals Containing *p-n* Junctions," Phys. Rev., *81*, 637 (1951).

3b. W. H. Pfann, "Principles of Zone-Refining," Trans. AIME, *194*, 747 (1952).

3c. M. Tanenbaum and D. E. Thomas, "Diffused Emitter and Base Silicon Transistor," Bell Syst. Tech. J., *35*, 1 (1956).

3d. C. A. Lee, "A High Frequency Diffused Base Germanium Transistor," Bell Syst. Tech. J., *35*, 23 (1956).

3e. C. J. Frosch and L. Derrick, "Surface Protection and Selective Masking During Diffusion in Silicon," J. Electrochem. Soc., *104*, 547 (1957).

4. H. C. Theuerer, J. J. Kleimack, H. H. Loar, and H. Christenson, "Epitaxial Diffused Transistors," Proc. IRE, *48*, 1642 (1960).

5. J. A. Hoerni, "Planar Silicon Transistor and Diodes," IRE Electron Devices Meeting, Washington, D. C. (1960).

6. M. P. Lepselter and R. W. MacDonald, "Beam-Lead Devices," also M. P. Lepselter, H. A. Waggener, and R. E. Davis, "Beam-Leaded and Intraconnected Integrated Circuits," IEEE Electron Device Meeting, Washington, D.C. (1964); and M. P. Lepselter, "Beam-Lead Technology," Bell Syst. Tech. J., *45*, 233 (1966).

7. W. Shockley, U.S. Patent 2,787,564 (1954). For a view on ion implantation see J. F. Gibbons, "Ion Implantation in Semiconductors—Part I, Range Distribution Theory and Experiments," Proc. IEEE, *56*, 295 (1968).

8. F. J. Biondi, Ed. *Transistor Technology*, Vol. I and II, D. Van Nostrand Co., Inc., Princeton, N.J. (1958).

9. J. N. Shive, *The Properties, Physics and Design of Semiconductor Devices*, D. Van Nostrand Co., Inc., Princeton, N.J. (1959).

10. A. B. Phillips, *Transistor Engineering*, McGraw Hill Book Co., Inc., New York (1962).

11. W. W. Gartner, *Transistors, Principle, Design and Application*, D. Van Nostrand Co., Inc., Princeton, N.J. (1960).

11a. R. L. Pritchard, *Electrical Characteristics of Transistors*, McGraw Hill Book Co., (1967).

12. SEEC (Semiconductor Electronics Education Committee), Vol. 1-4
    (1) R. B. Adler, A. C. Smith, and R. L. Longini, *Introduction to Semiconductor Physics*, SEEC Vol. 1, John Wiley & Sons, Inc., New York (1966).

    (2) P. E. Gray, D. DeWitt, A. R. Boothroyd, and J. F. Gibbons, *Physical Electronics and Circuit Models of Transistors*, SEEC Vol. 2, John Wiley & Sons, Inc., New York (1966).

    (3) C. L. Searle, A. R. Boothroyd, E. J. Angelo, P. E. Gray, and D. O. Pederson, *Elementary Circuit Properties of Transistors*, SEEC Vol. 3, John Wiley & Sons, Inc., New York (1966).

    (4) R. D. Thornton, D. DeWitt, E. R. Chenette, and P. E. Gray, *Characteristics and Limitations of Transistors*, SEEC Vol. 4, John Wiley & Sons, Inc., New York (1966).

13. E. O. Johnson, "Physical Limitations on Frequency and Power Parameters of Transistors," IEEE Interc. Conv. Record, Pt. 5, p. 27, (1965).

14. J. L. Moll and I. M. Ross, "The Dependence of Transistor Parameters on the Distribution of Base Layer Resistivity," Proc. IRE, *44*, 72 (1956).

15. J. E. Iwersen, A. R. Bray, and J. J. Kleimack, "Low-Current Alpha in Silicon Transistors," IRE Trans. Electron Devices, *ED-9*, 474 (1962).

16. H. K. Gummel, "Measurement of the Number of Impurities in the Base Layer of a Transistor," Proc. IRE, *49*, 834 (1961).

17. C. T. Sah, R. N. Noyce, and W. Shockley, "Carrier Generation and Recombination in *p-n* Junction and *p-n* Junction Characteristics," Proc. IRE, *45*, 1228 (1957).

18. W. M. Webster, "On the Variation of Junction-Transistor Current Amplification Factor with Emitter Current," Proc. IRE, *42*, 914 (1954).

19. M. J. Morant, *Introduction to Semiconductor Devices*, Addison-Wesley Publishing Co. Inc., Reading (1964).

20. J. M. Early, "Effects of Space-Charge Layer Widening in Junction Transistors," Proc. IRE, *40*, 1401 (1952).

21. R. L. Pritchard, "Frequency Response of Grounded-Base and Grounded-Emitter Transistors," AIEE Winter Meeting (Jan. 1954).

22. R. L. Pritchard, J. B. Angell, R. B. Adler, J. M. Early, and W. M. Webster, "Transistor Internal Parameters for Small-Signal Representation," Proc. IRE *49*, 725 (1961).

23. J. L. Moll, *Physics of Semiconductors*, McGraw-Hill Book Co. (1964).

24. H. Kroemer, "Transistor-I," RCA Laboratories, p. 202 (1956).

25. A. N. Daw, R. N. Mitra, and N. K. D. Choudhury, "Cutoff Frequency of a Drift Transistor," Solid State Electron., *10*, 359 (1967).

26. C. T. Kirk, "A Theory of Transistor Cutoff Frequency (f_T) Fall-Off at High Current Density," IEEE Trans. Electron Devices, *ED-9*, 164 (1962).

27. H. C. Poon, H. K. Gummel, and D. L. Scharfetter, "High Injection in Epitaxial Transistors," to be published.

28. R. Edwards and R. L. Pritchett, "Planar Germanium Microwave Transistors," NEREM Record, p. 246 (1965).

29. R. Edwards, "Fabrication Control Is Key to Microwave Performance," Electronics, pp. 109-113 (Feb. 1968).

29a. C. S. Fuller and K. B. Wolfstrin, "Cu-Doubling Effect in Gallium Arsenide," J. Phys. Chem. Solids, *27*, 1889 (1966).

30. S. M. Sze and H. K. Gummel, "Appraisal of Semiconductor-Metal-Semiconductor Transistors," Solid State Electron., *9*, 751 (1966).

30a. E. G. Nielson, "Behavior of Noise Figure in Junction Transistors," Proc. IRE, *45*, 957 (1957).

31. D. R. Carley, "A Worthy Challenger for *rf* Power Honors," Electronics, pp. 98-102 (Feb. 1968).

32. H. Becke, D. Flatley, and D. Stolnitz, "Double Diffused Gallium Arsenide Transistors," Solid State Electron., *8*, 255 (1965).

33. H. Statz, "Double Diffused *p-n-p* GaAs Transistor," Solid State Electron., *8*, 827 (1965).

34. W. Von Munch, H. Statz, and A. E. Blakeslee, "Isolated GaAs Transistors on High-Resistivity GaAs Substrate," Solid State Electron., *9*, 826 (1966).

35. W. Von Munch, "Gallium Arsenide Planar Technology," IBM Journal, p. 348 (Nov. 1966).

36. S. M. Sze and G. Gibbons, "Avalanche Breakdown Voltages of Abrupt and Linearly Graded *p-n* Junctions in Ge, Si, GaAs, and GaP," Appl. Phys. Letters, *8*, 111 (1966).

37. C. G. Thornton and C. D. Simmons, "A New High Current Mode of Transistor Operation," IRE Trans. Electron Devices, *ED-5*, 6 (1958).

38. H. A. Schafft, "Second-Breakdown—A Comprehensive Review," Proc. IEEE, *55*, 1272 (1967).

39. N. Klein, "Electrical Breakdown in Solids," a chapter in *Advances in Electronics* and *Electron Physics*, Ed. by L. Marton, Academic Press (1968).

40. H. Melchior and M. J. O. Strutt, "Secondary Breakdown in Transistors," Proc. IEEE, *52*, 439 (1964).

41. L. L. Lehner, "A Discrete Transistor That Is a Powerhouse," Electronics, pp. 105-108 (Feb. 1968).

42. W. D. Roehr and D. Thorpe, Editors, *Switching Transistor Handbook*, Motorola Semiconductor Product Inc., Phoenix (1966).

43. J. L. Moll, "Large-Signal Transient Response of Junction Transistors," Proc. IRE, *42*, 1773 (1954).

44. J. J. Ebers and J. L. Moll, "Large-Signal Behavior of Junction Transistors," Proc. IRE, *42*, 1761 (1954).

45. J. Sevick, unpublished results.

46. W. Shockley, G. L. Pearson, and J. R. Haynes, "Hole Injection in Germanium-Quantitative Studies and Filamentary Transistors," Bell Syst. Tech. J., *28*, 344 (1949).

47. V. A. Bluhm and T. P. Sylvan, "A High Performance Unijunction Transistor Using Conductivity Modulation of Spreading Resistance," Solid State Design, *5*, pp. 26-31 (1964).

48. L. E. Clark, "Unijunction Transistor," Patent 3,325,705 (1967).

49. L. S. Senhouse, "A Unique Filamentary-Transistor Structure," IEEE Electron Device Meeting, Paper 23.6, Washington D.C. (Oct. 1967).

50. I. A. Lesk and V. P. Mathis, "The Double-Base Diode—A New Semiconductor Device," IRE Conv. Rec., Part 6, p. 2 (1953).

51. F. N. Trofimenkoff and G. J. Huff, "DC Theory of the Unijunction Transistor," Int. J. Electronics, *20*, pp. 217-225 (1966).

52. L. E. Clark, "Now, New Unijunction Geometries," Electronics, *38*, pp. 93-97 (1965).

53. D. L. Scharfetter and A. G. Jordan, "Reactive Effects in Semiconductor Filaments Due to Conductivity Modulation and an Extension of the Theory of the Double-Base Diode," IRE Trans. Electron Devices, *ED-9*, 461 (1962).

54. T. P. Sylvan *The Unijunction Transistor Characteristics and Applications*, Application Note, Semiconductor Products Department, General Electric Co. (May 1965).

■ INTRODUCTION

■ SHOCKLEY DIODE AND
SEMICONDUCTOR-CONTROLLED
RECTIFIER

■ JUNCTION FIELD-EFFECT
TRANSISTOR
AND CURRENT LIMITER

7

p-n-p-n and
Junction Field-Effect Devices

I INTRODUCTION

The operations of four-layer *p-n-p-n* devices are intimately related to the junction transistor (also called the bipolar transistor) in which both electrons and holes are involved in the transport processes. This is in contrast to the junction field-effect devices which are unipolar devices with predominantly majority carriers participating in the current conduction mechanism. Although these two classes of devices differ from each other in their fundamental operation principles, they do have several aspects in common: both were invented by W. Shockley[1,2] in the early 50's, and both can be operated as two-, three-, or four-terminal devices.

Following Shockley's concept of the "hook" collector,[1] Ebers[3] developed a two-transistor analogue to explain the *p-n-p-n* characteristics. The detailed device principles and the first working *p-n-p-n* devices were reported by Moll et al.[4] This work has since served as the basis for all succeeding work in this field. As a two-terminal device, the Shockley diode (or *p-n-p-n* diode) possesses the properties of a classical switch which can change from a high-impedance off state to a low-impedance on state or vice versa. The three-terminal *p-n-p-n* device is called the semiconductor-controlled rectifier (SCR) or thyristor, since in many respects the electrical characteristics are similar to those of the gas thyratron. Because of their two stable states (on and off) and their low power dissipation in these two states, *p-n-p-n* devices have found unique usefulness in applications requiring latching action and power-handling capability. They have been extensively used as static switches, phase controllers, power inverters, and dc choppers. These *p-n-p-n* devices are now available with current ratings ranging from a few milliamperes to hundreds of amperes, and voltage ratings extending above 1000 volts. A comprehensive and authoritative treatment on *p-n-p-n* device principles and applications has been written recently by Gentry et al.[5]

Based on Shockley's theoretical treatment of the "unipolar" transistor,[2] the first working junction field-effect transistor (JFET), has been reported by Dacey and Ross[6] who have also considered the effect of mobility saturation. The generalized theory for JFET with arbitrary charge distributions has been developed by Bockemuehl[7] whose analysis has contributed significantly to the understanding of JFET behavior. As a matter of fact, the insulated-gate field-effect transistor (IGFET), to be considered in Chapter 10, is essentially a JFET with a delta function charge distribution. Because of their high input and output impedance and their square-law transfer characteristics, the JFET's are useful circuit components, particularly for low-noise, low-distortion amplification of low-frequency signals from high-impedance sources. The basic JFET circuit applications have been discussed in detail by Sevin.[8]

In this chapter we shall consider the current-voltage characteristics of *p-n-p-n* and junction field-effect devices under various operating conditions. In addition we shall present graphical solutions for both classes of devices whereby a family of *I-V* curves can be generated when a single curve is measured.

2 SHOCKLEY DIODE AND SEMICONDUCTOR-CONTROLLED RECTIFIER

(1) Basic Characteristics

A simple four-layer *p-n-p-n* structure is shown in Fig. 1. There are three *p-n* junctions, J1, J2, and J3 in series. The contact electrode to the outer *p* layer is called the anode and that to the outer *n* layer is called the cathode. For a general *p-n-p-n* device there may be two gate electrodes (also referred to as base) connected to the inner *n* and *p* layers. If there is no gate electrode, the device is operated as a two-terminal *p-n-p-n*, or Shockley, diode. With one gate electrode the device then has three terminals and is commonly called the semiconductor-controlled rectifier (SCR) or thyristor.

A typical doping profile of an alloy-diffused *p-n-p-n* device is shown in Fig. 2(a). An *n*-type wafer is chosen as the starting material. Then a diffusion step is used to form the *p*1 and *p*2 layers simultaneously. Finally an *n*-type layer is alloyed (or diffused) into one side of the wafer to form the *n*2 layer. A cross-sectional view[5] of a medium-current SCR is shown in Fig. 2(b). The copper block serves as the heat sink.

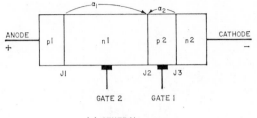

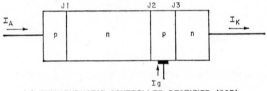

Fig. 1

(a) General four-layer *p-n-p-n* structure with anode, cathode, and two gate electrodes. There are three *p-n* junctions in the series J1, J2, and J3. Current gain α_1 is for the *p-n-p* transistor and α_2 is for the *n-p-n*. Under normal bias conditions, as shown, the center junction J2 is reverse-biased and serves as a common collector for the *p-n-p* and *n-p-n* transistors.

(b) Two-terminal *p-n-p-n* structure (Shockley diode).

(c) Three terminal *p-n-p-n* structure (semiconductor-controlled rectifier, SCR).

The basic current-voltage characteristic of a *p-n-p-n* device (with or without any gate electrodes) is shown in Fig. 3 which has a number of complex regions. In region 0–1 the device is in the forward blocking or "off" state with very high impedance. The forward breakover (or switching) occurs at

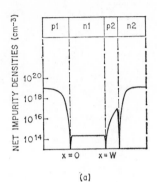

(a)

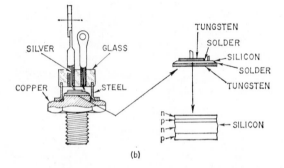

(b)

Fig. 2
(a) Typical doping profile. The most important parameters are the doping concentration N_D and the width W of the $n1$ layer.
(b) A cross-sectional view of a medium-current SCR.
(After Gentry et al., Ref. 5.)

the place where $dV/dI = 0$, and we define a switching voltage V_s and a switching current I_s. Region 1–2 is the negative resistance region, while 2–3 is the forward conducting or "on" state. At point (2) where again $dV/dI = 0$, we define the holding current I_h and holding voltage V_h. Region 0–4 is in the reverse blocking state, while 4–5 is the reverse breakdown region.

A *p-n-p-n* device when operated in the forward region is thus a bistable device which can switch from a high-impedance low-current state to a low-impedance high-current state or vice versa. To understand the basic switching phenomena, we shall use the method of the two-transistor analogue.[3] From Fig. 1 we see that the device can be considered as a *p-n-p* and an *n-p-n* transistor connected with the collector of one transistor attached to the base of

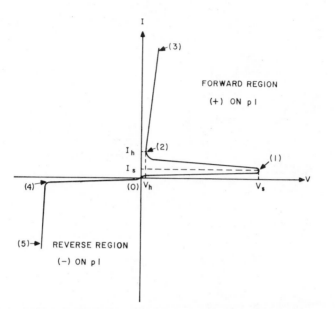

Fig. 3   Current-voltage characteristics of a p-n-p-n device. In the forward region (anode positively biased) 0–1 is the forward blocking region. V_S and I_S are the switching voltage and switching current respectively. 1–2 is the negative resistance region. V_h and I_h are the holding voltage and holding current respectively. 2–3 is the forward conduction region. 0–4 is the reverse blocking region and 4–5 is the reverse breakdown region.

the other and vice versa, as shown in Fig. 4(a) and 4(b) for a three-terminal p-n-p-n device. The center junction acts as the collector of holes from J1 and of electrons from J3.

The relationship between emitter, collector, and base currents (I_E, I_C, and I_B respectively) and the dc common-base current gain α_1 for a p-n-p transistor is shown in Fig. 4(c) where I_{CO} is the collector-base reverse saturation current. Similar relationships can be obtained for an n-p-n transistor, except that the currents are reversed. From Fig. 4(b) it is evident that the collector of the n-p-n transistor provides the base drive for the p-n-p transistor. Also the collector of the p-n-p transistor along with gate current I_g supplies the base drive for the n-p-n transistor. Thus a regeneration situation results when the total loop gain exceeds unity.

The base current of the p-n-p transistor is

$$I_{B1} = (1 - \alpha_1)I_A - I_{CO1} \tag{1}$$

which is supplied by the collector of the n-p-n transistor. The collector current

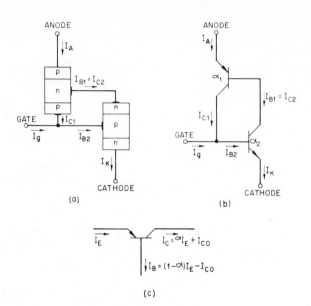

Fig. 4
(a) Two-transistor approximation of a three-terminal *p-n-p-n* device.
(b) Same as (a) using transistor notations.
(c) Current relationships in a *p-n-p* transistor.
(After Ebers, Ref. 3.)

of the *n-p-n* transistor with a dc common-base current gain α_2 is given by

$$I_{C2} = \alpha_2 I_K + I_{CO2}. \tag{2}$$

By equating I_{B1} and I_{C2}, we obtain

$$(I - \alpha_1)I_A - I_{CO1} = \alpha_2 I_K + I_{CO2}.$$

Since $I_K = I_A + I_g$, from Eq. (2) we obtain

$$I_A = \frac{\alpha_2 I_g + I_{CO1} + I_{CO2}}{1 - (\alpha_1 + \alpha_2)}. \tag{3}$$

It will be shown later that both α_1 and α_2 are functions of the current I_A and generally increase with increasing current. The above equation gives the static characteristic of the device. We note that all the current components in the numerator of Eq. (3) are small, hence I_A is small unless $(\alpha_1 + \alpha_2)$ approaches unity. At this point the denominator of the equation approaches zero and switching will occur. It is worthwhile to point out that if the polarities of the anode and cathode are reversed, then junctions J1 and J3 are reverse-biased while J2 is forward-biased. Under this condition there is no switching action, since only the center junction acts as an emitter, and no regenerative process can take place.

The depletion-layer widths and the corresponding energy-band diagrams for the equilibrium, forward off state, and forward on state are shown in Fig. 5(a), (b), and (c) respectively. In equilibrium there is at each junction a depletion region with a built-in potential which is determined by the impurity doping profile. When a positive voltage is applied to the anode, junction J2 will tend to become reverse-biased, while J1 and J3 will be forward-biased.

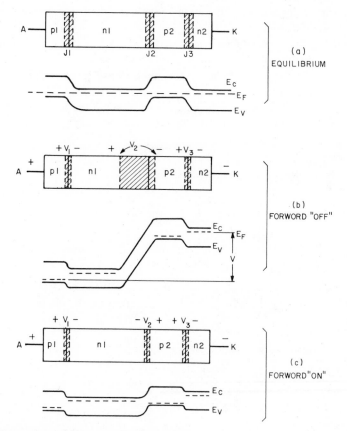

Fig. 5 Energy band diagrams for forward regions
(a) equilibrium condition,
(b) forward "off" condition where most of the voltage drops across the center junction J2, and
(c) forward "on" condition where all three junctions are forward-biased.

The anode-to-cathode voltage drop is approximately equal to the algebraic sum of the junction drops:

$$V_{AK} = V_1 + V_2 + V_3. \tag{4}$$

As the voltage increases, the current will increase. This in turn causes α_1 and α_2 to increase. Because of the regenerative nature of these processes, switching eventually occurs and the device is in its on state. Upon switching, the current through the device must be limited by an external load resistance; otherwise, the device would destroy itself if the supply voltage were sufficiently high. In this on state, J2 is also forward-biased, as shown in Fig. 5(c), and the voltage drop V_{AK} is given by $(V_1 - |V_2| + V_3)$ which is approximately equal to the voltage drop across one forward-biased *p-n* junction plus a saturated transistor.

Switching of a *p-n-p-n* device occurs when $dV_{AK}/dI_A = 0$. This condition is generally reached before $(\alpha_1 + \alpha_2) = 1$. We shall now show that the switching will begin when the sum of the small-signal alphas reaches unity.[9] Let us consider the situation which results when the gate current I_g is increased by a small amount ΔI_g. As a consequence of this increase the anode current will increase by an amount ΔI_A, so the incremental cathode current is given by

$$\Delta I_K = \Delta I_A + \Delta I_g. \tag{5}$$

The small-signal alphas are defined as

$$\tilde{\alpha}_1 \equiv \frac{\partial I_C}{\partial I_A} = \lim_{\Delta I_A \to 0} \frac{\Delta I_C}{\Delta I_A}, \tag{6a}$$

$$\tilde{\alpha}_2 \equiv \frac{\partial I_C}{\partial I_K} = \lim_{\Delta I_K \to 0} \frac{\Delta I_C}{\Delta I_K}. \tag{6b}$$

The hole current collected by J2 will be $\tilde{\alpha}_1 \Delta I_A$ and the electron current will be $\tilde{\alpha}_2 \Delta I_K$. Equating the change in anode current to the change in current across J2, we obtain

$$\Delta I_A = \tilde{\alpha}_1 \Delta I_A + \tilde{\alpha}_2 \Delta I_K. \tag{7}$$

Substitution of Eq. (7) into (5) yields

$$\frac{\Delta I_A}{\Delta I_g} = \frac{\tilde{\alpha}_2}{1 - (\tilde{\alpha}_1 + \tilde{\alpha}_2)}. \tag{8}$$

When $(\tilde{\alpha}_1 + \tilde{\alpha}_2)$ becomes unity, any small increase in I_g will cause the device to become unstable, since from Eq. (8) a small increase in I_g will cause an infinite increase in I_A. Although gate current was used in the analysis, the same effect can be obtained with a slight increase in temperature or voltage.

(2) Variation of Alphas With Current*

The dc common-base current gain α_1 of a transistor is given by

$$\alpha_1 = \alpha_T \gamma \tag{9}$$

where α_T is the transport factor defined as the ratio of the injected current reaching the collector junction to the injected current and γ is the injection efficiency defined as the ratio of the injected current to the total emitter current. From Fig. 4(c) we have the relationship

$$I_C = \alpha_1 I_E + I_{CO}. \tag{10}$$

By differentiating Eq. (10) with respect to emitter current, we obtain the small-signal alpha:

$$\tilde{\alpha}_1 \equiv \frac{\partial I_C}{\partial I_E} = \alpha_1 + I_E \frac{\partial \alpha_1}{\partial I_E}. \tag{11}$$

Substitution of Eq. (9) into Eq. (11) yields

$$\tilde{\alpha}_1 = \gamma \left(\alpha_T + I_E \frac{\partial \alpha_T}{\partial I_E} \right) + \alpha_T I_E \frac{\partial \gamma}{\partial I_E}. \tag{12}$$

The simplest approximations for α_T and γ are given by

$$\alpha_T = \frac{1}{\cosh\left(\dfrac{W}{\sqrt{D\tau}} \right)} \simeq 1 - \frac{W^2}{2D\tau}, \tag{12a}$$

$$\gamma \simeq \frac{1}{1 + \dfrac{N_B W}{N_E L_E}} \tag{12b}$$

where W is the base width, D and τ are, respectively, the diffusion coefficient and lifetime of minority carriers in the base, N_B and N_E are the base and emitter concentrations respectively, and L_E is the diffusion length in the emitter. To obtain large values of alpha, one must use small values of $W/\sqrt{D\tau}$ and N_B/N_E.

To investigate the dependence of dc alphas and small-signal alphas on current we must use the more detailed calculation, considering both diffusion

---

* This subsection is similar to that presented in 2(2) of Chapter 6. However, here we are primarily concerned with a common-base current gain that is substantially less than unity. We are also interested in the wide variation in current gain.

and drift current components. The hole currents at junctions J1 and J2 can be calculated from the equation

$$I_p(x) = qA_s\left(p_n\mu_p\mathscr{E} - D_p\frac{\partial p_n}{\partial x}\right) \tag{13}$$

where A_s is the area of the junction. The continuity equation for the $n1$ region as shown in Fig. 2(a) is given by

$$\frac{\partial p_n}{\partial t} = -\frac{p_n - p_{no}}{\tau_p} - \mu_p\mathscr{E}\frac{\partial p_n}{\partial x} + D_p\frac{\partial^2 p_n}{\partial x^2}. \tag{14}$$

And the boundary conditions are $p_n(x = 0) = p_{no}\exp(\beta V)$ where $\beta \equiv q/kT$, and $p_n(x = W) = 0$. The steady-state solution of the above equation subject to the boundary conditions is

$$p_n(x) = p_{no}\exp(\beta V)\exp[(C_1 + C_2)x] - p_{no}[\exp(\beta V)\exp(C_2 W)$$
$$+ \exp(-C_1 W)]\exp(C_1 x)\operatorname{csch}(C_2 W)\sinh(C_2 W) \tag{15}$$

where

$$C_1 \pm C_2 = \frac{\mu_p\mathscr{E}}{2D_p} \pm \left[\left(\frac{\mu_p\mathscr{E}}{2D_p}\right)^2 + \frac{1}{D_p\tau_p}\right]^{1/2}.$$

From Eqs. (13), (14), and (15) we obtain for the transport factor

$$\alpha_T = \frac{C_2\exp(C_1 W)}{C_1\sinh(C_2 W) + C_2\cosh(C_2 W)}. \tag{16}$$

The injection efficiency is given by

$$\gamma \equiv \frac{I_p}{I_p + I_n + I_r} \simeq \frac{I_p}{I_p + I_r} = \frac{I_{po}\exp(\beta V)}{I_{po}\exp(\beta V) + I_R\exp(\beta V/\eta)} \tag{17}$$

where I_p and I_n are the injected current flowing into the base and emitter regions respectively, I_r is the space-charge recombination current given by $I_R\exp(\beta V/\eta)$ where I_R and η are constants (generally $1 < \eta < 2$), and $I_{po} = qD_p A_s p_n(C_1 + C_2\coth C_2 W)$. For the doping profile of Fig. 2(a), $p_{po}(p1) \gg n_{no}(n1)$, the current I_n can be neglected in Eq. (17).

We can now calculate α_1 from Eqs. (16) and (17) as a function of the emitter current and the base layer width (W). In addition we can combine Eqs. (12), (16), and (17) to give the small-signal alpha. The results are shown in Fig. 6 for the doping profile shown in Fig. 2(a) and for some typical parameters of silicon.[10] We note that, for the current range shown, the small-signal alpha is always greater than the dc alpha. The ratio of the base width to diffusion length, W/L, is an important device parameter in determining the variation of gain with current. For small values of W/L, the transport factor

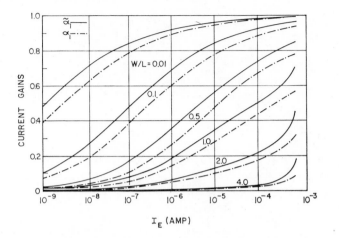

Fig. 6 Small-signal alpha and dc alpha as functions of current and base width for a transistor with the following parameters $n_{no} = 3 \times 10^{14}$ cm^{-3}, $p_{no} = 7.5 \times 10^5$ cm^{-3}, $A_s = 0.16$ mm^2, $\mu_n = 1400$ cm^2/V-sec, $\mu_p = 500$ cm^2/V-sec, $D_p = 13$ cm^2/sec, $\tau_p = 0.5$ μsec, $L_p = 25.5$ μm, $I_R = 2.5 \times 10^{-10}$ amp, and $\eta = 1.5$.
(After Yang and Voulgaris, Ref. 10.)

is independent of current and the gain varies with current only through the injection efficiency. This condition applies to the narrow base-width section of the devices (n-p-n section). For larger values of W/L both the transport factor and the injection efficiency are functions of current (p-n-p section). Thus the value of gain can, in principle, be tailored to the desired range by choosing the proper diffusion length and doping profile.

(3) Generalized Current-Voltage Characteristics

We shall now use a graphical method[11,12] to analyze the I-V characteristics of generalized p-n-p-n devices. Figure 7 shows a general device with leads connected to all four layers. Reference directions for voltages and currents are shown in the figure. We assume that the center junction of the device remains reverse-biased. We also assume that the voltage drop V_2 across this junction is sufficient to produce avalanche multiplication of carriers as they travel across the depletion region. We denote the multiplication factor for electrons by M_n and that for holes by M_p; both are functions of V_2. Because of the multiplication, a steady hole current $I_p(x_1)$ entering the depletion region at x_1 becomes $M_p I_p(x_1)$ at $x = x_2$. A similar result will be obtained for an electron current $I_n(x_2)$ entering the depletion layer at x_2. The total

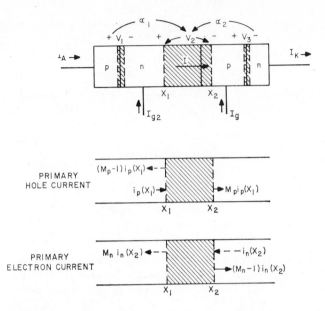

Fig. 7 Generalized *p-n-p-n* device. The current flowing through the center junction is *I*. There are electron, i_n, and hole, i_p, primary currents which generate, respectively, $M_n i_n$ and $M_p i_p$ current under avalanche multiplication conditions.

current I is given by

$$I = M_p I_p(x_1) + M_n I_n(x_2). \tag{18}$$

Since $I_p(x_1)$ is actually the collector current of the *p-n-p* transistor, we can express $I_p(x_1)$ as Fig. 4(c),

$$I_p(x) = \alpha_1(I_A)I_A + I_{CO1}. \tag{19}$$

Similarly, we can express the primary electron current $I_n(x_2)$ as

$$I_n(x) = \alpha_2(I_K)I_K + I_{CO2}. \tag{20}$$

Substitution of Eqs. (19) and (20) into Eq. (18) yields

$$I = M_p[\alpha_1(I_A)I_A + I_{CO1}] + M_n[\alpha_2(I_K)I_K + I_{CO2}]. \tag{21}$$

If we assume that $M_p = M_n = M$, Eq. (21) reduces to

$$\frac{1}{M(V_2)} = \frac{\alpha_1(I_A)I_A}{I} + \frac{\alpha_2(I_K)I_K}{I} + \frac{I_0}{I} \tag{22}$$

where $I_0 \equiv I_{CO1} + I_{CO2}$. This assumption of an equal multiplication factor is valid for Ge, GaAs, and GaP, since for each semiconductor the ionization rates of electrons and holes are about equal (refer to Fig. 30 of Chapter 2). For Si the assumption is not generally valid; however, the basic I-V characteristics still can be predicted reasonably well.

A. Shockley Diode. Based on Eq. (22) we shall first derive the I-V characteristics of a Shockley diode. Since $I_g = I_{g2} = 0$, and $I = I_A = I_K$, Eq. (22) reduces to

$$\frac{1}{M(V_2)} = \alpha_1(I) + \alpha_2(I) + I_0/I = f(I). \tag{23}$$

The graphical solution of Eq. (23) is illustrated in Fig. 8. We shall make the following assumptions: I_0 is some known constant, α_1 and α_2 are known functions of current similar to those shown in Fig. 6, and M can be expressed as

$$M(V_2) = \frac{1}{1 - \left(\dfrac{V_2}{V_{BD}}\right)^n} \tag{24}$$

where V_{BD} is the breakdown voltage and n is a constant (~ 3 for silicon). We first obtain the function $f(I)$ by adding the three curves as shown in Fig. 8(a). Figure 8(b) is a plot of $1/M$ versus M with the vertical scale identical to that in Fig. 8(a) and the horizontal scale identical to that in Fig. 8(c) which is a plot of Eq. (24). We now choose a value I in Fig. 8(d) at which we want to know the voltage drop. We project vertically upward to find the corresponding value of $f(I)$, point (1). Then we project horizontally to the left and locate a point on the $1/M$ versus M plot, point (2). With known M we can project vertically downward to find the required value (V_2/V_{BD}) in Fig. 8(c), point (3), and horizontally back to point (4). This gives the normalized voltage drop required to sustain the given current I. The entire I-V characteristic can thus be obtained by repeating this geometrical construction process. The result is shown in Fig. 8(d).

We note from Fig. 8 that the switching point (I_s, V_s) occurs at the location where the function $f(I)$ reaches its minimum. The holding point is defined as the low-voltage, high-current point at which $dV/dI = 0$. The above analysis does not enable us to find this point. However, the holding point can be approximately related to the coordinates $(I_1, 0)$ at which $f(I) = 1$. At $f(I) = 1$, $M(V_2) = 1$, this means that voltage V_2 is zero. If $V_2 = 0$, the saturation current I_0 of the center junction goes to zero. Then from Eq. (23) we have $\alpha_1(I) + \alpha_2(I) = 1$. For known α_1 and α_2 the current at which the center junction reaches zero bias can be determined. The voltage across the entire

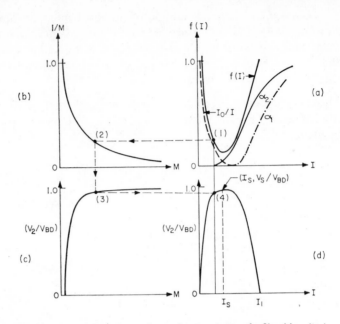

Fig. 8 Graphical solution of current-voltage characteristics of a Shockley diode. (After Gibbons, Ref. 12.)

device at this point will simply be the sum of forward voltage drops across the two outer junctions (about 1.2 to 1.4 volts for silicon devices.)

For current larger than I_1 the entire junction becomes forward-biased. (The above analysis cannot be extended to the forward case, since we have assumed that the junction J2 remains reverse-biased.) As the current increases beyond I_1, the voltage drop across the entire device continues to decrease and the device continues to exhibit negative dynamic resistance up to current I_h. Above this current the center junction voltage drop is comparable to the emitter junction voltage, and the dynamic resistance of the entire device again becomes positive.[13] Beyond the point (I_h, V_h) the device exhibits an *I-V* characteristic which is essentially identical to a single forward-biased *p-n* junction diode, $I \sim \exp(qV/kT)$.

B. Thyristor. For a three-terminal *p-n-p-n* device with one gate electrode, Eq. (22) can now be expressed as

$$\frac{1}{M(V_2)} = \alpha_1(I) + \alpha_2(I + I_g) + \frac{\alpha_2(I + I_g)}{I}I_g + \frac{I_0}{I}$$

$$= f(I, I_g). \tag{25}$$

In Eq. (25) the current I_K is replaced by $I + I_g$ and $I_{g2} = 0$. This equation would be identical to Eq. (23) if I_g were equal to zero. The $f(I, I_g)$ curve and I-V characteristics of the structure for $I_g = 0$ are shown in Fig. 9. The I-V characteristics for various values of I_g are obtained by replotting $\alpha_2(I + I_g)$ for each value of I_g and including the term $\alpha_2(I + I_g)/I$ in $f(I, I_g)$. This generates a set of $f(I, I_g)$ curves. We note that as I_g increases, the switching voltage decreases. This gives rise to the gate turn-on property of the structure.

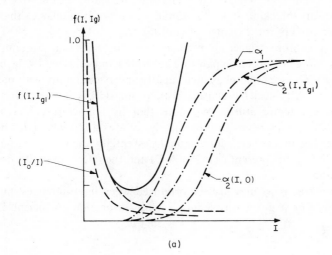

(a)

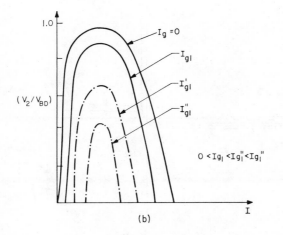

(b)

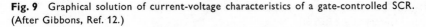

Fig. 9 Graphical solution of current-voltage characteristics of a gate-controlled SCR. (After Gibbons, Ref. 12.)

(4) Triggering Methods

To switch a *p-n-p-n* device from the off to the on state requires that the current be raised to a level high enough to satisfy the condition $\tilde{\alpha}_1 = \tilde{\alpha}_2 = 1$ (or $\alpha_1 + \alpha_2 \approx 1$). There are a number of methods whereby *p-n-p-n* devices can be triggered from the off to the on state: gate, temperature, radiation, and voltage triggering.

The gate-triggering method for an SCR has already been discussed in the last section. The complete *I-V* characteristics are shown in Fig. 10. Note that in the forward blocking state, the family of curves is similar to that shown in Fig. 9(b) except for a change of coordinates.

The temperature triggering method uses the fact that as temperature increases, more hole-electron pairs are generated and collected by the blocking junction J2. The sum of alphas then rapidly approaches unity with increasing temperature. A typical *I-V* characteristic of an SCR is shown in Fig. 11 for four different temperatures. We note that as temperature increases the blocking current increases approximately as $K \exp(aT)$ where K and a are constants. The breakover voltage remains essentially the same up to a certain temperature (in the present case $\sim 100°C$) and then decreases with increasing temperature.

Electron-hole pairs may also be produced when light impinges on a semiconductor. For a given wavelength λ with absorption coefficient $\alpha(\lambda)$, the

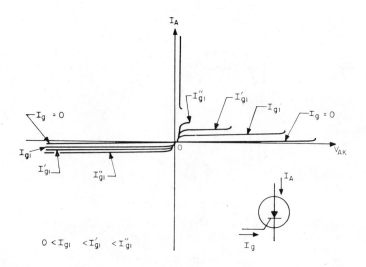

Fig. 10 Effect of gate current on current-voltage characteristics of an SCR. (After Gentry et al., Ref. 5.)

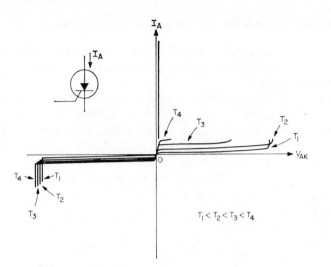

Fig. 11   Effect of temperature on the current-voltage characteristics of an SCR. (Ref. 5.)

generated current is proportional to $\Phi_0\{1 - \exp[-\alpha(\lambda)W_T]\}$ where Φ_0 is the light intensity and W_T is approximately the total device thickness. The basic switching mechanism and the *I-V* characteristic due to light triggering are essentially the same as temperature triggering. The *p-n-p-n* devices employing the light-triggering method are called light-activated SCR, or LASCR. Other forms of radiation, such as γ-rays, electrons, and x-rays, can be used to trigger a *p-n-p-n* device. However, great care must be exercised in their use, since the radiation may cause permanent damage to the semiconductor crystal lattice.

Voltage triggering is the most important method of switching a *p-n-p-n* device, especially for the Shockley diode. Voltage triggering can be accomplished in two ways: by slowly raising the forward voltage until the breakover voltage is reached, or by applying the anode voltage rapidly (referred to as dV/dt triggering). The breakover voltage, V_{BO}, can be obtained from Eqs. (23) and (24) under the condition that $I \gg I_0$, and we obtain

$$V_{BO} = V_{BD}(1 - \alpha_1 - \alpha_2)^{1/n} \tag{26}$$

where V_{BD} is the breakdown voltage to be discussed in the next section. In the dV/dt triggering, the rapidly varying anode voltage gives rise to a displacement current $d(CV)/dt$ where C is the capacitance of the collector junction J2. This current in turn can cause $(\tilde{\alpha}_1 + \tilde{\alpha}_2) \rightarrow 1$; then switching occurs.

(5) Forward and Reverse Blocking Characteristics

There are basically two factors which limit the reverse breakdown voltage and the forward breakover voltage: avalanche multiplication or depletion-layer punch-through.[14, 14a]

Consider the reverse blocking characteristics first. For the doping profile shown in Fig. 2(a), most of the applied reverse voltage will drop across the junction J1 as indicated in Fig. 12(a). Depending on the thickness of the $n1$ layer, W_n, the breakdown will be due to avalanche multiplication if the depletion-layer width at breakdown is less than W_n, or due to punch-through if the whole width W_n is filled out first by the depletion layer at which the junction J1 is effectively shorted to J2. Using the one-sided abrupt junction

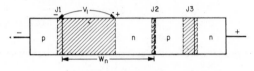

(a) REVERSE BLOCKING

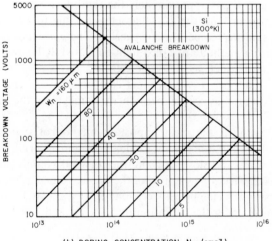

(b) DOPING CONCENTRATION N_D (cm⁻³)

Fig. 12 Reverse blocking capability of a *p-n-p-n* device. The avalanche breakdown line indicates the maximum voltage attainable in the $n1$ layer with doping concentration as a parameter. The parallel lines indicate the breakdown conditions due to punch-through of $n1$ layer for various layer widths.
(After Herlet, Ref. 14.)

approximation, the punch-through voltage is given by

$$V_{PT} \simeq \frac{qN_D W_n^2}{2\varepsilon_s} \qquad (27)$$

when N_D is the n-layer doping concentration and ε_s is the permittivity. The avalanche breakdown voltages have been considered previously in Chapter 3. Figure 12(b) shows the fundamental limit of the blocking capability of silicon p-n-p-n devices. For example, for $W_n = 80\mu$m, the maximum breakdown voltage is about 1000 volts which occurs at $N_D = 2 \times 10^{14}$ cm^{-3}; for lower dopings the breakdown voltage is limited by punch-through and for higher dopings by avalanche breakdown.[15]

For forward blocking, most of the applied voltage will drop across the center junction J2, as shown in Fig. 13(a). The breakover voltage can be obtained by combining Eq. (26) and Fig. 12(b). A typical result is shown in Fig. 13(b) for $W_n = 80\mu$m, $n \sim 3$, and various values of $(\alpha_1 + \alpha_2)$. For small values of $(\alpha_1 + \alpha_2)$, the breakover voltage is essentially the same as the reverse breakdown voltage. For values of $(\alpha_1 + \alpha_2)$ close to 1, the breakover voltages are substantially less than V_{BD}.

It has been shown in Chapter 3 that the breakdown voltage increases with increasing temperature. This fact coupled with Fig. 13(b) can help to explain the I-V characteristics shown in Fig. 11. The reverse characteristics are simply due to the temperature effect on breakdown voltage. For the forward characteristics, with temperatures slightly higher than room temperature, the temperature effect dominates, thus the breakover voltage increases slightly with temperature.[16] At higher temperatures, however, the variation of alphas with temperature dominates, and the breakover voltage decreases.

(6) Transient Operations

The anode current through a p-n-p-n device does not respond immediately to the application of gate current. The anode current can be characterized by a turn-on time as shown in Fig. 14(a). Because of the regenerative nature of a p-n-p-n device, the turn-on time is approximately the geometric mean value of the diffusion times in the $n1$ to $p2$ regions, or

$$t_{on} = \sqrt{t_1 t_2} \qquad (28)$$

where $t_1 \equiv W_n^2/2D_p$, $t_2 \equiv W_p^2/2D_n$, W_n and W_p are the layer widths of the $n1$ and $p2$ regions respectively, and D_p and D_n are the hole and electron diffusion coefficients respectively.

The above result can be derived from Fig. 4(b) with the help of the charge-control approach. We shall assume the stored charges in the p-n-p and n-p-n transistors are Q_1 and Q_2 respectively. The collector currents in the tran-

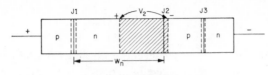

(a) FORWARD BLOCKING

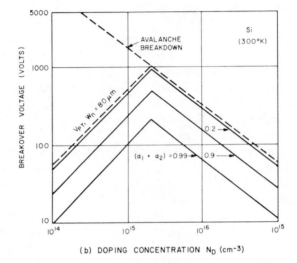

(b) DOPING CONCENTRATION N_D (cm^{-3})

Fig. 13 Forward breakover voltage as a function of the $n1$ layer doping concentration and width for various values of $(\alpha_1 + \alpha_2)$. For small $(\alpha_1 + \alpha_2)$ the breakover voltage is essentially the same as that for reverse breakdown. For large values of $(\alpha_1 + \alpha_2)$, however, the breakover voltages are substantially reduced.

sistors are then given by $I_{c2} \simeq Q_1/t_1$ and $I_{c1} \simeq Q_2/t_2$ respectively. Under the ideal condition that $dQ_1/dt = I_{c2}$ and $dQ_2/dt = I_g + I_{c1}$, we obtain the following equation:

$$\frac{d^2Q_1}{dt^2} - \frac{Q_1}{t_1 t_2} = \frac{I_g}{t_2}. \tag{28a}$$

It is then obvious that the solution of the above equation is of the form $\exp(-t/t_{on})$ with the time constant, t_{on}, given by Eq. (28). In order to reduce the turn-on time, one must employ devices with narrow $n1$ and $p2$ layer widths. This requirement, however, is in contrast to that for large break-

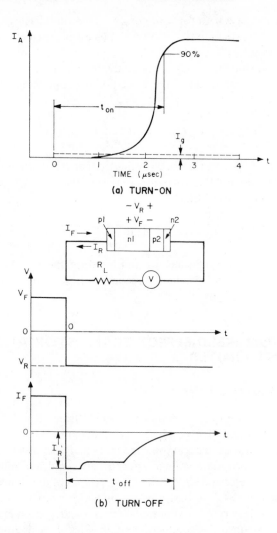

Fig 14
(a) Turn-on characteristic when a current step I_g is applied to an SCR.
(b) Turn-off characteristics where the voltage suddenly changes polarity.
(After Gentry et al., Ref. 5.)

down voltage. This is the reason that high-power, high-voltage thyristors have long turn-on times.

When a *p-n-p-n* device is in the on state, all three junctions are forward-biased. Consequently, in the device there are excess minority and majority carriers which increase with forward current. To switch back to the blocking state, these excess carriers must be swept out by an electric field or must decay by recombination.[17,18] A typical turn-off current waveform is shown in Fig. 14(b). The major time delay is due to the recombination time in layer $n1$. Since the hole current through the structure is proportional to the excess charge in $n1$, we can write

$$I = I_F \exp\left(-\frac{t}{\tau_p}\right) \tag{29}$$

where $I = I_F$ at $t = 0$, and τ_p is the minority-carrier lifetime. This current must drop below the holding current I_h to permit the device to block forward voltage. Thus, the turn-off time is

$$t_{off} = \tau_p \ln \frac{I_F}{I_h}. \tag{29a}$$

To obtain a small turn-off time, we must reduce the lifetime in layer $n1$. This can be achieved by introducing recombination centers such as gold in silicon during the diffusion process.

3 JUNCTION FIELD-EFFECT TRANSISTOR AND CURRENT LIMITER

(1) Basic Characteristics

A schematic diagram of a junction field-effect transistor (JFET) is shown in Fig. 15. It consists of a conductive channel provided with two ohmic contacts, one acting as the cathode (source) and the other as the anode (drain) with an appropriate voltage applied between drain and source. The third electrode (or electrodes), the gate, forms a rectifying junction with the channel. Thus the JFET is basically a voltage-controlled resistor and its resistance can be varied by varying the width of the depletion layer extending into the channel. Since the conduction process involves predominantly one kind of carrier, the JFET is also called a " unipolar " transistor in order to distinguish it from the bipolar (or junction) transistors in which both types of carriers are involved.

In Fig. 15 the basic dimensions of the device are the channel length L, channel width Z, half-channel depth at zero bias a, and depletion-layer width h. Three biasing conditions are shown in Fig. 16 for a p-channel JFET. The polarities are inverted for an n-channel JFET. The source electrode is generally grounded, and the gate and drain voltages are measured with respect to the source. When $V_G = V_D = 0$, the device is in equilibrium condition and

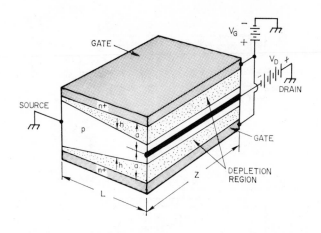

Fig. 15 Schematic diagram of a p-channel junction field-effect transistor (JFET) with channel length L, channel width Z, and channel depth $2a$. The source electrode is taken as the reference. Under normal operation the gate voltage has opposite polarity as compared to that of the drain.
(After Shockley, Ref. 2.)

no current flows. For a given V_G (zero or positive values), as the magnitude of the drain voltage increases (negative with respect to source), more current flows from source to drain. For sufficiently large V_D, the depletion layers near the drain end penetrate the entire p region. This is the pinch-off condition corresponding to $V_D = V_{D\text{sat}}$ at which the channel depth at L is reduced to zero and the conducting path between source and drain is pinched off. For even larger V_D, the point P will move towards the source end. The current will, however, remain essentially the same as the saturation current. This current flows because of carriers injected into the depletion region from the channel at the point P where the depletion regions touch. This is similar to a bipolar transistor in which carriers are injected from the emitter to a reverse-biased collector.

The basic current-voltage characteristics of a JFET are shown in Fig. 17 where the drain current is plotted against the drain voltage for various gate voltages. We can divide the characteristic into three regions: the linear region where the drain voltage is small and I_D is proportional to V_D, the saturation region where the current remains essentially constant and is independent of V_D, and the breakdown region where the drain current increases rapidly with a slight increase of V_D. As V_G increases, both the saturation current and

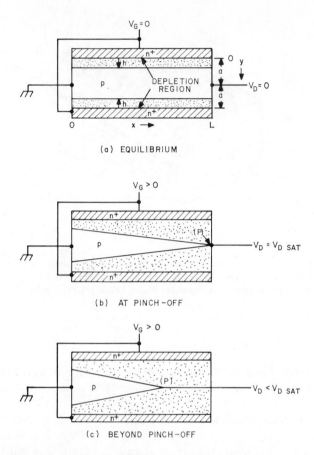

Fig. 16 Cross-sectional views of a JFET
(a) equilibrium condition,
(b) at pinch-off point where the depletion layers penetrate into the channel and meet at the drain end, and
(c) beyond pinch-off, the point (P) moves toward the source.

the saturation voltage decrease. This is because of reduced initial channel width which results in larger initial resistance.

We shall now consider the detailed current-voltage characteristics for a JFET with an arbitrary charge distribution. Because of the symmetry shown in Fig. 16, we need only consider the upper half of the device as shown in Fig. 18 where h is the depletion-layer width at an arbitrary point and y_1 and y_2 are the widths at the source and drain electrodes respectively.

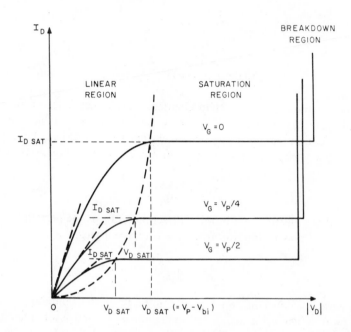

Fig. 17 Basic *I–V* characteristics of a JFET which includes the linear region, saturation region, and breakdown region. V_p is the pinch-off voltage. For a given V_G, the current and voltage at the point where saturation occurs are designated by $I_{D\,sat}$ and $V_{D\,sat}$ respectively.

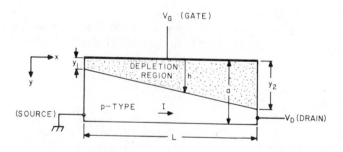

Fig. 18 Cross-sectional view of the upper-half portion of a JFET. y_1 and y_2 are the depletion layer widths at the source and drain ends. (After Bockemuehl, Ref. 7.)

We shall define an integral form of the charge density $Q(Y)$ as

$$Q(Y) \equiv \int_0^Y \rho(y)\, dy \qquad \text{Coul/cm}^2, \qquad (30a)$$

or

$$Q(h) \equiv \int_0^h \rho(y)\, dy \qquad \text{Coul/cm}^2 \qquad (30b)$$

where $\rho(y)$ is the charge density in Coul/cm^3. The dependence of the reverse bias voltage $V(h)$ on h and $\rho(y)$ can be derived from Poisson's equation. Under the conditions of gradual-channel approximation, i.e., the depletion layer width h varies only gradually along the channel (x-direction), we have

$$-\frac{\partial^2 V}{\partial y^2} = \frac{\partial \mathscr{E}_y}{\partial y} = \frac{\rho(y)}{\varepsilon_s} \qquad (31)$$

where $\mathscr{E}_y$ is the electric field in the y direction and ε_s is the permittivity. Integration of the above equation from $y = 0$ to $y = h$ yields

$$\mathscr{E}_y \equiv -\frac{\partial V}{\partial y} = \frac{1}{\varepsilon_s} \int_0^y \rho(y)\, dy + \text{const.} \qquad (32)$$

The integration constant can be determined from the boundary condition that $\mathscr{E}_y = 0$ at $y = h$ for an abrupt depletion layer, and is obtainable from Eq. (32) to be $-\dfrac{1}{\varepsilon_s} \displaystyle\int_0^h \rho(y)\, dy$. We thus have from Eqs. (30) and (32)

$$\frac{\partial V}{\partial y} = \frac{1}{\varepsilon_s} \left[\int_0^h \rho(y)\, dy - \int_0^y \rho(y)\, dy \right]$$

$$= \frac{1}{\varepsilon_s} [Q(h) - Q(y)]. \qquad (33)$$

Integrating once more from $y = 0$ to $y = h$, we obtain the voltage $V(h)$ across the depletion layer which includes both the applied and the built-in voltages:

$$V(h) = \frac{1}{\varepsilon_s} \left[Q(h) \int_0^h dy - \int_0^h Q(y)\, dy \right]$$

$$= \frac{1}{\varepsilon_s} \left[hQ(h) - \int_0^h Q(y)\, dy \right], \qquad (34)$$

or

$$V(h) = \frac{1}{\varepsilon_s} \int_0^h y\rho(y)\, dy. \qquad (35)$$

The right-hand expression in Eq. (35) can be readily shown to be identical to that of Eq. (34) using integration by parts.

The maximum value for the upper limit of the integration occurs at the point where $h = a$ and the corresponding voltage is called the pinch-off voltage as defined previously. Beyond this point the channel current remains essentially at a constant value. The pinch-off voltage is given by

$$V_P(\text{pinch-off voltage}) = V(h = a) = \frac{1}{\varepsilon_s} \int_0^a y\rho(y) \, dy. \tag{36}$$

Differentiation of Eq. (35) yields

$$dV/dh = h\rho(h)/\varepsilon_s \tag{37}$$

which shows that the voltage change required to move the depletion boundary a given distance, increases with the value h and is proportional to the space-charge density at that boundary. The junction capacitance per unit area is given by

$$C \equiv dQ(h)/dV = \left(\frac{dQ}{dh}\right)\left(\frac{dh}{dV}\right) \qquad f/\text{cm}^2. \tag{38}$$

The depletion layer thus acts as a plane capacitor with plate distance h, and the capacitance is independent of the charge distribution profile.

We next consider the current-voltage characteristics and the transconductance. The current density in the x-direction (transport along the channel) is given by the simple ohmic-law equation

$$J_x = \sigma(x)\mathscr{E}_x \tag{39}$$

where J_x is the current density, $\sigma(x)$ the conductivity, and $\mathscr{E}_x$ the electrical field along x direction $(-\partial V/\partial x)$. The channel (or drain) current is then given by

$$I_D = 2Z\mu \frac{dV}{dx} \int_h^a \rho(y) \, dy, \tag{40}$$

or

$$I_D \, dx = 2Z\mu(dV/dh) \, dh \int_h^a \rho(y) \, dy \tag{40a}$$

where μ is the drift mobility in the channel and the factor 2 is from the contribution of the lower half of the device. Substituting Eq. (37) and integrating with the boundary conditions $h = y_1$ at $x = 0$ and $h = y_2$ at $x = L$, we obtain

$$\int_0^L I_D \, dx = I_D \cdot L = \frac{2Z\mu}{\varepsilon_s} \int_0^L h\rho(h) \, dh \int_h^a \rho(y) \, dy \tag{41}$$

or

$$I_D = \frac{2Z\mu}{\varepsilon_s L} \int_{y_1}^{y_2} [Q(a) - Q(h)] h\rho(h) \, dh. \tag{42}$$

Equation (42) is the fundamental equation of the JFET.

From Eq. (42) we can derive two important device parameters, namely the transconductance, g_m, and the channel conductance (also called the drain conductance), g_D:

$$g_m \equiv \frac{\partial I_D}{\partial V_G} = \frac{\partial I_D}{\partial y_1} \frac{\partial y_1}{\partial V_G} + \frac{\partial I_D}{\partial y_2} \frac{\partial y_2}{\partial V_G}. \tag{43}$$

The partial derivatives are obtained from Eqs. (37) and (42):

$$g_m = \frac{2Z\mu}{L} [Q(y_2) - Q(y_1)] \tag{44}$$

which shows that g_m is equal to the conductance of the rectangular section of the semiconductor extending from $y = y_1$ to $y = y_2$. The channel conductance can be obtained from Eqs. (37) and (42) in a similar manner:

$$g_D = \frac{\partial I_D}{\partial V_D} = \frac{2Z\mu}{L} [Q(a) - Q(y_2)]. \tag{45}$$

This value approaches zero as $y_2 \approx a$ at which $V_D + V_G = V_P - V_{bi}$, where V_{bi} is the built-in potential. It is interesting to compare Eq. (44) with Eq. (45). For $V_D \to 0$, we have $y_2 \to y_1$ and g_D is proportional to $[Q(a) - Q(y_1)]$. On the other hand, when $V_D + V_G \geq V_P$, $y_2 \to a$ then g_m is also proportional to the same charge difference. Hence we obtain the following useful expression for any arbitrary charge distribution:

$$g_{DO}(V_D \to 0) = g_{ms}(|V_D| \gg |V_P|) = \frac{2Z\mu}{L} [Q(a) - Q(y_1)]$$

$$\equiv g_{max} \left[1 - \frac{Q(y_1)}{Q(a)} \right] \tag{46}$$

where

$$g_{max} \equiv \frac{2Z\mu}{L} Q(a). \tag{47}$$

(2) Specific Charge Distributions

We shall use the above equations to derive the current-voltage characteristics for some specific functions of $\rho(y)$, in particular for the uniformly doped charge distribution. For a uniformly doped semiconductor with doping N_D the function $Q(h)$ is given by $qN_A h$. Equation (42) can then be written as

$$
I_D = \frac{2Z\mu}{\varepsilon_s L} \int_{y_1}^{y_2} qN_A(a - h)hqN_A \, dh
$$

$$
= \frac{2Z\mu q^2 N_A^2}{\varepsilon_s L} \left[\frac{a}{2}\left(y_2^{\,2} - y_1^{\,2}\right) - \frac{1}{3}\left(y_2^{\,3} - y_1^{\,3}\right) \right]. \tag{48}
$$

The depletion layer widths y_2 and y_1 for an abrupt junction are given by

$$
y_2 = [2\varepsilon_s(V_D + V_G + V_{bi})/qN_A]^{1/2} \tag{49a}
$$

$$
y_1 = [2\varepsilon_s(V_{bi} + V_G)/qN_A]^{1/2} \tag{49b}
$$

where V_{bi} is the built-in potential given by $kT/q \ln(N_A/n_i)$ for an abrupt n^+p junction. Note that under normal operation of a p-channel device, V_G is positive, and V_D is negative. We shall use the absolute value of V_D in Eq. (49a) and also in the subsequent equations. Substitution of Eqs. (49a) and (49b) into Eq. (48) yields

$$
I_D = g_{max}\left\{ V_D - \frac{2}{3a}\sqrt{\frac{2\varepsilon_s}{qN_A}} \left[(V_D + V_G + V_{bi})^{3/2} - (V_{bi} + V_G)^{3/2} \right] \right\} \tag{50}
$$

where $g_{max} \equiv 2Z\mu qN_A a/L$ as given from Eq. (47). The maximum current, $I_{D\,sat}$, for a given V_G occurs at the point where the channel is pinched off. This current is obtainable from Eq. (48) by setting $y_2 = a$:

$$
I_{Dsat} = \frac{2Z\mu q^2 N_A^2}{\varepsilon_s L} \left[\frac{a}{2}\left(a^2 - y_1^{\,2}\right) - \frac{1}{3}\left(a^3 - y_1^{\,3}\right) \right]
$$

$$
= \frac{2Z\mu q^2 N_A^2}{\varepsilon_s L} \left[\frac{a^3}{6} - \frac{ay_1^{\,2}}{2} + \frac{1}{3}y_1^{\,3} \right]
$$

$$
= \frac{2Z\mu q^2 N_A^2 a^3}{6\varepsilon_s L} \left[1 - \frac{3y_1^{\,2}}{a^2} + 2\frac{y_1^{\,3}}{a^3} \right]
$$

$$
= I_P\left[1 - 3\left(\frac{V_{bi} + V_G}{V_P}\right) + 2\left(\frac{V_{bi} + V_G}{V_P}\right)^{3/2} \right] \tag{51}
$$

where I_P and V_P are the pinch-off current and the pinch-off voltage [which includes the built-in potential, Eq. (36)] which are given by

$$
I_P \equiv \frac{2Z\mu q^2 N_A^2 a^3}{6\varepsilon_s L}, \tag{52a}
$$

$$
V_P = \frac{qN_A a^2}{2\varepsilon_s}. \tag{52b}
$$

It is clear from Eq. (52a) that in order to increase the current I_P, one can employ device geometry with a large ratio of channel width to channel length (Z/L), high mobility, and low doping concentration. The last requirement comes about because of the fact that the depletion-layer width at breakdown is approximately proportional to $(N_A)^{-1}$; this makes the terms $N_A{}^2a^3$ in I_P proportional to $(N_A)^{-1}$. From Eq. (52b), the pinch-off voltage V_P will also vary as $(N_A)^{-1}$.

The transconductance and drain conductance can be directly obtained from Eqs. (44) and (45):

$$g_m = \frac{2Z\mu q N_A}{L}(y_2 - y_1) = \frac{2Z\mu}{L}\sqrt{2\varepsilon_s q N_A}\left(\sqrt{V_D + V_G + V_{bi}} - \sqrt{V_{bi} + V_G}\right).$$

(53a)

$$g_D = \frac{2Z\mu q N_A}{L}(a - y_2) = \frac{2Z\mu}{L}\sqrt{2\varepsilon_s q N_A}\left(\sqrt{V_P} - \sqrt{V_D + V_G + V_{bi}}\right).$$

(53b)

The current-voltage characteristics calculated from Eq. (50) are shown in Fig. 19 where the saturation current $I_{D\,sat}$ is given by Eq. (51) and the saturation voltage is given by

$$V_{D\,sat} = V_P - V_{bi} - V_G = \frac{2N_A a^2}{2\varepsilon_s} - \frac{kT}{q}\ln(N_A/n_i) - V_G.$$

(54)

In the linear region ($V_D \to 0$) the drain conductance is given from Eq. (46) or (53) as

$$g_{D0}(V_D \to 0) = g_{max}\left[1 - \sqrt{\frac{2\varepsilon_s(V_{bi} + V_G)}{qN_A a^2}}\right].$$

And this value is also equal to the transconductance in the saturation region as pointed out in Eq. (46).

Other charge distribution profiles can be calculated in a similar manner. Table 7.1 summarizes the results for three cases.[7] Case B represents the uniform charge distribution discussed previously. Cases A and C represent two limiting cases in which the charge distributions are delta functions located at $y = 2a$ and $y = 0$ respectively. We note that the dimensionless ratio $g_{max} V_P/I_P$ depends only on the charge distribution parameters. However, this ratio is limited to the range of 2 to 4. It can also be shown that for case A, the current-voltage relationship is given by

$$I_{D\,sat} = I_P\left[1 - \left(\frac{V_G + V_{bi}}{V_P}\right)\right]^2.$$

(55)

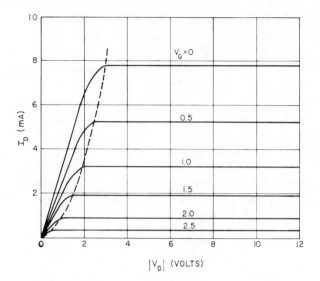

Fig 19 Theoretical I–V output characteristics of a silicon p-channel JFET with $Z/L = 200$, $\mu = 750$ cm^2/V-sec, $a = 2\mu$m, and $N_A = 10^{15}$ cm^{-3}.

TABLE 7.1

FIELD-EFFECT EQUATIONS FOR SPECIFIC CHARGE DISTRIBUTIONS
IN A RECTANGULAR STRUCTURE FOR REFLECTED-TYPE
JFET (TOTAL CHANNEL DEPTH 2a)

Parameter	Common Factor	Multiplying Factors for Specific Distributions		
		A. All Charge at $y = 2a$	B. Uniform	C. All Charge at $y = 0$
$g_{\max}$	$\dfrac{2Z\mu\rho a}{L}$	1	1	1
V_P	$\dfrac{4\rho a^2}{\varepsilon_s}$	$\dfrac{1}{4}$	$\dfrac{1}{8}$	0
I_P	$\dfrac{8Z\mu\rho^2 a^3}{\varepsilon_s L}$	$\dfrac{1}{8}$	$\dfrac{1}{24}$	0
$\dfrac{g_{\max} V_P}{I_P}$	1	2	3	4

A comparison of Eqs. (55) and (51) is shown in Fig. 20. We note that there is a surprisingly narrow range of possible transfer characteristics as indicated by the shaded region. It can be shown that the shaded region can be represented by

$$\left[1 - \left(\frac{V_G + V_{bi}}{V_P}\right)\right]^n$$

where n varies[19] between 2 and 2.25. Because of this narrow range of n (~ 2) the square-law approximation to the JFET's transfer characteristic can be very useful for many circuit applications.

It is also interesting to note that the current-voltage characteristics for one value of V_G are sufficient to completely describe the device behavior for any channel impurity profile. This can be shown easily from Eq. (42) which is now rewritten as[20]

$$I_D(V_G, V_D) = \int_{y_1}^{y_2} F(h)\, dh = \int_{y(V_G)}^{y(V_D + V_G)} F(h)\, dh. \tag{56}$$

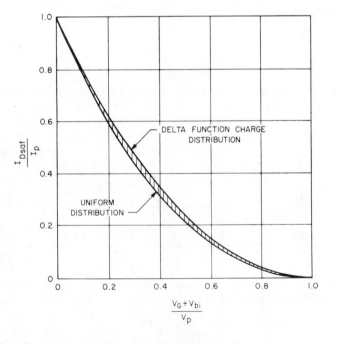

Fig. 20 Transfer characteristics for two specific charge distributions. (After Middlebrook and Richer, Ref. 19.)

The integral can be split into two terms to yield

$$I_D(V_G, V_D) = \int_{y(0)}^{y(V_D+V_G)} F(h)\,dh + \int_{y(V_G)}^{y(0)} F(h)\,dh \tag{57}$$

or

$$I_D(V_G, V_D) = I_D(0, V_D + V_G) - I_D(0, V_G). \tag{57a}$$

Equation (57a) states that if the relationship between I_D and V_D is known for the case of $V_G = 0$, the right-hand side of Eq. (57a) permits the calculation of I_D for any combination of V_D and V_G. One can thus use a graphical method to generate the complete I-V characteristics. An example is shown in Fig. 21. For a given $I_D(0, V_D)$, curve (1), the first term on the right-hand side of Eq. (57) is the curve $I_D(0, V_D)$ displaced by an amount V_G along the V_D axis [Fig. 21(a) is for an n channel so that V_G has negative values]. The second term is a constant in the I_D-V_D plane and represents a displacement along the I_D axis such that the resulting characteristic curve always passes through the origin of the coordinates. The graphical steps are illustrated in Fig. 21(a) for an n-channel JFET. A comparison between theoretical and experimental results is shown in Fig. 21(b). The agreement is better than 5 % at all data points.

From the above discussion we conclude that the basic characteristic of a JFET is essentially independent of the charge distributions in the channel. The JFET has approximately a square-law transfer characteristic. In addition, the family of I-V curves of a JFET can be generated by knowing the I_D-V_D characteristic for only one value of V_G. We shall now consider some modification of the basic static characteristics and the dynamic behavior at high frequencies.

(3) Static Characteristics

A. Channel Conductance. A planar junction field-effect transistor, shown in Fig. 22, is generally made by the use of two successive diffusion processes. A typical output characteristic (I_D vs V_D) is shown in Fig. 23. The constant-current approximation in the pinch-off region is reasonably valid. The slight upward tilt of the current (corresponding to a nonzero channel conductance) is mainly due to the reduction of the effective channel length as indicated in Fig. 16. The potential at the end of the channel (point P) is fixed at the value $V_{D\,\text{sat}}$ as given by Eq. (54). As the drain voltage increases, the depletion-layer width also increases. This causes the point P to move towards the source.[21] The voltage at P remains at the same value ($V_{D\,\text{sat}}$), but the effective distance from the source to the drain is reduced. This results in the finite value of the channel conductance in the saturation region.

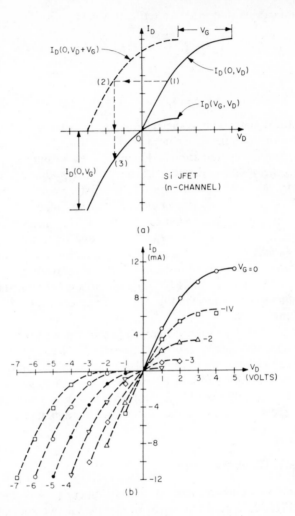

Fig. 21
(a) Graphical solution of *I–V* characteristics of a JFET.
(b) Comparison between theory and experiment.
(After Wedlock, Ref. 20.)

B. Breakdown Voltage. As the drain voltage increases, eventually avalanche breakdown of the gate-to-channel diode occurs, and the drain current will suddenly increase. The breakdown occurs at the drain end of the channel where the reverse voltage is highest:

$$V_B(\text{breakdown voltage}) = |V_D| + V_G. \tag{58}$$

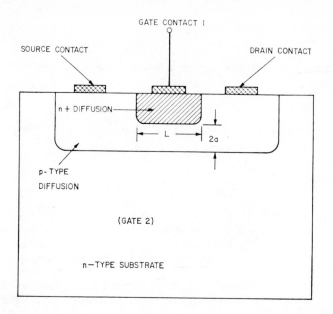

Fig. 22 Typical cross-sectional view of a double diffused JFET. *P*-type impurities (e.g., boron) are first diffused into the *n*-type substrate. Then *n*-type impurities (e.g., phosphorous) are diffused into the *p*-type layer. The channel length and depth are indicated by *L* and $2a$ respectively.

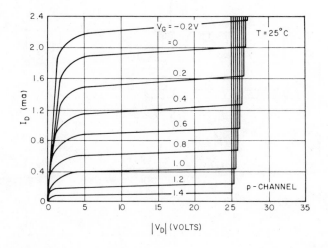

Fig. 23 *I–V* characteristics of a *p*-channel JFET. (After Sevin, Ref. 8.)

For example, in Fig. 23 the breakdown voltage is 27 V for $V_G = 0$. At $V_G = 1$ V, the breakdown voltage is still 27 V and the drain voltage at breakdown is $(V_B - V_G)$ or 26 volts.

C. Input Resistance. The current in the reverse-biased gate-to-channel junction can be expressed as

$$I_G = I_S \left[\exp(qV_G/\eta kT) - 1 \right] \tag{59}$$

where η is a constant which equals 1 for ideal current and 2 for generation-recombination current. The input resistance is given by

$$R_{in} \equiv \frac{1}{g_{in}} = \left(\frac{\partial I_G}{\partial V_G} \right)^{-1} = \frac{\eta kT}{q(I_G - I_S)}. \tag{60}$$

At $I_G = 0$, the input resistance at room temperature is about 250 $M\Omega$ for $I_S = 10^{-10}$ amp, and will increase with the reverse bias. It is obvious that the JFET has a very high input resistance.

D. Series Resistances. Because of the finite distances between the gate and the source and drain contacts as shown in Fig. 24, there are series resistances R_S (source resistance) and R_D (drain resistance) outside the channel region. These resistances, which cannot be modulated by the gate voltage,

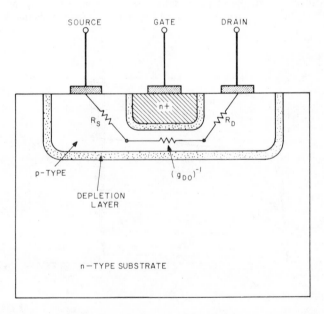

Fig. 24 Schematic diagram to show the series resistances at the source end (R_S) and drain end (R_D). The gate and drain voltages can only modulate the channel conductance g_{DO}.

introduce an IR drop between the gate and the source and drain contacts and reduce the drain conductance as well as the transconductance. The voltages V_D and V_G in Eq. (50) should then be replaced by $V_D - I_D(R_S + R_D)$ and $(V_G - I_D R_S)$ respectively. In the linear region the resistances R_S, $1/g_{DO}$, and R_D are in series, and the measured drain conductance is given by $g_{DO}/[1 + (R_S + R_D)g_{DO}]$. A similar result can be obtained for the measured transconductance. In the saturation region, however, the transconductance is affected only by the source resistance. The drain resistance R_D will cause an increase of the drain voltage at which current saturation occurs. Beyond that voltage, $V_D > V_{D\,sat}$, the magnitude of V_D has no effect on the drain current, thus R_D has no further effect on g_m, so the transconductance measured is equal to $g_m/(1 + R_S g_m)$.

E. Effects of Temperature and Mobility. If the density of the ionized donors remains essentially the same over a certain temperature range, then the voltage V_P, Eq. (52b), will be constant. However, the current I_P, Eq. (52a), will vary with temperature because of mobility variation. A typical result is shown in Fig. 25 where I_P decreases with increasing temperature, and varies as T^{-2}. The effect of temperature on I_D can be obtained from Eq. (50) and is mainly due to two quantities, μ and V_{bi}. Both quantities cause I_D to decrease with increasing temperature.

In the previous derivations it is assumed that the channel mobility is a constant. However, as the electric field increases, the mobility tends to decrease. Equation (42) should be slightly modified by placing μ within the integral sign. This results in a decrease of the drain current.[6, 22, 23] The general behavior of the I_D versus V_D, however, remains essentially the same.

(4) Dynamic Characteristics

A. Cutoff Frequency. Under high-frequency operation there are two factors which limit the frequency response of a JFET: the transit time and the RC time constant. The transit time effect is due to the finite time required for carriers to travel from source to drain. For the constant mobility case the transit time is given by

$$\tau \equiv \frac{L}{\mu \mathscr{E}_x} \approx \frac{L^2}{\mu V_D}. \tag{61}$$

And for the constant velocity case (under a large electric field)

$$\tau \simeq L/v_{sl} \tag{62}$$

where v_{sl} is the scattering-limited velocity. Usually the above transit time is

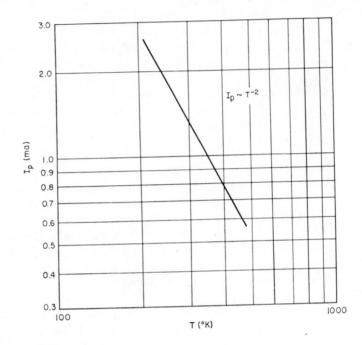

$$I_p \sim T^{-2}$$

Fig. 25 Temperature dependence of the maximum drain current I_p.
(After Sevin, Ref. 8.)

small compared to the RC time constant resulting from the input capacitance C_{in} and the transconductance.

A schematic common-source configuration of a p-channel JFET is shown in Fig. 26(a). For an n-channel JFET the arrow is inverted. The corresponding equivalent circuit is shown in Fig. 26(b) where C_{in} and g_{in} are the input capacitance and input conductance respectively. C_{out} and g_{out} are the output capacitance and conductance respectively, g_{GD} is a feedback capacitance, and $g_m V_G$ is the current generator. Under normal operating conditions the two most important terms are C_{in} and $g_m V_G$.

The maximum operating frequency can then be defined as the frequency at which the current through C_{in} is equal to the current of the current generator $g_m V_G$:

$$f_m = \frac{g_m}{2\pi C_{in}} \le \frac{\left(\dfrac{2Z\mu q N_A a}{L}\right)}{2\pi\left(\dfrac{\varepsilon_s}{a} ZL\right)} = \frac{q N_A \mu a^2}{\pi L^2 \varepsilon_s}. \tag{63}$$

For high operating frequencies one should reduce the channel length L. Equation (63) also indicates that an n-channel Si JFET has a higher maximum operating frequency than a p-channel Si JFET with the same device geometry. This is because the electron drift mobility is higher than that of holes in silicon.

B. Noise.[23-27] There are three noise sources in JFET: shot noise from the gate leakage current, thermal noise generated in the conductive channel, and the generation-recombination type noise due to surface effect.

The shot-noise current is given by

$$I_{sh}^2 = 2q I_G B \tag{64}$$

where B is the bandwidth and I_G is the gate current, Eq. (59). Since the gate current under reverse bias is very small, of the order of 10^{-10} amp, this noise source makes only a small contribution to the total noise. The approximated thermal noise is given as an equivalent noise voltage:

$$\bar{V}_{th}^2 = 4kT(g_m)^{-1}B \tag{65}$$

where g_m is the transconductance. The generation-recombination noise is an additional noise source which is given by the noise voltage modified by a $1/f$ frequency dependence, $\bar{V}_{th}^2 f_c/f$, where f_c is the corner frequency at the low end of the frequency spectrum. It is clear that for high-frequency operation, the dominant noise source is the thermal noise which can be substantially reduced by increasing the transconductance.

(5) Current Limiter

In this section we shall consider two classes of current-regulator diodes which are two-terminal field-effect devices. They are the field-effect diode and the limiting-velocity diode.

The operation principle of a field-effect diode (FED)[28, 29] is the same as that of a shorted-gate-to-source junction field-effect transistor. Its I-V characteristic is shown in Fig. 27 which is similar to the curve shown in Fig. 23 for $V_G = 0$. We shall consider four important parameters pertinent to the operation of current limiters: the limiting current I_l, the saturation voltage V_{sat}, the slope g_l in the limiting range, and the breakdown voltage. For an FED

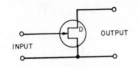

(a) COMMON−SOURCE CONFIGURATION
(p−CHANNEL)

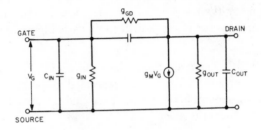

(b) EQUIVALENT CIRCUIT OF JFET

Fig. 26
(a) Common-source configuration.
(b) Equivalent circuit of a JFET.

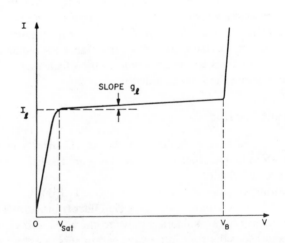

Fig. 27 Basic characteristics of a current limiter where I_l is the limiting current, V_{sat} the saturation voltage, g_l the conductance in the limiting region, and V_B the breakdown voltage.

all the above parameters have already been discussed in connection with the junction field-effect transistor. To reduce V_{sat}, one can use devices with a small channel depth a, and a low channel doping concentration Eq. (52b). To reduce I_l, one can reduce N_A and a, or increase the channel length L, Eq. (52a). To decrease g_l, one must increase L. To increase V_B, a low channel doping concentration should be used.

For the limiting-velocity diode[30] the current-limiting characteristic is obtained by employing the effect that at high electric fields the carrier drift velocity saturates. A schematic geometry is shown in Fig. 28 where a p-type Ge substrate is used. The high-field region is a shallow n-type diffused layer about 0.5 μm deep and 3 μm long. The reason for choosing Ge is that its velocity saturation characteristics occur at low field (~ 4 kV/cm) in contrast to that in Si (~ 30 kV/cm).

The four important parameters as shown in Fig. 27 will now be discussed. The limiting current is given by

$$I_l = qN_D v_{sl} A + I_S \tag{66}$$

where v_{sl} is the scattering-limited velocity, A the area, and I_S the p-n junction reverse-biased saturation current. The current I_S increases with increasing temperature while the velocity v_{sl} decreases with temperature. A minimum in I_l is thus expected when these two competing mechanisms cancel.

The saturation voltage V_{sat} is given by

$$V_{sat} = \mathscr{E}_s L + I_l R_c \tag{67}$$

where $\mathscr{E}_s$ is the electric field at the onset of velocity saturation, and R_c is the residual resistance associated with the contacts. In an ideal limiter V_{sat} is zero; in a practical limiter V_{sat} should be as small as possible.

The slope g_l results from two effects: first, the electron drift velocity in the saturation range is not completely field independent, and secondly there is a space-charge-limited resistance due to injected carriers. This resistance is given by $(L^2/2\varepsilon_s v_{sl} A)$ as considered in Chapter 5.

The breakdown voltage depends on the impact ionization in the conduction channel. The electric field will approximately increase linearly from the negative contact to a peak value $\mathscr{E}_p$ at the positive contact, and the breakdown voltage is given by

$$V_B \approx \frac{\mathscr{E}_p}{2} L \tag{68}$$

where $(\mathscr{E}_p/2)$ is the average field in the channel. For Ge., $\mathscr{E}_p$ is about 1.5×10^5 V/cm at breakdown. The breakdown voltage is then expected to be about 20 volts for a channel length of 3 μm.

The experimental results are shown in Fig. 29 for a Ge current limiter

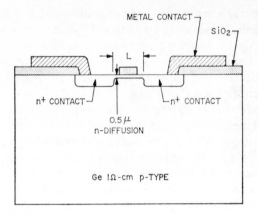

Fig. 28 Schematic geometry of a current limiter (limiting-velocity diode). (After Boll et al., Ref. 30.)

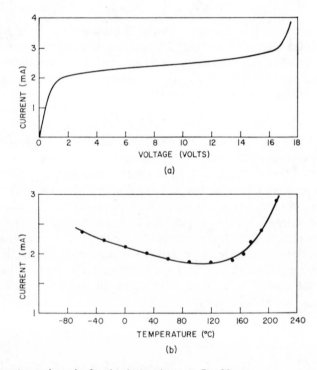

(a)

(b)

Fig. 29 Experimental results for the device shown in Fig. 28
(a) *I–V* characteristics and
(b) temperature dependence of the limiting current I_l (at 2 volts).
(Ref. 30.)

with geometries shown in Fig. 28. Figure 29(a) shows the current-voltage characteristic at room temperature. We note that the current is reasonably saturated at about 2 volts, and the breakdown occurs near 18 volts in agreement with the expected values. Figure 29(b) shows the temperature dependence of the limiting current I_l (at 2 V) with minimum current occurring at about 120°C. The limiting-velocity diode can be operated at higher speed than the field-effect diode because, for the same current level I_l, the depletion-layer with the former can be made much greater than that of the corresponding FED so that the shunt-capacitance of a limiting-velocity diode is smaller than the input capacitance of the field-effect diode.

REFERENCES

1. W. Shockley, *Electrons and Holes in Semiconductors*, D. Van Nostrand Co. Inc., p. 112 (1950).

2. W. Shockley, "A Unipolar Field-Effect Transistor," Proc. IRE, *40*, 1365 (1952).

3. J. J. Ebers, "Four-Terminal *p-n-p-n* Transistors," Proc. IRE, *40*, 1361 (1952).

4. J. L. Moll, M. Tanenbaum, J. M. Goldey, and N. Holonyak, "*p-n-p-n* Transistor Switches," Proc. IRE, *44*, 1174 (1956).

5. F. E. Gentry, F. W. Gutzwieler, N. H. Holonyak, and E. E. Von Zastrow, *Semiconductor Controlled Rectifiers*, Prentice-Hall, Inc., Englewood Cliffs, N. J. (1964).

6. G. C. Dacey and I. M. Ross, "The Field-Effect Transistor," Bell Syst. Tech. J., *34*, 1149 (1955), and "Unipolar Field-Effect Transistor," Proc. IRE, *41*, 970 (1953).

7. R. R. Bockemuehl, "Analysis of Field-Effect Transistors With Arbitrary Charge Distribution," IEEE Trans. Electron Devices, *ED-10*, 31 (1963).

8. L. J. Sevin, *Field Effect Transistors*, McGraw-Hill Book Co., New York (1965).

9. F. E. Gentry, "Turn-on Criterion for *p-n-p-n* Devices," IEEE Trans. Electron Devices, *ED-11*, 74 (1964).

10. E. S. Yang and N. C. Voulgaris, "On the Variation of Small-Signal Alphas of a *p-n-p-n* Device With Current," Solid State Electron., *10*, 641 (1967).

11. Y. Kawana and T. Misawa, "A Silicon *p-n-p-n* Power Triode," J. Electronics and Control, *6*, 324 (1959).

12. J. F. Gibbons, "Graphical Analysis of the *I-V* Characteristics of Generalized *p-n-p-n* Devices," Proc. IEEE, *55*, 1366 (1967).

13. J. F. Gibbons, "A Critique of the Theory of *p-n-p-n* Devices," IEEE Trans. Electron Devices, *ED-11*, 406 (1964).

14. A. Herlet, "The Maximum Blocking Capability of Silicon Thyristors," Solid State Electron, *8*, 655 (1965).

14a. W. Fulop, "Three Therminal Measurement of Current Amplification Factors of Controlled Rectifiers," IEEE Trans. Electron Devices, *ED-10*, 120 (1963).

15. S. M. Sze and G. Gibbons, "Avalanche Breakdown Voltages of Abrupt and Linearly Graded *p-n* Junctions in Ge, Si, GaAs, and GaP," Appl. Phys. Letters, *8*, 111 (1966).

16. C. R. Crowell and S. M. Sze, "Temperature Dependence of Avalanche Multiplication in Semiconductors," Appl. Phys. Letters, *9*, 242 (1966).

17. E. S. Yang, "Turn-off Characteristics of *p-n-p-n* Devices," Solid State Electron., *10*, 927 (1967).

18. T. S. Sundresh, "Reverse Transient in *p-n-p-n* Triodes," IEEE Trans. Electron Devices, *ED-14*, 400 (1967).

19. R. D. Middlebrook and I. Richer, "Limits on the Power-Law Exponent for Field-Effect Transistor Transfer Characteristics," Solid State Electron., *6*, 542 (1963).

20. B. D. Wedlock, "On the Field-Effect Transistor Characteristics," IEEE Trans. Electron Devices, *ED-15*, 181 (1968).

21. J. R. Hauser, "Characteristics of Junction Field-Effect Devices with Small Channel Length-to-Width Ratios," Solid State Electron., *10*, 577 (1967).

22. J. M. LeMée, "Influence of Carrier Mobility and Design Parameters on Field Effect Transistor Characteristics," IEEE Trans. Electron Devices, *ED-15*, 110 (1968).

23. H. E. Halladay and A. Van Der Ziel, "DC Characteristics of Junction Field-Effect Transistors," IEEE Trans. Electron Devices, *ED-13*, 531 (1966).

24. C. T. Sah, "Theory of Low-Frequency Generation Noise in Junction-Gate Field-Effect Transistors," Proc. IEEE, *52*, 795 (1964).

25. F. M. Klaassen, "High-Frequency Noise of the Junction Field-Effect Transistor," IEEE Trans. Electron Devices, *ED-14*, 368 (1967).

26. W. C. Bruncke and A. Van Der Ziel, "Thermal Noise in Junction-Gate Field-Effect Transistors," IEEE Trans. Electron Devices, *ED-13*, 323 (1966).

27. H. E. Halladay, and A. Van Der Ziel, "Field-Dependent Mobility Effects in the Excess Noise of Junction-Gate Field-Effect Transistors," IEEE Trans. Electron Devices, *ED-14*,10 (1967).

28. R. M. Warner, W. H. Jackson, E. I. Doucette, and H. A. Stone, "A Semiconductor Current Limiter," Proc. IRE, *47*, 45 (1959).

29. H. Lawrence, "A Diffused Field-Effect Current Limiter," IRE Trans. Electron Devices, *ED-9*, 82 (1962).

30. H. J. Boll, J. E. Iwersen, and E. W. Perry, "High-Speed Current Limiters," IEEE Trans. Electron Devices, *ED-13*, 904 (1966).

PART **III**

INTERFACE AND
THIN-FILM DEVICES

- Metal-Semiconductor Devices
- Metal-Insulator-Semiconductor (MIS) Diodes
- IGFET and Related Surface Field Effects
- Thin-Film Devices

■ INTRODUCTION

■ SCHOTTKY EFFECT

■ ENERGY BAND RELATION
AT METAL-SEMICONDUCTOR
CONTACT

■ CURRENT TRANSPORT THEORY
IN SCHOTTKY BARRIERS

■ MEASUREMENT OF
SCHOTTKY BARRIER HEIGHT

■ CLAMPED TRANSISTOR,
SCHOTTKY GATE FET,
AND METAL-SEMICONDUCTOR
IMPATT DIODE

■ MOTT BARRIER,
POINT-CONTACT RECTIFIER,
AND OHMIC CONTACT

■ SPACE-CHARGE-LIMITED DIODE

8

Metal-Semiconductor Devices

1 INTRODUCTION

The earliest systematic investigation on metal-semiconductor rectifying systems is generally attributed to Braun[1] who in 1874 noted the dependence of the total resistance on the polarity of the applied voltage and on the detailed surface conditions. The point-contact rectifier in various forms found practical applications[2] beginning in 1904. In 1931 Wilson[3] formulated the transport theory of semiconductors based on the band theory of solids. This theory was then applied to the metal-semiconductor contacts. In 1938 Schottky[4] suggested that the potential barrier could arise from stable space charges in the semiconductor alone without the presence of a chemical layer. The model arising from this consideration is known as the Schottky barrier. In 1938 Mott[5] also devised an appropriate theoretical model for swept-out metal-semiconductor contacts that is known as the Mott barrier. The basic theory and the historical development of rectifying metal-semiconductor contacts were summarized by Henisch[6] in 1957 in his book *Rectifying Semiconductor Contacts.*

Because of their importance in direct current and microwave applications and as tools in the analysis of other fundamental physical parameters, metal-semiconductor contacts have been studied extensively. Recently, reproducible and near-ideal metal-semiconductor contacts have been fabricated with the help of modern transistor technology[29a] and improved vacuum technology. In Section 2 we shall first consider the Schottky effect. The formation of an energy barrier between a metal and a semiconductor is discussed in Section 3. Section 4 presents the current transport theory. The measurements of the barrier heights using current-voltage, capacitance-voltage, and photoelectric methods are considered in Section 5. Section 6 considers a variety of applications using metal-semiconductor contacts including the clamped transistor, field-effect transistor, and IMPATT diode. In Section 7 the Mott barrier, point-contact rectifier, and the ohmic contact are discussed. Finally, the space-charge-limited diodes are briefly considered.

2 SCHOTTKY EFFECT

In a metal-vacuum system the minimum energy necessary for an electron to escape into vacuum from an initial energy at the Fermi level is defined as the work function. This quantity is denoted by $q\phi_m$ (ϕ_m in volts), as shown in Fig. 1. For metals, $q\phi_m$ is of the order of a few electron volts and varies from 2 to 6 eV. The values of $q\phi_m$ are generally very sensitive to surface contamination. The most reliable values[7] for clean surfaces are given in Fig. 2.

When an electron is at a distance x from the metal, a positive charge will be induced on the metal surface. The force of attraction between the electron and the induced positive charge is equivalent to the force which would exist between the electron and an equal positive charge located at $(-x)$. This positive charge is referred to as the image charge. The attractive force, called the image force, is given by

$$F = \frac{-q^2}{4\pi(2x)^2\varepsilon_0} = \frac{-q^2}{16\pi\varepsilon_0 x^2} \tag{1}$$

where ε_0 is the permittivity of free space. The work done by an electron in the course of its transfer from infinity to the point x is given by

$$E(x) = \int_\infty^x F\,dx = \frac{q^2}{16\pi\varepsilon_0 x}. \tag{2}$$

The above energy corresponds to the potential energy of an electron at a

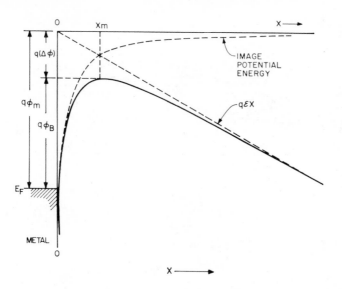

Fig. 1 Energy band diagram between a metal surface and a vacuum. The metal work function is $q\phi_m$. The effective work function (or barrier) is lowered when an electric field is applied to the surface. The lowering is due to the combined effects of the field and the image force.

distance x from the metal surface, shown in Fig. 1, and is measured downwards from the x-axis.

When an external field $\mathscr{E}$ is applied, the total potential energy, PE, as a function of distance (measured downwards from the x-axis) is given by the sum

$$PE(x) = \frac{q^2}{16\pi\varepsilon_0 x} + q\mathscr{E}x \qquad \text{eV.} \tag{3}$$

The Schottky barrier lowering (also referred to as image force lowering), $\Delta\phi$, and the location of the lowering, x_m (as shown in Fig. 1), are given by the condition $d[PE(x)]/dx = 0$, or

$$x_m = \sqrt{\frac{q}{16\pi\varepsilon_0 \mathscr{E}}} \qquad \text{cm} \tag{4}$$

$$\Delta\phi = \sqrt{\frac{q\mathscr{E}}{4\pi\varepsilon_0}} = 2\mathscr{E}x_m \qquad \text{volts.} \tag{5}$$

The lowering of the metal work function by an amount $\Delta\phi$ as a result of the

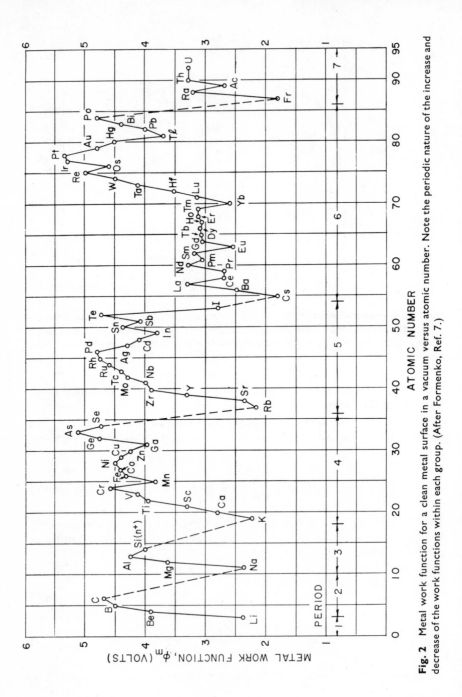

Fig. 2 Metal work function for a clean metal surface in a vacuum versus atomic number. Note the periodic nature of the increase and decrease of the work functions within each group. (After Formenko, Ref. 7.)

image force and the electric field is called the Schottky effect. From Eqs. (4) and (5) we obtain $\Delta\phi = 0.12$ V and $x_m \simeq 60$ Å for $\mathscr{E} = 10^5$ V/cm; and $\Delta\phi = 1.2$ V and $x_m \simeq 10$ Å for $\mathscr{E} = 10^7$ V/cm. Thus at high fields there is considerable Schottky barrier lowering, and the effective metal work function for thermionic emission ($q\phi_B$) is reduced.

The above results can also be applied to metal-semiconductor systems. However, the field should be replaced by the maximum field at the interface, and the free-space permittivity ε_0 should be replaced by an appropriate permittivity ε_s characterizing the semiconductor medium. This value may be different from the semiconductor static permittivity. This is because, during the emission process, if the electron transit time from the metal-semiconductor interface to the barrier maximum x_m is shorter than the dielectric relaxation time, the semiconductor medium does not have enough time to be polarized, and smaller permittivity than the static value is expected. It will be shown, however, that for Ge and Si the appropriate permittivities are about the same as their corresponding static values.

Because of the larger values of ε_s in a metal-semiconductor system the barrier lowering and the location of the potential maximum are smaller than those for a corresponding metal-vacuum system. For example, for $\varepsilon_s = 16\varepsilon_0$, $\Delta\phi$ as obtained from Eq. (5) is only 0.03 V for $\mathscr{E} = 10^5$ V/cm and even smaller for smaller fields. Although the barrier lowering is small, it does have a profound effect on current transport processes in metal-semiconductor systems. These will be considered in Section 4.

The dielectric constant ($\varepsilon_s/\varepsilon_0$) in gold-silicon barriers has been obtained from the photoelectric measurement which will be discussed in Section 5. The experimental result[8] is shown in Fig. 3 where the measured barrier lowering is plotted as a function of the square root of the electric field. From Eq. (5) the image-force dielectric constant is determined to be 12 ± 0.5. For $\varepsilon_s/\varepsilon_0 = 12$, the distance x_m varies between 10 Å and 50 Å as in the field range shown in Fig. 3. Assuming a carrier velocity of the order of 10^7 cm/sec, the transit time for these distances should be between 1×10^{-14} sec and 5×10^{-14} sec. The image-force dielectric constant should thus be comparable to the dielectric constant of approximately 12 for electromagnetic radiation of roughly these periods (wavelengths between 3 μm and 15 μm).[9] The dielectric constant of silicon is essentially constant (11.7) from dc to $\lambda = 1$ μm, therefore the lattice has time to polarize while the electron is traversing the depletion layer. There is thus excellent agreement between the photoelectric measurements and data deduced from the optical constants. For Ge and GaAs the dependence of the optical dielectric constant on wavelength is similar to that of Si. It is thus expected that the image-force permittivities of these semiconductors in the above field range are approximately the same as the corresponding static values.

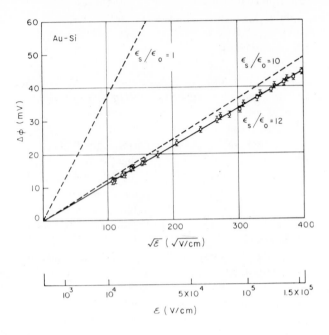

Fig. 3 Measurement of barrier lowering as a function of the electric field in a Au-Si diode. (After Sze et al., Ref. 8.)

3 ENERGY BAND RELATION AT METAL-SEMI-CONDUCTOR CONTACT

(1) Ideal Condition and Surface States

When a metal is making intimate contact with a semiconductor, the Fermi levels in the two materials must be coincident at thermal equilibrium. We will first consider two limiting cases;[6] a more general result will be considered later.[10] The two limiting cases are shown in Fig. 4. Figure 4(a) shows the electronic energy relations at an ideal contact between a metal and an *n*-type semiconductor in the absence of surface states. At far left, the metal and semiconductor are not in contact and the system is not in thermal equilibrium. If a wire is connected between the semiconductor and the metal so that charge will flow from the semiconductor to the metal and electronic equilibrium is established, the Fermi levels on both sides line up. Relative to the Fermi level in the metal, the Fermi level in the semiconductor has lowered by an amount equal to the difference between the two work functions. This potential

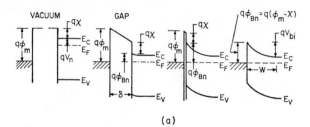

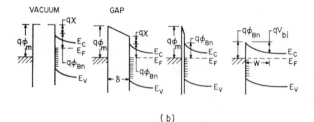

(b)

Fig. 4   Energy band diagrams of metal-semiconductor contacts.
(After Henisch, Ref. 6.)

difference, $q\phi_m - q(\chi + V_n)$, is called the contact potential where $q\chi$ is the electron affinity measured from the bottom of the conduction band to the vacuum level. As the distance δ decreases, an increasing negative charge is built up at the metal surface. An equal and opposite charge (positive) must exist in the semiconductor. Because of the relatively low carrier concentration, this positive charge is distributed over a barrier region near the semiconductor surface. When δ is small enough to be comparable with interatomic distances, the gap becomes transparent to electrons, and we obtain the limiting case as shown in the far right (Fig. 4). It is clear that the limiting value of the barrier height, $q\phi_{Bn}$ (neglecting the Schottky lowering), is given by

$$q\phi_{Bn} = q(\phi_m - \chi). \qquad (6)$$

The barrier height is simply the difference between the metal work function and the electron affinity of the semiconductor. For an ideal contact between a metal and a p-type semiconductor, the barrier height, $q\phi_{Bp}$, is given by

$$q\phi_{Bp} = E_g - q(\phi_m - \chi). \qquad (7)$$

For a given semiconductor and for any metals, the sum of the barrier heights on n-type and p-type substrates is thus expected to be equal to the band gap, or

$$q(\phi_{Bn} + \phi_{Bp}) = E_g. \tag{8}$$

The second limiting case is shown in Fig. 4(b) where a large density of surface states is present on the semiconductor surface. At far left, the figure shows equilibrium between the surface states and the bulk of the semiconductor but nonequilibrium between metal and the semiconductor. In this case the surface states are occupied to a level E_F. When the metal-semiconductor system is in equilibrium, the Fermi level of the semiconductor relative to that of the metal must fall an amount equal to the contact potential and, as a result, an electric field is produced in the gap δ. If the density of the surface states is sufficiently large to accommodate any additional surface charges resulting from diminishing δ without appreciably altering the occupation level E_F, the space charge in the semiconductor will remain unaffected. As a result the barrier height is determined by the property of the semiconductor surface and is independent of the metal work function.

(2) Depletion Layer

It is clear from the above discussion that when a metal is brought into intimate contact with a semiconductor, the conduction and valence bands of the semiconductor are brought into a definite energy relationship with the Fermi level in the metal. Once this relationship is known, it serves as a boundary condition on the solution of the Poisson equation in the semiconductor, which proceeds in exactly the same manner as in the p-n junctions. The energy band diagrams for metals on both n-type and p-type materials are shown, under different biasing conditions, in Fig. 5.

Under the abrupt approximation that $\rho \simeq qN_D$ for $x < W$, and $\rho \simeq 0$, $dV/dx \simeq 0$ for $x > W$ where W is the depletion width, the results for the metal-semiconductor barrier are identical to those of the one-sided abrupt p-n junction and we obtain

$$W(\text{depletion width}) = \sqrt{\frac{2\varepsilon_s}{qN_D}\left(V_{bi} - V - \frac{kT}{q}\right)}, \tag{9}$$

$$|\mathscr{E}(x)| = \frac{qN_D}{\varepsilon_s}(W - x) = \mathscr{E}_m - \frac{qN_D}{\varepsilon_s}x, \tag{10}$$

$$V(x) = \frac{qN_D}{\varepsilon_s}\left(Wx - \frac{1}{2}x^2\right) - \phi_{Bn}, \tag{11}$$

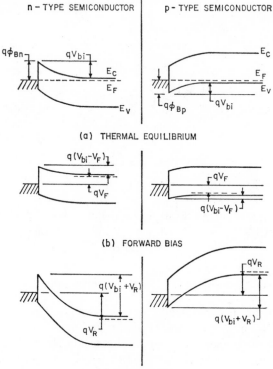

n – TYPE SEMICONDUCTOR | p – TYPE SEMICONDUCTOR

(a) THERMAL EQUILIBRIUM

(b) FORWARD BIAS

(c) REVERSE BIAS

Fig. 5 Energy band diagram of metal n-type and metal p-type semiconductors under different biasing conditions.

where the term kT/q arises from the contribution of the mobile carriers to the electric field and $\mathscr{E}_m$ is the maximum field strength which occurs at $x = 0$:

$$\mathscr{E}_m = \mathscr{E}(x = 0) = \sqrt{\frac{2qN_D}{\varepsilon_s}\left(V_{bi} - V - \frac{kT}{q}\right)} = 2\left(V_{bi} - V - \frac{kT}{q}\right)\Big/W. \quad (12)$$

The space charge Q_{sc} per unit area of the semiconductor and the depletion-layer capacitance C per unit area are given by

$$Q_{sc} = qN_D W = \sqrt{2q\varepsilon_s N_D\left(V_{bi} - V - \frac{kT}{q}\right)} \qquad \text{coul/cm}^2 \qquad (13)$$

$$C \equiv \frac{\partial Q_{sc}}{\partial V} = \sqrt{\frac{q\varepsilon_s N_D}{2\left(V_{bi} - V - \frac{kT}{q}\right)}} = \frac{\varepsilon_s}{W} \qquad \text{farad/cm}^2. \qquad (14)$$

Equation (14) can be written in the form

$$\frac{1}{C^2} = \frac{2\left(V_{bi} - V - \dfrac{kT}{q}\right)}{q\varepsilon_s N_D} \tag{15a}$$

$$\frac{d\left(\dfrac{1}{C^2}\right)}{(-dV)} = \frac{2}{q\varepsilon_s N_D} \tag{15b}$$

or

$$N_D = \frac{2}{q\varepsilon_s} \frac{(-dV)}{d\left(\dfrac{1}{C^2}\right)}. \tag{15c}$$

If N_D is constant throughout the depletion region, one should obtain a straight line by plotting $1/C^2$ versus V. If N_D is not a constant, one can use the differential capacitance method to determine the doping profile from Eq. (15c).

(3) General Expression for the Barrier Height

The barrier heights of metal-semiconductor systems are, in general, determined by both the metal work function and the surface states. A general expression[10] of the barrier height can be obtained on the basis of the following two assumptions: (1) with intimate contact between the metal and the semiconductor and with an interfacial layer of atomic dimensions, this layer will be transparent to electrons and can withstand potential across it, and (2) the surface states per unit area per electron volt at the interface are a property of the semiconductor surface and are independent of the metal.

A more detailed energy band diagram of a metal n-type semiconductor contact is shown in Fig. 6. The various quantities used in the derivation which follows are defined in this figure. The first quantity which is of interest is the energy level $q\phi_0$. This was the energy difference between the Fermi level and the valence-band edge at the surface before the metal-semiconductor contact was formed. It specified the level below which all surface states must have been filled for charge neutrality at the surface.[11] The second quantity is $q\phi_{Bn}$, the barrier height of the metal-semiconductor contact; it is this barrier which must be surmounted by electrons flowing from the metal into the semiconductor. The interfacial layer will be assumed to have a thickness of a few angstroms and will therefore be essentially transparent to electrons.

We consider a semiconductor with acceptor surface states whose density is D_s states/cm^2/eV, and D_s is a constant over the energy range from $q\phi_0$ to the Fermi level. The surface-state charge density on the semiconductor Q_{ss}

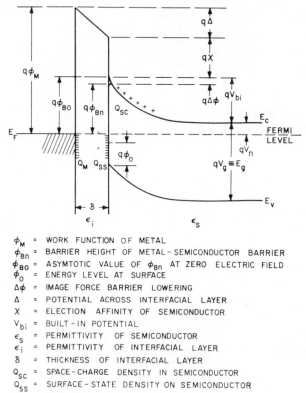

ϕ_M = WORK FUNCTION OF METAL
ϕ_{Bn} = BARRIER HEIGHT OF METAL – SEMICONDUCTOR BARRIER
ϕ_{BO} = ASYMTOTIC VALUE OF ϕ_{Bn} AT ZERO ELECTRIC FIELD
ϕ_O = ENERGY LEVEL AT SURFACE
$\Delta\phi$ = IMAGE FORCE BARRIER LOWERING
Δ = POTENTIAL ACROSS INTERFACIAL LAYER
X = ELECTION AFFINITY OF SEMICONDUCTOR
V_{bi} = BUILT – IN POTENTIAL
ϵ_s = PERMITTIVITY OF SEMICONDUCTOR
ϵ_i = PERMITTIVITY OF INTERFACIAL LAYER
δ = THICKNESS OF INTERFACIAL LAYER
Q_{sc} = SPACE – CHARGE DENSITY IN SEMICONDUCTOR
Q_{ss} = SURFACE – STATE DENSITY ON SEMICONDUCTOR
Q_M = SURFACE – CHARGE DENSITY ON METAL

Fig. 6 Detailed energy band diagram of a metal n-type semiconductor contact with an interfacial layer of the order of atomic distance. (After Cowley and Sze, Ref. 10.)

is given by

$$Q_{ss} = -qD_s(E_g - q\phi_0 - q\phi_{Bn} - q\Delta\phi) \qquad \text{coul/cm}^2 \qquad (16)$$

where $q\Delta\phi$ is the Schottky barrier lowering. The quantity in parentheses is simply the difference between the Fermi level at the surface and $q\phi_0$. D_s times this quantity yields the number of surface states above $q\phi_0$ which are full.

The space charge which forms in the depletion layer of the semiconductor at thermal equilibrium is given in Eq. (13) and is rewritten as

$$Q_{sc} = \sqrt{2q\varepsilon_s N_D\left(\phi_{Bn} - V_n + \Delta\phi - \frac{kT}{q}\right)} \qquad \text{coul/cm}^2. \qquad (17)$$

The total equivalent surface charge density on the semiconductor surface is given by the sum of Eqs. (16) and (17). In the absence of any space charge effects in the interfacial layer, an exactly equal and opposite charge, Q_M (coul/cm^2), develops on the metal surface. For thin interfacial layers such effects are negligible, and Q_M can be written as

$$Q_M = -(Q_{ss} + Q_{sc}).$$ (18)

The potential Δ across the interfacial layer can be obtained by the application of Gauss' law to the surface charge on the metal and semiconductor:

$$\Delta = -\delta \frac{Q_M}{\varepsilon_i}$$ (19)

where ε_i is the permittivity of the interfacial layer and δ its thickness. Another relation for Δ can be obtained by inspection of the energy band diagram of Fig. 6:

$$\Delta = \phi_m - (\chi + \phi_{Bn} + \Delta\phi).$$ (20)

This results from the fact that the Fermi level must be constant throughout this system at thermal equilibrium.

If Δ is eliminated from Eqs. (19) and (20), and Eq. (18) is used to substitute for Q_M, we obtain

$$(\phi_m - \chi) - (\phi_{Bn} + \Delta\phi) = \sqrt{\frac{2q\varepsilon_s N_D \delta^2}{\varepsilon_i^2}\left(\phi_{Bn} + \Delta\phi - V_n - \frac{kT}{q}\right)}$$
$$- \frac{qD_s\delta}{\varepsilon_i}(E_g - q\phi_0 - q\phi_{Bn} - q\Delta\phi).$$ (21)

Equation (21) can now be solved for ϕ_{Bn}. Introducing the quantities

$$c_1 \equiv \frac{2q\varepsilon_s N_D \delta^2}{\varepsilon_i^2}$$ (22a)

$$c_2 \equiv \frac{\varepsilon_i}{(\varepsilon_i + q^2\delta D_s)}$$ (22b)

we can write the solution to Eq. (21) as

$$\phi_{Bn} = \left[c_2(\phi_m - \chi) + (1 - c_2)\left(\frac{E_g}{q} - \phi_0\right) - \Delta\phi\right]$$
$$+ \left\{\frac{c_2^2 c_1}{2} - c_2^{3/2}\left[c_1(\phi_m - \chi) + (1 - c_2)\left(\frac{E_g}{q} - \phi_0\right)\frac{c_1}{c_2} - \frac{c_1}{c_2}\left(V_n + \frac{kT}{q}\right)\right.\right.$$
$$\left.\left. + \frac{c_2 c_1^2}{4}\right]^{1/2}\right\}.$$ (23)

Equation (22a) can be used to calculate c_1 if values of δ and ε_i are estimated; for vacuum-cleaved or well-cleaned semiconductor substrates the interfacial layer will have a thickness of atomic dimensions, i.e., 4 or 5 Å. The permittivity of such a thin layer can be well approximated by the free space value, and since this approximation represents a lower limit for ε_i, it leads to an overestimation of c_2. For $\varepsilon_s \approx 10\varepsilon_0$, $\varepsilon_i = \varepsilon_0$, and $N_D < 10^{18}$ cm^{-3}, c_1 is small, of the order of 0.01 V, and the { } term in Eq. (23) is estimated to be less than 0.04 V. Neglecting the { } term in Eq. (23) reduces the equation to

$$\phi_{Bn} = c_2(\phi_m - \chi) + (1 - c_2)\left(\frac{E_g}{q} - \phi_0\right) - \Delta\phi \equiv c_2\,\phi_m + c_3. \qquad (24)$$

If c_2 and c_3 can be determined experimentally and if χ is known, then

$$\phi_0 = \frac{E_g}{q} - \frac{(c_2\chi + c_3 + \Delta\phi)}{(1 - c_2)} \qquad (25)$$

and from (22b)

$$D_s = \frac{(1 - c_2)\varepsilon_i}{c_2\,\delta q^2}. \qquad (26)$$

Using the previous assumptions for δ and ε_i, we obtain

$$D_s \simeq 1.1 \times 10^{13}(1 - c_2)/c_2 \qquad \text{states/cm}^2/\text{eV}. \qquad (26a)$$

The two limiting cases considered previously can be obtained directly from Eq. (24):

(A) When $D_s \to \infty$, $c_2 \to 0$ then,

$$q\phi_{Bn} = (E_g - q\phi_0) - q\Delta\phi. \qquad (27)$$

In this case the Fermi level at the interface is "pinned" by the surface states at the value $q\phi_0$ above the valence band. The barrier height is independent of the metal work function and is determined entirely by the doping and surface properties of the semiconductor.

(B) When $D_s \to 0$, $c_2 \to 1$ then,

$$q\phi_{Bn} = q(\phi_m - \chi) - q\Delta\phi. \qquad (28)$$

This is identical to Eq. (6) (except for the Schottky lowering term) and is for the barrier height of an ideal Schottky barrier where surface-state effects are neglected.

The experimental results of the metal n-type silicon system are shown in Fig. 7. A least-square straight line fit to the data yields

$$\phi_{Bn} = 0.235\phi_m - 0.352.$$

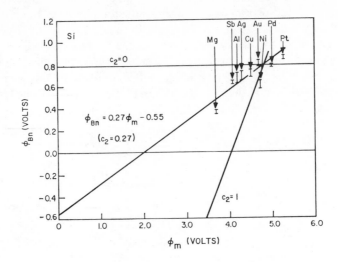

Fig. 7 Experimental results of barrier heights for metal *n*-type silicon system. (Ref. 10.)

Comparing this expression with Eq. (24) and using Eqs. (25) and (26a), we obtain $c_2 = 0.235$, $q\phi_0 = 0.33$ eV, and $D_s = 4 \times 10^{13}$ states/cm^2/eV. Similar results are obtained for GaAs, GaP, and CdS which are shown in Fig. 8 and listed in Table 8.1.

We note that the values of $q\phi_0$ for Si, GaAs, and GaP are very close to one-third of the band gap. Similar results are obtained for other semiconductors. Figure 9 shows a plot of $(E_C - q\phi_0)$ versus the band gap for gold contacts on various semiconductors. The solid line is for $(E_C - q\phi_0) = 2E_g/3$ which passes through most of the experimental points.[12] This fact indicates

TABLE 8.1

SUMMARY OF BARRIER HEIGHT DATA AND CALCULATIONS FOR
Si, GaP, GaAs, AND CdS

Semi-conductor	c_2	$c_3(V)$	$\chi(V)$	$D_s \times 10^{-13}$ (eV^{-1} cm^{-2})	$q\phi_0$(eV)	$q\phi_0/E_g$
Si	0.27 ± 0.05	-0.55 ± 0.22	4.05	2.7 ± 0.7	0.30 ± 0.36	0.27
GaP	0.27 ± 0.03	-0.01 ± 0.13	4.0	2.7 ± 0.4	0.66 ± 0.2	0.294
GaAs	0.07 ± 0.05	$+0.49 \pm 0.24$	4.07	12.5 ± 10.0	0.53 ± 0.33	0.38
CdS	0.38 ± 0.16	-1.20 ± 0.77	4.8	1.6 ± 1.1	1.5 ± 1.5	0.6

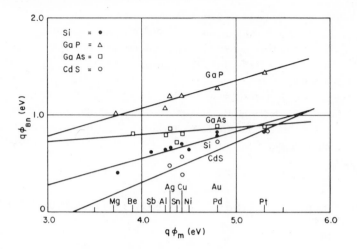

Fig. 8 Similar results for other metal-semiconductor systems. (Ref. 10.)

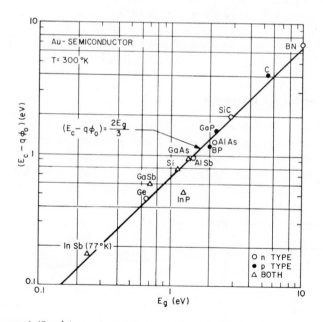

Fig. 9 Measured $(E_c - q\phi_0)$ versus the band gap for Au contacts on various *n*-type semiconductors.
(After Mead and Spitzer, Ref. 12.)

that most semiconductor surfaces have a high peak density of surface states near one-third of the gap from the valence band edge. The theoretical calculation by Pugh[13] for $\langle 111 \rangle$ diamond indeed gives a narrow band of surface states slightly below the center of the forbidden gap. It is thus expected that a similar situation may exist for other semiconductors.

For CdS, however, the value of $q\phi_0$ is very large ($\sim 0.8\ E_g$) and the metal-CdS system behaves as if there were a low density of surface states. This may be explained by the fact that the surface states are very close to the band edge, and the Fermi level at the interface can move up or down in energy over a relatively large range without requiring any surface charge to fill or empty these states.

4 CURRENT TRANSPORT THEORY IN SCHOTTKY BARRIERS

The current transport in metal-semiconductor barriers is mainly due to majority carriers in contrast to p-n junctions where the minority carriers are responsible. We shall present in this section three different approaches: (1) the simple isothermal thermionic emission theory by Bethe,[14] (2) the simple isothermal diffusion theory by Schottky,[4] and (3) the more general theory which incorporates the above two theories into a single thermionic emission-diffusion theory by Crowell and Sze.[16] Also included in this section is a discussion of the minority carrier injection ratio.

(1) Thermionic Emission Theory

The thermionic emission theory is derived from the assumptions that (1) the barrier height $q\phi_{Bn}$ is much larger than kT, (2) electron collisions within the depletion region are neglected, and (3) the effect of image force is also neglected. Because of the above assumptions the shape of the barrier profile is immaterial and the current flow depends solely on the barrier height. The current density $J_{s \to m}$ from the semiconductor to the metal is then given by the standard thermionic emission equation[14]

$$
\begin{aligned}
J_{s \to m} &= \frac{qn(m^*)^{3/2}}{(2\pi kT)^{3/2}} \int_{-\infty}^{\infty} dv_y \int_{-\infty}^{\infty} dv_z \int_{v_{0x}}^{\infty} v_x \exp\left[-\frac{m^*(v_x^2 + v_y^2 + v_z^2)}{2kT} \right] dv_x \\
&= qn\left(\frac{m^*}{2\pi kT}\right)^{1/2} \int_{v_{0x}}^{\infty} v_x \exp\left(-\frac{m^* v_x^2}{2kT} \right) dv_x \\
&= qn\left(\frac{kT}{2\pi m^*}\right)^{1/2} \exp\left(-\frac{m^* v_{0x}^2}{2kT} \right).
\end{aligned}
\tag{29}
$$

The velocity v_{0x} is the minimum velocity required in x-direction to surmount the barrier and is given by the relation

$$\tfrac{1}{2}m^*v_{0x}^2 = q(V_{bi} - V) \tag{30}$$

where V_{bi} and V are the built-in potential and the applied voltage respectively (V is positive for forward bias). The electron concentration n is given by

$$n = N_C \exp\left(-\frac{E_C - E_F}{kT}\right) = 2\left(\frac{2\pi m^*kT}{h^2}\right)^{3/2} \exp\left(-\frac{qV_n}{kT}\right). \tag{31}$$

Substitution of Eq. (30) and (31) into Eq. (29) yields

$$J_{s \to m} = A^*T^2 \exp\left(-\frac{q\phi_{Bn}}{kT}\right)\exp\left(\frac{qV}{kT}\right)$$

$$A^* \equiv \frac{4\pi q m^* k^2}{h^3}. \tag{32}$$

For free electrons, $A^* = 120$ amp/cm^2/$^\circ$K$^2 \equiv A$, which is the Richardson constant for thermionic emission into a vacuum. For semiconductors with isotropic effective mass in the lowest minimum of the conduction band such as n-type GaAs, $A^*/A = m^*/m_0$ where m^* and m_0 are the effective mass and the free-electron mass respectively. For multiple-valley semiconductors the appropriate Richardson constant A^* associated with a single energy minimum is given by[15]

$$\frac{A_1^*}{A} = \frac{1}{m_0}(l_1^2 m_y^* m_z^* + l_2^2 m_z^* m_x^* + l_3^2 m_x^* m_y^*)^{1/2} \tag{33}$$

where l_1, l_2, and l_3 are the direction cosines of the normal to the emitting plane relative to the principal axes of the ellipsoid, and m_x^*, m_y^*, and m_z^* are the components of the effective mass tensor. For Ge the emission in the conduction band arises from minima at the edge of the Brillouin zone in the $\langle 111 \rangle$ direction. These minima are equivalent to four ellipsoids with longitudinal mass $m_l^* = 1.6\, m_0$ and transverse mass $m_t^* = 0.082\, m_0$. The sum of all the A_1^* values has a minimum in the $\langle 111 \rangle$ direction

$$\left(\frac{A^*}{A}\right)_{n\text{-Ge}\langle 111 \rangle} = m_t^*/m_0 + [(m_t^*)^2 + 8m_l^* m_t^*]^{1/2}/m_0 = 1.11. \tag{34}$$

The maximum A^* occurs for the $\langle 100 \rangle$ direction:

$$\left(\frac{A^*}{A}\right)_{n\text{-Ge}\langle 100 \rangle} = \frac{4}{m_0}\left[\frac{(m_t^*)^2 + 2m_t^* m_l^*}{3}\right]^{1/2} = 1.19. \tag{35}$$

For Si the conduction band minima occur in the $\langle 100 \rangle$ directions and $m_l^* = 0.97\ m_0$, $m_t^* = 0.19\ m_0$. All minima contribute equally to the current in the $\langle 111 \rangle$ direction, yielding the maximum A^*:

$$\left(\frac{A^*}{A}\right)_{n\text{-Si}\langle 111\rangle} = \frac{6}{m_0}\left[\frac{(m_t^*)^2 + 2m_t^* m_l^*}{3}\right]^{1/2} = 2.2. \tag{36}$$

The maximum value of A^* occurs for the $\langle 100 \rangle$ direction

$$\left(\frac{A^*}{A}\right)_{n\text{-Si}\langle 100\rangle} = 2m_t^*/m_0 + 4(m_l^* m_t^*)^{1/2}/m_0 = 2.1. \tag{37}$$

For holes in Ge, Si, and GaAs the two energy maxima at $\mathbf{k} = 0$ give rise to approximately isotropic current flow from both the light and heavy holes. Adding the currents due to these carriers, we obtain

$$\left(\frac{A^*}{A}\right)_{p\text{-type}} = (m_{lh}^* + m_{hh}^*)/m_0. \tag{38}$$

A summary of the values[15] of (A^*/A) is given in Table 8.2.

Since the barrier height for electrons moving from the metal into the semiconductor remains the same, the current flowing into the semiconductor is thus unaffected by the applied voltage. It must therefore be equal to the current flowing from the semiconductor into the metal when thermal equilibrium prevails, i.e., when $V = 0$. The corresponding current density is obtained from Eq. (32) by setting $V = 0$,

$$J_{m\rightarrow s} = -A^* T^2 \exp\left(-\frac{q\phi_{Bn}}{kT}\right). \tag{39}$$

TABLE 8.2

VALUES OF A*/A

Semiconductor	Ge	Si	GaAs
p-type	0.34	0.66	0.62
n-type $\langle 111 \rangle$	1.11	2.2	0.068 (low field)
n-type $\langle 100 \rangle$	1.19	2.1	1.2 (high field)

The total current density is given by the sum of Eqs. (32) and (39),

$$J_n = \left\{ A^* T^2 \exp\left(-\frac{q\phi_{Bn}}{kT} \right) \right\} \left[\exp\left(\frac{qV}{kT} \right) - 1 \right]$$

$$= J_{ST} \left[\exp\left(\frac{qV}{kT} \right) - 1 \right] \tag{40}$$

where

$$J_{ST} \equiv A^* T^2 \exp\left(-\frac{q\phi_{Bn}}{kT} \right). \tag{41}$$

Equation (40) is similar to the Shockley equation for $p\text{-}n$ junctions. However, the expressions for the saturation current densities are quite different.

(2) Diffusion Theory[4]

The diffusion theory is derived from the assumptions that (1) the barrier height is much larger than kT; (2) the effect of electron collisions within the depletion region is included; (3) the carrier concentrations at $x = 0$, and and $x = W$ are unaffected by the current flow, i.e., they have their equilibrium values; and (4) the impurity concentration of the semiconductor is nondegenerate.

Since the current in the depletion region depends on the local field and the concentration gradient, we must use the current density equation:

$$J_x = J_n = q \left[n(x)\mu\mathscr{E} + D_n \frac{\partial n}{\partial x} \right]$$

$$= qD_n \left[-\frac{qn(x)}{kT} \frac{\partial V(x)}{\partial x} + \frac{\partial n}{\partial x} \right]. \tag{42}$$

Under the steady-state condition, the current density is independent of x, and Eq. (42) can be integrated using $\exp[-qV(x)/kT]$ as an integrating factor. We then have

$$J_n \int_0^W \exp\left[-\frac{qV(x)}{kT} \right] dx = qD_n \left\{ n(x)\exp\left[-\frac{qV(x)}{kT} \right] \right\}_0^W, \tag{43}$$

and the boundary conditions

$$\left. \begin{array}{l} qV(0) = -q(V_n + V_{bi}) = -q\phi_{Bn} \\[2mm] qV(W) = -qV_n - qV \\[2mm] n(0) = N_C \exp\left[-\frac{E_C(0) - E_F}{kT} \right] = N_C \exp\left(-\frac{q\phi_{Bn}}{kT} \right) \\[2mm] n(W) = n = N_C \exp\left(-\frac{qV_n}{kT} \right). \end{array} \right\} \tag{44}$$

Substitution of Eq. (44) into Eq. (43) yields

$$J_n = qN_C D_n \left[\exp\left(\frac{qV}{kT}\right) - 1 \right] \bigg/ \int_0^W \exp\left[-\frac{qV(x)}{kT} \right] dx. \tag{45}$$

For Schottky barriers, neglecting image-force effect, the potential distribution is given by Eq. (11), or

$$qV(x) = \frac{q^2 N_D}{\varepsilon_s} \left(Wx - \frac{x^2}{2} \right) - q\phi_{Bn}.$$

Substituting into Eq. (45) and expressing W in terms of $V_{bi} + V$, leads to

$$J_n \simeq \frac{q^2 D_n N_C}{kT} \left[\frac{q(V_{bi} - V)2N_D}{\varepsilon_s} \right]^{1/2} \exp\left(-\frac{q\phi_{Bn}}{kT} \right) \left\{ \frac{\exp\left(\frac{qV}{kT}\right) - 1}{1 - \exp\left[-\frac{2q(V_{bi} - V)}{kT} \right]} \right\} \tag{46}$$

where V has positive values for forward bias, and negative values for reverse bias. Since $qV_{bi} \gg kT$ is one of the conditions on which the present theory is based, the exponential term in the denominator can be neglected for all reverse voltages and for small forward voltages, Eq. (46) reduces to

$$J_n = \left\{ \frac{q^2 D_n N_C}{kT} \left[\frac{q(V_{bi} - V)2N_D}{\varepsilon_s} \right]^{1/2} \exp\left(-\frac{q\phi_{Bn}}{kT} \right) \right\} \left[\exp\left(\frac{qV}{kT}\right) - 1 \right]$$

$$= J_{SD} \left[\exp\left(\frac{qV}{kT}\right) - 1 \right]. \tag{47}$$

The current density expressions of the diffusion and thermionic emission theories, Eqs. (40) and (47), are basically very similar. However, the " saturation current density," J_{SD}, for the diffusion theory varies more rapidly with the voltage but is less sensitive to temperature in comparison with the " saturation current density," J_{ST}, of the thermionic emission theory.

(3) Thermionic Emission-Diffusion Theory[16]

A synthesis of the above thermionic emission and diffusion approaches will now be considered. This approach is derived from the boundary condition of a thermionic recombination velocity, v_R, near the metal-semiconductor interface. In addition, effects of electron optical-phonon scattering and quantum-mechanical reflection at the metal-semiconductor interface are incorporated. The electron optical-phonon scattering between the barrier energy maximum (x_m) and the metal predicts a low-field limit for applying the thermionic emission theory, i.e., for assuming that the metal acts as a perfect sink for carriers which cross the potential maximum in the direction of the metal. The effect of quantum-mechanical reflection and quantum tunneling on the recombination velocity predicts the high-field limit of validity of the thermionic emission theory and the onset of thermionic-field emission.

Since the diffusion of carriers is strongly affected by the potential configuration in the region through which the diffusion occurs, we consider the electron potential energy, $q\psi(x)$, versus distance as shown in Fig. 10 for a metal-semiconductor barrier. The origin of the barrier itself has been considered in the previous section and is due mainly to a combination of the effects of surface states and the metal work function. We will consider the case where the barrier height is large enough that the charge density between the metal surface and $x = W$ is essentially that of the ionized donors, i.e., W is the edge of the electron depletion layer. The rounding of $q\psi$ near the metal-semiconductor interface is due to the superimposed effects of the electric field associated with the ionized donors (shown by the dotted extrapolation of ψ) and the attractive image force experienced by an electron when it approaches the metal. As drawn, the applied voltage V between the metal and the semiconductor bulk would give rise to a flow of electrons into the metal. The imref $(-q\phi_n)$ associated with the electron current density J in the barrier is also shown schematically as a function of distance in Fig. 10. Throughout

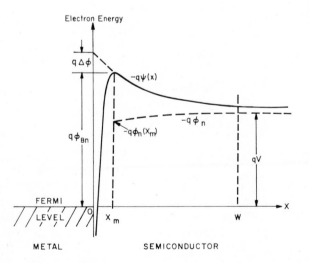

Fig. 10 Electron potential energy $(q\psi)$ versus distance for a metal-semiconductor barrier.

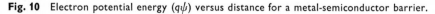

the region between x_m and W,

$$J = -q\mu n \frac{d\phi_n}{dx} \tag{48}$$

where the electron density at the point x is given by

$$n = N_C e^{-q(\phi_n - \psi)/kT} \tag{49}$$

where N_C is the effective density of states in the conduction band, and T is the electron temperature. We will assume that the region between x_m and W is isothermal and that the electron temperature is equal to the lattice temperature. Equations (48) and (49) will not be applicable between x_m and the interface ($x = 0$) since there the potential energy changes rapidly in distances comparable to the electron mean free path. In this region the distribution of carriers cannot be described by an imref ϕ_n nor be associated with an effective density of states. If this portion of the barrier acts as a sink for electrons, however, we can describe the current flow in terms of an effective recombination velocity v_R at the potential energy maximum:

$$J = q(n_m - n_0)v_R \tag{50}$$

where n_m is the electron density at x_m when the current is flowing. n_0 is a quasi-equilibrium electron density at x_m, the density which would occur if it were possible to reach equilibrium without altering the magnitude or position of the potential energy maximum. It is convenient to measure both ϕ and ψ with respect to the Fermi level in the metal. Then

$$\phi_n(W) = -V,$$

$$n_0 = N_C e^{-q\phi_{Bn}/kT}$$

and

$$n_m = N_C \exp\left[\frac{-q\phi_n(x_m) - q\phi_{Bn}}{kT}\right] \tag{51}$$

where $q\phi_{Bn}$ is the barrier height, and $q\phi_n(x_m)$ is the imref at x_m.

If n is eliminated from Eqs. (48) and (49) and the resulting expression for ϕ_n is integrated between x_m and W,

$$\exp\left[\frac{q\phi_n(x_m)}{kT}\right] - \exp\left(\frac{qV}{kT}\right) = -\frac{J}{\mu N_C kT} \int_{x_m}^{W} \exp\left(\frac{-q\psi}{kT}\right) dx. \tag{52}$$

Then from Eqs. (50), (51), and (52),

$$J = \frac{qN_C v_R}{1 + \dfrac{v_R}{v_D}} \exp\left[-\frac{q\phi_{Bn}}{kT}\right]\left[\exp\left(\frac{-qV}{kT}\right) - 1\right] \tag{53}$$

where

$$v_D \equiv \left[\int_{x_m}^{W} \frac{q}{\mu k T} \exp\left[-\frac{q}{kT}(\phi_{Bn} + \psi) \right] dx \right]^{-1} \tag{54}$$

is an effective diffusion velocity associated with the transport of electrons from the edge of the depletion layer at W to the potential energy maximum. If the electron distribution is Maxwellian for $x \geq x_m$, and if no electrons return from the metal other than those associated with the current density $q n_0 v_R$, the semiconductor acts as a thermionic emitter. Then

$$v_R = \frac{A^* T^2}{q N_C}, \tag{55}$$

where A^* is the effective Richardson constant, as shown in Table 8.2. At 300°K, v_R is 7.0×10^6, 5.2×10^6, and 1.0×10^7 cm/sec for $\langle 111 \rangle$ oriented n-type Ge, $\langle 111 \rangle$ n-type Si, and n-type GaAs respectively. If $v_D \gg v_R$, the pre-exponential term in Eq. (53) is dominated by v_R and the thermionic emission theory most nearly applies. If, however, $v_D \ll v_R$, the diffusion process is dominant. If we were to neglect image-force effects, and if the electron mobility were independent of the electric field, v_D would be equal to $\mu \mathscr{E}$, where $\mathscr{E}$ is the electric field in the semiconductor near the boundary. The standard Schottky diffusion result as given by Eq. (46) would then be obtained, and

$$J \simeq q N_C \mu \mathscr{E} \exp\left(-\frac{q\phi_{Bn}}{kT} \right) \left[\exp\left(\frac{qV}{kT} \right) - 1 \right]. \tag{56}$$

To include image-force effects on the calculation of v_D, the appropriate expression for ψ in Eq. (54) is

$$\psi = \phi_{Bn} + \Delta\phi - \mathscr{E}x - \frac{q}{16\pi\varepsilon_s x} \tag{57}$$

where $\Delta\phi$ is the barrier lowering as given by Eq. (5) (assuming the electric field is constant for $x < x_m$). Substitution of Eq. (57) into Eq. (54) yields the results that $v_D \simeq \mu \mathscr{E}$ for $\Delta\phi < kT/q$, and v_D reduces to 0.3 $\mu \mathscr{E}$ when $\Delta\phi$ increases to 20 kT/q.

In summary, Eq. (53) gives a result which is a synthesis of Schottky's diffusion theory and Bethe's thermionic emission theory, and which predicts currents in essential agreement with the thermionic emission theory if $\mu\mathscr{E}(x_m) > v_R$. The latter criterion is more rigorous than Bethe's condition $\mathscr{E}(x_m) > kT/q\lambda$ where λ is the carrier mean free path.

In the previous section a recombination velocity v_R associated with thermionic emission was introduced as a boundary condition to describe the collecting action of the metal in a Schottky barrier. In many cases there is an appreciable probability that an electron which crosses the potential energy maximum will be backscattered by electron optical-phonon scattering with a subsequent reduction in the net current over the barrier. Provided the backscattered electrons are a small fraction of the total electron flux, this effect can be viewed as a small perturbation. The optical-phonon mean free paths (λ) of Ge, Si, and GaAs have been listed in Table 2.4 of Chapter 2. As a first approximation, the probability of electron emission over the potential maximum can be given by[17,18]

$$f_p \simeq \exp\left(-\frac{x_m}{\lambda}\right) = \exp\left\{-\left(\frac{q}{16\pi\varepsilon_s\,\varepsilon}\right)^{1/2}\Big/\left[\lambda_0\tanh\left(\frac{E_p}{2kT}\right)\right]\right\}. \tag{58}$$

The probability of more detailed calculation over a Maxwellian distribution of electrons at the potential maximum of metal-Ge, metal-Si, and metal-GaAs Schottky diodes is shown in Fig. 11 for four different temperatures. We note that for small fields (corresponding to large x_m) and high temperatures (corresponding to small λ) there is a considerable reduction of the emission probability as expected from Eq. (58). A value of f_p less than unity implies that v_R should be replaced by a smaller recombination velocity $f_p v_R$ in Eqs. (50) and (53). If we assume that values of f_p less than 0.7 are indicative of failure of the thermionic emission boundary condition, then at room temperature the minimum electric fields are 2×10^3, 2×10^2, and 9×10^3 V/cm for Schottky barriers in Ge, Si, and GaAs respectively.

In addition to effects of phonon scattering the energy distribution of carriers should be further distorted from a Maxwellian distribution because

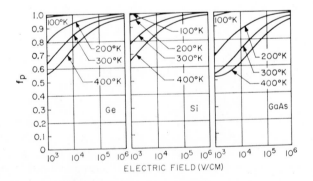

Fig. 11 Calculated probability of electron emission over the potential energy maximum for Au-Ge, Au-Si and Au-GaAs diodes as a function of the electric field at the interface and for various lattice temperatures.
(After Crowell and Sze, Ref. 18.)

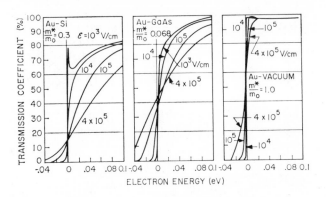

Fig. 12 Calculated quantum-mechanical transmission coefficient for Au-Si, Au-GaAs, and Au-vacuum systems as a function of electron energy measured from the potential maximum. (After Crowell and Sze, Ref. 19.)

of quantum-mechanical reflection of electrons by the Schottky barrier and also because of tunneling of electrons through the barrier. The predicted quantum-mechanical transmission coefficients (P_Q) for Au-Si, Au-GaAs, and Au-vacuum system are shown in Fig. 12. The electron energy is measured with respect to the potential energy maximum. We note that in a metal-semiconductor barrier for a given field and a given energy below the potential maximum, the tunneling probability increases with decreasing effective mass. The theoretical ratio f_Q of the total current flow considering tunneling and quantum-mechanical reflection, to the current flow neglecting these effects is shown in Fig. 13 as a function of electric field,[19]

$$f_Q \equiv \int_{-\infty}^{\infty} P_Q\, e^{-E/kT}\, \frac{dE}{kT}. \tag{59}$$

The field at which f_Q starts to rise rapidly marks the transition between thermionic and thermionic-field (T-F) emission, since at this point the field-enhanced tunneling process becomes the dominant mechanism. These fields are listed in Table 8.3 for Ge, Si, and GaAs. We notice that f_Q (or the current) in a metal-GaAs barrier increases rapidly near 10^5 V/cm at room temperature. This is mainly due to the contribution of the tunneling component at large fields.[16a] For Si and Ge, however, f_Q (or the currents) remains essentially constant in the field range shown in Fig. 13.

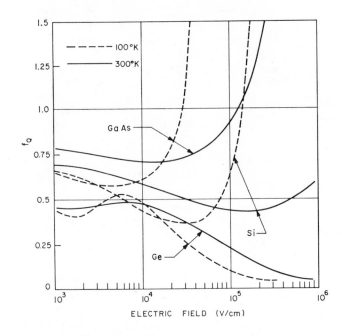

Fig. 13 The ratio f_Q of the predicted total current flow considering tunneling and quantum-mechanical reflection to the predicted current neglecting these effects. The field at which f_Q starts to increase rapidly marks the transition between thermionic and thermionic-field emission. For GaAs because of its small effective mass the transition occurs at about 10^5 V/cm. (Ref. 19).

TABLE 8.3

FIELD LIMITATIONS FOR THE
THERMIONIC-EMISSION MODEL AT 300°K

Semiconductor	Field (V/cm)	
	$\mathscr{E}_{D-T}$	$\mathscr{E}_{T-F}$
GaAs	9×10^3	1×10^5
Si	2×10^2	4×10^5
Ge	2×10^3	4×10^5

The complete expression of the J-V characteristics taking into account f_p and f_Q is thus

$$J = J_S(e^{qV/kT} - 1) \tag{60}$$

$$J_S = A^{**}T^2 \exp\left(-\frac{q\phi_{Bn}}{kT}\right) \tag{61}$$

where

$$A^{**} = \frac{f_p f_Q A^*}{(1 + f_p f_Q v_R/v_D)}. \tag{61a}$$

Figure 14 shows[19a] the calculated room-temperature values of the effective Richardson constant, A^{**}, for metal-Si systems with an impurity concentration of 10^{16} cm^{-3}. We note that for electrons (n-type Si), A^{**} in the field range 10^4 to 2×10^5 V/cm remains essentially at a constant value of about 110 amp/cm^2/°K^2. For holes (p-type Si), A^{**} in the above field range also

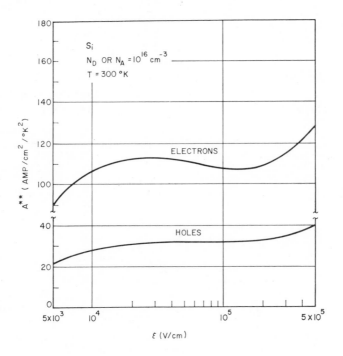

Fig. 14 Calculated A^{**} vs. electric field for metal Si barriers. (After Andrews and Lepselter, Ref. 19a.)

remains essentially constant but at a considerably lower value (~ 30 amp/cm^2/°K^2).

We conclude from the above discussions that at room temperature in the electric field range of 10^4 V/cm to about 10^5 V/cm, the current transport mechanism in most Ge, Si, and GaAs Schottky barrier diodes is due to thermionic emission of majority carriers.

(4) Minority Carrier Injection Ratio[20]

The Schottky barrier diode is a majority carrier device under low-injection conditions. At sufficiently large forward bias, the minority carrier injection ratio, γ, (ratio of minority carrier current to total current) increases with current due to the enhancement of the drift-field component which becomes much larger than the diffusion current.

At steady state, the one-dimensional continuity and current density equations for the minority carriers are given by

$$0 = -\frac{p_n - p_{n0}}{\tau_p} - \frac{1}{q}\frac{\partial J_p}{\partial x}, \tag{62}$$

$$J_p = q\mu_p p_n \mathscr{E} - qD_p \frac{\partial p_n}{\partial x}. \tag{63}$$

We consider the energy band diagram as shown in Fig. 15 where x_1 is the boundary of the depletion layer, and x_2 occurs at the interface between the n-type epitaxial layer and the n^+ substrate. From the rectifying theory as

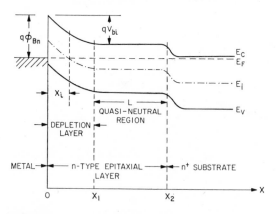

Fig. 15 Energy band diagram of an epitaxial Schottky barrier. (After Scharfetter, Ref. 20.)

discussed in Chapter 3, the minority carrier density at x_1 is

$$p_n(x_1) = p_{n0}\left[\exp\left(\frac{qV}{kT}\right) - 1\right] \simeq \frac{n_i^2}{N_D}\left[\exp\left(\frac{qV}{kT}\right) - 1\right] \qquad (64)$$

where N_D is the n-type donor concentration. The quantity $p_n(x)$ at $x = x_1$ expressed as a function of the forward current density can be obtained from Eqs. (60) and (64):

$$p_n(x_1) = \frac{n_i^2}{N_D}\frac{J}{J_S}. \qquad (65)$$

The boundary condition on $p_n(x)$ at $x = x_2$ can be stated in terms of a transport velocity, $v_T = D_p/L_p$, for the minority carriers

$$J_p(x_2) = qv_T p_n = q\left(\frac{D_p}{L_p}\right)p_{n0}\left[\exp\left(\frac{qV}{kT}\right) - 1\right] \quad \text{for} \quad L \ll L_p \qquad (66)$$

where D_p and L_p are the minority carrier diffusion constant and diffusion length respectively, and L is the distance of the quasi-neutral region.

For the low-injection conditions, the minority carrier drift component in Eq. (63) is negligible in comparison with the diffusion term, and the injection ratio, γ, is given by

$$\gamma \equiv \frac{J_p}{J_p + J_n} \simeq \frac{J_p}{J_n} = \frac{qn_i^2 D_p}{N_D L_p A^{**}T^2 \exp\left(-\dfrac{q\phi_{Bn}}{kT}\right)}. \qquad (67)$$

For gold-n-type silicon diodes ($\phi_{Bn} = 0.8$ V), the ratio is generally much less than 0.1 % at room temperature.

For sufficiently large forward bias, however, the electric field causes a significant carrier drift current component which eventually dominates the minority carrier current. From Eqs. (56), (63), and (65) we obtain for the high-current-limiting condition

$$\gamma \simeq \frac{J_p}{J_n} \simeq \frac{n_i^2}{N_D^2}\left(\frac{\mu_p}{\mu_n}\right)\frac{J}{J_S}. \qquad (68)$$

For example, a gold-n-type silicon diode with $N_D = 10^{15}$ cm^{-3}, and $J_S = 5 \times 10^{-7}$ amp/cm^2 would be expected to have an injection ratio of about 5 % at a current density of 350 amp/cm^2. The intermediate cases have been considered by Scharfetter, and the computed results are shown in Fig. 16(a) where the normalization factors are given by

$$J_0 \equiv \frac{qD_n N_D}{L}, \qquad (69a)$$

$$\gamma_0 \equiv \frac{qD_p n_i^2}{N_D L J_S}. \qquad (69b)$$

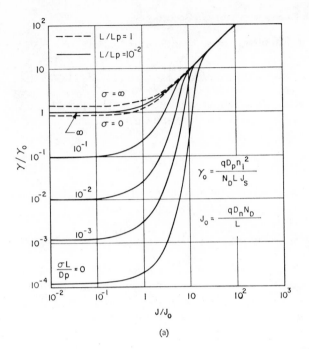

(a)

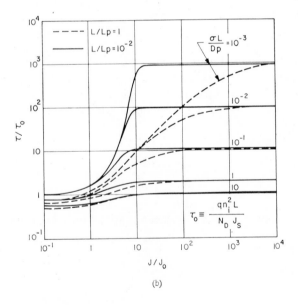

(b)

Fig. 16
(a) Normalized minority carrier injection ratio versus normalized diode current density.
(b) Normalized minority carrier storage time versus normalized current density.
(Ref. 20.)

392

It is clear from Fig. 16(a) that in order to reduce the minority carrier injection ratio one must use a metal-semiconductor system with large N_D (corresponding to low resistivity material), large J_S (corresponding to small barrier height), and small n_i (corresponding to large band gap).

Another quantity associated with the injection ratio is the minority storage time τ_s, which is defined as the minority carrier stored in the quasi-neutral region per unit current density:

$$\tau_s \equiv \frac{\int_{x_1}^{x_2} q p(x)\, dx}{J}. \tag{70}$$

For high current limit, τ_s is given by

$$\tau_s \simeq \frac{q n_i^2 L_p}{N_D J_S}. \tag{71}$$

The results for τ_s versus the current density are shown in Fig. 16(b) where similar parameters as in Fig. 16(a) are used. For example, in a Au-Si diode with $N_D = 1.5 \times 10^{14}$ cm^{-3}, $L = 7\ \mu$m, and $D_p/L_p = 2000$ cm/sec, the storage time for $J = 10$ amp/cm^2 is about 1 ns. If N_D is increased to 1.5×10^{16} cm^{-3}, the storage time would decrease to 0.01 ns even at a current density of 1000 amp/cm^2.

5 MEASUREMENT OF SCHOTTKY BARRIER HEIGHT

(1) Current-Voltage Measurement

A. Forward Characteristics. From Eq. (60) one can predict the ideal forward and reverse I-V characteristics of a Schottky barrier diode. In the forward direction with $V > 3kT/q$, we can rewrite Eq. (60) as

$$J = A^{**} T^2 \exp\left(-\frac{q\phi_{B0}}{kT}\right)\exp\left[\frac{q(\Delta\phi + V)}{kT}\right] \tag{72}$$

where ϕ_{B0} is the zero-field asymtotic barrier height as shown in Fig. 6, A^{**} is the effective Richardson constant, and $\Delta\phi$ is the Schottky barrier lowering. Since both A^{**} and $\Delta\phi$ are functions of the applied voltage, the forward J-V characteristic (for $V > 3kT/q$) is not represented by $J \sim \exp(qV/kT)$ but rather by

$$J \sim \exp\left(\frac{qV}{nkT}\right) \tag{72a}$$

where the parameter n is given by

$$n \equiv \frac{q}{kT} \frac{\partial V}{\partial(\ln J)}$$

$$= \left[1 + \frac{\partial \Delta \phi}{\partial V} + \frac{kT}{q} \frac{\partial(\ln A^{**})}{\partial V}\right]^{-1}. \tag{73}$$

Typical examples are shown in Fig. 17 where $n = 1.02$ for the W-Si diode and $n = 1.04$ for the W-GaAs diode.[21] The extrapolated value of current density to zero voltage gives the saturation current J_S, and the barrier height can be

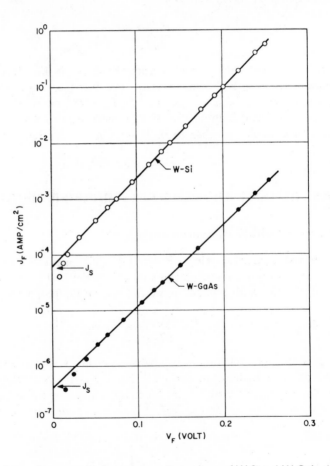

Fig. 17 Forward current density versus applied voltage of W-Si and W-GaAs diodes. (After Crowell et al., Ref. 21.)

obtained from the following equation:

$$\phi_{Bn} = \frac{kT}{q} \ln\left(\frac{A^{**}T^2}{J_S}\right). \tag{74}$$

The value of ϕ_{Bn} is not very sensitive to the choice of A^{**}, since at room temperature, a 100% increase in A^{**} will cause an increase of only 0.018 volt in ϕ_{Bn}. The theoretical relationship between J_S and ϕ_{Bn} (or ϕ_{Bp}) at room temperature is plotted in Fig. 18 for $A^{**} = 120$ amp/cm^2/°K. For other values of A^{**}, parallel lines can be drawn on this plot to obtain the proper relationship. The experimental values of the barrier heights obtained from the current-voltage measurements are listed in Table 8.4 (compiled by Mead[22]).

From Eq. (72a) we can obtain the junction resistance R_j

$$R_j \equiv \frac{\partial V}{\partial I} = \frac{nkT}{qJA_j} \tag{75}$$

where A_j is the junction area. Typical experimental results of R_j versus I are shown in Fig. 19 for Au-Si and Au-GaAs diodes. Also shown is the result for Si point contact to be discussed later. We note for sufficiently high forward bias the junction resistance does not decrease to zero as predicted by Eq. (75) but instead approaches a constant value. This value is the series resistance R_S given, see Fig. 15, by

$$R_S = \frac{1}{A_j} \int_{x_1}^{x_2} \rho(x)\,dx + \frac{\rho_B}{4r} + R_c \tag{76}$$

where the first term on the right-hand side is the series resistance of the quasi-neutral region, and x_1 and x_2 are the depletion layer edge and the epitaxial layer-substrate boundary respectively. The second term is the spreading resistance of the metal-semiconductor barrier substrate with a resistivity ρ_B and a circular area of radius r ($A_j = \pi r^2$). The last term R_c is the resistance due to the ohmic contact with the substrate.

An important figure of merit for microwave application of the Schottky diodes is the forward bias cutoff frequency, f_{c0}, which is defined as

$$f_{c0} \equiv \frac{1}{2\pi R_F C_F} \tag{77}$$

where R_F and C_F are the resistance and capacitance, respectively, at a forward bias of 0.1 V to the flat-band condition.[23] The value of f_{c0} is considerably smaller than the corresponding cutoff frequency using zero-bias capacitance, and can be used as a lower limit for practical consideration. A typical result[24] is shown in Fig. 20. We note that for a given doping and a given junction diameter (e.g., 10 μm), the Schottky diode of n-type GaAs gives the highest cutoff frequency. This is mainly due to the fact that the electron mobility is considerably higher in GaAs.

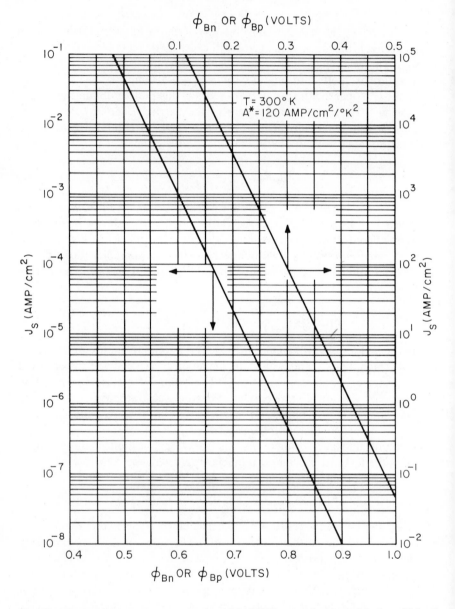

Fig. 18 Theoretical saturation current density at 300°K versus barrier height for a Richardson constant of 120 amp/cm²/°K².

TABLE 8.4

MEASURED SCHOTTKY BARRIER HEIGHTS (300°K)

Semiconductor	Metal	Barrier Height (V)		
		I-V	C-V	Photo
n-AlAs (Vac Cleave)	Au			1.2
	Pt			1.0
p-AlSb (Vac Cleave)	Au		0.53	0.55
	Au (77°K)		0.59	
p-BN (Chem)	Au		3.1	
p-BP	Au			0.87
n-CdS (Vac Cleave)	Au		0.80	0.78
	Cu		0.35	0.36
	Ni			0.45
	Al (77°)			
	Ag		0.58	0.56
	Pt		0.86	0.85
n-CdS (Chem)	Pt	1.2	1.2	1.1
	Au	0.68	0.66	0.68
	Pd	0.61	0.59	0.62
	Cu	0.47	0.41	0.50
	Ag		0.35	
n-CdSe (Vac Cleave)	Pt			0.37
	Au			0.49
	Ag			0.43
	Cu			0.33
n-CdTe (Vac Cleave)	Au		0.71	0.60
	Pt		0.76	0.58
	Ag		0.81	0.66
	Al			0.76
p Diamond	Au		1.35	
n-GaAs (Vac Cleave)	Au		0.95	0.90
	Pt		0.94	0.86
	Be		0.82	0.81
	Ag		0.93	0.88
	Cu		0.87	0.82
	Al		0.80	0.80
	Al (77°)		0.88	
p-GaAs (Vac Cleave)	Au		0.48	0.42
	Au (77°)		0.46	
	Pt (77°)		0.48	

TABLE 8.4 (Cont.)

Semiconductor	Metal	Barrier Height (V)		
		I-V	C-V	Photo
p-GaAs (Vac Cleave)	Ag (77°)		0.44	
(cont.)	Cu (77°)		0.52	
	Al		0.63	0.50
	Al (77°)		0.61	
n-GaAs	W	0.71	0.77	0.80
p-GaP (Chem)	Au			0.715
p-GaP	Au	0.68	0.75	0.72
n-GaP	Au	1.1	1.3	1.3
n-GaP (Chem)	Cu		1.34	1.20
	Al		1.14	1.05
	Au		1.34	1.28
	Pt		1.52	1.45
	Mg		1.09	1.04
	Ag			1.20
n-GaSb (Vac Cleave)	Au		0.61	0.60
	Au (77°)		0.75	
p-GaSb (Vac Cleave)	Au (77°)	ohmic		
n-Ge (Vac Cleave)	Au		0.45	
	Au (77°)		0.50	
	Al		0.48	
n-Ge	W	0.48		
n-InAs (Vac Cleave and Chem)	Au, Ag, Al (77°)	ohmic		
p-InAs (Vac Cleave and Chem)	Au (77°)		0.47	
n-InP (Vac Cleave)	Au		0.49	0.52
	Au (77°)		0.56	
	Cu (77°)		0.5	
	Ag		0.54	0.57
	Ag (77°)		0.50	
p-InP (Vac Cleave)	Au		0.76	
	Au (77°)		0.78	
n-InSb (Vac Cleave and Chem)	Au (77°)		0.17	
	Ag (77°)		0.18	
p-InSb (Vac Cleave and Chem)	Au (77°)	ohmic		

TABLE 8.4 (Cont.)

Semiconductor	Metal	Barrier Height (V)		
		I-V	C-V	Photo
n-PbO	Ag			0.95
	Bi			0.94
	Ni			0.96
	Pb			0.95
	In			0.93
n-Si (Chem)	Au	0.79	0.80	0.78
p-Si (Chem)	Au	0.25		
n-Si (Chem)(200)	Au			0.82
n-Si (Chem)	Mo	0.59	0.57	0.56
n-Si (Back Sputtering)	PtSi	0.85	0.86	0.85
p-Si (Back Sputtering)	PtSi	0.20		
n-Si (Chem)	W	0.67	0.65	0.65
n-SiC (hexag) (Chem)	Au		1.95	
	Al		2.0	
n-SnO$_2$	Au			0.98
	Ag			0.65
	Cu			0.47
n-ZnO (Vac Cleave)	Au			0.65
	Pt			0.75
	Pd			0.68
	Ag			0.68
	Cu	0.45		
	In	<0.3		
	Al	ohmic		
	Ti	<0.3		
n-ZnS (Vac Cleave)	Au			2.0
	Pd			1.87
	Pt			1.84
	Cu			1.75
	Ag			1.65
	In			1.50
	Al			0.8
	Ti			1.1
	Mg			0.82
n-ZnSe (Vac Cleave)	Au			1.36
	Pt			1.40
	Cu			1.10
	Mg	<0.4		0.70

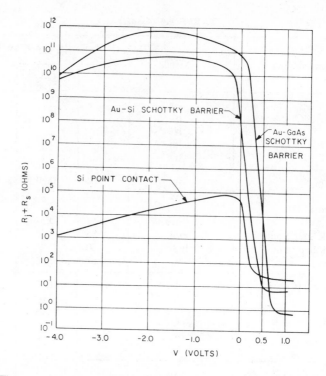

Fig. 19 The sum of the junction resistance and the series resistance versus applied voltage for Au-Si, Au-GaAs, and point-contact diodes.
(After Irvin and Vanderwal, Ref. 24.)

B. Reverse Characteristics. In the reverse direction the dominant effect is due to the Schottky barrier lowering, or

$$J_R \simeq J_S \,(\text{for } V_R > 3kT/q)$$

$$= A^{**}T^2 \exp\!\left(-\frac{q\phi_{B0}}{kT}\right)\!\exp\!\left(+\frac{q\sqrt{q\mathscr{E}/4\pi\varepsilon_s}}{kT}\right) \tag{78}$$

where

$$\mathscr{E} = \sqrt{\frac{2qN_D}{\varepsilon_s}\left(V + V_{bi} - \frac{kT}{q}\right)}.$$

If the barrier height $q\phi_{Bn}$ is reasonably smaller than the band gap such that the depletion-layer generation-recombination current is small in comparison

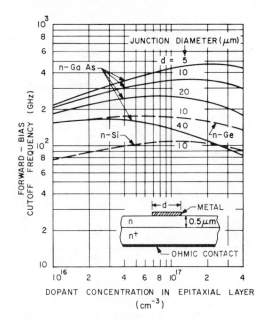

Fig. 20 The forward-bias cutoff frequency versus doping concentration in the epitaxial layer for a 0.5 μm epitaxial layer and with various junction diameters. (Ref. 24.)

with the Schottky emission current, then the reverse current will increase gradually with the reverse bias as given by Eq. (78).

For most of the practical Schottky diodes, however, the dominant reverse current component is the edge leakage current which is caused by the sharp edge around the periphery of the metal plate. This is similar to the junction curvature effect (with $r_j \to 0$) as discussed in Chapter 3. To eliminate this effect, metal-semiconductor diodes have been fabricated with a diffused guard ring as shown[25] in Fig. 21(a). The guard ring is a deep p-type diffusion, and the doping profile is tailored to give the p-n junction a higher breakdown voltage than that of the metal-semiconductor contact. Because of the elimination of the sharp-edge effect, near-ideal forward and reverse I-V characteristics have been obtained. Figure 21(b) shows a comparison between experimental measurement from a PtSi-Si diode with guard ring and theoretical calculation based on Eq. (78). The agreement is excellent. The sharp increase of current near 30 V is due to avalanche breakdown and is expected for the diode with a donor concentration of 2.5×10^{16} cm^{-3}.

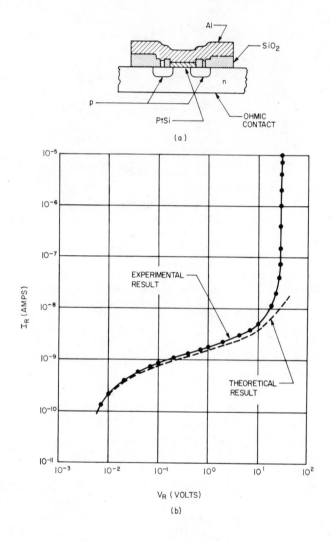

Fig. 21
(a) PtSi-Si diode with a diffused guard ring.
(b) Comparison of experiment with theoretical prediction of Eq. (78) for a PtSi-Si diode.
(After Lepselter and Sze, Ref. 25.)

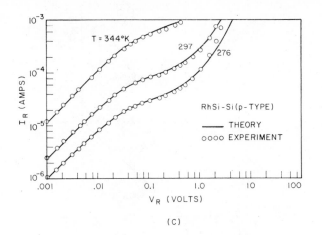

Fig. 21
(c) Comparison of experiment with theoretical prediction.
(After Andrews and Lepselter, Ref. 19a.)

In some Schottky diodes there is an additional effect due to intrinsic barrier lowering, i.e., $\partial\phi_{B0}/\partial\mathscr{E} \neq 0$. In other words, in addition to the image-force lowering effect the intrinsic barrier height ϕ_{B0} is also lowered as the electric field increases.[28b] Figure 21(c) shows a comparison of theory and experiment of the reverse characteristics for a RhSi-Si diode. The theory is calculated based on a value of $\partial\phi_{B0}/\partial\mathscr{E} = 17$ Å. We note that there is general agreement particularly in the temperature dependence of the reverse characteristics.[19a]

(2) Capacitance-Voltage Measurement

The barrier height can also be determined by the capacitance measurement. When a small ac voltage is superimposed upon a dc bias, charges of one sign are induced on the metal surface and charges of the opposite sign in the semiconductor. The relationship between C and V is given by Eq. (14). Figure 22 shows some typical results where $1/C^2$ is plotted against the applied voltage. From the intercept on the voltage axis the barrier height can be determined:[21,26]

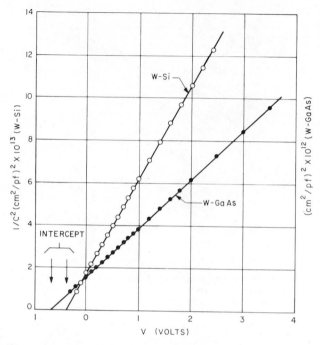

Fig. 22 $1/C^2$ versus applied voltage for W-Si and W-GaAs diodes. (After Crowell et al., Ref. 21.)

$$\phi_{Bn} = V_i + V_n + \frac{kT}{q} - \Delta\phi \tag{79}$$

where V_i is the voltage intercept, and V_n the depth of the Fermi level below the conduction band which can be computed if the doping concentration is known. From the slope one can determine the carrier density, Eq. (15c). (This method can also be used to measure the doping variation in an epitaxial layer.) Table 8.4 lists some results of barrier heights measured by the capacitance method.[22]

(3) Photoelectric Measurement

The photoelectric measurement is the most accurate and most direct method of determining the barrier height.[27] When a monochromatic light is incident upon a metal surface, photocurrent may be generated. The basic setup is shown in Fig. 23. For the front illumination the light can generate excited electrons in the metal, process (1), if $hv > q\phi_{Bn}$, and also can generate electron-hole pairs in the semiconductor, process (2), if the metal film is thin enough and $hv > E_g$. For the back illumination, photoelectrons can be generated, process (1), if $hv > q\phi_{Bn}$; however, when $hv > E_g$, the light will be strongly absorbed at the back semiconductor surface, and the photo-excited electron-hole pairs have very small probability of reaching the metal-semiconductor interface.

The photocurrent per absorbed photon, R, as a function of the photon energy, hv, is given by the Fowler theory:[28]

$$R \sim \frac{T^2}{\sqrt{E_s - hv}} \left[\frac{x^2}{2} + \frac{\pi^2}{6} - \left(e^{-x} - \frac{e^{-2x}}{4} + \frac{e^{-3x}}{9} - \cdots \right) \right] \quad \text{for} \quad x \ge 0 \quad (80)$$

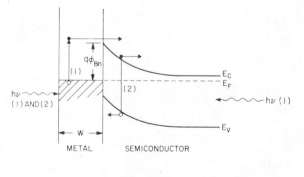

(a)

(b)

Fig. 23
(a) Schematic setup for photoelectric measurement.
(b) Energy band diagram for photoexcitation processes.

where hv_0 is the barrier height ($q\phi_{Bn}$), E_s the sum of hv_0 and the Fermi energy measured from the bottom of the metal conduction band, and $x \equiv h(v - v_0)/kT$. Under the condition that $E_s \gg hv$, and $x > 3$, Eq. (80) reduces to

$$R \sim (hv - hv_0)^2 \quad \text{for} \quad h(v - v_0) > 3kT \quad (81)$$

or

$$\sqrt{R} \sim h(v - v_0). \quad (81a)$$

When the square root of the photoresponse is plotted as a function of photon energy, a straight line should be obtained, and the extrapolated value on the energy axis should give directly the barrier height. Figure 24 shows the photoresponse of W-Si and W-GaAs diodes, with barrier heights of 0.65 eV and 0.80 eV respectively. Similar results are also listed in Table 8.4 for other metal-semiconductor systems.

The photoelectric measurement has been used to determine the image-force dielectric constant of Au-Si diodes.[8] The photoresponse is shown in Fig. 25 for three different applied voltages. From the values of voltage and doping concentration the electric field can be determined, Eq. (12); and the shift of the photothreshold gives directly the barrier lowering. A plot of $\Delta\phi$ versus

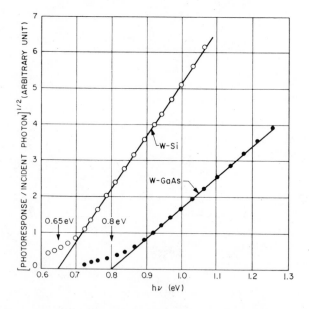

Fig. 24 Square root of the photoresponse per incident photon versus photon energy for W-Si and W-GaAs diodes. The extrapolated values are the corresponding barrier heights. (Ref. 21.)

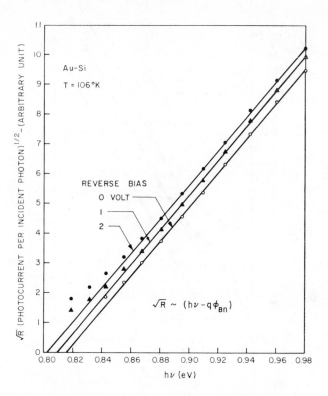

Fig. 25 Square root of the photoresponse per incident photon, $\sqrt{R}$, versus photon energy for three different biasing conditions.
(After Sze et al., Ref. 8.)

$\sqrt{\mathscr{E}}$ can be made and the image-force dielectric constant $(\varepsilon_s/\varepsilon_0)$ can be determined (Fig. 3). It is important to note that if the surface states at the metal-semiconductor interface also play a role in the potential distribution, then a larger barrier lowering will result.[28a]

The photoelectric measurement can also be used to study the temperature dependence of the barrier heights.[29] Figure 26(a) shows the photoresponse of Au-Si diodes at three different temperatures. The shift correlates reasonably well with the temperature dependence of the silicon band gap as shown in Fig. 26(b). This result implies that the Fermi level at the Au-Si interface is pinned in relation to the valence band edge. Another interesting example is to use the photoelectric method to measure the direct and indirect band gaps of semiconductors. Figure 27(a) shows the photoresponse (of front

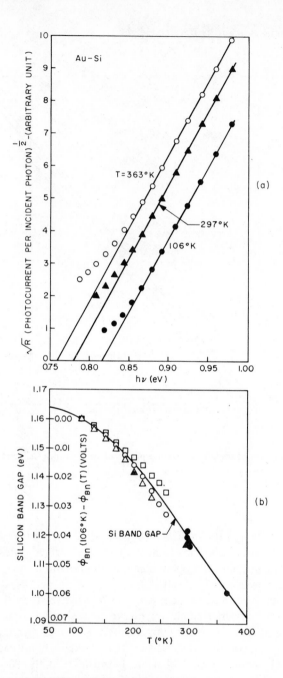

Fig. 26
(a) $\sqrt{R}$ versus $h\nu$ at three different lattice temperatures.
(b) Variation of the Au-Si barrier height with temperature.
(After Crowell et al., Ref. 29.)

408

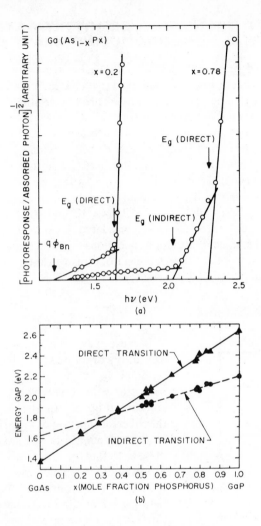

Fig. 27
(a) $\sqrt{R}$ versus $h\nu$ for Au-Ga(As$_{1-x}$P$_x$) diodes.
(b) Measured direct and indirect band gaps from (a) as a function of the phosphorus mole fraction.
(After Spitzer and Mead, Ref. 30.)

illumination) for two samples of $\text{Ga}(\text{As}_{1-x}P_x)$ at room temperature.[30] The breaking points of the curves indicate the direct and indirect band-to-band transitions. The threshold energies as a function of the mole fraction of phosphorus are shown in Fig. 27(b).

6 CLAMPED TRANSISTOR, SCHOTTKY GATE FET, AND METAL-SEMICONDUCTOR IMPATT DIODE

As mentioned previously, Schottky diode behavior is electrically similar to a one-sided abrupt *p-n* junction, and yet it can be operated as a majority-carrier device with inherent fast response. Thus the terminal functions of a *p-n* junction diode as considered in Chapter 3 can also be performed by a Schottky diode with one exception for the charge-storage diode. This is because the charge-storage time in a majority-carrier device is extremely small.

In addition, Schottky diodes can be used (1) as the drain and source contacts in an insulated-gate field-effect transistor to be discussed in Chapter 10, (2) as the emitter and collector junctions in a hot-electron transistor in Chapter 11, (3) as photodetectors in Chapter 12, and (4) as the third terminal electrode in a Gunn oscillator to be considered in Chapter 14. In this section we shall consider three important applications of the Schottky diode: (1) in a clamped transistor, (2) as the gate electrode of a field-effect transistor, and (3) as a passivated IMPATT oscillator.

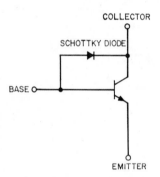

Fig. 28 Composite transistor with a Schottky diode connected between the base and collector terminals.
(After Tada and Laraya, Ref. 32.)

(1) Clamped Transistor[31,32]

Because of its fast response a Schottky diode can be incorporated into the base-collector terminal to form a clamped (composite) transistor with a very short saturation time constant, Fig. 28. In the saturation region the collector junction of the original transistor is slightly forward-biased, instead of reverse-biased. If the forward voltage drop in the Schottky diode is much lower than the base-collector "on" voltage of the original transistor, most of the excess base current flows through the diode in which minority carriers are not stored. Thus the saturation time is reduced markedly as compared with that of the original transistor. The measured satura-

tion time[32] can be reduced to about 10% of that of the original transistor and can be shorter than 1 ns.

(2) Schottky Barrier Gate Field-Effect Transistor (FET)

The feasibility of a field-effect transistor employing a Schottky barrier gate was first demonstrated by Mead.[33] Figure 29(a) shows a schematic

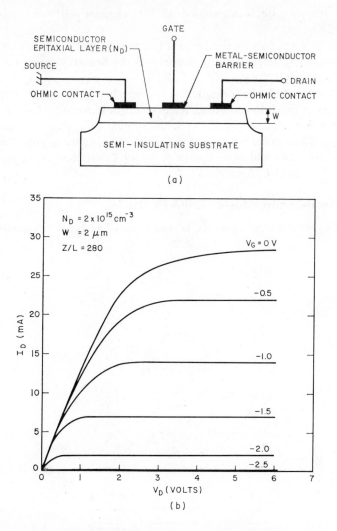

Fig. 29
(a) Schematic diagram of a Schottky barrier gate field-effect transistor.
(After Mead, Ref. 33.)
(b) Output characteristics of a GaAs Schottky gate FET.
(After Hooper and Lehrer, Ref. 34.)

device structure where the notations are identical to those defined in Chapter 7.
Figure 29(b) shows[34] the output (drain) characteristic of a GaAs Schottky
barrier gate FET. The device is fabricated using an n-type GaAs epitaxial
layer 2 μm thick with doping 2×10^{15} cm^{-3} on a semi-insulating GaAs
substrate Source and drain ohmic contacts are alloyed Ag-In-Ge, and the

Schottky barrier gate is evaporated Al. The channel width-to-length ratio, (Z/L), is 280.

The *I-V* characteristic is similar to a junction field-effect transistor. There are, however, two important differences: (1) the Schottky-type FET can be made in semiconductors (such as CdS) in which *p*-type doping cannot be easily formed and (2) the formation of the metal-semiconductor contact can be achieved at much lower temperatures than those required for a *p-n* junction. At the present time the GaAs Schottky gate FET gives the best power and noise performances among various types of GaAs transistors. The maximum available gain and noise figures of the FET described above have been shown in Fig. 20 of Chapter 6.

(3) Metal-Semiconductor IMPATT Diode

The feasibility of microwave CW oscillation has been demonstrated in

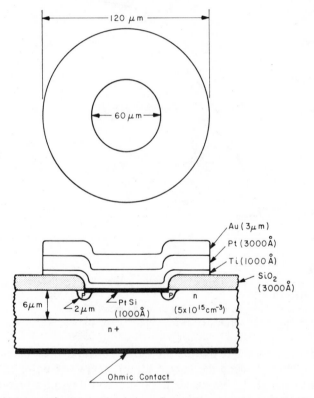

Fig. 30 Device geometry and cross-sectional view of a metal-semiconductor IMPATT diode. (After Sze et al., Ref. 35.)

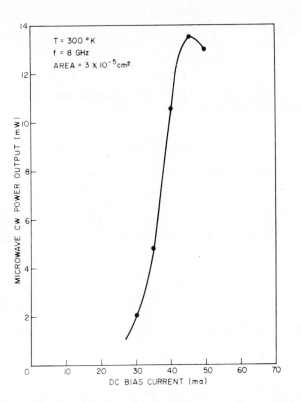

Fig. 31 Microwave CW power output versus dc bias current of a metal-semiconductor IMPATT diode.
(Ref. 35.)

passivated metal-semiconductor diodes.[35] The diodes were fabricated using. PtSi on epitaxial n-type Si substrate, including a diffused p-type guard ring. A typical device geometry is shown in Fig. 30, and the experimental result obtained from the device is shown in Fig. 31. Although the efficiency is low, the device does demonstrate the following interesting features: (1) For a background doping of 6×10^{15} cm^{-3}, the expected breakdown voltage and oscillation frequency of a p^+n diode are about 90 volts and 8 GHz, respectively. Thus the basic microwave characteristics of a metal-semiconductor IMPATT diode are similar to those of a p^+n IMPATT diode with the same background doping. (2) The device shown in Fig. 30 uses beam-lead sealed-junction technology and is completely passivated. (3) The input dc power density of this diode can be as high as half a million watts per cm^2, indicating that the device is capable of high-power operation and that the metal contact can effectively conduct excessive heat away. It is expected that by optimization of the guard-ring structure and the depletion-layer width, high-efficiency metal-semiconductor IMPATT diodes can be realized.

7 MOTT BARRIER, POINT-CONTACT RECTIFIER, AND OHMIC CONTACT

(1) Mott Barrier (or Punch-Through Barrier)[5,6]

When the thickness of the epitaxial layer is reduced (Fig. 15), the depletion layer will eventually reach the heavily doped substrate. A Mott barrier is defined as the limiting case in which the epitaxial layer is much narrower than the required depletion layer width such that the layer is swept out even under forward bias. The band diagrams of a Mott barrier are shown in Fig. 32. When a voltage is applied, the potential energy distribution is given by

$$qV(x) = -q\phi_{Bn} + \frac{q(V_{bi} - V)x}{W}.$$ (82)

Substitution of Eq. (82) into Eq. (45) yields the current-voltage relationship for Mott barriers:

$$J = \frac{q^2 D_n N_C(V_{bi} - V)}{WkT} \exp\left(-\frac{q\phi_{Bn}}{kT}\right)\left\{\frac{\exp\left(\frac{qV}{kT}\right) - 1}{1 - \exp\left[-\frac{q(V_{bi} - V)}{kT}\right]}\right\}.$$ (83)

Comparison of Eq. (83) with Eq. (46) indicates that the Mott barrier is more sensitive to the voltage variation than the Schottky barrier. Since the epitaxial layer is swept out at zero bias, the reverse depletion capacitance is independent of the bias and is given by $\varepsilon_s/W = $ constant.

The dc conductance per unit area of a Mott barrier tends to a limiting value as the reverse voltage increases. This follows from Eq. (83):

$$\lim_{V \to -\infty} |(J/V)| = \left(\frac{q^2 D_n N_C}{WkT}\right)\exp\left(-\frac{q\phi_{Bn}}{kT}\right).$$ (84)

For a Schottky barrier there is no such limiting value. It can be shown from Eqs. (46) and (83) that for a given set of N_D, ϕ_{Bn}, and W the junction resistance near the zero bias, as defined in Eq. (75), of a Schottky barrier is about one-half that of a Mott barrier.

(2) Point-Contact Rectifier[6]

When a small metal wire with a sharp point makes a contact with a semiconductor, one generally obtains a point-contact rectifier. The contact may be just a simple mechanical contact, or may be formed by electrical discharge processes which may result in a small alloyed *p-n* junction.

A point-contact rectifier usually has poor forward and reverse *I-V* characteristics in comparison with a planar Schottky diode. Its characteristics are also difficult to predict from theory, since the rectifiers are subject to wide variations such as the whisker pressure, contact area, crystal structure, surface treatment, whisker composition, and heat or forming processes.

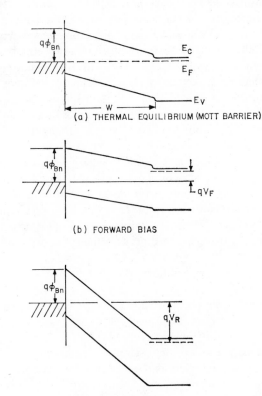

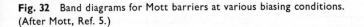

Fig. 32 Band diagrams for Mott barriers at various biasing conditions. (After Mott, Ref. 5.)

The advantage of a point-contact rectifier is its small area, which can give very small capacitance, a desirable feature for microwave application. The disadvantages are its large spreading resistance ($R_S = \rho/2\pi r_0$ where r_0 is the radius of the hemispheric point contact), its large leakage current due mainly to the surface effect which gives rise to poor rectification ratio, and its soft reverse breakdown characteristics due to a large concentration of field beneath the metal point.

(3) Ohmic Contact

There are various methods of forming a metal-semiconductor contact. The methods include thermal evaporation, chemical decomposition, electron-gun bombardment, sputtering, or plating of metals onto chemically etched, mechanically polished, vacuum-cleaved, back-sputtered, heat-treated, or ion-bombarded semiconductor surfaces. To our great surprise, a rectifying barrier is generally formed on n-type semiconductors. This is believed to be due to the surface effect. Surface states give rise to a depletion layer which is only weakly influenced by the metal. Most of the metal-semiconductor contacts are formed in a vacuum system.[36] One of the most important parameters concerning vacuum deposition of metals is the vapor pressure which is defined as the pressure exerted when a solid or liquid is in equilibrium with its own vapor. The vapor pressure[37] versus temperature for the more common elements is shown in Fig. 33(a) and 33(b).

An ohmic contact is defined as a contact which will not add a significant parasitic impedance to the structure on which it is used, and it will not sufficiently change the equilibrium carrier densities within the semiconductor to affect the device characteristics. In other words, an ohmic contact should have a linear and symmetrical current-voltage relationship; it is characterized by having no potential barrier (hence no asymmetry) and an infinite surface recombination velocity (hence linearity). At an ohmic contact the electrons and holes are at their thermal equilibrium values.

In practice, the above ideal ohmic contact can only be approximated. A metal-semiconductor contact is approximately ohmic if the semiconductor is very heavily doped. A direct contact between a metal and a semiconductor, however, does not generally give an ohmic contact, especially when the resistivity of the semiconductor is high. The most common approach is to use metal-n^+-n or metal-p^+-p contacts as ohmic contacts. For ohmic contacts on Ge and Si, one can first evaporate Au-Sb alloy (with 0.1 percent Sb) onto n-type semiconductors. These contacts are then alloyed at the corresponding eutectic temperature into the semiconductors under an inert gas (such as argon or nitrogen). For n-type GaAs, one can first evaporate 5000 Å indium onto a GaAs surface. This is followed by 5000 Å of nickel deposited by the electroless process. The contact is alloyed into GaAs under a pressure of 10^{-2} torr of forming gas (15 percent H_2 and 85 percent N_2) using an alloying cycle of about 30 seconds at 300°C. Similarly one can use Ni-Sn or In-Au combinations for the GaAs ohmic contacts.

8 SPACE-CHARGE-LIMITED DIODE

In this section we shall consider a related metal-semiconductor diode in which the impurity concentration of the semiconductor is very low.[38] Figure 34(a) shows[39] the energy band diagram of a space-charge-limited

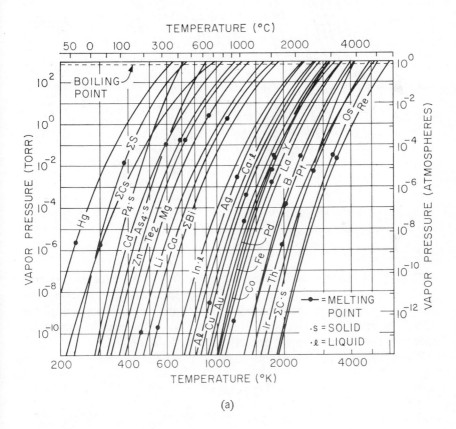

(a)

Fig. 33 Vapor pressure versus temperature for solid and liquid elements. (After Honig, Ref. 37.)

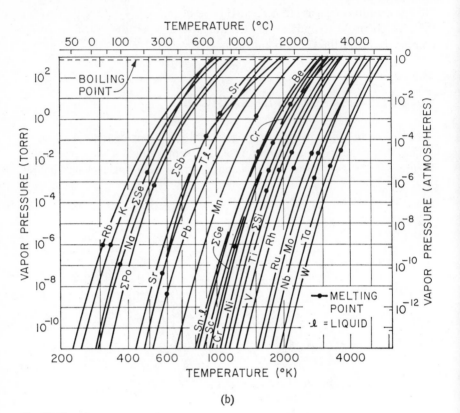

(b)

Fig. 33 (Cont.)

(SCL) diode at thermal equilibrium. Figure 34(b) and (c) are for the forward-bias and reverse-bias conditions. This is drawn specifically for the case when electrons are the charge carriers, but a similar diagram can be obtained for holes. The injecting contact (source) is shown as n^{++} material and the drain contact is shown as an abrupt potential energy discontinuity.

When current is carried by both electrons and holes we have

$$
\left.
\begin{aligned}
J_n &= q\mu_n n\mathscr{E} \\
J_p &= q\mu_p p\mathscr{E} \\
J_{tot} &= J_n + J_p
\end{aligned}
\right\}
\tag{85}
$$

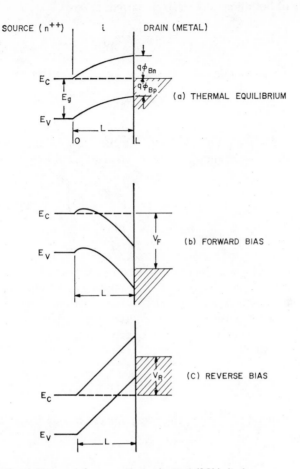

SOURCE (n^{++}) i DRAIN (METAL)

Fig. 34 Energy band diagrams of a space-charge-limited (SCL) diode. (After Buget and Wright, Ref. 39.)

where the symbols have their usual meaning and J_{tot} is the total current density. The Poisson equation is given by

$$\frac{\partial \mathscr{E}}{\partial x} = -\frac{q(n - p)}{\varepsilon_s}.$$ (86)

Note the above equation does not contain the impurity concentration which is assumed to be negligibly smaller than n and p.

Under the assumptions (1) low current density, (2) short carrier transit time across the i region, and (3) no hole-electron recombination, then the divergence of both hole and electron current is zero:

$$\frac{dJ_n}{dx} = \frac{dJ_p}{dx} = 0. \tag{87}$$

From Eq. (86) we obtain

$$\zeta \equiv \frac{p}{n} = \frac{\mu_n J_p}{\mu_p J_n}. \tag{88}$$

Substitution for p in Eq. (86) and integration with the usual boundary conditions that $n = \infty$ and $\mathscr{E} = 0$ at $x = 0$ gives the results

$$n = \frac{1}{q} \left[\frac{\varepsilon_s J_F}{2x(1 - \zeta)(\mu_n + \zeta\mu_p)} \right]^{1/2} \tag{89}$$

$$J_F = \frac{9\varepsilon_s}{8} \frac{(\mu_n + \zeta\mu_p)}{(1 - \zeta)} \frac{V_F^2}{L^3}. \tag{90}$$

In this equation V_F is the applied drain forward voltage and L is the source to drain spacing. The value of ζ can be found by consideration of conditions at the drain contact. At small applied voltages current is small. This justifies regarding the hole density at the drain as being in thermal equilibrium with the drain metal. Thus $p_D(x = L) = N_V \exp(-q\phi_{Bp}/kT)$, and from Eqs. (88) and (89) we have

$$\zeta = \frac{4q p_D L^2}{3\varepsilon_s V_D + 4q p_D L^2}. \tag{91}$$

Substitution in Eq. (90) gives the result for double-injection (or two-carrier injection):

$$J_F = \frac{3q p_D(\mu_n + \mu_p)V_F}{2L} + \frac{9\varepsilon_s \mu_n V_F^2}{8L^3}. \tag{92}$$

At large applied voltages the hole current across the drain contact will saturate at the value J_S given by the Richardson thermionic emission equation. In this case we have from Eqs. (85) and (88) that

$$\zeta = \frac{\mu_n J_S}{\mu_p(J_F - J_S)}. \tag{93}$$

Substitution in Eq. (90) gives the result

$$J_F = \left[1 + \frac{\mu_n}{\mu_p}\right]J_S + \frac{9\varepsilon_s \mu_n V_F^2}{8L^3}. \tag{94}$$

If p_D and J_S are sufficiently small, both Eqs. (92) and (94) reduce to the standard Mott-Gurney Law[40] for one-carrier injection:

$$J_F = \frac{9\varepsilon_s \mu_n V_F^2}{8L^3}.$$ (95)

Equations (92), (94), and (95) are illustrated in Fig. 35 where (1) is the ideal Mott-Gurney square law of Eq. (95); (2) is the linear current term of Eq. (92); and (3) is the saturation current term of Eq. (94). The experimental data show the resultant I-V characteristic of an Au-Si SCL diode with resistivity of about 25,000 Ω-cm and $L \simeq 10$ μm. It is clear that a variety of I-V characteristics may be expected depending on the relative magnitude of the terms involved.

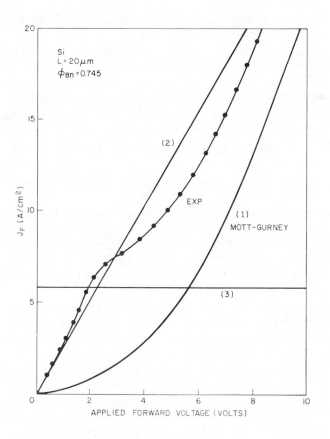

Fig. 35 Theoretical and experimental results for SCL diode. (Ref. 39.)

REFERENCES

1. F. Braun, "Über die Stromleitung durch Schwefelmetalle," Ann. Physik Chem., *153*, 556 (1874).

2. J. C. Bose, U.S. Patent 775,840 (1904).

3. A. H. Wilson, Proc. Roy. Soc., *A133*, 458 (1931).

4. W. Schottky, Natarwiss., *26*, 843 (1938).

5. N. F. Mott, "Note on the Contact Between a Metal and an Insulator or Semiconductor," Proc. Camb. Phil. Soc., *34*, 568 (1938).

6. H. K. Henisch, *Rectifying Semiconductor Contacts*, Oxford at the Clarendon Press., Oxford (1957).

7. V. S. Formenko, *Handbook of Thermionic Properties*, edited by G. V. Samsonov, Plenum Press Data Division, New York (1966).

8. S. M. Sze, C. R. Crowell, and D. Kahng, "Photoelectric Determination of the Image Force Dielectric Constant for Hot Electrons in Schottky Barriers," J. Appl. Phys., *35*, 2534 (1964).

9. C. D. Salzberg and G. G. Villa, J. Opt. Soc. Am., *47*, 244 (1957).

10. A. M. Cowley and S. M. Sze, "Surface States and Barrier Height of Metal-Semiconductor Systems," J. Appl. Phys., *36*, 3212 (1965).

11. J. Bardeen, "Surface States and Rectification at a Metal Semiconductor Contact," Phys. Rev., *71*, 717 (1947).

12. C. A. Mead and W. G. Spitzer, "Fermi-Level Position at Metal-Semiconductor Interfaces," Phys. Rev., *134*, A713 (1964).

13. D. Pugh, "Surface States on the ⟨111⟩ Surface of Diamond," Phys. Rev. Letters, *12*, 390 (1964).

14. H. A. Bethe, "Theory of the Boundary Layer of Crystal Rectifiers," MIT Radiation Laboratory, Report 43-12 (1942).

15. C. R. Crowell, "The Richardson Constant for Thermionic Emission in Schottky Barrier Diodes," Solid State Electron., *8*, 395 (1965).

16. C. R. Crowell and S. M. Sze, "Current Transport in Metal-Semiconductor Barriers," Solid State Electron., *9*, 1035 (1966).

16a. F. A. Padovani and R. Stratton, "Field and Thermionic-Field Emission in Schottky Barriers," Solid State Electron., *9*, 695 (1966).

17. C. R. Crowell and S. M. Sze, "Electron-Phonon Collector Backscattering in Hot Electron Transistors," Solid State Electron., *8*, 673 (1965).

18. C. R. Crowell and S. M. Sze, "Electron-Optical-Phonon Scattering in the Emitter and Collector Barriers of Semiconductor-Metal-Semiconductor Structures," Solid State Electron., *8*, 979 (1965).

19. C. R. Crowell and S. M. Sze, "Quantum Mechanical Reflection at Metal-Semiconductor Barriers," J. Appl. Phys., *37*, 2683 (1966).

19a. J. M. Andrews and M. P. Lepselter, " Reverse Current-Voltage Characteristics of Metal-Silicide Schottky Diodes," IEEE Solid State Device Conference, Washington, D.C. (Oct. 1968).

20. D. L. Scharfetter, "Minority Carrier Injection and Charge Storage in Epitaxial Schottky Barrier Diodes," Solid State Electron., *8*, 299 (1965).

21. C. R. Crowell, J. C. Sarace, and S. M. Sze, "Tungsten-Semiconductor Schottky Barrier Diodes," Trans. Met. Soc. AIME, *233*, 478 (1965).

22. C. A. Mead, "Metal-Semiconductor Surface Barriers," Solid State Electron., *9*, 1023 (1966).

23. N. C. Vanderwal, "A Microwave Schottky-Barrier Varistor Using GaAs for Low Series Resistance," IEEE Int. Elec. Devices Meeting, Washington, D.C. (Oct. 18-20, 1967).

24. J. C. Irvin and N. C. Vanderwal " Schottky-Barrier Devices," a chapter of *Microwave Semiconductor Devices and Their Circuit Applications*, Ed. H. A. Watson, McGraw-Hill Book Co. (1968).

25. M. P. Lepselter and S. M. Sze, "Silicon Schottky Barrier Diode with Near-Ideal *I-V* Characteristics," Bell Syst. Tech. J., *47*, 195 (1968).

26. A. M. Goodman, "Metal-Semiconductor Barrier Height Measurement by the Differential Capacitance Method—One Carrier System," J. Appl. Phys., *34*, 329 (1963).

27. C. R. Crowell, W. G. Spitzer, L. E. Howarth, and E. E. Labate, "Attenuation Length Measurements of Hot Electrons in Metal Films," Phys. Rev., *127*, 2006 (1962).

28. R. H. Fowler, Phys. Rev., *38*, 45 (1931).

28a. G. H. Parker, T. C. McGill, C. A. Mead, and D. Hoffman, "Electric Field Dependence of GaAs Schottky Barriers," Solid State Electron., *11*, 201 (1968).

28b. C. R. Crowell, H. B. Shore, and E. E. Labate, "Surface State and Interface Effect in Schottky Barriers at *n*-Type Silicon Surface," J. Appl. Phys., *36*, 3843 (1965).

29. C. R. Crowell, S. M. Sze, and W. G. Spitzer, "Equality of the Temperature Dependence of the Gold-Silicon Surface Barrier and the Silicon Energy Gap in an *n*-Type Si Diode," Appl. Phys. Letters, *4*, 91 (1964).

29a. D. Kahng, "Conduction Properties of the An-*n*-Type-Si Schottky Barrier," Solid State Electron., *6*, 281 (1963).

30. W. G. Spitzer and C. A. Mead, "Conduction Band Minima of $Ga(As_{1-x}P_x)$," Phys. Rev., *133*, A872 (1964).

31. R. H. Baker, "Maximum Efficiency Switching Circuit," MIT Lincoln Lab., Lexington, Mass., Report TR-110 (1956).

32. K. Tada and J. L. R. Laraya, "Reduction of the Storage Time of a Transistor Using a Schottky-Barrier Diode," Proc. IEEE, *55*, 2064 (1967).

33. C. A. Mead, "Schottky Barrier Gate Field-Effect Transistor," Proc. IEEE, *54*, 307 (1966).

34. W. W. Hooper and W. I. Lehrer, "An Epitaxial GaAs Field-Effect Transistor," Proc. IEEE, *55*, 1237 (1967).

35. S. M. Sze, M. P. Lepselter, and R. W. MacDonald, "Metal-Semiconductor IMPATT Diode," Solid State Electron., *11*, (1968).

36. For a general reference on vacuum deposition, see L. Holland, *Vacuum Deposition of Thin Films*, Chapman and Hall Ltd., London (1966).

37. R. E. Honig, "Vapor Pressure Data for the Solid and Liquid Elements," RCA Review, *23*, 567 (1962).

38. For a general review, see M. A. Lampert, *Injection Currents in Solids*, Academic Press, New York (1965).

39. U. Buget and G. T. Wright, "Space-Charge-Limited Current in Silicon," Solid State Electron., *10*, 199 (1967).

40. N. F. Mott and R. W. Gurney, *Electronic Processes in Ionic Crystals*, Clarendon Press, Oxford (1940).

■ INTRODUCTION

■ IDEAL METAL-INSULATOR-
SEMICONDUCTOR DIODE

■ SURFACE STATES,
SURFACE CHARGES,
AND SPACE CHARGES

■ EFFECTS OF METAL WORK FUNCTION,
CRYSTAL ORIENTATION,
TEMPERATURE, ILLUMINATION,
AND RADIATION
ON MIS CHARACTERISTICS

■ SURFACE VARACTOR, AVALANCHE,
TUNNELING, AND
ELECTROLUMINESCENT MIS DIODES

■ CARRIER TRANSPORT
IN INSULATING FILMS

9

Metal-Insulator-Semiconductor Diodes

I INTRODUCTION

The metal-insulator-semiconductor (MIS) diode is the most useful device in the study of semiconductor surfaces. Since the reliability and stability of all semiconductor devices are intimately related to their surface conditions, an understanding of the surface physics with the help of MIS diodes is of great importance to device operations. In this chapter we shall be mainly concerned with the metal-oxide-silicon (MOS) system. This system has been extensively studied because it is directly related to most of the planar devices and integrated circuits.

The MIS structure was first proposed as a voltage variable capacitor in 1959 by Moll[1] and by Pfann and Garrett.[2] Its characteristics were then analyzed by Frankl[3] and Lindner.[4] The MIS diode was first employed in the study of a thermally oxidized silicon surface by Terman[5] and by Lehovec and Slobodskoy.[6] A comprehensive treatment of the theory of the semiconductor surface can be found in *Semiconductor Surfaces* by Many et al.[7] In this chapter we will first consider the ideal MIS diode, which will serve as a basis for understanding of nonideal MIS diode characteristics. Section 3 presents the nonideal situations caused by surface states, surface charges, and space charges in the insulator. The ion transport as discovered by Snow et al,[8] the conductance method as proposed by Nicollian and Goetzberger,[9] the time constant dispersion, and $1/f$ noise will also be considered in Section 3. Section 4 is concerned with the effects on MIS characteristics due to metal work function, crystal orientation, lattice temperature, illumination, and γ-ray radiation. Section 5 considers some of the applications of MIS diodes as electronic and optical devices, and as a tool to study fundamental physical processes. Section 6 presents a brief discussion of the carrier transport and maximum dielectric strength of thin insulating films.

2 IDEAL METAL-INSULATOR-SEMICONDUCTOR (MIS) DIODE

The MIS structure is shown in Fig. 1 where d is the thickness of the insulator and V is the applied voltage on the metal field plate. This defines that the voltage, V, is positive when the metal plate is positively biased with respect to the ohmic contact, and V is negative when the metal plate is negatively biased with respect to the ohmic contact. This convention will be used throughout this chapter.

The energy band diagram of an ideal MIS structure for $V = 0$ is shown in Fig. 2, where Figs. 2(a) and 2(b) are for n-type and p-type semiconductors respectively. An ideal MIS diode is defined as follows: (1) at zero applied bias there is no energy difference between the metal work function ϕ_m and

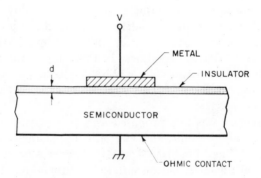

Fig. 1 Metal-insulator-semiconductor (MIS) structure.

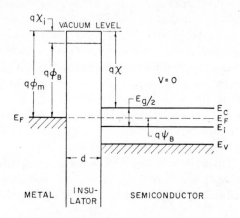

(a) IDEAL MIS DIODE (n-TYPE SEMICONDUCTOR)

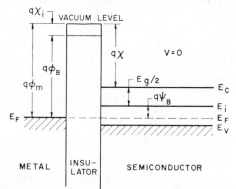

(b) IDEAL MIS DIODE (p-TYPE SEMICONDUCTOR)

Fig. 2 Energy band diagrams for ideal MIS structures at $V = 0$.

the semiconductor work function, or the work function difference ϕ_{ms} is zero:

$$\phi_{ms} \equiv \phi_m - \left(\chi + \frac{E_g}{2q} - \psi_B\right) = 0 \qquad \text{for } n\text{-type} \qquad (1a)$$

$$\phi_{ms} \equiv \phi_m - \left(\chi + \frac{E_g}{2q} + \psi_B\right) = 0 \qquad \text{for } p\text{-type} \qquad (1b)$$

where ϕ_m is the metal work function, χ the semiconductor electron affinity, χ_i the insulator electron affinity, E_g the band gap, ϕ_B the potential barrier between the metal and the insulator, and ψ_B the potential difference between the Fermi level E_F and the intrinsic Fermi level E_i; in other words, the band is flat (flat-band condition) when there is no applied voltage; (2) the only charges which can exist in the structure under any biasing conditions are those in the semiconductor and those with the equal but opposite sign on the metal surface adjacent to the insulator; and (3) there is no carrier transport through the insulator under dc biasing conditions, or the resistivity of the insulator is infinity. The ideal MIS diode theory to be considered in this section serves as a foundation to understand practical MIS structures and to explore the physics of semiconductor surfaces.

When an ideal MIS diode is biased with positive or negative voltages, there are basically three cases which may exist at the semiconductor surface. These cases are illustrated in Fig. 3. Consider the p-type semiconductor first. When

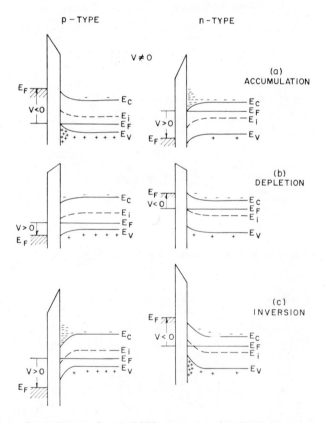

Fig. 3 Energy band diagrams for ideal MIS structures when $V \neq 0$ for both p-type and n-type semiconductors.

a negative voltage ($V < 0$) is applied to the metal plate, Fig. 3(a), the top of the valence band bends upward and is closer to the Fermi level. For an ideal MIS diode there is no current flow in the structure [or $\partial(\text{Imref})/\partial x = 0$], so the Fermi level remains constant in the semiconductor. Since the carrier density depends exponentially on the energy difference ($E_F - E_V$), this band bending causes an accumulation of majority carriers (holes) near the semiconductor surface. This is the case of "accumulation." When a small positive voltage ($V > 0$) is applied, Fig. 3(b), the bands bend downward, and the majority carriers are depleted. This is the case of "depletion." When a larger positive voltage is applied, Fig. 3(c), the bands bend even more downward such that the intrinsic level E_i at the surface crosses over the Fermi level E_F. At this point the number of electrons (minority carriers) at the surface is larger than that of the holes, the surface is thus inverted, and this is the case of "inversion." Similar results can be obtained for the n-type semiconductor. The polarity of the voltage, however, should be changed for the n-type semiconductor.

(1) Surface Space-Charge Region

We shall derive the relations between the surface potential, space charge, and electric field in this subsection. These relations will then be used to derive the capacitance-voltage characteristics of the ideal MIS structure in the next subsection.

Figure 4 shows a more detailed band diagram at the surface of a p-type semiconductor. The potential ψ is defined as zero in the bulk of the semiconductor, and is measured with respect to the intrinsic Fermi level E_i as shown. At the semiconductor surface $\psi = \psi_s$, and ψ_s is called the surface potential. The electron and hole concentrations as a function of ψ are given by the following relations:

$$n_p = n_{po} \exp(q\psi/kT) = n_{po} \exp(\beta\psi) \tag{2}$$

$$p_p = p_{po} \exp(-q\psi/kT) = p_{po} \exp(-\beta\psi) \tag{3}$$

where ψ is positive when the band is bent downward (as shown in Fig. 4), n_{po} and p_{po} are the equilibrium densities of electrons and holes respectively in the bulk of the semiconductor, and $\beta \equiv q/kT$. At the surface the densities are

$$n_s = n_{po} \exp(\beta\psi_s)$$
$$p_s = p_{po} \exp(-\beta\psi_s). \tag{4}$$

It is obvious from the previous discussions and with the help of Eq. (4) that the following regions of surface potential can be distinguished:

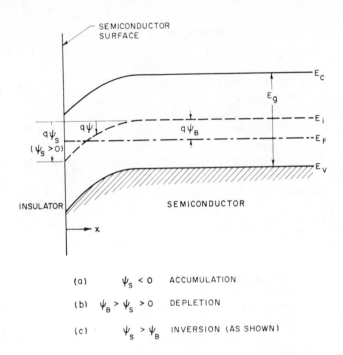

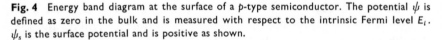

(a) $\psi_s < 0$ ACCUMULATION

(b) $\psi_B > \psi_s > 0$ DEPLETION

(c) $\psi_s > \psi_B$ INVERSION (AS SHOWN)

Fig. 4 Energy band diagram at the surface of a p-type semiconductor. The potential ψ is defined as zero in the bulk and is measured with respect to the intrinsic Fermi level E_i. ψ_s is the surface potential and is positive as shown.

$\psi_s < 0$ Accumulation of holes (bands bend upward)
$\psi_s = 0$ Flat-band condition
$\psi_B > \psi_s > 0$ Depletion of holes (bands bend downward)
$\psi_s = \psi_B$ Midgap with $n_s = p_s = n_i$ (intrinsic concentration)
$\psi_s > \psi_B$ Inversion (electron enhancement, bands bend downward).

The potential ψ as a function of distance can be obtained by using the one-dimensional Poisson equation

$$\frac{\partial^2 \psi}{\partial x^2} = -\frac{\rho(x)}{\varepsilon_s} \tag{5}$$

where ε_s is the permittivity of the semiconductor and $\rho(x)$ is the total space-charge density given by

$$\rho(x) = q(N_D^+ - N_A^- + p_p - n_p) \tag{6}$$

where N_D^+ and N_A^- are the densities of the ionized donors and acceptors respectively. Now, in the bulk of the semiconductor, far from the surface, charge neutrality must exist. Therefore $\rho(x) = 0$ and $\psi = 0$, and we have

$$N_D^+ - N_A^- = n_{po} - p_{po}. \tag{7}$$

In general for any value of ψ, we have from Eqs. (2) and (3)

$$p_p - n_p = p_{po} \exp(-\beta\psi) - n_{po} \exp(\beta\psi). \tag{8}$$

The resultant Poisson's equation to be solved is therefore

$$\frac{\partial^2 \psi}{\partial x^2} = -\frac{q}{\varepsilon_s} [p_{po}(e^{-\beta\psi} - 1) - n_{po}(e^{\beta\psi} - 1)]. \tag{9}$$

Integration of Eq. (9) from the bulk toward the surface[10]

$$\int_0^{\partial\psi/\partial x} \left(\frac{\partial\psi}{\partial x}\right) d\left(\frac{\partial\psi}{\partial x}\right) = -\frac{q}{\varepsilon_s} \int_0^\psi [p_{po}(e^{-\beta\psi} - 1) - n_{po}(e^{\beta\psi} - 1)] \, d\psi \tag{10}$$

gives the relation between the electric field ($\mathscr{E} \equiv -\partial\psi/\partial x$) and the potential ψ:

$$\mathscr{E}^2 = \left(\frac{2kT}{q}\right)^2 \left(\frac{qp_{po}\beta}{2\varepsilon_s}\right) \left[(e^{-\beta\psi} + \beta\psi - 1) + \frac{n_{po}}{p_{po}}(e^{\beta\psi} - \beta\psi - 1)\right]. \tag{11}$$

We shall introduce the following abbreviations:

$$L_D \equiv \sqrt{\frac{2kT\varepsilon_s}{p_{po}q^2}} \equiv \sqrt{\frac{2\varepsilon_s}{qp_{po}\beta}} \tag{12}$$

and

$$F\left(\beta\psi, \frac{n_{po}}{p_{po}}\right) \equiv \left[(e^{-\beta\psi} + \beta\psi - 1) + \frac{n_{po}}{p_{po}}(e^{\beta\psi} - \beta\psi - 1)\right]^{1/2} \geq 0 \tag{13}$$

where L_D is called the extrinsic Debye length for holes. Thus the electric field becomes

$$\mathscr{E} = -\frac{\partial\psi}{\partial x} = \pm \frac{2kT}{qL_D} F\left(\beta\psi, \frac{n_{po}}{p_{po}}\right) \tag{14}$$

with positive sign for $\psi > 0$ and negative sign for $\psi < 0$. To determine the electric field at the surface, we let $\psi = \psi_s$:

$$\mathscr{E}_s = \pm \frac{2kT}{qL_D} F\left(\beta\psi_s, \frac{n_{po}}{p_{po}}\right). \tag{15}$$

Similarly, by Gauss' law the space charge per unit area required to produce this field is

$$Q_s = \varepsilon_s \mathscr{E}_s = \mp \frac{2\varepsilon_s kT}{qL_D} F\left(\beta\psi_s, \frac{n_{po}}{p_{po}}\right). \tag{16}$$

To determine the change in hole density, Δp, and electron density, Δn, per unit area when the ψ at the surface is shifted from zero to a final value ψ_s, it is necessary to evaluate the following expressions:[11]

$$\Delta p = p_{po} \int_0^\infty (e^{-\beta\psi} - 1)\, dx$$

$$= \frac{q p_{po} L_D}{2kT} \int_{\psi_s}^0 \frac{(e^{-\beta\psi} - 1)}{F\left(\beta\psi, \dfrac{n_{po}}{p_{po}}\right)} d\psi \qquad (\text{cm}^{-2}), \qquad (17)$$

$$\Delta n = n_{po} \int_0^\infty (e^{\beta\psi} - 1)\, dx$$

$$= \frac{q n_{po} L_D}{2kT} \int_{\psi_s}^0 \frac{(e^{\beta\psi} - 1)}{F\left(\beta\psi, \dfrac{n_{po}}{p_{po}}\right)} d\psi \qquad (\text{cm}^{-2}). \qquad (18)$$

A typical variation of the space-charge density Q_s as a function of the surface potential ψ_s is shown in Fig. 5 for a p-type silicon with $N_A = 4 \times 10^{15} \text{ cm}^{-3}$ at room temperature. We note that for negative ψ_s, Q_s is positive and corresponds to the accumulation region. The function F is dominated by the first term in Eq. (13), i.e., $Q_s \sim \exp(q|\psi_s|/2kT)$. For $\psi_s = 0$, we have the flat-band condition and $Q_s = 0$. For $\psi_B > \psi_s > 0$, Q_s is negative and we have the depletion case. The function F is now dominated by the second term, i.e., $Q_s \sim \sqrt{\psi_s}$. For $\psi_s \gg \psi_B$, we have the inversion case with the function F dominated by the fourth term, i.e., $Q_s \sim -\exp(q\psi_s/2kT)$. We also note that the strong inversion begins at a surface potential,

$$\psi_s(\text{inv}) \simeq 2\psi_B \simeq \frac{2kT}{q} \ln\left(\frac{N_A}{n_i}\right). \qquad (19)$$

The differential capacitance of the semiconductor space-charge region is given by

$$C_D \equiv \frac{\partial Q_s}{\partial \psi_s} = \frac{\varepsilon_s}{L_D} \frac{\left[1 - e^{-\beta\psi_s} + \dfrac{n_{po}}{p_{po}}(e^{\beta\psi_s} - 1)\right]}{F\left(\beta\psi_s, \dfrac{n_{po}}{p_{po}}\right)} \qquad \text{farad/cm}^2. \qquad (20)$$

At flat-band condition, i.e., $\psi_s = 0$, C_D can be obtained by expanding the exponential terms into series, and we obtain

$$C_D(\text{flat-band}) = \sqrt{2}\varepsilon_s/L_D \qquad \text{farad/cm}^2. \qquad (21)$$

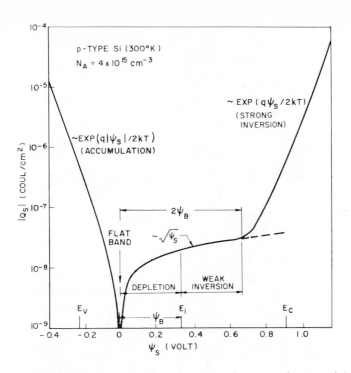

Fig. 5 Variation of space-charge density in the semiconductor as a function of the surface potential ψ_s for a p-type silicon with $N_A = 4 \times 10^{15}$ cm^{-3} at room temperature. ψ_B is the potential difference between the Fermi level and the intrinsic level of the bulk. (After Garrett and Brattain, Ref. 10.)

(2) Ideal MIS Curves

Figure 6(a) shows the band diagram of an ideal MIS structure with the band bending of the semiconductor identical to that shown in Fig. 4. The charge distribution is shown in Fig. 6(b). For charge neutrality of the system it is required that

$$Q_M = Q_n + qN_A W = Q_s \qquad (22)$$

where Q_M is charges per unit area on the metal, Q_n is the electrons per unit area in the inversion region, $qN_A W$ is the ionized acceptors per unit area in the space-charge region with space-charge width W, and Q_s is the total charges per unit area in the semiconductor. The electric field and the potential as obtained by first and second integrations of Poisson's equation are shown in Fig. 6(c) and 6(d) respectively.

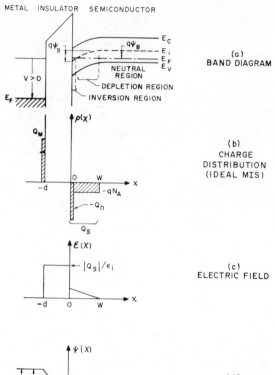

(a)
BAND DIAGRAM

(b)
CHARGE
DISTRIBUTION
(IDEAL MIS)

(c)
ELECTRIC FIELD

(d)
POTENTIAL

Fig. 6
(a) Band diagram of an ideal MIS structure.
(b) Charge distribution under inversion condition.
(c) Electric field distribution.
(d) Potential distribution.

It is clear that in the absence of any work-function differences, the applied voltage will partly appear across the insulator and partly across the silicon. Thus

$$V = V_i + \psi_s \tag{23}$$

where V_i is the potential across the insulator and is given, see Fig. 6(c), by

$$V_i = \frac{Q_s d}{\varepsilon_i} \left(\equiv \frac{Q_s}{C_i} \right). \tag{24}$$

The total capacitance, C, of the system is a series combination of the insulator capacitance, $C_i (= \varepsilon_i/d)$, and the silicon space-charge capacitance C_D:

$$C = \frac{C_i C_D}{C_i + C_D} \qquad \text{farad/cm}^2. \tag{25}$$

For a given insulator thickness d, the value of C_i is constant and corresponds to the maximum capacitance of the system. The silicon capacitance C_D as given by Eq. (20) depends on the voltage. Combination of Eqs. (20), (23), (24), and (25) gives the complete description of the ideal MIS curve as shown in Fig. 7 curve (a). Of particular interest is the total capacitance at flat-band condition, i.e., $\psi_s = 0$. From Eqs. (21) and (25), we obtain

$$C_{FB}(\psi_s = 0) = \frac{\varepsilon_i}{d + \frac{1}{\sqrt{2}} \left(\frac{\varepsilon_i}{\varepsilon_s}\right) L_D} = \frac{\varepsilon_i}{d + \left(\frac{\varepsilon_i}{\varepsilon_s}\right) \sqrt{\frac{kT\varepsilon_s}{p_{po} q^2}}} \tag{26}$$

where ε_i and ε_s are the permittivities of the insulator and the semiconductor respectively, and L_D is the extrinsic Debye length given by Eq. (12).

In describing this curve we begin at the left side (negative voltage) where we have an accumulation of holes and therefore a high differential capacitance of the semiconductor. As a result the total capacitance is close to the insulator capacitance. As the negative voltage is reduced sufficiently, a depletion region which acts as a dielectric in series with the insulator is formed near the semiconductor surface, and the total capacitance decreases. The capacitance goes through a minimum and then increases again as the inversion layer of electrons forms at the surface. The minimum capacitance and the corresponding minimum voltage are designated by C_{min} and V_{min} respectively (shown in Fig. 7). It should be pointed out that the increase of the capacitance is dependent on the ability of the electron concentration to follow the applied ac signal. This is only possible at low frequencies where the recombination-generation rates of minority carriers (in our example, electrons) can keep up with the small signal variation and lead to charge exchange with the inversion layer in step with the measurement signal. Experimentally it is found that for the Metal-SiO$_2$-Si system the frequency is between 5 to 100 Hz.[12,13] As a consequence, MIS curves measured at higher frequencies do not show the increase of capacitance on the right side, Fig. 7(b). Figure 7(c) shows the capacitance curve under nonequilibrium conditions (pulse condition) which will be discussed later in connection with the avalanche effect in MIS diode.[47,48]

The high-frequency curve can be obtained using an analogous approach as in a one-sided abrupt p-n junction.[14] When the semiconductor surface is depleted, the ionized acceptors in the depletion region are given by $(-qN_A W)$ where W is the depletion width. Integration of Poisson's equation yields the

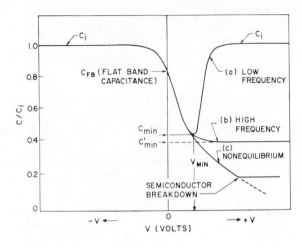

Fig. 7 MIS capacitance-voltage curves.
(a) Low frequency.
(b) High frequency.
(c) Nonequilibrium case.
(After Grove et al., Ref. 13.)

potential distribution in the depletion region:

$$\psi = \psi_s\left(1 - \frac{x}{W}\right)^2 \tag{27}$$

where the surface potential ψ_s is given by

$$\psi_s = \frac{qN_A W^2}{2\varepsilon_s}. \tag{27a}$$

When the applied voltage increases, ψ_s increases, so does W. Eventually strong inversion will occur. As shown in Fig. 5, strong inversion begins at $\psi_s(\text{inv}) \simeq 2\psi_B$. Once strong inversion occurs, the depletion-layer width reaches a maximum. When the bands are bent down far enough that $\psi_s = 2\psi_B$, the semiconductor is effectively shielded by further penetration of electric field by the inversion layer and even a very small increase in band bending (corresponding to a very small increase in the depletion-layer width) will result in a very large increase in the charge density within the inversion layer. Accordingly, the maximum width, W_m, of the surface depletion region can be obtained from Eqs. (19) and (27a)

$$W_m \simeq \sqrt{\frac{2\varepsilon_s \psi_s(\text{inv})}{qN_A}} = \sqrt{\frac{4\varepsilon_s kT \ln(N_A/n_i)}{q^2 N_A}}. \tag{28}$$

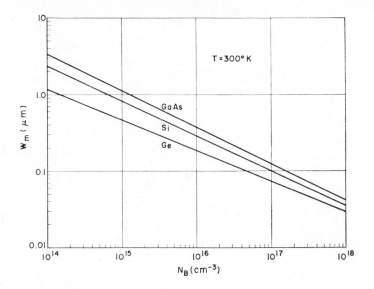

Fig. 8 Maximum depletion layer width versus impurity concentration of the semiconductors Ge, Si, and GaAs under heavy inversion condition.

The relationship between W_m and the impurity concentration is shown in Fig. 8 for Ge, Si, and GaAs, where N_B is equal to N_A for p-type and N_D for n-type semiconductors. Another quantity of interest is the so-called turn-on voltage, V_T, at which strong inversion occurs. From Eqs. (19) and (23), we obtain

$$V_T(\text{strong inversion}) = \frac{Q_s}{C_i} + 2\psi_B. \qquad (29)$$

And the corresponding total capacitance is given by

$$C'_{\min} = \frac{\varepsilon_i}{d + \left(\dfrac{\varepsilon_i}{\varepsilon_s}\right)W_m}. \qquad (30)$$

The high-frequency capacitance curve with its approximated segments (dotted curves) is shown in Fig. 9. Also shown in the insert are the measured MIS curves at different frequencies.[13] We note that the onset of the low-frequency curves occurs at $f \lesssim 100$ Hz.

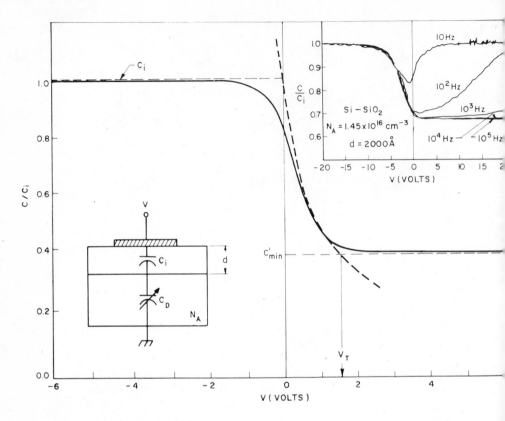

Fig. 9 High-frequency MIS capacitance-voltage curve showing its approximated segments (dotted lines). The inset shows the frequency effect. (After Grove et al., Ref. 13.)

The ideal MIS curves of the metal-SiO_2-Si system have been computed for various oxide thicknesses and semiconductor doping densities.[15] Figure 10 shows some typical examples for p-type silicon. We note that, as the oxide film becomes thinner, larger variation of the capacitance is obtained. Figure 11 shows the dependence of ψ_s on applied voltage for the same systems as in Fig. 10. Figures 12, 13, and 14 show respectively the normalized flat-band capacitance (C_{FB}/C_i), the normalized minimum capacitances (C_{min}/C_i) and (C'_{min}/C_i), and the minimum voltage (V_{min}) versus oxide thickness with silicon doping concentration as the parameter. The conversion to n-type silicon is achieved simply by changing the sign of the voltage axes, the conversion to other insulators requires scaling of the oxide thickness with the

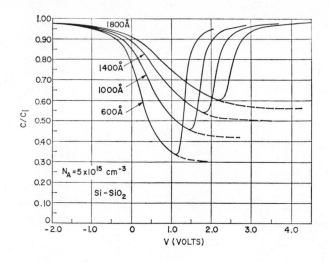

Fig. 10 Ideal MIS capacitance-voltage curve. Solid lines for low frequencies. Dotted lines for high frequencies.
(After Goetzberger, Ref. 15.)

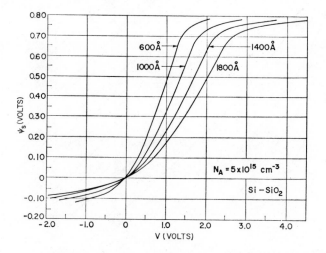

Fig. 11 Surface potential versus applied voltage for ideal MIS diodes.
(Ref. 15.)

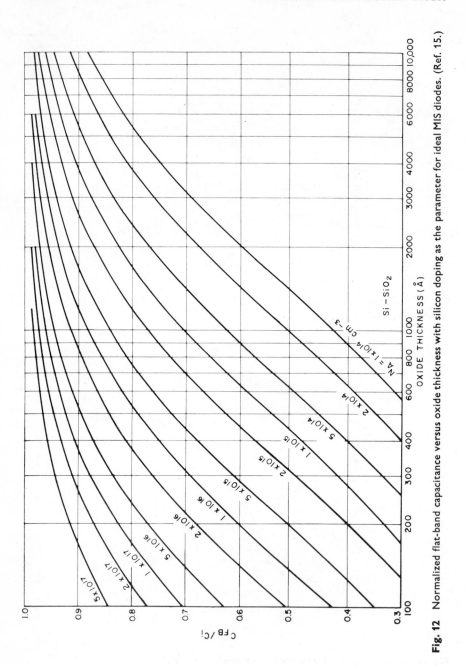

Fig. 12 Normalized flat-band capacitance versus oxide thickness with silicon doping as the parameter for ideal MIS diodes. (Ref. 15.)

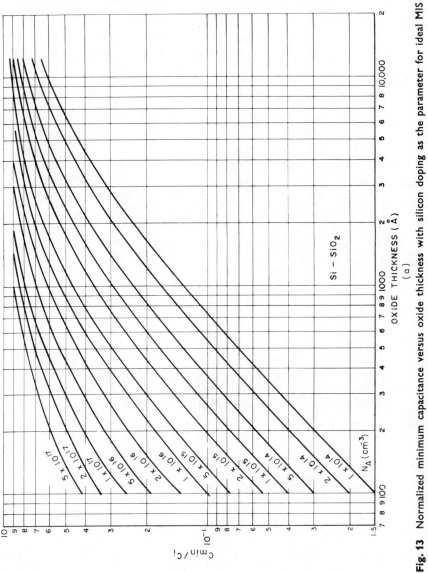

Fig. 13 Normalized minimum capacitance versus oxide thickness with silicon doping as the parameter for ideal MIS diodes under low-frequency condition. (Ref. 15.)

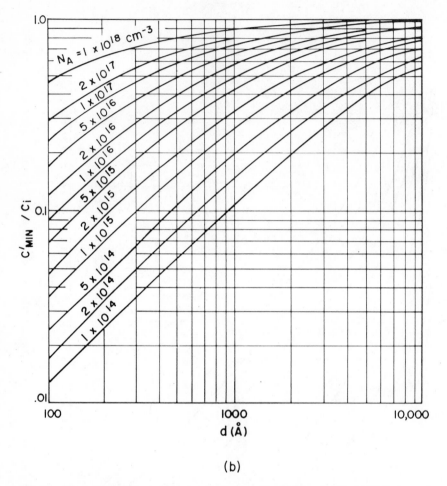

(b)

Fig. 13 Normalized minimum capacitance versus oxide thickness for ideal MIS diodes under high-frequency condition. (Ref. 15.)

ratio of the permittivities of SiO_2 and other insulator

$$d_c = d_i \frac{\varepsilon_i(SiO_2)}{\varepsilon_i(\text{insulator})} \tag{31}$$

where d_c is the thickness to be used in these curves, d_i is the actual thickness of the insulator, $\varepsilon_i(\text{insulator})$ is the permittivity of the new insulator, and $\varepsilon_i(SiO_2) = 3.4 \times 10^{-13}$ f/cm. For other semiconductors, the MIS curves can

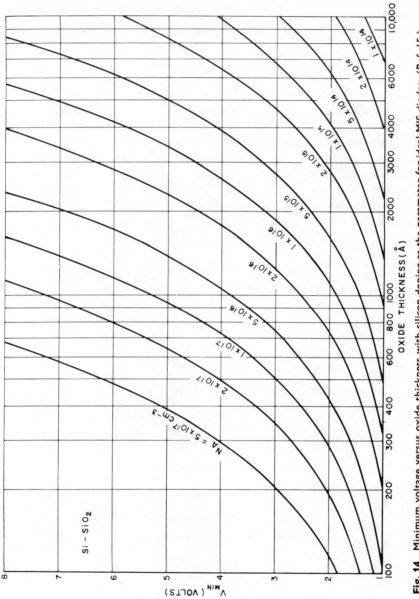

Fig. 14 Minimum voltage versus oxide thickness with silicon doping as the parameter for ideal MIS diodes. (Ref. 15.)

be constructed similarly to that as shown in Fig. 9 by using Eqs. (26), (27), (28), and (29).

The ideal MIS curves as shown in Fig. 10 through 14 will be used in subsequent sections to compare with the experimental results and to understand the practical MIS systems.

3 SURFACE STATES, SURFACE CHARGES, AND SPACE CHARGES

In a practical MIS diode there exist many other states and charges which will, in one way or another, affect the ideal MIS characteristic. The basic classifications of these states and charges are shown in Fig. 15. There are (1) surface states or interface states which are defined as energy levels within the forbidden band gap at the insulator-semiconductor interface which can exchange charges with the semiconductor in a short time, (2) fixed surface

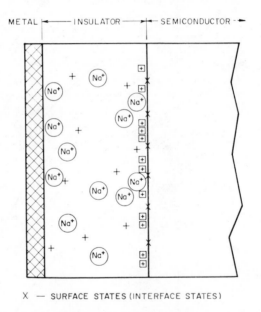

X — SURFACE STATES (INTERFACE STATES)

⊞ — FIXED SURFACE CHARGES

(Na⁺) — MOBILE IONS

+ — IONIZED TRAPS

Fig. 15 Basic classification of states and charges in a nonideal MIS diode.

charges which are located near or at the semiconductor surface (~ 200 Å) and are immobile under applied electric fields, (3) mobile ions such as sodium which are mobile within the insulator under bias-temperature aging conditions, and (4) ionized traps which can be created, for example, by x-ray radiation.

(1) Surface States

The surface states have been theoretically studied by Tamm,[16] Shockley,[17] and others [18,19] and have been shown to exist within the forbidden gap due to the interruption of the periodic lattice structures at the surface of a crystal. The existence of surface states was first found experimentally by Shockley and Pearson[20] in their surface conductance measurement. Measurement on clean surfaces in an ultrahigh vacuum system have confirmed that the density of surface states is very high—of the order of the density of surface atoms.[21] Historically, surface states have been classified into fast and slow states. The fast states exchange charge with the conduction or valence band rapidly, and are assumed to lie close to the interface between the semiconductor and the insulator. Slow states, on the other hand, exist at the interface of the air and insulator and require a longer time for charge exchange. For the present MIS diodes with thick insulating layers, the only states that we are concerned with are the surface states or interface states at the insulator-semiconductor interface. And these states are not necessarily fast surface states either, since at low temperatures, the time constant of these states is very long.[9,22]

A surface state is considered as a donor state if it can be neutral or it can become positive by donating (giving up) an electron. For an acceptor surface state, it can be neutral or it can become negative by accepting an electron. The distribution functions for the surface states are similar to those for the bulk impurity levels as discussed in Chapter 2, Section (4):

$$F_{SD}(E_t) = \left[1 - \frac{1}{1 + \dfrac{1}{g} \exp\left(\dfrac{E_t - E_F}{kT}\right)} \right] = \frac{1}{1 + g \exp\left(\dfrac{E_F - E_t}{kT}\right)} \qquad (32a)$$

for donor surface states; and

$$F_{SA}(E_t) = \frac{1}{1 + \dfrac{1}{g} \exp\left(\dfrac{E_t - E_F}{kT}\right)} \qquad (32b)$$

for acceptor surface states where E_t is the energy of the surface states and g is the ground state degeneracy which is 2 for donor and 4 for acceptor.

When a voltage is applied, the surface levels will move up or down with the valence and conductance bands while the Fermi level remains fixed.

A change of charge in the surface state occurs when it crosses the Fermi level. This change of charge will contribute to the MIS capacitance and alter the ideal MIS curve. The basic equivalent circuit[23] incorporating the surface states effect is shown in Fig. 16(a). In the figure, C_i and C_D are the insulator capacitance and the semiconductor depletion-region capacitance respectively and are identical to those shown in the insert of Fig. 9. C_s and R_s are the capacitance and resistance associated with the surface states and are functions of the surface potential. The product $C_s R_s$ is defined as the surface state lifetime which determines the frequency behavior of the surface states. The parallel branch of the equivalent circuit in Fig. 16(a) can be converted into a frequency-dependent capacitance C_p in parallel with a frequency-dependent conductance G_p as shown in Fig. 16(b) where

$$C_p = C_D + \frac{C_s}{1 + \omega^2 \tau^2} \tag{33}$$

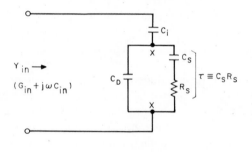

(a)

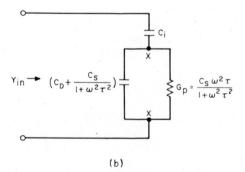

(b)

Fig. 16   Equivalent circuit including surface state effect where C_s and R_s are associated with surface state densities.
(After Nicollian and Goetzberger, Ref. 23.)

and

$$\frac{G_p}{\omega} = \frac{C_s \omega \tau}{1 + \omega^2 \tau^2} \tag{34}$$

with $\tau \equiv C_s R_s$. And the input admittance, Y_{in}, is given by

$$Y_{in} \equiv G_{in} + j\omega C_{in} \tag{35}$$

where

$$G_{in} = \frac{\omega^2 C_s \tau C_i^2}{(C_i + C_D + C_s)^2 + \omega^2 \tau^2 (C_i + C_D)^2} \tag{36a}$$

$$C_{in} = \frac{C_i}{C_i + C_D + C_s} \left[C_D + C_s \frac{(C_i + C_D + C_s)^2 + \omega^2 \tau^2 C_D (C_i + C_D)}{(C_i + C_D + C_s)^2 + \omega^2 \tau^2 (C_i + C_D)^2} \right]. \tag{36b}$$

(2) Capacitance Method

To evaluate the surface states density one can either use the capacitance measurement or the conductance measurement since, in Eqs. (36a) and (36b), both the input conductance and the input capacitance contain similar information about the surface states. It will be shown that the conductance technique can give the most accurate results especially for MIS diodes with relatively low surface-state densities ($\sim 10^{10}$ states/cm^2/eV). The evaluation of surface-state density using capacitance measurement can be achieved, for example, by the differentiation procedure, the integration procedure, or the temperature procedure.

(A) Differentiation Procedure. This method has been used first by Terman.[5] The capacitance is first measured at a high frequency ($\omega\tau \gg 1$) such that Eq. (36b) reduces to Eq. (25) which is free of capacitance due to the surface states. This yields the high-frequency curve as shown in Fig. 17 (dashed lines) for both n-type and p-type semiconductors. The influences of the surface states on the voltage, however, cause a shift of the ideal MIS curve along the voltage axis. This is because, when surface states are present, the electric field in the oxide is higher than the field in the semiconductor surface, and more charges on the metal are necessary to create a given surface field in the semiconductor. Comparison of Fig. 17 with the ideal MIS curves gives a curve of ΔV versus V where ΔV is the voltage shift. The total charge in the surface states (Q_{ss}) at a given surface potential is then given by

$$Q_{ss} = C_i(\Delta V) \qquad \text{coul/cm}^2. \tag{37}$$

The surface state density per unit energy (N_{ss}) is then obtained by graphical differentiation:

$$N_{ss} = \frac{1}{q} \left(\frac{\partial Q_{ss}}{\partial \psi_s} \right)_V \qquad \text{states/cm}^2/\text{eV}. \tag{38}$$

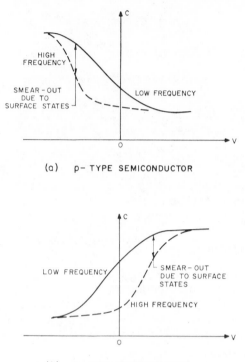

(a)   p– TYPE SEMICONDUCTOR

(b)   n – TYPE SEMICONDUCTOR

Fig. 17   High-frequency and low-frequency C-V curves for p-type and n-type semiconductor samples.

Measurements at various lower frequencies (Fig. 17, solid lines) can be used to determine the time constants (τ) of surface states. The above differentiation method is useful for MIS diodes with large surface state density. However, the facts that only the integral of N_{ss} is measured and there is uncertainty about the magnitude of the semiconductor depletion-layer capacitance make this method unreliable.[24]

(B)   Integration Procedure. This method proposed by Berglund[25] makes it possible to determine the semiconductor surface potential as a function of the applied voltage directly from low-frequency differential capacitance measurement, and no graphical differentiation is required to determine N_{ss}. When space-charge effects in the insulator can be neglected, from Eqs. (23) and (25) and the fact $dQ = C_i \, dV_i = C \, dV$, we obtain

$$\left(\frac{\partial \psi_s}{\partial V}\right) = 1 - \frac{C}{C_i} \tag{39a}$$

and

$$\frac{d\psi_s}{dV_i} = \frac{C_i}{C} - 1. \tag{39b}$$

Integration of Eq. (39a) from V_1 to V_2 yields

$$\psi_s(V_1) - \psi_s(V_2) = \int_{V_2}^{V_1} \left[1 - \frac{C}{C_i}\right] dV. \tag{40}$$

Equation (40) indicates that the surface potential at any applied voltage can be determined by integrating a curve of $(1 - C/C_i)$. It should be noted that Eq. (40) is valid only when the surface states are in equilibrium at all times during the measurement of $C(V)$; that is, the measurement frequency must be low enough that all surface states can follow both the dc bias and the ac signal. Another relationship can be obtained from the relation of charge neutrality in the MIS system. Referring to Fig. 6(b), in addition to the charge in the semiconductor (Q_s) we now have $N_{ss}(q\psi_s) = N_{SD}^+(q\psi_s) + N_{SA}^-(q\psi_s)$ where N_{SD}^+ and N_{SA}^- are respectively the donor and acceptor surface state densities. The charge neutrality requirement gives

$$\frac{\varepsilon_i V_i}{d} = q \int_0^\infty \left[N_{SD}^+ F_{SD}(E_t) - N_{SA}^- F_{SA}(E_t)\right] dE + Q_s. \tag{41}$$

Differentiation of Eq. (41) with respect to ψ_s gives

$$\frac{\partial \psi_s}{\partial V_i} = \frac{\varepsilon_i}{d} \bigg/ \left[\frac{dQ_s}{d\psi_s} + qN_{ss}(q\psi_s)\right]. \tag{42}$$

From Eqs. (39b) and (40), a curve of $\partial \psi_s / \partial V_i$ versus ψ_s can be obtained directly using low-frequency capacitance measurement of an MIS diode. If the doping density of the semiconductor and the temperature are known, this curve can be compared to that given by Eq. (42) to determine N_{ss}. This method does not require graphical differentiation and the errors introduced by uncertainties in the semiconductor doping are also reduced.

(C) Temperature Procedure. This method, as proposed by Gray and Brown,[26] can separate the effects of the space-charge in the insulator and the surface states on the surface potential ψ_s. Figure 18(a) is a schematic representation of how the $C(V)$ curves shift with decreasing temperature. For p-type samples (as shown), it requires larger and larger negative voltages on the metal electrode to reach the flat-band condition as the temperature

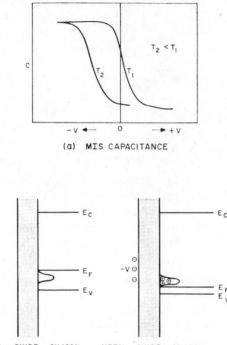

(a)   MIS CAPACITANCE

METAL   OXIDE   SILICON   METAL   OXIDE   SILICON
        $T = T_1$ $T = T_2 < T_1$

(b)   SIMPLIFIED BAND STRUCTURE

Fig. 18
(a) Schematic representation of the shift of the MIS C-V curves with temperature.
(b) Change of Fermi level due to temperature.
(After Gray and Brown, Ref. 26.)

decreases from T_1 to T_2, whereas for n-type samples, the flatband voltage increases in the positive direction as T decreases. (Note: For ideal MIS diodes the flat-band voltages are always at zero bias independent of temperature.) The change in Fermi level in the semiconductor due to temperature is shown schematically in Fig. 18(b). At T_1 the Fermi level is assumed to be above the surface states which are filled with electrons and are neutral. When the temperature is decreased to T_2, the Fermi level moves closer to the valence band, and some of the surface states lose electrons and become positively charged. In this instance the surface states control the surface potential, and it requires larger negative voltage on the metal electrode to deplete the surface states of their electrons in order to reach the flat-band condition. This group of states near the valence band edge (as shown) therefore are donors. A similar explanation can be given for the results on n-type samples where the surface states are acceptors (negatively charged when filled with electrons). The increment in flat-band voltage as the temperature is changed gives the change in surface-state charge directly, since there is no band bending for this condition, and the surface potential ψ_s is equal to the bulk Fermi level which is calculable. In addition this method makes the results independent of the oxide space charge since it exerts a constant effect on the surface charge for all temperatures. The experimental procedure consists of (1) varying the temperature while (2) maintaining the flat-band condition by observing the changes in capacitance and continuously adjusting the bias, (3) recording flat-band voltage versus temperature, and (4) converting these data to surface charge Q_{ss} versus surface potential, $\psi_s = (E_F - E_V)/q$. The surface-state density is given by

$$N_{ss} = \frac{1}{q} \frac{\partial Q_{ss}}{\partial \psi_s}.$$

The experimental results for the Si-SiO$_2$ system are shown in Fig. 19. We note the peaks of N_{ss} near the band edges, and the dependence of N_{ss} on the silicon surface orientation. The peaks will be discussed in Section 3(4), and the orientation dependence will be considered in the next section. It should be pointed out that in the above method, it is assumed that (1) the ac frequency is high enough that Q_{ss} does not contribute appreciably to the capacitance, (2) oxide-space charge and electron affinity differences are temperature-independent, and (3) N_{ss} is relatively constant over a small energy range in the forbidden gaps.[27]

(3) Conductance Method

A detailed and comprehensive discussion of the conductance method is given by Nicollian and Goetzberger.[9] In the capacitance measurement, the difficulty arises from the fact that the surface-state capacitance must be extracted from the measured capacitance which consists of oxide capacitance, depletion capacitance, and surface-state capacitance. Since, as mentioned previously, both the capacitance and conductance as functions of voltage and frequency contain identical information about surface states, greater inaccuracies arise in extracting this information from the measured capacitance. This difficulty does not apply to the measured conductance (which is measured simultaneously in a standard capacitance bridge) because it is directly related to the surface states.

Thus conductance measurements yield more accurate and reliable results, particularly when N_{ss} is low as in the thermally oxidized SiO$_2$-Si system. This

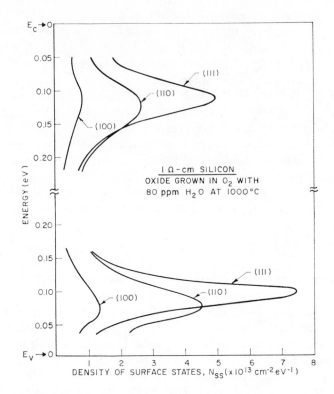

Fig. 19 Measured surface-state densities for the Si-SiO$_2$ system for three crystal orientations. (Ref. 26.)

is illustrated in Fig. 20 which shows the measured capacitance and measured conductance at 5 kHz and 100 kHz. The largest capacitance spread is only 14 percent while the magnitude of the conductance peak increases by over one order of magnitude in this frequency range.

The principle of the MIS conductance technique is easily illustrated by the simplified equivalent circuit shown previously in Fig. 16. The admittance of the MIS diode is measured by a bridge across the diode terminals. The insulator capacitance is measured in the region of strong accumulation. The admittance of the circuit is then converted into an impedance. The reactance of the insulator capacitance is subtracted from this impedance and the resulting impedance converted back into an admittance. This leaves C_D in parallel with the series $R_s C_s$ network of the surface states. The capacitance and equivalent parallel conductance divided by ω are given by Eqs. (33) and

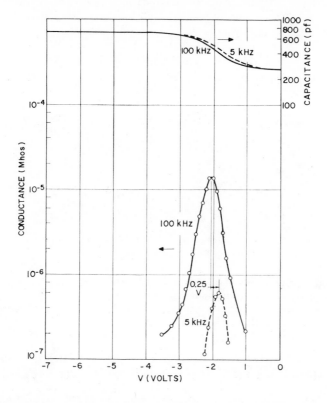

Fig. 20 Comparison of MIS capacitance measurement and conductance measurement at two frequencies.
(After Nicollian and Goetzberger, Ref. 9.)

(34). Equation (34), $G_p/\omega = C_s\omega\tau/(1 + \omega^2\tau^2)$, does not contain C_D and depends only on the surface-state branch of the equivalent circuit. At a given bias, G_p/ω can be measured as a function of frequency. A plot of G_p/ω versus $\omega\tau$ will go through a maximum when $\omega\tau = 1$. This gives τ directly. The value of G_p/ω at the maximum is $C_s/2$. Thus, equivalent parallel conductance corrected for C_i gives C_s and τ ($\equiv R_sC_s$) directly from the measured conductance. Once C_s is known, the surface-state density is obtained by using the relation $N_{ss} = C_s/qA$ where A is the metal plate area. A typical result of N_{ss} in a Si-SiO$_2$ system is shown[33] in Fig. 21. This is very similar to those shown in Fig. 19. The variation of the time constant τ versus the surface potential is shown in Fig. 22 for MIS diodes with steam-grown oxides on $\langle111\rangle$ silicon

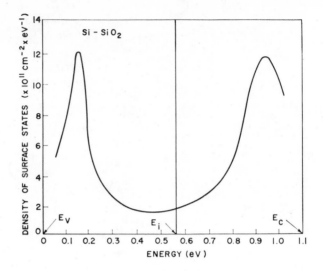

Fig. 21 Surface-state density of a Si-SiO$_2$. Near the band edges the temperature method is used. At the midgap the conductance method is used. (Ref. 9.)

substrates where ψ_B is the potential difference between intrinsic level and the Fermi level, and $\bar{\psi}_s$ is the average surface potential to be discussed later. It is seen that these curves can be fitted by the following expressions and are similar to those for the generation-recombination processes as discussed in Chapter 2:

$$\tau = \frac{1}{\bar{v}\sigma_p n_i} \exp\left[-\frac{q(\psi_B - \bar{\psi}_s)}{kT}\right] \qquad \text{for } p\text{-type}$$

$$\tau = \frac{1}{\bar{v}\sigma_n n_i} \exp\left[\frac{q(\psi_B - \bar{\psi}_s)}{kT}\right] \qquad \text{for } n\text{-type} \qquad (43)$$

where σ_p and σ_n are the capture cross sections of holes and electrons respectively, and $\bar{v}$ is the average thermal velocity. The above results indicate that the capture cross section is independent of energy. The capture cross sections obtained[9] from Fig. 22 are $\sigma_p = 2.2 \times 10^{-16}$ cm^2 and $\sigma_n = 5.9 \times 10^{-16}$ cm^2 where the values of $\bar{v} = 10^7$ cm/sec and $n_i = 1.6 \times 10^{10}$ cm^{-3} have been used.

As illustrated in Figs. 19 and 21, the surface states in the Si-SiO$_2$ system are comprised of many levels so closely spaced in energy that they cannot be distinguished as separate levels. They actually appear as a continuum over

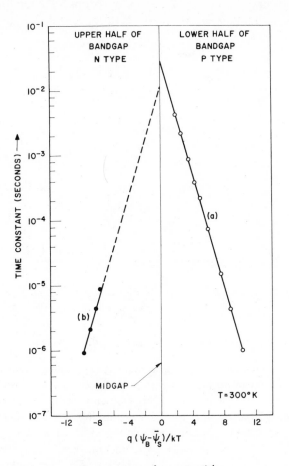

Fig. 22 Variation of time constant versus surface potentials. (Ref. 9.)

the band gap of the semiconductor. The equivalent circuit shown in Fig. 16, which is for an MIS diode with a single-level time constant, should therefore be modified. In addition to the effect due to the surface-state continuum, we must consider the statistical fluctuation of surface potential due to surface charges which include the fixed surface charges and the charged surface states, since from Eq. (43), a small fluctuation in $\bar{\psi}_s$ will cause large fluctuation in τ. If we assume the surface charges are randomly distributed in the plane of the interface, the electric field at the semiconductor surface will fluctuate over the plane of the interface. This is shown schematically in Fig. 23. The expression of G_p/ω and C_p due to time-constant dispersion resulting from the surface-state continuum and the statistical (Poisson) distribution of surface charges is given by[9]

$$\frac{G_p}{\omega} = \frac{1}{2} q N_{ss} [2\pi(\sigma_S^2 + \sigma_B^2)]^{-1/2} \int_{-\infty}^{\infty} \exp(-z - y)\ln(1 + e^{2y}) d\left(\frac{q\psi_s}{kT}\right) \quad (44)$$

$$C_p = C_D(\bar{\psi}_s) + q N_{ss} [2\pi(\sigma_S^2 + \sigma_B^2)]^{-1/2} \int_{-\infty}^{\infty} \exp(-z - y)\tan^{-1}(e^y) d\left(\frac{q\psi_s}{kT}\right)$$

$$(45)$$

where

$$\sigma_S \equiv \frac{qW}{kT(WC_i + \varepsilon_s)} \left(\frac{q\bar{Q}}{A_c}\right)^{1/2}$$

$$\sigma_B \equiv \frac{q^2(\bar{N}_A W_3)^{1/2}\left[1 - \exp\left(-\dfrac{q\bar{\psi}_s}{kT}\right)\right]}{2kT[WC_i + \varepsilon_s]}$$

$$y \equiv \ln\frac{\omega}{\bar{v}\sigma_p n_i} + \frac{q(\psi_s - \psi_B)}{kT}$$

$$z \equiv q^2(\psi_s - \bar{\psi}_s)^2/[2k^2T^2(\sigma_S^2 + \sigma_B^2)]$$

where $\bar{N}_A$ is the mean ionized acceptor density, $\bar{Q}$ the mean density of surface charges, $\bar{\psi}_s$ the mean surface potential, W the depletion-layer width at the potential $\bar{\psi}_s$, and A_c a characteristic area which is proportional to the square of W. Figure 24 shows plots of G_p/ω versus $\ln f$ for a Si-SiO$_2$ MIS diode biased in the depletion region (broad curve) and in the weak inversion region (narrow curve). The circles are the experimental results, and the solid broad curve is the theoretical calculation based on Eq. (44). The excellent agreement indicates the importance of the statistical model. For the case of weak inversion, however, one requires only a single time constant to fit the experimental results.

The above results can be explained with the help of the modified equivalent circuits shown in Fig. 25. Figure 25(a) shows the time-constant dispersion caused primarily by statistical fluctuation of surface potential. Each subnetwork consisting of C_s and R_s in series represents a time constant of the continuum of surface states in a characteristic area A_c. This circuit corresponds to the case of depletion-accumulation. Figure 25(b) is for the case of the midgap region where $q\psi_s = q\psi_B \pm$ a few kT. R_{ns} and R_{ps} represent the capture resistances for the electrons and holes respectively. We now have two resistances for each subnetwork; when the surface potential is equal to

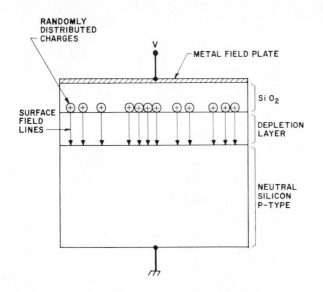

Fig. 23 Randomly distributed surface charge at the semiconductor-insulator interface causes time-constant dispersion.
(Ref. 9.)

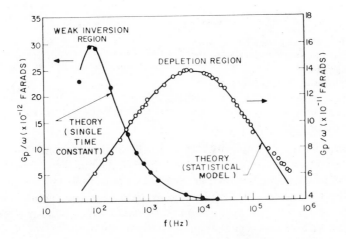

Fig. 24 G_p/ω versus frequency for a Si-SiO$_2$ MIS diode biased in the depletion region (broad curve) and in the weak inversion region (narrow curve). Circles are experimental results. Lines are the theoretical calculations.
(Ref. 9.)

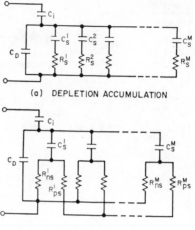

(a) DEPLETION ACCUMULATION

(b) MIDGAP

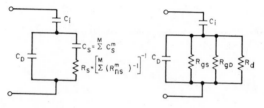

(c) WEAK INVERSION   (d) HEAVY INVERSION

Fig. 25 Modified equivalent circuits for
(a) depletion-accumulation region,
(b) midgap region,
(c) weak inversion region, and
(d) heavy inversion region.
(After Nicollian and Goetzberger, Ref. 9, and Hoftsein and Warfield, Ref. 12.)

or greater than ψ_B, the minority carrier density at the semiconductor surface will be equal to or greater than the magnitude of the majority carrier density at the surface. The minority carrier density can thus no longer be ignored as in the depletion case, Fig. 25(a). Figure 25(c) represents weak inversion $(2\psi_B > \psi_s > \psi_B)$; whereas Fig. 25(b) is simplified by the fast response (small R_{ns}) of minority carriers such that all R_{ns}'s are shorted together, the equivalent surface-state capacitance is given by the sum of each individual C_s, and the resistance by the sum of the capture resistances for the majority carriers. Therefore, in weak inversions good agreement is obtained between the experimental result and the single time constant theory. Figure 25(d) shows the circuit for heavy inversion [12] ($\psi_s > 2\psi_B$) where C_i and C_D are the insulator capacitance and the depletion capacitance respectively; R_d is the resistance associated with the diffusion current of the minority carriers from the bulk to the edge of the depletion region and through the depletion region to the surface; R_{gs} is the resistance associated with the current flow of holes between the valence band and the surface states, with the current flow of electrons between the conductance band and the surface states, and with the flow of the majority carriers from the bulk to the surface; and R_{gD} is the resistance associated with the finite generation and recombination within the depletion layer which is found to be the dominant factor in controlling the frequency response of the inversion layer. The surface state capacitance C_s can be assumed to be essentially zero ($C_s \ll C_D$) under heavy inversion condition. It is clear from Fig. 25(d) that, in the heavy inversion region, the surface states have only minor effect on the MIS characteristics.

The conductance method also serves as a powerful tool to study the $1/f$-type noise. It has long been recognized that surface states at the Si-SiO$_2$ interfaces which exchange charge with the silicon can give rise to $1/f$-type noise.[29] It has been shown experimentally that the $1/f$ noise of an MIS diode is directly related to surface-state density and capture conductance over the energy gap.[28] The mean square open circuit noise voltage across the MIS diode terminals is given by [29]

$$\langle V_n^2 \rangle = \frac{4kTBG_p(\omega)}{G_p^2(\omega) + \omega^2 C_p^2(\omega)} \tag{46}$$

where B is the bandwidth in Hz; G_p and C_p are given by Eqs. (44) and (45) respectively, and the noise spectral distribution is given by[29]

$$S(\omega) = \left(\frac{2kT}{q}\right) BN_{ss}[2\pi(\sigma_S^2 + \sigma_B^2)]^{-1/2}$$

$$\times \int_{-\infty}^{\infty} \exp(-z-y)\omega^{-1}\ln(1 + e^{2y})d\left(\frac{q\psi_s}{kT}\right). \tag{47}$$

Figure 26 shows a typical theoretical noise spectrum versus frequency calculated from Eq. (47) for an MIS diode with a surface-state density of 3×10^{11} cm^{-2}eV^{-1} at 300°K. We note that, in the intermediate frequency range, the noise has a $1/f$-type frequency dependence. The frequency at which the noise spectrum is tangent to the line $1/f$ gives the maximum value of $G_p(\omega)/\omega$. From Eqs. (46) and (47) it can be shown that $1/f$-type noise due to surface states can be reduced by decreasing N_{ss}, and that $1/f$-type noise can be calculated from the electrical properties of the interface obtained by measuring the admittance of an MIS diode.

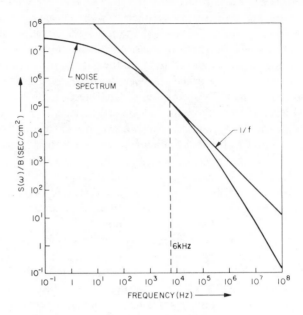

Fig. 26 Theoretical noise spectrum versus frequency for an MIS diode with surface-state density 3×10^{11} cm^{-2} eV^{-1} at 300°K.
(After Nicollian and Melchoir, Ref. 29.)

(4) Surface Charges and Space Charges in the Insulator

The surface charges include the fixed surface charge, the mobile ions, and the ionized traps which are located near or at the insulator-semiconductor interface. The space charges in the insulator include the mobile ions and the ionized traps inside the insulator. The fixed surface charge has the following properties: it is fixed and cannot be charged or discharged over a wide variation of ψ_s; it is located within the order of 100 Å of the insulator-semiconductor interface;[30] its density Q_{fc} is not greatly affected by the insulator thickness or by the type or concentration of impurities in the semiconductor; Q_{fc} depends on the formation (oxidation) and annealing conditions, and on the semiconductor orientation. It has been suggested that the excess ionic silicon in the oxide is the origin of the fixed surface charge in the Si-SiO$_2$ system.[30] The effect of the fixed charge on the MIS capacitance curve is a parallel shift of the curve along the voltage axis, and the amount of shift ΔV is given by

$$\Delta V = \frac{Q_{fc}}{C_i}. \tag{48}$$

This can be explained with the help of Fig. 27(a). When a positive fixed surface charge is present, the electric field $\mathscr{E}_i$ in the insulator is higher than the field on the semiconductor surface. Therefore more charges in the metal electrode are required to create a given surface field $\mathscr{E}_s$. Consequently a larger voltage is required to set up the surface potential ψ_s. An interesting example is shown in Fig. 27(b) where the capacitance and conductance of an unstable Si-SiO$_2$ system are plotted as functions of applied voltage before and after temperature aging. The shift of the C-V curve is caused by the increase in surface charges at the Si-SiO$_2$ interface.[9,23]

It was first demonstrated by Snow et al.[8] that alkali ions such as sodium in the thermally grown SiO$_2$ films are mainly responsible for the instability of

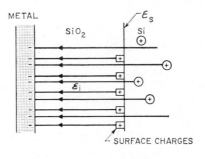

(a)

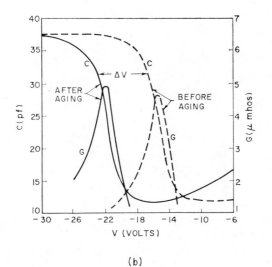

(b)

Fig. 27 Effect of surface charge on MIS curves.
(After Nicollian and Goetzberger, Ref. 23.)

the oxide-passivated device structures. The reliability problems in semiconductor devices operated at high temperatures and voltages may be related to trace contamination by alkali ions, since, under those conditions, the ions can move through the oxide film and give rise to voltage shift. A typical example[8] of the effect due to mobile ions is shown in Fig. 28(a). The initial *C-V* curve (1) of a Si-SiO$_2$ system is annealed at 127°C for 30 minutes with +10 V on the metal electrode to give curve (2). Curve (3) is obtained by annealing the sample at the same temperature for the same time but with −10 V. We note that there is a partial recovery of the *C-V* curve. This drift phenomena can be explained using Fig. 28(b) where the charge distributions correspond to the capacitance-voltage curves in (a). Initially all the alkali ions are located near the metal electrode. When a positive voltage is applied to

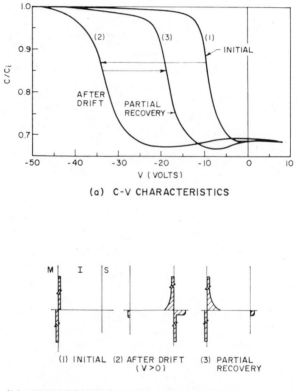

(a) C-V CHARACTERISTICS

(b) CHARGE DISTRIBUTIONS CORRESPONDING TO (a)

Fig. 28 Effect of mobile ions on MIS curves.
(After Snow et al., Ref. 8.)

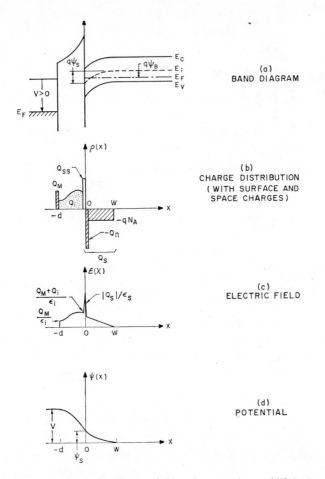

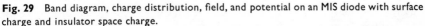

Fig. 29 Band diagram, charge distribution, field, and potential on an MIS diode with surface charge and insulator space charge.

the metal electrode (at 127°C), the positive ions will move towards the semiconductor and eventually reach the Si-SiO$_2$ interface causing a large voltage shift. When a negative voltage is applied (at 127°C), most of the ions will drift backwards to the metal electrode and give rise to the partial recovery. Other ions such as proton (H$^+$) can also give a similar drift effect.[31]

The space charge in the insulator will also cause a voltage shift on the MIS C-V curve. Figure 29 shows the band diagram, the charge distribution, the electric field, and the potential for an MIS diode with both surface states and insulator space charge. Comparing this figure with Fig. 6 of the ideal MIS

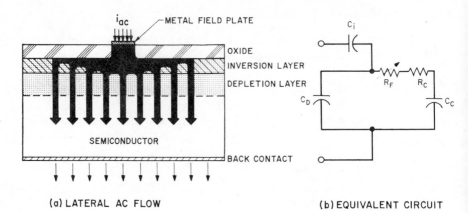

(a) LATERAL AC FLOW   (b) EQUIVALENT CIRCUIT

Fig. 30
(a) Lateral ac flow pattern.
(b) Equivalent circuit.
(After Nicollian and Goetzberger, Ref. 32.)

diode, we note that for the same surface potential ψ_s, the applied voltage V is reduced indicating a voltage shift of the $C\text{-}V$ curve toward negative voltage. And the shift due to the space charge density $\rho_i(x)$ is given by

$$\Delta V = \frac{1}{C_i d} \int_0^d x \rho_i(x)\, dx. \tag{49}$$

In addition to the above mentioned departures from the ideal MIS characteristics, there is the lateral alternating current flow which occurs in surfaces having a permanent inversion layer due to charges in the insulator and/or on the surface. In this case the inversion layer is essentially connected with the minority carriers on the entire surface.[32] The flow lines of the alternating current are shown in Fig. 30(a). The current flows across the insulator as a displacement current, then spreads out in the high-conductivity inversion region (channel region) and finally flows across the depletion capacitance in an area much larger than the metal plate. This effect increases the semiconductor capacitance without necessity of recombination processes. The equivalent circuit is shown in Fig. 30(b) which is for a nonideal MIS diode in the inversion region. The depletion capacitance C_D is now shunted by a branch consisting of R_F and R_C in series with C_C where R_F is the resistance of the inversion layer and where R_C and C_C are the lumped resistance and capacitance of the channel outside the metal plate. This channel connection greatly enhances the response time of minority carriers, and the cutoff frequency of the channel is considerably higher than that for minority carriers under equilibrium condition. This is demonstrated in Fig. 31 which shows the C-V curves of a metal-SiO$_2$-Si diode at three frequencies. The cutoff frequency in this case is about 50 MHz which is considerably higher than the 100 Hz shown in Fig. 9.

The experimentally observed distributions of surface-state density versus energy in MIS diodes with thermally grown oxides on silicon substrates have been correlated to the existence of charges in the oxide layer by Goetzberger et al.[33] This is because (1) in ion-free oxides a reduction of surface charges by annealing leads also to a reduction of surface states, (2) ionizing radiation creates both surface charges and surface states, (3) incorporation of either holes or electrons by the surface avalanche effect (to be discussed in Section 5) into the oxides leads to a surface charge of the proper sign and a proportional density of surface states. A schematic illustration of the correlation is shown in Fig. 32 where it is assumed that the oxide contains approximately equal densities of positive and negative charge centers, that all those charges within some distance D of the interface give rise to surface states, and that each positive (negative) charge in D gives rise to one localized surface state near the conduction (valence) band edge. These states are not the same as those found on clean or cleaved surfaces. Because the binding energy of the states is at least 0.1 eV, their size cannot be greater than the radius $R \approx 30$ Å of the scaled hydrogen atom model of a bulk donor state. However, a typical density of 10^{12} cm^{-2} of surface states means there is only one surface state in an area 100 Å square. A single positive charge at the interface will give rise to a bound donor state analogous to that formed in the bulk. Using the effective mass approximation for shallow states we obtain for the energy, from Eq. (21) of Chapter 2,

$$E_S = \left[\frac{\varepsilon_s}{\frac{(\varepsilon_i + \varepsilon_s)}{2}} \right]^2 \left(\frac{m^*}{m_0} \right) E_H = Z^2 E_D \qquad (50)$$

where E_D is bulk donor ionization energy and $Z^2 = [2\varepsilon_s/(\varepsilon_i + \varepsilon_s)]^2$ which is 2.25 for the Si-SiO$_2$ system. A similar result can be obtained for the acceptor states. Since $E_D = 0.025$ eV and $E_A = 0.05$ eV in silicon, the expected surface energies are about 0.06 eV for the surface donors and 0.12 eV for the surface acceptors. These values are in reasonable agreement with the peak positions in Figs. 19 and 21.

For a charge located at some distance x in the oxide, the energy level, which is independent of x, remains unchanged. This explains the existence and the sharpness of the peaks. Because of the effective-mass approximation, a

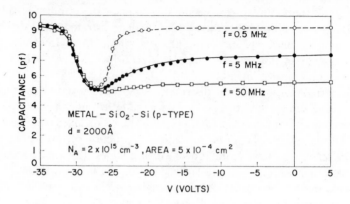

Fig. 31 Capacitance-voltage curves of a metal-SiO₂-Si diode at three frequencies. (Ref. 32.)

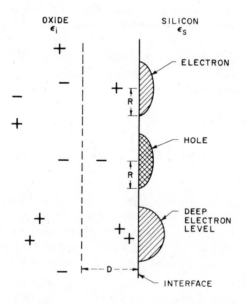

Fig. 32 Schematic diagram of correlation between surface state and surface charge. (After Goetzberger et al., Ref. 33.)

reasonable upper limit of x is R ($D \simeq R$) the radius of the hydrogen-like orbit. The surface states near the midgap are due to charge clusters, e.g., two positive charges situated close together within the depth D act effectively as a single center of double charge, quadrupling the energy to 0.24 eV. Larger clusters can give even deeper surface energy levels but with smaller surface-state density because the charges are randomly distributed, and the chance that many charges will be located close together is very small.

From the above discussions it is clear that to obtain stability of MIS device parameters under operational stresses such as high electric field and elevated temperature, we must control both the ion drift instability and the surface-charge instability.[34] The basic approaches to the control of the ion drift are (1) modification of the insulator, e.g., phosphorus doping on the Si-SiO$_2$ system or use of an insulator such as silicon nitride which is immune to ion drift and (2) clean fabrication so as to avoid contamination of sodium and other impurities. To control the surface-charge instability, we must use (1) proper annealing, e.g., high-temperature dry-hydrogen annealing which can effectively reduce the surface charge density and (2) semiconductor substrates which are free from foreign impurities other than the desired dopants.[35] Because of the correlation between the surface charge and surface states, if we can control the surface charge instability, we automatically control the surface-state instability.

4 EFFECTS OF METAL WORK FUNCTION, CRYSTAL ORIENTATION, TEMPERATURE, ILLUMINATION, AND RADIATION ON MIS CHARACTERISTICS

(1) Metal Work Function Effect

For an ideal MIS diode it has been assumed that the work function difference (Fig. 2) for a p-type semiconductor,

$$\phi_{ms} \equiv \phi_m - \left(\chi + \frac{E_g}{2q} + \psi_B \right), \tag{51}$$

is zero. If the value of ϕ_{ms} is not zero, and if a fixed surface charge density Q_{fc} exists at the insulator-semiconductor interface (assuming negligible surface states and other charges), the experimental capacitance-voltage curve will be displaced from the ideal theoretical curve by an amount

$$V_{FB} = \frac{Q_{fc}}{C_i} - \phi_{ms} \tag{52}$$

where V_{FB} is the shift of voltage corresponding to the flat-band capacitance, and C_i is the insulator capacitance per unit area (ε_i/d). The fixed charge density is obtained from Eq. (52):

$$Q_{fc} = C_i(V_{FB} + \phi_{ms}) \qquad \text{coul/cm}^2 \tag{53a}$$

$$= \frac{C_i}{q}(V_{FB} + \phi_{ms}) \qquad \text{charges/cm}^2. \tag{53b}$$

The energy band diagram for the interface between silicon and thermally grown SiO_2 has been obtained from electron photoemission measurement[36] and is shown in Fig. 33. For SiO_2 the band gap is found to be about 8 eV and the electron affinity ($q\chi_i$) 0.9 eV. It is also found that deep electron traps are present at a level of 2 eV below the oxide conduction-band edge with a capture cross section of about 10^{-12} cm^2. The mobility in the oxide is estimated to be either 34 or 17 cm^2/V-second depending on whether the trapping center is singly or doubly charged. Photoemission of holes from silicon to silicon dioxide has also been studied,[37] and it is obtained that for holes in the oxide valence band, the mobility mean-free-time product is about 10^{-12} cm^2/V.

The effect of the metal work function on MIS diodes can be studied using the photoresponse measurement and the MIS capacitance-voltage measurement. Figure 34(a) shows the cube root of photoresponse versus photon energy for MIS diodes using various metals.[38] The intercept on the $h\nu$-axis corresponds to the metal-SiO_2 barrier energy $q\phi_B$. The metal work function is given by the sum of ϕ_B and χ_i where χ_i is the electron affinity of the insulator (refer to Fig. 2). Similar results are obtained from the MIS capacitance measurements as shown in Fig. 34(b). From Eq. (52), if two different metals

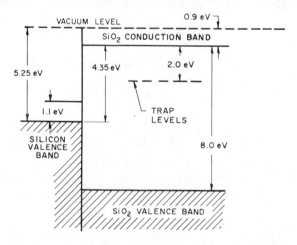

Fig. 33 Energy band diagram of Si-SiO$_2$ system obtained from photoemission measurement. (After Williams, Ref. 36.)

are deposited as field plates on the same oxidized silicon sample, the displacement between the two experimental MIS curves will represent the difference in metal work functions ($\phi_{m1} - \phi_{m2}$) or ($\phi_{B1} - \phi_{B2}$). Hence if the value of ϕ_m for any one metal is known (e.g., Al with $\phi_m = 3.2$ V from photoresponse), then the ϕ_m values for other metal can be determined. The results are shown in Table 9.1; also shown are the vacuum work functions. The metal work

TABLE 9.1 METAL WORK FUNCTION (VOLTS)

Metal	ϕ_m (C − V)	ϕ_m (Photoresponse)	ϕ_m (Vacuum Work Function)
Mg	3.35	3.15	3.7
Al	4.1*	4.1	4.25
Ni	4.55	4.6	4.5
Cu	4.7	4.7	4.25
Au	5.0	5.0	4.8
Ag	5.1	5.05	4.3

* Value of ϕ_m for Al (4.1 V) is the sum of the barrier height (3.2 V) and the SiO$_2$ electron affinity (0.9 V).

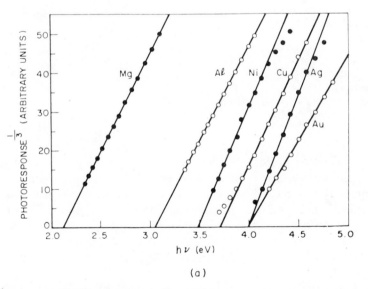

Fig. 34

(a) Cube root of photoresponse versus photon energy for MIS diodes using various metals.

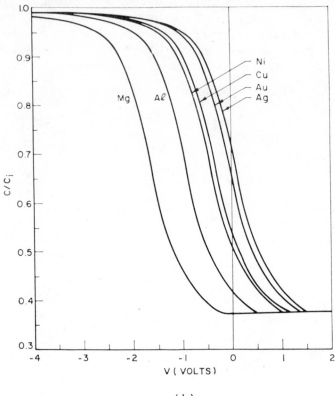

Fig. 34
(b) The corresponding MIS capacitance curves.
(After Deal et al., Ref. 38.)

functions as obtained from the photoresponse and the capacitance curve are in excellent agreement. These values are, however, different from that of the vacuum work function. This is of no surprise, since the deposited metal films are polycrystalline and the insulator-metal interface is quite different from the vacuum-metal (single crystal) interface as used in the vacuum work function measurement. It has also been found[38] using the above methods that the silicon-silicon dioxide barrier is independent of silicon orientation to within 0.1 V.

The above results show that ϕ_{ms} can significantly affect the silicon surface potential, and that in evaluating the surface charge densities from C-V curves, the voltage displacement must be corrected for ϕ_{ms}, Eq. (53). Figure 35(a) and (b) show the band diagrams[38] for two common electrode metals (Al and Au) on n- and p-type Si (10^{16} cm^{-3}) respectively for the case of an oxide thickness of 500 Å and zero surface charge. It can be seen that by appropriate choice of the electrode metal, the n-type surface can be varied from accumulation to depletion, and the p-type surface can be varied from flat-band to inversion. In Fig. 35(c) the required ϕ_{ms} correction is shown as a function of silicon doping and type for both gold and aluminum electrodes.

(2) Crystal Orientation Effect

It has been shown in Fig. 19 that the surface state density of a Si-SiO$_2$ system under a given oxidation condition has its largest value for $\langle 111 \rangle$

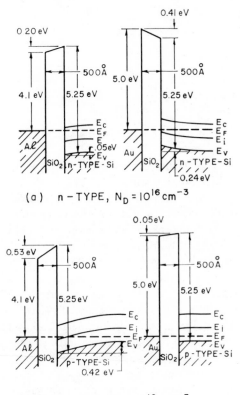

(a) n – TYPE, $N_D = 10^{16}$ cm^{-3}

(b) p – TYPE, $N_A = 10^{16}$ cm^{-3}

Fig. 35
(a) Band diagram for Au and Al on n-type silicon,
(b) Band diagram for Au and Al on p-type silicon, and

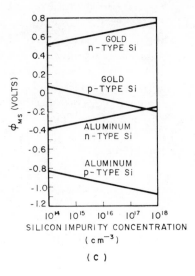

Fig. 35
(c) ϕ_{ms} versus silicon doping.
(Ref. 38.)

oriented substrate. This result can be correlated with the effect of the crystal orientation on oxidation rates.[30,39] The oxidation rate is given by $C \exp(-E_a/kT)$ where C is proportional to the available silicon bonds per cm^2 to react with the oxygen or water molecule, and E_a is the activation energy. Table 9.2 shows the properties of silicon crystal planes oriented along $\langle 111 \rangle$, $\langle 110 \rangle$, and $\langle 100 \rangle$. It is apparent that the (111) surface has the largest number of available bonds per cm^2 and the (100) surface has the smallest. The activation energy also is expected to depend on the orientation. Bonds parallel to the surface are expected to react most readily. Those at an angle to the surface plane will react less easily because of a steric hindrance presented by the position of the silicon atoms in the neighborhood of the bond. On examination of surfaces with different orientations, one notes that $E_a(100) > E_a(111) > E_a(110)$. Since the (100) surface has the largest E_a and the smallest available bonds, the oxidation rate of (100) would therefore be the smallest. If the activation energies for the (111) and (110) surfaces are reasonably close, then one would expect that the (111) surface has the largest oxidation rate because it has 23 percent more available bonds per cm^2. If we assume that the origin of the fixed surface charge is due to the excess ionic silicon in the oxide,[30] then the larger the oxidation rate, the larger the amount of the

TABLE 9.2 PROPERTIES OF SILICON CRYSTAL PLANES

Orientation	Plane Area of Unit Cell (cm^2)	Atoms in Area	Available Bonds in Area	Atoms/cm^2	Available Bonds/cm^2
$\langle 111 \rangle$	$\sqrt{3}a^2/2$	2	3	7.85×10^{14}	11.8×10^{14}
$\langle 110 \rangle$	$\sqrt{2}a^2$	4	4	9.6×10^{14}	9.6×10^{14}
$\langle 100 \rangle$	a^2	2	2	6.8×10^{14}	6.8×10^{14}

excess silicon ions; thus the (111) surface should have the highest fixed charge density. This in turn gives the highest surface states density on the (111) surface because of the correlation between surface charge and surface states as discussed in Section 3(4).

(3) Temperature Effect

For an MIS diode, the charge in the inversion layer can communicate with the bulk under steady-state conditions only by means of generation-recombination processes. At room temperature the inversion-layer cutoff frequencies in Si-SiO$_2$ systems are normally below 100 Hz, sometimes below 1 Hz. At lower temperatures the build up of the inversion charge is very slow. It is found that a forward bias of about 0.25 V must be reached across the space charge region before a noticeable injection of the inversion charge occurs.[22] The true inversion capacitance-voltage curve can only be measured by allowing a long time for equilibration at each bias point. At elevated temperatures, however, the generation is more rapid, and the temperature effect on the MIS characteristics especially the generation mechanisms can be easily studied. A typical example is shown in Fig. 36 where (a) is for a family of capacitance versus voltage curves and (b) for a family of conductance versus voltage curves at 6 kHz for an n-type sample.[40] It is seen that both capacitance and conductance saturate in the inversion region at negative voltage. Due to the residual surface states small humps appear in the depletion region. The appropriate equivalent circuit for the heavy inversion region has been shown previously in Fig. 25(d). The total conductance G, $1/R_{gs} + 1/R_{gD} + 1/R_d$, can be obtained from Fig. 36(b), and is plotted versus $1/T$ in (c). As discussed previously in Chapter 3, the space-charge recombination process is proportional to n_i with an activation energy of $E_g/2$, and the diffusion process is proportional to n_i^2 with an activation energy of E_g. From Fig. 36(c) it is seen that the space charge recombination is the dominant effect ($1/R_{gD} \sim n_i$) up to 140°C. In this range the experimental activation energy is 0.56 eV, line

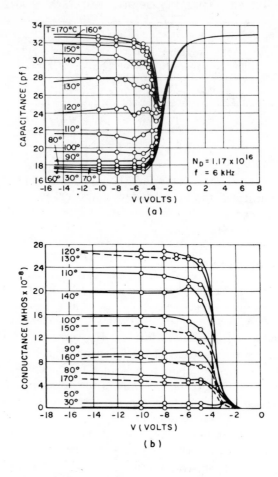

Fig. 36
(a) MIS capacitance versus voltage curves at various temperatures.
(b) Conductance versus voltage curves at various temperatures.

(a), in excellent agreement with the expected activation energy of n_i or $(E_g/2)$. Above 140°C a new process dominates as shown by the break in the $1/T$ plot. The high-temperature region, line (c), is obtained by subtraction of the influence of space-charge generation from the total conductance, line (b)-line (a). The activation energy is found to be 1.17 eV (E_g) corresponding to the expected result for the diffusion process $(1/R_d \sim n_i^2)$. The above results demonstrate the validity of the equivalent circuit shown in Fig. 25(d) for the heavy inversion region.[12]

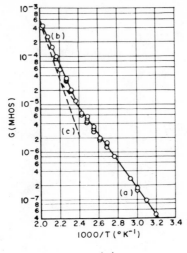

(c)

Fig. 36
(c) Conductance versus $1/T$.
(After Goetzberger and Nicollian, Ref. 40.)

(4)   Illumination Effect

The main effect of illumination on the MIS capacitance-voltage curves is that the capacitance in the heavy inversion region approaches the low-frequency value as the intensity of illumination is increased. This effect is illustrated in the insert of Fig. 37 which shows the normalized capacitance measured at 100 kHz versus applied voltage when white light with different intensities is illuminated on the MIS device.[13] There are two basic mechanisms which are responsible for this effect. The first is the decrease in the time constant, τ_{inv}, of minority carrier generation in the inversion layer.[13] The second is the generation of electron-hole pairs by photons which causes a decrease of the surface potential ψ_s under constant applied voltage.[41] This decrease of ψ_s results in a reduction of the width of the space-charge layer, with a corresponding increase of the capacitance. The second mechanism is preponderant when the measurement frequency is high.

From the equivalent circuit shown in Fig. 25(d), the MIS capacitance under the heavy inversion condition is given by

$$\frac{C}{C_i - C} = \frac{C_D}{C_i}\frac{1 + \omega^2\tau_{inv}^2}{\omega^2\tau_{inv}^2} \tag{54}$$

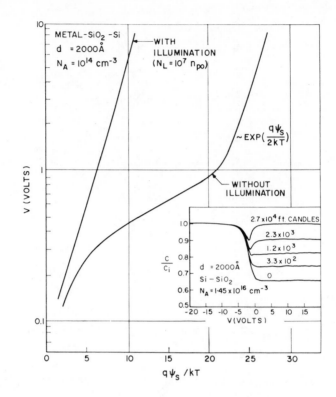

Fig. 37 Effect due to illumination.
(After Grove et al., Ref. 13, and Grosvalet and Jund, Ref. 41.)

where τ_{inv} is the time constant of the inversion layer given by $C_D(\Sigma 1/R)^{-1}$. The ratio of the time constants with or without illumination is given by

$$\frac{\tau_{\text{inv}}(\text{with illumination})}{\tau_{\text{inv}}(\text{without illumination})} = \frac{n_i}{n_i + G\tau} \qquad (55)$$

where n_i is the intrinsic carrier concentration, G the generation rate of electron-hole pairs by photons, and τ the lifetime of minority carriers. It is clear from Eqs. (54) and (55) that as G increases, the lifetime τ_{inv} decreases, which in turn causes an increase in C. This explains the first mechanism. However, in order to obtain a large increase in C, one must use very strong illumination.

Consider now the second mechanism. Without illumination and under strong inversion, the applied voltage [from Eqs. (15), (23), and (24)] is given by

$$V = \psi_s + \frac{\varepsilon_s}{\varepsilon_i} d\mathscr{E}_s \approx \frac{\varepsilon_s}{\varepsilon_i} d\mathscr{E}_s$$

$$\approx \frac{\varepsilon_s d}{\varepsilon_i} \sqrt{\frac{2kT n_{po}}{\varepsilon_s}} \exp\left(\frac{q\psi_s}{2kT}\right) \tag{56}$$

where d is the oxide thickness, and $\mathscr{E}_s$ is the surface electric field. With uniform illumination, the equilibrium concentrations in the bulk become (for a p-type semiconductor)

$$N_L = n_{po} + \Delta n, \tag{57}$$
$$P_L = p_{po} + \Delta n.$$

And the approximate expression for the applied voltage is given by Eq. (56) with N_L replacing n_{po}. Figure 37 shows a typical example of V versus ψ_s for silicon with $N_A = 10^{14}$ cm^{-3} and SiO$_2$ layer of 2000 Å. Without illumination, the curve is similar to that shown in Fig. 5. With illumination corresponding to generation such that $N_L = 10^7 n_{po}$ ($\approx 10^{13}$ cm^{-3}), the curve shifts toward the left. Thus the surface potential ψ_s at a given bias voltage is considerably reduced and results in an increase in C.

(5) Radiation Effect

As shown[42] in Fig. 38(a), the MIS structure is under ionization radiation such as electrons,[42] x-ray,[43] or γ-ray.[44] As the radiation passes through the insulator, hole-electron pairs are generated. Since the holes are relatively immobile,[37] they are trapped or recombine with electrons before they can leave the insulator. The electrons, on the other hand, are more mobile and drift toward the positive electrode. Since the Si is unable to supply electrons to the SiO$_2$ (because of a large potential barrier at the Si-SiO$_2$ interface) a net positive space charge, Q_R, builds up near the Si-SiO$_2$ interface. Figure 38(b) illustrates this buildup for an MIS diode with positive voltage applied to the metal electrode.[44] As the positive space charge grows, the electric field in the oxide between the space charge and the positive electrode decreases. When the field in this region is reduced to zero, no further charge will accumulate unless the applied voltage is increased. It is thus expected that the charge Q_R will depend on V. Based on the above model, it is predicted[45] that (1) the dependence of the charge buildup on radiation dose, D_R, is approximately of the form [$1 - \exp(-\text{constant} \cdot D_R)$], (2) the charge

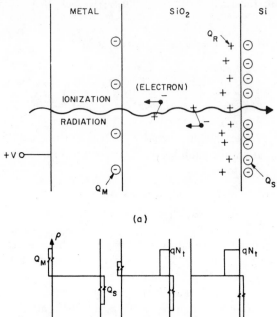

(a)

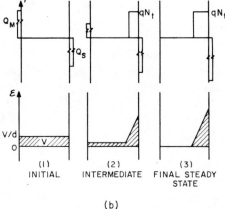

(b)

Fig. 38
(a) MIS structure under ionization radiation.
(After Zianinger, Ref. 42.)
(b) Buildup of the positive space charge near the Si-SiO$_2$ interface.
(After Snow et al., Ref. 44.)

buildup increases linearly with the voltage for both polarities, (3) the charge buildup depends on the total dose absorbed and not on the rate at which the dose was received.

The charge Q_R will cause a voltage shift of the MIS capacitance curve (Fig. 39, insert). For $V < 0$, however, the accumulation of the trapped charge

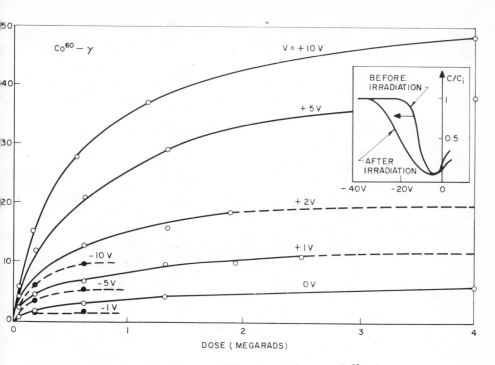

Fig. 39 Experimental results of measured MIS voltage shift versus Co^{60}-γ dose. (After Mitchell, Ref. 45.)

will take place near the metal-insulator interface rather than near the insulator-semiconductor interface. In this case the space charge is further removed from the semiconductor surface and hence will have less effect on the MIS curve. Figure 39 shows some typical experimental results[45] where the voltage shifts are plotted as a function of the dose of Co^{60} γ-rays. We note, as expected, that the voltage shifts for positive bias values are considerably larger for the voltage shifts for the corresponding negative bias values. And the exponential buildup is also in good agreement with the prediction.

5 SURFACE VARACTOR, AVALANCHE, TUNNELING, AND ELECTROLUMINESCENT MIS DIODES

As mentioned in the Introduction, at the present time the most useful application of MIS diodes is in the study of semiconductor surfaces and in the passivation of semiconductor planar devices. The MIS diodes, however, have other potentials per se as useful devices both for electronic application and for studies of fundamental physical parameters and physical processes. In this section, we will consider briefly some of the interesting potentials of MIS diodes.

(1) MIS Surface Varactor

Actually the first MIS structure was proposed in 1959 as a surface varactor by Moll,[1] and by Pfann and Garrett.[2] The basic characteristics of this device are its capacitance-voltage dependence and its negligible dc conduction. Two quantities of importance for the performance of a varactor are the ratio between the maximum and minimum capacitances and the cutoff frequency. The MIS varactor approaches the ideal requirement of an abrupt change from one constant value of capacitance to another (for a frequency sufficiently high that a high-frequency curve is obtained). The ratio of the maximum ($\approx C_i$) and the minimum (C'_{min} in the inversion region) capacitance can be varied by varying the insulator thickness and the semiconductor doping density. For large ratios, thin insulating films and low doping densities (or high resistivities) should be used [as illustrated in Fig. 13(b)]. The cutoff frequency for small ac signal operation is defined as[3,4]

$$f_c = \frac{1}{2\pi ARC} \tag{58}$$

where A is the area, R the resistance, and C the capacitance per unit area at dc bias voltage. For large ac signals, the capacitance will change appreciably during a cycle, and a minimum cutoff frequency is defined as

$$f_c(\text{min}) = \frac{1}{2\pi ARC_{max}} \tag{59}$$

when C_{max} is the highest possible capacitance under operating conditions. To reduce the resistance, an epitaxial layer should be used. The thickness of the layer cannot be less than the width of the space-charge layer at maximum voltage, $(\varepsilon_s\, d/\varepsilon_i)(C_{max}/C'_{min} - 1)$, to obtain the desired capacitance change. It can be shown that comparable performance can be obtained using either a step p-n junction varactor or an MIS surface varactor. It is expected that stable operation of an MIS varactor with good efficiency will ultimately be achieved at frequencies exceeding 300 GHz by using thin epitaxial layers and thin insulating films.[46] At medium frequencies the performance is equivalent to the junction varactor, and at still lower frequencies it is superior.

The latter result follows from the larger capacitance ratio available in the MIS varactor, since an equivalent capacitance ratio can be obtained with the junction varactor only at the expense of a large voltage swing, or by operation in the forward-biased direction where the junction is a lossy element. In addition, the MIS varactor is easy to fabricate. It can be made with a small active area and low capacitance, and without the need of a very low-resistance ohmic contact to the semiconductor. Thus for high-frequency use, in some cases where high contact resistance to the semiconductor or high capacitance is a limiting factor in the performance of a junction varactor, the MIS surface varactor will prove to be superior.

(2) Avalanche and Tunneling at Semiconductor Surface and Optical MIS Diodes

When a bias voltage is applied so rapidly to an MIS diode that buildup of an inversion is impossible, the capacitance-voltage curve will show the nonequilibrium characteristic of a continuously decreasing capacity, Fig. 7(c). In this case the space-charge region extends into the semiconductor as in a step p-n junction. When a voltage pulse is applied to the MIS diode with a sufficiently high amplitude to make the field at the surface reach the avalanche field, $\mathscr{E}_{sa}$, impact ionization occurs at this point and prevents the surface field from exceeding $\mathscr{E}_{sa}$. This results in a constant capacitance of low value at pulse voltages above breakdown voltage. The influence of surface states on capacitance can be neglected here because the time constants are much too long.

The relation between the applied voltage V, the avalanche field $\mathscr{E}_{sa}$, and the breakdown surface potential ψ_{sa} (identical to the breakdown voltage in the one-sided p-n junction as discussed in Chapter 3) can be obtained from Poisson's equation and is given by

$$V - V_{FB} = \frac{qN_D W}{C_i} + \psi_{sa} \tag{60}$$

where W is the space-charge layer width, and V_{FB} is the flat-band voltage shift due to work function difference and surface charges. Since $\psi_{sa} = (qN_D/2\varepsilon_s)W^2$, W can be eliminated from Eq. (60), and we obtain[47]

$$\psi_{sa} = (\sqrt{\alpha^2 + (V - V_{FB})} - \alpha)^2 \tag{61}$$

$$\mathscr{E}_{sa} = \sqrt{\frac{2qN_D \psi_{sa}}{\varepsilon_s}} \tag{62}$$

where

$$\alpha \equiv \frac{1}{C_i}\left(\frac{qN_D \varepsilon_s}{2}\right)^{1/2}.$$

We also obtain the relation between the measured capacitance and the voltage:

$$\left(\frac{C_i}{C}\right)^2 = 1 + \frac{(V - V_{FB})}{\alpha^2}. \tag{63}$$

An experimental result[48] is shown in Fig. 40 where $(C_i/C)^2$ is plotted against pulse voltage for an MIS diode (Si-SiO$_2$) with $d = 600$ Å and $N_D = 9.3 \times 10^{16}$ cm^{-3}. For lower voltages the plot follows the relation given by Eq. (62). At the avalanche point the silicon surface breaks down and a constant field, $\mathscr{E}_{sa}$, is reached, which results in a constant capacitance. Beyond the avalanche point any additional voltage will be applied across the insulator layer only. The result of ψ_{sa} (or V_B) as obtained from Eq. (61) is plotted in Fig. 41 as a function of the silicon doping concentration.[48] For lower dopings (Region I) the space-charge region is wide, as shown in Fig. 42(a), and the fringing field is high. This breakdown is similar to the junction edge effect[49] (Chapter 3), and gives rise to lower voltages than those predicted for plane junctions[50] (solid line in Fig. 41). For intermediate dopings (Region II) where the space-charge region is narrow, there is no edge effect, Fig. 42(b). The breakdown voltages agree with the theoretical values of abrupt p-n junctions. At very high dopings ($> 10^{18}$ cm^{-3}, Region III), tunneling occurs as shown in Fig. 42(c). Since the band bending has to be equal to the band gap, the breakdown voltage is always about 1.1 volts.[48]

The above effects can be used to obtain electroluminescence in MIS diodes.[51] A setup for observation of light emission is shown in Fig. 43(a). For a p-type sample, when a large positive voltage is applied, the bands in the semiconductor will bend down as shown in Fig. 43(b), and minority carriers (electrons) will accumulate at the semiconductor-insulator interface. If a large negative voltage is suddenly applied, the bands will bend up slightly as shown in Fig. 43(c), holes will accumulate at the interface, and the electrons previously generated at the interface may diffuse and drift into the semiconductor bulk and recombine with the holes resulting in light emission. To produce minority carriers fast enough in the positive half-cycle of the voltage, we must use either avalanche multiplication or the tunneling effect. The minority carrier charge density Q accumulated at the interface on positive half-cycles and the average minority carrier current J are given approximately by[51]

$$Q \simeq \varepsilon_i(\mathscr{E}_i - \mathscr{E}_{sa}) \tag{64a}$$

$$J \simeq Qf \tag{64b}$$

where $\mathscr{E}_i$ is the peak electric field in the insulator and f is the frequency. Figure 44(a) shows the observed spectrum of a p-type GaAs (4×10^{18} cm^{-3})

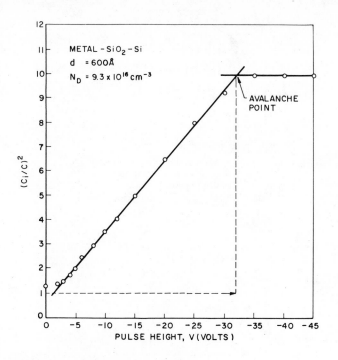

Fig. 40 $(C_i/C)^2$ versus pulse voltage for an MIS diode.
(After Goetzberger and Nicollian, Ref. 47.)

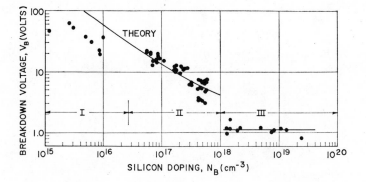

Fig. 41 Breakdown voltage in the silicon versus silicon doping concentration.
(After Goetzberger and Nicollian, Ref. 48.)

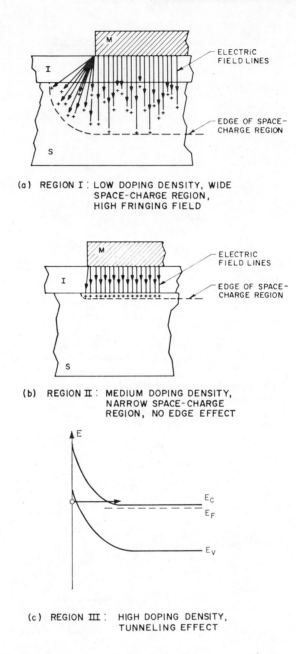

(a) REGION I : LOW DOPING DENSITY, WIDE
        SPACE-CHARGE REGION,
        HIGH FRINGING FIELD

(b) REGION II : MEDIUM DOPING DENSITY,
        NARROW SPACE-CHARGE
        REGION, NO EDGE EFFECT

(c) REGION III : HIGH DOPING DENSITY,
        TUNNELING EFFECT

Fig. 42 Schematic diagrams corresponding to the three regions in Fig. 41.
(Ref. 48.)

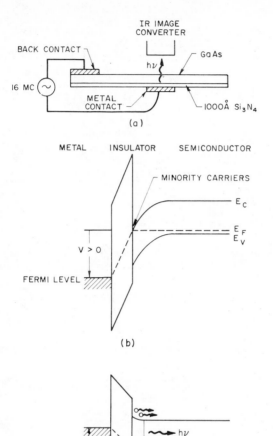

Fig. 43
(a) Schematic setup for observation of light emission.
(b) Band bending due to large positive voltage.
(c) Band bending due to negative voltage.
(After Berglund, Ref. 51.)

sample coated with 1000 Å Si_3N_4. The primary means of minority carrier generation is tunneling because of the large doping of the GaAs. Figure 44(b) shows the spectrum of a similar MIS structure with more lightly doped GaAs (5×10^{17} cm^{-3}). The means of minority generation in this case is

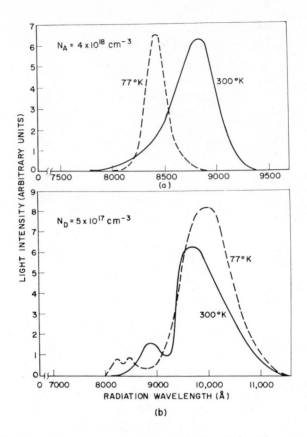

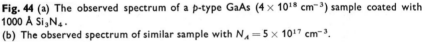

Fig. 44 (a) The observed spectrum of a *p*-type GaAs (4×10^{18} cm^{-3}) sample coated with 1000 Å Si$_3$N$_4$.
(b) The observed spectrum of similar sample with $N_A = 5 \times 10^{17}$ cm^{-3}.
(Ref. 51.)

avalanche multiplication. The minority carrier injection scheme described above may have significant advantages for materials that cannot be made both *n* and *p* type (e.g., some II-VI compounds). In connection with electroluminescence, the MIS diodes can also be used for optical detection such as the InSb MIS infrared detector which exhibits high quantum efficiency and high detectivity.[52] The MIS optical detectors can be made using a semitransparent metal electrode on the insulator. Light can be transmitted through the metal and the insulator layer ($hv < E_g$ of the insulator) and incident upon the semiconductor. If $hv > E_g$ of the semiconductor electron-hole pairs will be generated at the semiconductor surface. The MIS optical detector should be very useful because of the fact that the depletion region is at the semiconductor surface, and the absorption of light occurs in a high-field region, thereby yielding high collection efficiency.

(3) Tunneling in Insulator and Tunnel MIS Diodes

It can be shown that, under certain conditions, the current-voltage characteristic of an MIS diode with a very thin insulating layer on a degenerate semiconductor substrate exhibits negative resistance similar to that of a tunnel diode (Chapter 4). The simplified band diagrams[53] including surface states for MIS structures with p^{++} and n^{++} semiconductor substrates are shown in Fig. 45. The band bending, image forces, and potential drops across the oxide layer at equilibrium are neglected for simplicity. Consider the p^{++} type first. Application of a positive voltage to the metal, Fig. 45(b), causes electron tunneling from the valence band to the metal. This tunneling current increases monotonically with the increasing energy range between the Fermi levels; it further increases with the decreasing insulator barrier height. Application of a small negative voltage to the metal, Fig. 45(c), results in electron tunneling from the metal to the unoccupied semiconductor valence band. An increase of the voltage $-V$, according to Fig. 45(d), implies an increase in the effective barrier height for electrons tunneling from the metal to the unoccupied states of the valence band, i.e., a negative I-V characteristic [if $qV_C < qV_V$ as shown in Fig. 45(a)]. However, electrons in the metal with

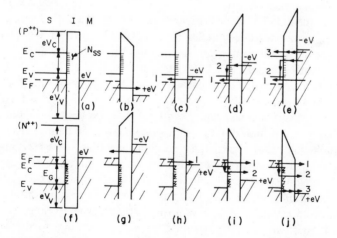

Fig. 45 Simplified band diagrams including surface states of tunneling MIS diode. (After Dahlke and Sze, Ref. 53.)

higher energies can tunnel simultaneously into the empty surface states and momentarily recombine with holes in the valence band, resulting in another current component. Since the insulator barrier decreases with bias, this current component has a positive I-V characteristic. Finally, an additional increase of the bias, Fig. 45(e), results in a third very fast-growing tunnel current component from the metal into the conduction band of the semiconductor.

Next consider the n^{++} type semiconductors. As shown in Fig. 45(f), the effective insulator barriers for the n^{++} type are expected to be smaller than those of the p^{++} type samples; hence for a given bias, there will be larger tunnel currents. For a negative bias on the metal Fig. 45(g), electrons tunnel from the metal into the empty states of the semiconductor conduction band, resulting in a large, rapidly increasing current. A small positive voltage on the metal, Fig. 45(h), leads to increasing electron tunneling from the conduction band of the semiconductor into the metal. If the surface states are filled with conduction electrons by recombination, a further increase in bias will give rise to a second current component caused by the tunneling of electrons from the surface states into the metal. This current component increases with increasing bias, Fig. 45(i), since the effective insulator barrier decreases. For larger voltage, Fig. 45(j), additional tunneling from the valence band to the metal is possible, but its influence on the total I-V characteristic is comparatively small because of the related high oxide barrier. Thus, the band structure of the semiconductor has a much smaller influence on the tunneling characteristics of the n^{++} type compared to p^{++} type structures.

Figure 46(a) shows the measured I-V characteristics at room temperature (solid lines) and liquid nitrogen temperature (dotted lines) for three p^{++} silicon samples with oxide layers (20 Å) treated in different ways. It should be noted that the small influence of temperature on the I-V characteristics is typical for tunneling. The samples have an oxide grown in dry oxygen (top one), in steam (middle one), and in steam with 30-minute annealing in H_2 at 350°C (bottom one). The band structure of the semiconductor is most distinctly reflected by the I-V characteristic of the bottom one; for negative voltages the current increases gradually with bias until $-V \approx 1$ volt where the current starts to increase rapidly with bias. This voltage corresponds to the silicon band gap at high doping (with some band-edge tailing effect).

The predicted negative resistance at small negative voltages is apparently masked by the tunneling of electrons from the metal into the surface states as discussed previously. The I-V characteristics of the top and the middle samples show in principle the same trend as the bottom one but exhibit considerably increased currents especially in the forbidden energy range ($-1.1\ V < V < 0$). If these currents are assumed to be proportional to the density of surface states, Fig. 46(a) leads to the conclusion that there is an increase of 1 or 2 orders of magnitude in surface state density when changing from annealed

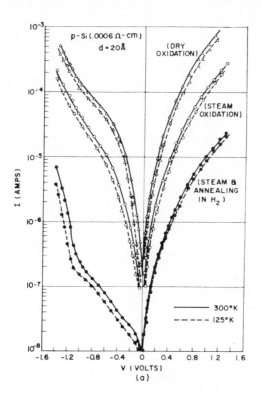

Fig 46

(a) Measured *I-V* characteristics at room temperature (solid lines) and liquid nitrogen temperature (dotted lines) for three p^{++} silicon samples with oxide layer (20 Å) treated in different ways.

to steam-grown and finally to dry-oxygen-grown oxide layers. This is in qualitative agreement with an experimentally determined increase of surface states for annealed-steam or oxygen-grown oxide layers of larger thickness ($d \approx 1000$ Å).[9] The effects of the semiconductor band structure and the density of surface states on the dc tunnel characteristics are even more distinctly reflected by the conductance-voltage curves, as shown in Fig. 46(b) which are obtained by differentiating the measured curves for Fig. 46(a). The solid lines and dotted lines are for measurements at room temperature and liquid nitrogen temperature respectively. The left branches of the curves for $V < -1.1$ V in Fig. 46(b) represent electron tunneling from the metal into the conduction band; the right branches, $V > 0$, represent tunneling from the

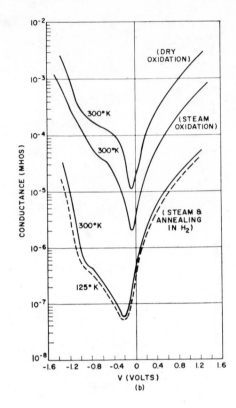

Fig. 46
(b) The corresponding conductance (dI/dV) curves.
(Ref. 53.)

valence band into the metal. The minimum conductance at small negative voltages is a result of the superposition of two current components as shown in Fig. 45(d). The expected negative conductance of the first component is apparently compensated for by the larger positive conductance of the surface states.

The measured ac capacitance and conductance as a function of applied voltage and frequency are presented in Fig. 47(a) and (b) respectively. The capacitance curves of the unannealed sample, as expected, are higher than the annealed ones, since the former has higher surface-state densities. Much larger frequency dependence is observed on the ac conductance curves. For $f < 5$ kHz, the conductance curves are virtually identical to the curves shown in Fig. 46(b). The basic equivalent circuit including the tunneling effect is shown in the insert of Fig. 47. The RC circuit to the left of the line AA is identical to that shown in Fig. 16(a) for thicker insulator films. R_T is the equivalent resistance for the tunneling of electrons into the valence band and/or conductance band of the p^{++} semiconductor, [corresponding to current components 1 and 3 in Fig. 45(c) and (e)]. R_{TS} is the equivalent resistance for the tunneling of electrons into the surface states and the recombining of them with holes in the valence band [corresponding to current component 2 in Fig. 45(d) or (e)]. Both R_T and R_{TS} are functions of the applied voltage. The complete circuit can be simplified as a capacitor $C(\omega)$ in parallel with a conductor $G(\omega)$, both frequency dependent.

It can be shown that $d[C(\omega)]/d\omega \le 0$; and that $d[G(\omega)]/d\omega \ge 0$. For thin oxide layers and/or highly doped semiconductor substrates, the capacitance will increase from its high-frequency value[53]

$$C(\infty) = \frac{C_i C_D}{(C_i + C_D)} \tag{65}$$

to

$$C(0) \simeq C(\infty) + AC_s \tag{66}$$

as $\omega \to 0$. And the conductance will increase from its low-frequency value

$$G(0) = \frac{1}{R_T} + \frac{1}{(R_{TS} + R_s)} \tag{67}$$

to

$$G(\infty) = G(0) + \frac{A(R_{TS} + R_s)}{R_{TS} R_s} \tag{68}$$

as $\omega \to \infty$, where

$$A \equiv \frac{(C_i R_{TS} - C_D R_s)^2}{(C_i + C_D)^2 (R_{TS} + R_s)^2}.$$

The experimental results as shown in Fig. 47 are in good agreement with the above discussion.

The predicted negative resistance[54] is obtained in an MIS tunnel diode of Al-Al$_2$O$_3$-SnTe. The SnTe is highly doped p-type with a concentration of 8×10^{20} cm^{-3}; and the Al$_2$O$_3$ is about 50 Å thick. The current-voltage characteristics at three different temperatures are shown in Fig. 48(a) where the negative resistance occurs between 0.6 to 0.8 volt. The corresponding conductance curve at 4.2°K is shown in Fig. 48(b). Reasonable agreement is obtained between the experimental result (solid line) and the prediction (dotted line) based on the WKB approximation.

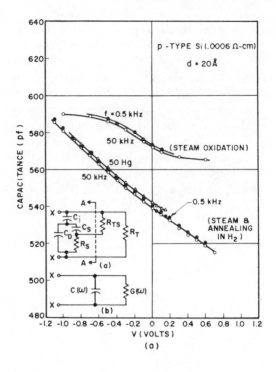

Fig. 47
(a) MIS capacitance-voltage curves measured at different frequencies.
The insert shows the equivalent circuits for tunnel MIS diode.

6 CARRIER TRANSPORT IN INSULATING FILMS

In an ideal MIS diode it is assumed that the conductance of the insulating film is zero. Real insulators, however, show carrier conduction at a field of 10^6 V/cm or lower. To estimate the electric field in an insulator under biasing conditions, we obtain from Eqs. (16) and (24) that

$$\mathscr{E}_i = \mathscr{E}_s \left(\frac{\varepsilon_s}{\varepsilon_i} \right) \tag{69}$$

where $\mathscr{E}_i$ and $\mathscr{E}_s$ are the electric fields in the insulator and the semi-conductor respectively, and ε_i and ε_s are the corresponding permittivities. For a Si-SiO$_2$ system the field for silicon at avalanche breakdown[49] is about 5×10^5 V/cm; the corresponding field in the insulator is then 3 times larger

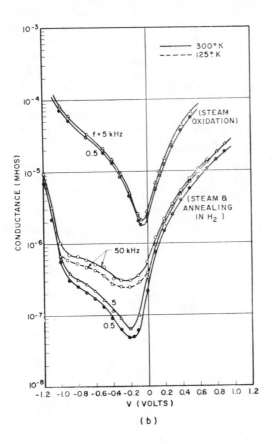

Fig. 47
(b) MIS conductance-voltage curves measured at different frequencies.
(Ref. 53.)

$(\varepsilon_{Si}/\varepsilon_{SiO_2} = 12/3.9)$, i.e., about 1.5×10^6 V/cm. At this field the electron and hole conductions in SiO_2 are negligible even at elevated temperatures; however, mobile ions such as sodium will transport through the insulator and give rise to device instability and the hysteresis effect.[8] To improve the device stability and to fully develop the potential usefulness of MIS structures, we should investigate other systems than the Si-SiO$_2$. Also, we should study the composite system, which combines several of them to form a new system incorporating the desirable features of each.

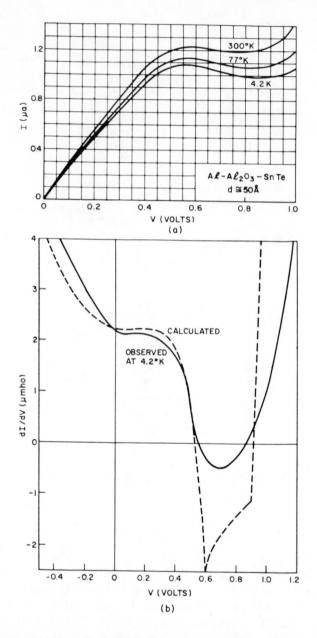

Fig. 48
(a) Tunnel MIS diode (Al-Al$_2$O$_3$-SnTe) *I-V* characteristics at three temperatures.
(b) The conductance curve at 4.2°K. Solid line is for experimental result. Dotted line from theoretical WKB approximation. (After Esaki and Stiles, Ref. 54.)

The composite systems that have been studied include the metal-(Al_2O_3-SiO_2)-Si system,[55] and the metal-(phospho-silicate glass-SiO_2)-Si system.[56,57] It has been found that the former gives a voltage shift in the C-V curve toward positive voltage. This implies that a p-type semiconductor surface can be inverted at zero or negative bias voltages. The latter system, the phospho-silicate glass which is SiO_2 rich in P_2O_5, can be formed on the outside of a SiO_2 layer during the phosphorus diffusion process and can substantially reduce the instability of a contaminated MIS diode, since the sodium has much greater solubility in glass than in SiO_2.

Another composite system is the metal-Si_3N_4-SiO_2-Si system.[58] The silicon nitride films are found to have much smaller diffusion coefficients of various impurities,[59] particularly sodium, in comparison with SiO_2 films. It is thus expected that this system will have the desired features of low surface states because of the clean SiO_2-Si interface, and will be immune to ion drift because of the outer Si_3N_4 film. In the visible range, Si_3N_4 films have a slightly larger dielectric constant than the SiO_2 films. A color comparison[60] of SiO_2 and Si_3N_4 films is listed in Table 9.3 which also shows film thickness.

TABLE 9.3
COLOR COMPARISON OF SiO_2 AND Si_3N_4 FILMS

Order	Color	SiO_2 Thickness Range* (μm)	Si_3N_4 Thickness Range (μm)
	Silicon	0–0.027	0–0.020
	Brown	0.027–0.053	0.020–0.040
	Golden Brown	0.053–0.073	0.040–0.055
	Red	0.073–0.097	0.055–0.073
	Deep Blue	0.097–0.010	0.073–0.077
1st	Blue	0.10–0.12	0.077–0.093
	Pale Blue	0.12–0.13	0.093–0.10
	Very Pale Blue	0.13–0.15	0.10–0.11
	Silicon	0.15–0.16	0.11–0.12
	Light Yellow	0.16–0.17	0.12–0.13
	Yellow	0.17–0.20	0.13–0.15
	Orange Red	0.20–0.24	0.15–0.18
1st	Red	0.24–0.25	0.18–0.19
	Dark Red	0.25–0.28	0.19–0.21
2nd	Blue	0.28–0.31	0.21–0.23
	Blue-Green	0.31–0.33	0.23–0.25
	Light Green	0.33–0.37	0.25–0.28
	Orange Yellow	0.37–0.40	0.28–0.30
2nd	Red	0.40–0.44	0.30–0.33

* The ratio of refractive index $\dfrac{\bar{n}(Si_3N_4)}{\bar{n}(SiO_2)} = \dfrac{1.97}{1.48} = 1.33 = \dfrac{SiO_2 \text{ thickness}}{Si_3N_4 \text{ thickness}}$.

The basic conduction processes in insulators are summarized in Table 9.4. The Schottky emission process is similar to that discussed previously in Chapter 8, where thermionic emissions across the metal-insulator interface or the insulator-semiconductor interface are responsible for the carrier transport. A plot of $\ln(J/T^2)$ versus $1/T$ should yield a straight line with a slope determined by the permittivity ε_i of the insulator. The Frenkel-Poole emission[65a] is due to field-enhanced thermal excitation of trapped electrons into the conduction band. For trap states with coulomb potentials, the expression is virtually identical to that of the Schottky emission. The barrier height, however, is the depth of the trap potential well, and the quantity $\sqrt{q/\pi\varepsilon_i}$ is larger than in the case of Schottky emission by a factor of 2 since the barrier lowering is twice as large due to the immobility of the positive charge. The tunnel emission is due to field ionization of trapped electrons into the conduction band or due to electrons tunneling from the metal Fermi energy into the insulator conduction band. The tunnel emission has the strongest dependence on the applied voltage but is essentially independent of the temperature. The space-charge-limited current results from carrier

TABLE 9.4

BASIC CONDUCTION PROCESSES IN INSULATORS

Process	Expression†	Voltage and Temperature Dependence‡
Schottky Emission	$J = A^{*}T^2 \exp\left[\dfrac{-q(\phi_B - \sqrt{q\mathscr{E}/4\pi\varepsilon_i})}{kT}\right]$	$\sim T^2 \exp(+ a\sqrt{V}/T - q(\phi_B/kT)$
Frenkel-Poole Emission	$J \sim \mathscr{E} \exp\left[\dfrac{-q(\phi_B - \sqrt{q\mathscr{E}/\pi\varepsilon_i})}{kT}\right]$	$\sim V \exp(+2a\sqrt{V}/T - q(\phi_B/kT)$
Tunnel or Field Emission	$J \sim \mathscr{E}^2 \exp\left[-\dfrac{4\sqrt{2m^{*}}(q\phi_B)^{3/2}}{3q\hbar\mathscr{E}}\right]$	$\sim V^2 \exp(-b/V)$
Space-Charge-Limited	$J = \dfrac{8\varepsilon_i\mu V^2}{9d^3}$	$\sim V^2$
Ohmic	$J \sim \mathscr{E} \exp(-\Delta E_{ae}/kT)$	$\sim V \exp(-c/T)$
Ionic Conduction	$J \sim \dfrac{\mathscr{E}}{T} \exp(-\Delta E_{ai}/kT)$	$\sim \dfrac{V}{T}\exp(-d'/T)$

† A^{*} = effective Richardson constant, ϕ_B = barrier height, $\mathscr{E}$ = electric field, ε_i = insulator dynamic permittivity, m^{*} = effective mass, d = insulator thickness, ΔE_{ae} = activation energy of electrons, ΔE_{ai} = activation energy of ions, and a $\equiv \sqrt{q/(4\pi\varepsilon_i d)}$.

‡ $V = \mathscr{E}d$. Positive constants independent of V or T are b, c, and d'.

injected into the insulator where there is no compensating charge present. The current for the unipolar trap-free case is proportional to the square of the applied voltage. At low voltage and high temperatures, current is carried by thermally excited electrons hopping from one isolated state to the next. This mechanism yields an ohmic characteristic exponentially dependent on temperature. The ionic conduction is similar to a diffusion process. Generally, the dc ionic conductivity decreases during the time the electric field is applied, since ions cannot be readily injected into or extracted from the insulator. After an initial current flow, positive and negative space charges will build up near the metal-insulator and the semiconductor-insulator interfaces. This causes a distortion of the potential distribution. When the applied field is removed, large internal fields remain which cause some, but not all, of the ions to flow back toward their equilibrium position and hysteresis effects result.

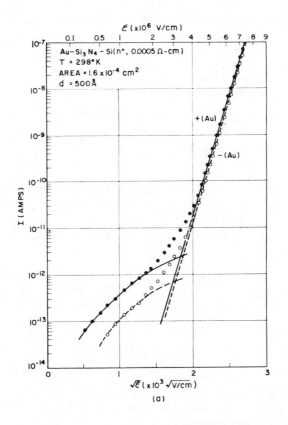

Fig. 49
(a) Current-voltage characteristics of Au-Si$_3$N$_4$-Si diode at room temperature.

For a given insulator, each conduction process may dominate in certain temperature and voltage ranges. The processes are also not exactly independent of one another and should be carefully examined. For example, for the large space-charge effect, the tunneling characteristic is found to be very similar to the Schottky-type emission.[61]

An example of the conduction processes for silicon nitride films[62] is shown in Fig. 49. The films are deposited on degenerate silicon substrate (0.0005 Ω-cm n-type) by the process of reaction of $SiCl_4$ and NH_3 at 1000°C. An MIS diode is made by evaporation of Au onto the Si_3N_4 film. Figure 49(a) shows ln I versus $\sqrt{\mathscr{E}}$. The +(Au) curve is for the gold-electrode positive and the −(Au) is for the gold-electrode negative. It will be noted that the two curves are

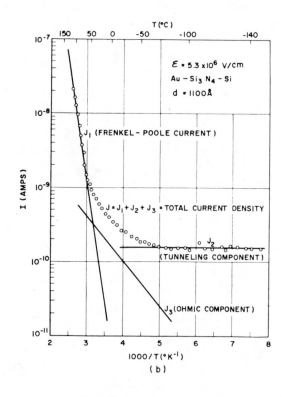

Fig. 49
(b) Current versus $1/T$ of an Au-Si_3N_4-Si diode at a field of 5.3×10^6 V/cm. The total current density is separated into three components: J_1 (Frenkel-Poole current), J_2 (tunnel current), and J_3 (ohmic conduction current). (Ref. 62.)

virtually identical. The slight difference (especially at low fields) is believed to be mainly due to the difference in barrier heights at the gold-nitride and nitride-silicon interfaces. One also notices that there are two distinct regions. In high electric fields the current varies exponentially with the square root of the field;[62, 63] at low fields, the characteristic is ohmic. It has been found that at room temperature for a given field, the characteristics of current density versus field are essentially independent of the film thickness, the device area, the electrode materials, and the polarity of the electrodes. These results strongly suggest that the current is bulk-controlled rather than electrode-controlled as in Schottky-barrier diodes. At low temperatures the current-voltage characteristic becomes nearly independent of the temperature.

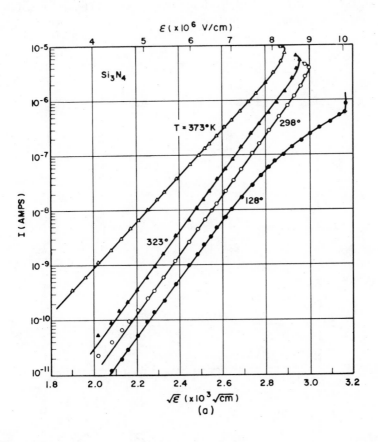

Fig. 50

(a) Current-voltage curves of a typical nitride film 1100Å thick and 0.2 mm in diameter at four different ambient temperatures.

Figure 49(b) shows plots of the current versus $1/T$ for a field of 5.3 MV/cm. One notes that the conduction current J in the nitride films can be separated into three components: J_1, J_2, and J_3. The current J_1 is due to the Frenkel-Poole emission which dominates at high temperatures and high fields. The dynamic dielectric constant, $\varepsilon_i/\varepsilon_0$, obtained from the slopes of Fig. 49 is found to be 5.5, in agreement with the optical measurement.[60] The current J_2 is due to tunnel emission of trapped electrons into the conductance band, which dominates at low temperatures and high fields. The current J_3 is the ohmic component which contributes at low fields and moderate temperatures.

Another important parameter in the insulator is the maximum dielectric strength. Figure 50(a) shows $I\text{-}V$ curves of a typical nitride film 1100 Å thick and 0.2 mm in diameter at four different ambient temperatures.[62] As the ramp voltage increases, the conduction current increases accordingly. Eventually the voltage reaches a maximum value beyond which it decreases

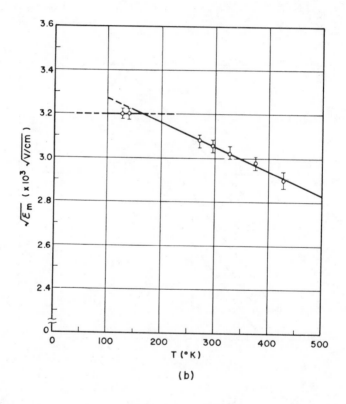

(b)

Fig. 50

(b) Square root of maximum dielectric strength versus temperature of silicon nitride films. (Ref. 62.)

while the current continues to increase and destructive breakdown occurs. The above maximum voltage normalized to the film thickness is defined as the maximum dielectric strength.

Based on the assumption that at high temperatures, the maximum dielectric strength, $\mathscr{E}_m$, is limited by the thermal instability of the structure, the value of $\mathscr{E}_m$ can be determined by equating the Joule heat with that lost by heat transfer, and is given by[62]

$$\mathscr{E}_m \simeq \left(\frac{\pi \varepsilon_i}{q}\right)[\phi_B - CT]^2 \tag{70}$$

where T is the temperature, ϕ_B the barrier height, and C a slowly varying function of T. At low temperatures, an intrinsic dielectric strength is expected and $\mathscr{E}_m$ approaches a constant value.[64] Figure 50(b) shows the plot of $\mathscr{E}_m^{1/2}$ versus T. It will be noted that at high temperatures $\mathscr{E}_m$ varies approximately as Eq. (70); and at low temperatures $\mathscr{E}_m$ becomes independent of temperature.

Similar results have been obtained in other insulating films. For example, the current transports in tantalum oxide[65] and silicon oxide[66] films have the general behavior of nitride films. Maximum dielectric strength as large as 4×10^7 V/cm has been observed in silicon dioxide films.[67]

REFERENCES

1. J. L. Moll, "Variable Capacitance with Large Capacity Change," Wescon Convention Record, Part 3, p. 32 (1959).

2. W. G. Pfann and C. G. B. Garrett, "Semiconductor Varactor Using Space-Charge Layers," Proc. IRE, 47, 2011 (1959).

3. D. R. Frankl, "Some Effects of Material Parameters on the Design of Surface Space-Charge Varactors," Solid State Electron., 2, 71 (1961).

4. R. Lindner, "Semiconductor Surface Varactor," Bell Syst. Tech. J., 41, 803 (1962).

5. L. M. Terman, "An Investigation of Surface States at a Silicon/Silicon Dioxide Interface Employing Metal-Oxide-Silicon Diodes," Solid State Electron., 5, 285 (1962).

6. K. Lehovec and A. Slobodskoy, "Field-Effect Capacitance Analysis of Surface States on Silicon," Phys. Stat. Solidi, 3, 447 (1963).

7. A. Many, Y. Goldstein, and N. B. Grover, Semiconductor Surfaces, John Wiley & Sons, Inc., New York (1965).

8. E. H. Snow, A. S. Grove, B. E. Deal, and C. T. Sah, "Ion Transport Phenomena in Insulating Films," J. Appl. Phys., 36, 1664 (1965).

9. E. H. Nicollian and A. Goetzberger, "The Si-SiO₂ Interface-Electrical Properties as Determined by the MIS Conductance Technique," Bell Syst. Tech. J., 46, 1055 (1967).

10. C. G. B. Garrett and W. H. Brattain, "Physical Theory of Semiconductor Surfaces," Phys. Rev., 99, 376 (1955).

11. R. H. Kingston and S. F. Neustadter, "Calculation of the Space Charge, Electric Field, and Free Carrier Concentration at the Surface of a Semiconductor," J. Appl. Phys., *26*, 718 (1955).

11a. C. F. Young, "Extended Curves of the Space Charge, Electric Filed, and Free Carrier Concentration at the Surface of a Semiconductor and Curves of the Electrostatic Potential Inside a Semiconductor," J. Appl. Phys., *32*, 329 (1961).

12. S. R. Hofstein and G. Warfield, "Physical Limitation on the Frequency Response of a Semiconductor Surface Inversion Layer," Solid State Electron., *8*, 321 (1965).

13. A. S. Grove, B. E. Deal, E. H. Snow, and C. T. Sah, "Investigation of Thermally Oxidized Silicon Surfaces Using Metal-Oxide-Semiconductor Structures," Solid State Electron., *8*, 145 (1965).

14. A. S. Grove, E. H. Snow, B. E. Deal, and C. T. Sah, "Simple Physical Model for the Space-Charge Capacitance of Metal-Oxide-Semiconductor Structures," J. Appl. Phys., *33*, 2458 (1964).

15. A. Goetzberger, "Ideal MOS Curves for Silicon," Bell Syst. Tech. J., *45*, 1097 (1966).

16. I. Tamm, Physik. Z. Sowjetunion, *1*, 733 (1933).

17. W. Shockley, "On the Surface States Associated with a Periodic Potential," Phys. Rev., *56*, 317 (1939).

18. J. Kontecky, "On the Theory of Surface States," J. Phys. Chem. Solids, *14*, 233 (1960).

19. D. Pugh, "Surface States on the (111) Surface of Diamond," Phys. Rev. Letters, *12*, 390 (1964).

20. W. Shockley and G. L. Pearson, "Modulation of Conductance of Thin Films of Semiconductors by Surface Charges," Phys. Rev., *74*, 232 (1948).

21. F. G. Allen and G. W. Gobeli, "Work Function, Photoelectric Threshold and Surface States of Atomically Clean Silicon," Phys. Rev., *127*, 150 (1962).

22. A. Goetzberger, "Behavior of MOS Inversion Layers at Low Temperature," IEEE Trans. on Electron Devices, *ED-14*, 787 (1967).

23. E. H. Nicollian and A. Goetzberger, "MOS Conductance Technique for Measuring Surface State Parameters," Appl. Phys. Letters, *7*, 216 (1965).

24. K. H. Zaininger and G. Warfield, "Limitation of the MOS Capacitance Method for the Determination of Semiconductor Surface Properties, "IEEE Trans. on Electron Devices, *ED-12*, 179 (1965).

25. C. N. Berglund, "Surface States at Steam-Grown Silicon-Silicon Dioxide Interface," IEEE Trans. on Electron Devices, *ED-13*, 701 (1966).

26. P. V. Gray and D. M. Brown, "Density of SiO_2-Si Interface States," Appl. Phys. Letters, *8*, 31 (1966).

27. D. R. Frankl, "Comment on Density of SiO_2-Si Interface States by Gray and Brown," J. Appl. Phys., *38*, 1996 (1967).

28. C. T. Sah and F. H. Hielscher, "Evidence of the Surface Origin of the $1/f$ Noise," Phys. Rev. Letters, *17*, 956 (1966).

29. E. H. Nicollian and H. Melchior, "A Quantitative Theory of $1/f$ Type Noise Due to Interface States in Thermally Oxidized Silicon," Bell Syst. Tech. J., *46*, 1935 (1967).

30. B. E. Deal, M. Sklar, A. S. Grove, and E. H. Snow, "Characteristics of the Surface-State Charge (Q_{ss}) of Thermally Oxidized Silicon," J. Electrochemical Soc., *114* (March 1967).

31. S. R. Hofstein, "Proton and Sodium Transport on SiO_2 Films," Trans. on Electron Devices, *ED-14*, 749 (1967).

32. E. H. Nicollian and A. Goetzberger, "Lateral AC Current Flow Model for Metal-Insulator-Semiconductor Capacitors," IEEE Trans. Electron Devices, *ED-12*, 108 (1965).

33. A. Goetzberger, V. Heine, and E. H. Nicollian, "Surface States in Silicon from Charges in the Oxide Coating," Appl. Phys. Letters, *12*, 95 (1968).

34. S. R. Hofstein, "Stabilization of MOS Devices," Solid State Electron., *10*, 657 (1967).

35. R. Schmidt, private communication.

36. R. Williams, "Photoemission of Electrons From Silicon Into Silicon Dioxide," Phys. Rev., *140*, A569 (1965).

37. A. M. Goodman, "Photoemission of Electrons From Silicon and Gold Into Silicon Dioxide, "Phys. Rev., *144*, 588 (1966).

38. B. E. Deal, E. H. Snow, and C. A. Mead, "Barrier Energies in Metal-Silicon Dioxide-Silicon Structures," J. Phys. Chem. Solids. *27*, 1873 (1966).

39. J. R. Ligenza, "Effect of Crystal Orientation on Oxidation Rates of Silicon in High Pressure Steam," J. Phys. Chem., *65*, 2011 (1961).

40. A. Goetzberger and E. H. Nicollian, "Temperature Dependence of Inversion Layer Frequency Response in Silicon," Bell Syst. Tech. J., *46*, 513 (1967).

41. J. Grosvalet and C. Jund, "Influence of Illumination on MIS Capacitance in the Strong Inversion Region," IEEE Trans. on Electron Devices, *ED-14*, 777 (1967).

42. K. H. Zaininger, "Electron Bombardment of MOS Capacitors," Appl. Phys. Letters, *8*, 140 (1966).

43. D. R. Collins and C. T. Sah, "Effects of X-Ray Irradiation on the Characteristics of MOS Structures," Appl. Phys. Letters, *8*, 124 (1966).

44. E. H. Snow, A. S. Grove and D. J. Fitzgerald, "Effect of Ionization Radiation on Oxidized Silicon Surfaces and Planar Devices," Proc. IEEE, *55*, 1168 (1967).

45. J. P. Mitchell, "Radiation-Induced Space-Charge Buildup in MOS Structures," IEEE Trans. on Electron Devices, *ED-14*, 764 (1967).

46. D. P. Howson, B. Owen, and G. T. Wright, "The Space-Charge Varactor," Solid State Electron., *8*, 913 (1965).

47. A. Goetzberger and E. H. Nicollian, "Transient Voltage Breakdown Due to Avalanche in MIS Capacitors," Appl. Phys. Letters, *9*, 444 (1966).

48. A. Goetzberger and E. H. Nicollian, "MOS Avalanche and Tunneling Effects in Silicon Surfaces," J. Appl. Phys., *38*, 4582 (1967).

49. S. M. Sze and G. Gibbons, "Effect of Junction Curvature on Breakdown Voltage in Semiconductors," Solid State Electron., *9*, 831 (1966).

50. S. M. Sze and G. Gibbons, "Avalanche Breakdown Voltages in Abrupt and Linearly Graded Ge, Si, GaAs, and GaP *p-n* Junctions," Appl. Phys. Letters, *8*, 111 (1966).

51. C. N. Berglund, "Electroluminescence Using GaAs MIS Structures," Appl. Phys. Letters, *9*, 441 (1966).

52. R. J. Phelan, Jr., and J. O. Dimmock, "InSb MOS Infrared Detector," Appl. Phys. Letters, *10*, 55 (1967).

53. W. E. Dahlke and S. M. Sze, "Tunneling in Metal-Oxide-Silicon Structures," Solid State Electron., *10*, 865 (1967).

54. L. Esaki and P. J. Stiles, "New Type of Negative Resistance in Barrier Tunneling," Phys. Rev. Letters, *16*, 1108 (1966).

55. G. T. Cheney, R. M. Jacobs, H. W. Korb, H. E. Nigh, and J. Stack, "Al_2O_3-SiO_2 IGFET Integrated Circuits," paper 2.2, IEEE Device Meeting, Washington, D.C. (Oct. 18-21, 1967).

56. D. R. Kerr, J. S. Logan, P. J. Burkhardt, and W. A. Pliskin," Stabilization of SiO_2 Passivation Layers with P_2O_5," IBM J., *8*, 376 (1964).

57. E. H. Snow and B. E. Deal, "Polarization Phenomena and Other Properties of Phosphosilicate Glass Films on Silicon," J. Elect. Soc., *113*, 2631 (1966).

58. T. L. Chu, J. R. Szedon, and C. H. Lee, "The Preparation and C-V Characteristics of Si-Si_3N_4 and Si-SiO_2-Si_3N_4 Structure," Solid State Electron., *10*, 897 (1967).

59. J. V. Dalton, "Sodium Drift and Diffusion in Silicon Nitride Films," J. Electrochem. Soc., *113*, 1650 (1966).

60. F. Reizman and W. Van Gelder, "Optical Thickness Measurement of SiO_2-Si_3N_4 Films on Silicon," Solid State Electron., *10*, 625 (1967).

61. J. J. O'Dwyer, "Current-Voltage Characteristics of Dielectric Films," J. Appl. Phys., *37*, 599 (1966).

62. S. M. Sze, "Current Transport and Maximum Dielectric Strength of Silicon Nitride Films," J. Appl. Phys., *38*, 2951 (1967).

63. S. M. Hu, "Properties of Amorphous Silicon Nitride Films," J. Electrochemical Soc., *113*, 693 (1966).

64. J. J. O'Dwyer, *The Theory of Dielectric Breakdown of Solids*, Clarendon Press, Oxford, England (1964).

65. C. A. Mead, "Electron Transport Mechanisms in Thin Insulating Films," Phys. Rev., *128*, 2088 (1962).

65a. J. Frenkel, "On the Theory of Electric Breakdown of Dielectrics and Electronic Semi-conductors," Tech. Phys. USSR, *5*, 685 (1938); also "On Pre-Breakdown Phenomena in Insulators and Electronic Semiconductors," Phys. Rev., *54*, 647 (1938).

66. T. E. Hartman, J. C. Blair, and R. Bauer, "Electrical Conduction Through SiO_2 Films," J. Appl. Phys., *37*, 2468 (1966).

67. N. Klein, "The Mechanism of Self-Healing Electrical Breakdown in MOS Structures," IEEE Trans. on Electron Devices, *ED-13*, 788 (1966).

- INTRODUCTION
- SURFACE-SPACE-CHARGE REGION UNDER NONEQUILIBRIUM CONDITION
- CHANNEL CONDUCTANCE
- BASIC DEVICE CHARACTERISTICS
- GENERAL CHARACTERISTICS
- IGFET WITH SCHOTTKY BARRIER CONTACTS FOR SOURCE AND DRAIN
- IGFET WITH A FLOATING GATE —A MEMORY DEVICE
- SURFACE FIELD EFFECTS ON p-n JUNCTIONS AND METAL-SEMICONDUCTOR DEVICES

10

IGFET and Related Surface Field Effects

1 INTRODUCTION

The word IGFET stands for the *I*nsulated-*G*ate *F*ield-*E*ffect *T*ransistor. Since the gate structures for most IGFET are of the MOS type, this device has also been called MOSFET (metal-oxide-semiconductor field-effect transistor) or MOST (metal-oxide-semiconductor transistor). The principle of the surface field-effect transistor was first demonstrated back in the early 1930's when Lilienfeld[1] and Heil[2] proposed using the surface field effect to achieve a solid-state amplifier. It was subsequently studied by Shockley and Pearson[3] in the late 1940's. In 1960 Kahng and Attala[4] proposed and fabricated the first IGFET using a thermally oxidized silicon structure. The basic device characteristics have been subsequently studied by Ihantola and Moll,[5,6] Sah,[7] and Hofstein and Heiman.[8] The technology and applications of IGFET have been reviewed recently by Wallmark and Johnson.[9]

Since the current in an IGFET is transported by carriers of one polarity only (e.g., electrons in an *n*-channel device), the IGFET is usually referred to as a unipolar device in contrast to the *p-n* junction transistor which is a bipolar device involving both types of carriers. The IGFET is a member of the family of unipolar transistors. The first one, the junction field-effect transistor as proposed by Shockley[10] in 1952, has already been considered in Chapter 7. We shall consider in the next chapter the thin-film-type field-effect transistor as proposed by Weimer[11] in 1961. The behaviors of the various unipolar transistors are basically quite similar. The IGFET, however, is particularly important because of its application in linear circuits and digital circuits.[9] Although the IGFET has been made on various semiconductors such as Ge,[12] Si,[13] and GaAs[14] and uses various insulators such as SiO_2, Si_3N_4, and Al_2O_3, the most important system is the Si-SiO_2 combination. Hence most of the experimental results in this chapter are obtained from the Si-SiO_2 system.

In this chapter we shall first consider the MIS device under a nonequilibrium condition in Section 2. This serves as a foundation to an understanding of the IGFET operation. The channel conductance and the basic device characteristics are considered in Sections 3 and 4 respectively. The general device characteristics such as the environmental effects and physical limitations are discussed in Section 5. Section 6 considers an IGFET using Schottky barrier contacts for source and drain. At room temperature the device characteristics are comparable to conventional IGFET's with similar electrode geometry. At lower temperatures the current transport is by the tunneling of carriers from the metal across the Schottky barrier to the semiconductor inversion layer. Section 7 presents an interesting memory device which is obtained by adding a floating gate to a regular IGFET. Finally, the surface field effects on junction breakdown and reverse current are considered in Section 8.

2 SURFACE-SPACE-CHARGE REGION UNDER NON-EQUILIBRIUM CONDITION

The basic structure of an insulated-gate field-effect transistor (IGFET) is illustrated in Fig. 1. This device consists of a p-type semiconductor substrate into which two n^+ regions, the source and the drain, are formed (e.g., by diffusion or by ion-implantation). The metal contact on the insulator is called the gate electrode. When there is no voltage applied to the gate, the source-to-drain electrodes correspond to two p-n junctions connected back to back. The only current that can flow from the source to the drain is the reverse leakage current (or saturation current).* When a sufficiently large positive bias is applied to the gate such that a surface inversion layer (or channel) is formed between the two n^+ regions, the source and the drain are thus connected by

---

* This is the n–channel enhancement-type IGFET. Other types will be discussed later.

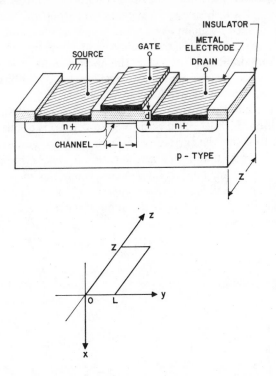

Fig. 1 Schematic diagram of an insulated-gate field-effect transistor. The important parameters are the channel length (L), the channel width (Z), and the insulator thickness (d). (After Kahng and Atalla, Ref. 4.)

a conducting-surface n channel through which a large current can flow. The conductance of this channel can be modulated by varying the gate voltage. One may readily extend the discussion to a p-channel device by exchanging p for n and reversing the polarity of the voltages. The important parameters of an IGFET are the channel length L, the channel width Z, the insulator thickness d (with permittivity ε_i), and the channel conductance.

Before we calculate the channel conductance, we must consider the gate MIS structure under nonequilibrium conditions, i.e., the situation when a voltage is applied across the source-drain contacts, and the imref of the minority carriers (electrons, in the present case) is lowered from the equilibrium Fermi level. In order to show more clearly the band bending across the

n^+p junction under the nonequilibrium condition, the IGFET in Fig. 1 is turned 90 degrees and is shown in Fig. 2(a). The portion of the surface enclosed by the dashed frame in Fig. 2(a) is shown in Fig. 2(b).

The idealized junction and the energy band diagrams pertaining to the nonequilibrium conditions (i.e., when the applied voltage $V_D \neq 0$) are illustrated in Fig. 3. In this representation the energy band is shown as a function of the two directions x and y. When no gate voltage (V_G) is applied, the only variation of the band is in the y direction and is due to the applied reverse bias on the drain. In Fig. 3(b) a positive gate voltage is applied, but it is not large enough to invert the surface of the p region. The gate voltage required for inversion is larger than in the zero junction bias case, in which $\psi_s(\text{inv}) \simeq 2\psi_B$. This is because the applied reverse bias lowers the imref of the minority carriers (electrons), and an inversion layer can be formed only when the potential at the surface crosses over the imref of the minority carrier. Thus the

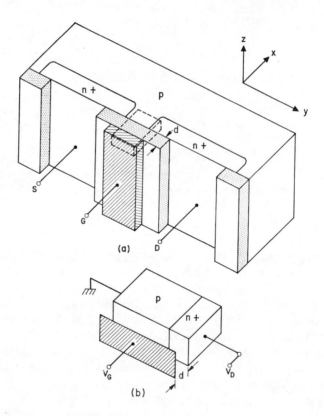

Fig. 2
(a) Same as in Fig. 1. The IGFET is turned 90°.
(b) The portion of the surface enclosed by the dashed frame in (a).

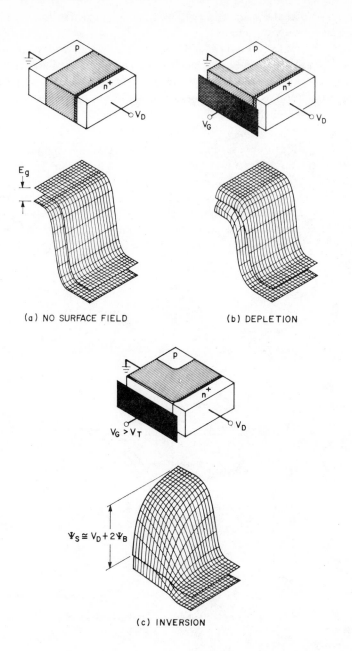

(a) NO SURFACE FIELD   (b) DEPLETION

(c) INVERSION

Fig. 3 The idealized junction and energy band diagrams under nonequilibrium conditions. (After Grove and Fitzgerald, Ref. 15.)

band bending at the surface in Fig. 3(b) is still insufficient for inversion, and the surface is only depleted.

Figure 3(c) illustrates the condition at which the gate voltage is large enough to bring the conduction band sufficiently near the imref for electrons to cause an inversion at the surface. The surface potential ψ_s at the onset of strong inversion is given, to a good approximation, by

$$\psi_s(\text{inv}) \simeq V_D + 2\psi_B. \tag{1}$$

As in the equilibrium case discussed in the preceding chapter, the surface depletion region reaches a maximum width W_m at inversion. The width is now a function of the bias V_D. Figure 4 shows a comparison of the charge distribution and energy band variation of an inverted p region for (a) the equilibrium case and (b) the nonequilibrium case.

The characteristic of the surface-space charge under the nonequilibrium condition[15] can be derived similarly to that in Chapter 9. The two assumptions

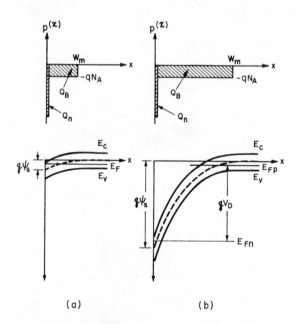

Fig. 4 Comparison of the charge distribution and energy band variation of an inverted p region for
(a) the equilibrium case and
(b) the nonequilibrium case.
(After Grove and Fitzgerald, Ref. 15.)

are that (1) the imref for the majority carriers of the substrate does not vary with distance from the bulk to the surface and (2) the imref for the minority carriers of the substrate is separated by the applied junction bias V_D from the imref for the majority carriers, i.e., $E_{Fp} = E_{Fn} + qV_D$ for a p substrate. The first assumption introduces little error when the surface is inverted, since majority carriers are then only a negligible part of the surface-space charge; the second assumption is correct under the inversion condition, since the only condition under which minority carriers are an important part of the surface-space-charge region is that when the surface is inverted.

Based on the above assumptions, the one-dimensional Poisson equation for the surface-space-charge region is given by

$$\frac{\partial^2 \psi}{\partial x^2} = -\frac{q}{\varepsilon_s}(N_D^+ - N_A^- + p - n) \tag{2}$$

where

$$\left. \begin{array}{c} N_D^+ - N_A^- = n_{po} - p_{po} \\[6pt] p = p_{po}\, e^{-\beta\psi} \\[6pt] n = n_{po}\, e^{\beta\psi - \beta V_D}. \qquad \beta \equiv q/kT. \end{array} \right\} \tag{3}$$

Following the same approach as in Chapter 9, we obtain

$$\mathscr{E} \equiv -\frac{\partial \psi}{\partial x} = \pm \frac{2kT}{qL_D} F\!\left(\beta\psi,\, V_D,\, \frac{n_{po}}{p_{po}}\right) \tag{4}$$

and

$$Q_s = \varepsilon_s \mathscr{E}_s = \mp \frac{2\varepsilon_s kT}{qL_D} F\!\left(\beta\psi_s,\, V_D,\, \frac{n_{po}}{p_{po}}\right) \tag{5}$$

where

$$F\!\left(\beta\psi,\, V_D,\, \frac{n_{po}}{p_{po}}\right) \equiv \left[e^{-\beta\psi} + \beta\psi - 1 + \frac{n_{po}}{p_{po}} e^{-\beta V_D}\!\left(e^{\beta\psi} - \beta\psi e^{\beta V_D} - 1\right) \right]^{1/2} \tag{6a}$$

and

$$L_D \equiv \left[\frac{2kT\varepsilon_s}{p_{po}\, q^2}\right]^{1/2}. \tag{6b}$$

The charge due to minority carriers within the inversion layer is given by

$$|Q_n| \equiv q \int_0^{x_i} n(x)\, dx = q \int_{\psi_s}^{\psi_B} \frac{n(\psi)\, d\psi}{d\psi/dx}$$

$$= q \int_{\psi_s}^{\psi_B} \frac{n_{po}\, e^{(\beta\psi - \beta V_D)}\, d\psi}{\left(\dfrac{2kT}{qL_D}\right) F\!\left(\beta\psi,\, V_D,\, \dfrac{n_{po}}{p_{po}}\right)} \tag{7}$$

where x_i denotes the point at which the intrinsic Fermi level intersects the imref for electrons.

Under strong inversion the fourth term in Eq. (6a) dominates, and we obtain

$$F\left(\beta\psi, V_D, \frac{n_{po}}{p_{po}}\right) \simeq \sqrt{\frac{n_{po}}{p_{po}}} e^{(\beta\psi - \beta V_D)}. \tag{8}$$

The potential is given from Eq. (4):

$$-\frac{\partial\psi}{\partial x} \simeq \frac{2kT}{qL_D} \sqrt{\left(\frac{n_{po}}{p_{po}}\right)} e^{\beta(\psi - V_D)/2}. \tag{9}$$

Integrating Eq. (9), we have

$$x = \sqrt{\frac{2\varepsilon_s kT}{n_{po} q^2}} \left[\exp\left(\frac{-\beta\psi + \beta V_D}{2}\right) - \exp\left(\frac{-\beta\psi_s + \beta V_D}{2}\right)\right] \tag{10}$$

where the boundary condition ($x = 0$, $\psi = \psi_s$) is applied. The quantity x_i is then given by

$$x_i \simeq \sqrt{\frac{2\varepsilon_s kT}{n_{po} q^2}} \exp(-\beta\psi_B). \tag{10a}$$

Comparing the above results with those derived in Chapter 9, we note the main difference is that, under a strong inversion condition, the surface potential $\psi_s(\text{inv})$ is approximately given by $(V_D + 2\psi_B)$ for the nonequilibrium case and by $2\psi_B$ for the equilibrium case. The charge per unit area after strong inversion is then given by

$$Q_s = Q_n + Q_B \tag{11}$$

$$Q_B = -qN_A W_m = -\sqrt{2qN_A \varepsilon_s (V_D + 2\psi_B)}. \tag{12}$$

3 CHANNEL CONDUCTANCE

The conductivity of a semiconductor in a one-dimensional case is given by

$$\sigma(x) = q[p(x)\mu_p(x) + n(x)\mu_n(x)] \tag{13}$$

where $\mu_p(x)$ and $\mu_n(x)$ are the hole and electron mobilities, respectively, and $p(x)$ and $n(x)$ are the hole and electron concentrations respectively. For n-channel devices under normal operating conditions $[n(x) \gg p(x)]$ the conductivity of the channel can be approximated by

$$\sigma(x) \simeq qn(x)\mu_n(x). \tag{13a}$$

The channel conductance is then given by

$$g \equiv \frac{Z}{L} \int_0^{x_i} \sigma(x) \, dx = \frac{Z}{L} q \int_0^{x_i} \mu_n(x) n(x) \, dx. \tag{14}$$

From Eqs. (7) and (14) we have for the channel conductance

$$g = \frac{Z}{L} \mu_{eff} |Q_n| \tag{15}$$

where the effective mobility μ_{eff} is given by

$$\mu_{eff} \equiv \frac{q \int_0^{x_i} \mu_n(x) n(x) \, dx}{|Q_n|}. \tag{16}$$

The charge per unit area Q_n that results from electrons within the inversion layer can be calculated as a function of the total charge per unit area induced in the semiconductor surface Q_s, $(Q_n = Q_s - Q_B)$. With known values of Q_n and g at a given gate voltage, the effective mobility can be readily determined from Eq. (15).

The experimental effective mobilities of electrons and holes in inversion layers on oxidized silicon are shown in Fig. 5(a) and (b) as a function of the total charge per unit area induced in the semiconductor.[16] We note that up to approximately $|Q_s/q| = 10^{12}$ cm^{-2} (corresponding to $\mathscr{E}_s \equiv Q_s/\varepsilon_s \cong 10^5$ V/cm), the mobility is practically constant (μ_{const}). Beyond this point the mobility decreases with increasing field. The values of the mobilities are a factor of approximately two lower than the corresponding bulk mobilities. This is due to the effect of the surface scattering and the effect of the redistribution of impurities during thermal oxidation which can cause the impurity concentration near the oxide-silicon interface to deviate from that in the bulk.

The surface-scattering effect was first considered by Schrieffer[17] and has recently been reviewed by Frankl.[18] The result for a constant surface field is given by

$$\frac{\mu_{eff}}{\mu_{bulk}} = 1 - \exp(\alpha^2) \operatorname{erfc}(\alpha) \tag{17}$$

where

$$\alpha \equiv \frac{1}{\mathscr{E}_s} \frac{1}{\mu_{bulk}} \sqrt{\frac{2kT}{m^*}}. $$

It is thus expected that, as the surface field $\mathscr{E}_s$ increases, the values of both α and μ_{eff} will decrease. Figure 5(c) shows a comparison between typical

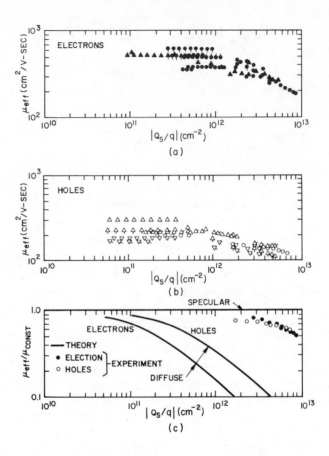

Fig. 5
(a) Effective channel mobility of electrons as a function of total surface charges.
(b) Effective channel mobility of holes as a function of total surface charges.
(c) Comparison between the experimental channel mobilities and the theoretical curves.
(After Leistiko et al., Ref. 16.)

experimental inversion layer mobilities and the theoretical curves calculated from Eq. (17) using μ_{const} as the bulk mobility μ_{bulk}. The result indicates that the carriers are essentially specularly reflected at the surface. And the diffuse surface-scattering process has significant effect only for large surface fields. Although the general behaviors are quite similar, it appears that the redistribution of the impurities is the more important effect in the $Si\text{-}SiO_2$ system.

The dependence of the effective mobility on temperature[16] is shown in Fig. 6. It appears that both electron and hole mobilities in inversion layers have a $T^{-3/2}$ power dependence at higher temperatures. The behavior is indicative of a scattering mechanism similar to lattice scattering.

4 BASIC DEVICE CHARACTERISTICS

We shall first present a qualitative discussion of the device operation principle. Let us consider first that a voltage is applied to the gate to cause an inversion at the semiconductor surface, Fig. 7(a). If a small drain voltage is applied, a current will flow from the source to the drain through the conducting channel. Thus the channel acts as a resistance, and the drain current I_D is proportional to the drain voltage V_D. This is the linear region. As the drain voltage increases, it eventually reaches a point at which the channel depth x_i at $y = L$ is reduced to zero, Fig. 7(b). This is the pinch-off point. Beyond the pinch-off point the drain current remains essentially the same. This is because, for $V_D > V_{D\,sat}$, the depletion region near the drain will increase and the point marked by Y will move toward the source. The voltage

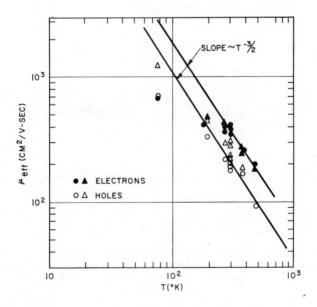

Fig. 6 Dependence of the effective mobilities on temperature. (After Leistiko et al., Ref. 16.)

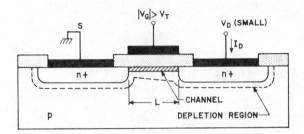

(a) LINEAR REGION

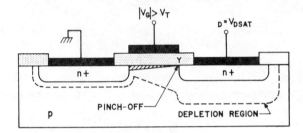

(b) ONSET OF SATURATION

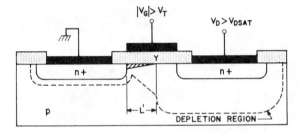

(c) BEYOND SATURATION

Fig. 7
(a) IGFET operated in linear region (low drain voltage).
(b) IGFET operated at onset of saturation. The point Y indicates the pinch-off point.
(c) IGFET operated beyond saturation and the effective channel length reduced to L'.

at Y, however, remains the same, $V_{D\,sat}$. Thus the number of carriers arriving at the point Y from the source, and hence the current flowing from source to drain, will remain the same. Carrier injection from point Y into the drain depletion region is quite similar to the case of carrier injection from an emitter-base junction to the base-collector depletion region of a junction transistor.

We shall now derive the basic IGFET characteristics under the following idealized conditions: (1) the gate MIS structure corresponds to an ideal MIS diode as defined in Chapter 9, i.e., no surface states, no surface charge, no work function difference, etc., (2) the carrier mobility in the inversion layer is constant, (3) the reverse leakage current is negligibly small, (4) only the drift current will be considered, and (5) the transverse field ($\mathscr{E}_x$ in x direction) in the channel is much larger than the longitudinal field ($\mathscr{E}_y$ in y direction). The last condition corresponds to the so-called gradual channel approximation.

Under the above idealized conditions the total charge induced in the semiconductor per unit area Q_s at a distance y from the source is given by

$$Q_s(y) = [-V_G + \psi_s(y)]C_i \qquad (18)$$

where $C_i \equiv \varepsilon_i/d$ is the capacitance per unit area. The charge in the inversion layer is given by

$$Q_n(y) = Q_s(y) - Q_B(y)$$
$$= -[V_G - \psi_s(y)]C_i - Q_B(y). \qquad (19)$$

The surface potential $\psi_s(y)$ at inversion can be approximated by $2\psi_B + V(y)$ where $V(y)$ is the reverse bias between the point y and the source electrode (which is assumed to be grounded, see Fig. 8). The charge within the surface depletion region $Q_B(y)$ is given previously as

$$Q_B(y) = -qN_A W_m = -\sqrt{2\varepsilon_s qN_A[V(y) + 2\psi_B]}. \qquad (20)$$

Substitution of Eq. (20) into Eq. (19) yields

$$Q_n(y) = -[V_G - V(y) - 2\psi_B]C_i + \sqrt{2\varepsilon_s qN_A[V(y) + 2\psi_B]}. \qquad (21)$$

The channel resistance of an elemental section (dy) is given by

$$dR = \frac{dy}{gL} = \frac{dy}{Z\mu_n|Q_n(y)|}. \qquad (22)$$

And the voltage drop across this elemental section is given by

$$dV = I_D\,dR = \frac{I_D\,dy}{Z\mu_n|Q_n(y)|} \qquad (23)$$

where I_D is the drain current and is a constant independent of y. Substitution

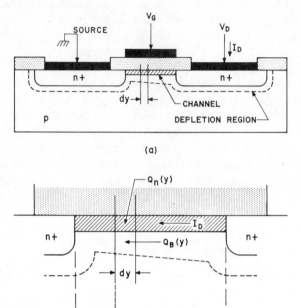

(a)

Fig. 8 Detailed view of the channel. The drain current is I_D which is a constant. At a distance y measured from the source (which is grounded) the charge densities per unit area in an incremental element dy are $Q_n(y)$ and $Q_B(y)$ for the inversion layer and the depletion layer respectively.

of Eq. (21) into Eq. (23) and integration from the source ($y = 0$, $V = 0$) to the drain ($y = L$, $V = V_D$) yield

$$I_D = \frac{Z}{L} \mu_n C_i \left\{ \left(V_G - 2\psi_B - \frac{V_D}{2} \right) V_D - \frac{2}{3} \frac{\sqrt{2\varepsilon_s q N_A}}{C_i} \left[(V_D + 2\psi_B)^{3/2} - (2\psi_B)^{3/2} \right] \right\}$$

(24)

for the present idealized case (no surface state, surface charge, etc.).

Equation (24) predicts that for a given V_G the drain current will first increase linearly with drain voltage (the linear region), then gradually will level off, approaching a saturated value (the saturation region). The basic output characteristic of an idealized IGFET is shown in Fig. 9. The dotted line indicates the locus of the drain voltage ($V_{D\,\text{sat}}$) at which the current reaches a maximum value.

We shall now consider the above mentioned two regions. For the case of small V_D, Eq. (24) reduces to

$$I_D \simeq \frac{Z}{L}\mu_n C_i\left[(V_G - V_T)V_D - \frac{V_D{}^2}{2}\right] \tag{25}$$

or

$$I_D \simeq \left(\frac{Z}{L}\right)\mu_n C_i(V_G - V_T)\,V_D \quad \text{for} \quad V_D \ll (V_G - V_T) \tag{25a}$$

where V_T is called the turn-on or threshold voltage and is given by

$$V_T = 2\psi_B + \frac{\sqrt{2\varepsilon_s q N_A(2\psi_B)}}{C_i}. \tag{26}$$

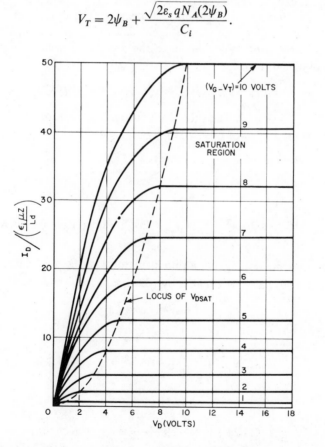

Fig. 9 Idealized output characteristics (I_D vs. V_D) of an IGFET. The dotted line indicates the locus of the saturation drain voltage ($V_{D\,\text{sat}}$). For $V_D > V_{D\,\text{sat}}$ the drain current is constant.

The calculated values of V_T as a function of semiconductor doping density and insulator thickness are shown in Fig. 10 for the Si-SiO$_2$ system. Since for $V_D \ll (V_G - V_T)$, I_D is proportional to V_D, this is the linear region. In this region, channel conductance g_D and transconductance g_m are given as

$$g_D \equiv \left.\frac{\partial I_D}{\partial V_D}\right|_{V_G = \text{const}} = \frac{Z}{L}\mu_n C_i(V_G - V_T) \tag{27a}$$

$$g_m \equiv \left.\frac{\partial I_D}{\partial V_G}\right|_{V_D = \text{const}} = \frac{Z}{L}\mu_n C_i V_D. \tag{27b}$$

When the drain voltage is increased to a point such that the charge in the inversion layer $Q(y)$ at $y = L$ becomes zero, the number of mobile electrons at the drain experiences a drastic fall-off. This point, called the pinch-off, is analogous to the junction field-effect transistor. The drain voltage and the drain current at this point are designated as $V_{D\,\text{sat}}$ and $I_{D\,\text{sat}}$ respectively. Beyond the pinch-off point we have the saturation region. The value of $V_{D\,\text{sat}}$ is obtained from Eq. (21) under the condition $Q_n(L) = 0$:

$$V_{D\,\text{sat}} = V_G - 2\psi_B + K^2(1 - \sqrt{1 + 2V_G/K^2}) \tag{28}$$

where $K \equiv \sqrt{\varepsilon_s q N_A}/C_i$. The saturation current $I_{D\,\text{sat}}$ can be obtained by substitution of Eq. (28) into Eq. (24):

$$I_{D\,\text{sat}} = \frac{Z\mu_n C_i}{6L} V_{\text{sum}}^2 \tag{29}$$

where

$$V_{\text{sum}}^2 \equiv [(V_{D\,\text{sat}} + 2\psi_B)^2 + V_G(V_{D\,\text{sat}} + 2\psi_B) - 12\psi_B(V_G - \psi_B - \tfrac{4}{3}K\sqrt{\psi_B})]. \tag{29a}$$

For low substrate doping and a thin insulator layer, $K \ll 1$, and the last term of Eq. (28) may be approximated by $\sqrt{2\varepsilon_s q N_A V_G}/C_i$, and we obtain

$$V_{D\,\text{sat}} \simeq V_G - V_T' \tag{30}$$

and

$$I_{D\,\text{sat}} \simeq \frac{Z}{2L}\mu_n C_i(V_G - V_T')^2 = \frac{Z}{2L}\mu_n C_i V_{D\,\text{sat}}^2 \tag{31}$$

where

$$V_T' \equiv 2\psi_B + \frac{\sqrt{2\varepsilon_s q N_A V_G}}{C_i} \equiv 2\psi_B + K\sqrt{2V_G}. \tag{31a}$$

If $K \ll 1$, we obtain $V_T \approx V_T' \approx 2\psi_B$.

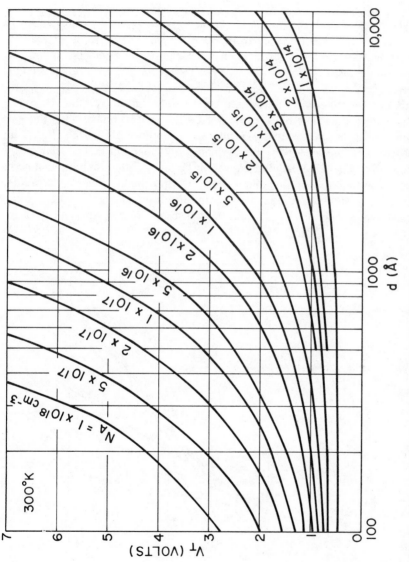

Fig. 10 Turn-on voltage as a function of thickness of the SiO₂ layer for various Si substrate dopings.

521

The transconductance in the saturation region is given by

$$g_m \equiv \left.\frac{\partial I_D}{\partial V_G}\right|_{V_D=\text{const}} = \frac{Z}{L}\mu_n C_i(V_G - V_T'). \tag{32}$$

The transfer characteristics,[7] I_D versus V_G with V_D as a parameter for an n-channel IGFET as obtained from Eq. (25), are shown in Fig. 11(a). They are straight lines below saturation when the channel is not pinched off and the gate voltage satisfies $V_G - V_T > V_D$. The intercepts of these straight lines are given by $V_G - V_T = V_D/2$. Between $V_G - V_T = 0$ and $V_G - V_T = V_D$, the drain current follows the saturation current locus given by Eq. (31). Figure 11(b) shows the transconductance versus the drain voltage with the gate voltage as a parameter. It is evident that below saturation $V_D < V_{D\text{sat}}$, the transconductance is proportional to the drain voltage but independent of the gate voltage Eq. (27b). Beyond saturation $V_D > V_{D\text{sat}}$ the transconductance is independent of the drain voltage but is proportional to the gate voltage Eq. (32). A similar representation with the drain and gate voltage interchanged is shown in Fig. 11(c). It is also of interest to note that for

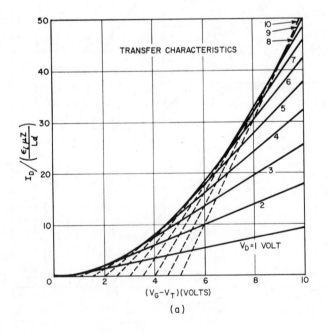

Fig. 11
(a) Idealized transfer characteristics (I_D versus V_G) of an IGFET.

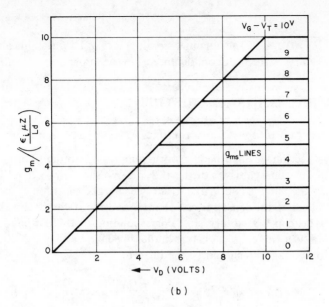

Fig. 11
(b) Transconductance versus drain voltage with gate voltage as a parameter.

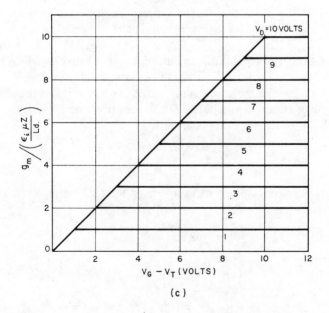

Fig. 11
(c) Transconductance versus gate voltage with drain voltage as a parameter.
(After Sah, Ref. 7.)

K $(\equiv \sqrt{\varepsilon_s q N_A}/C_i) \ll 1$, the channel conductance g_D in the linear region Eq. (27a) is equal to the transconductance g_m in the saturation region Eq. (32).

5 GENERAL CHARACTERISTICS

(1) Current-Voltage Characteristics

In the last section we have made many assumptions in order to bring out the most important characteristics of the IGFET. We shall now extend the consideration to include the effects due to surface charges, diffusion currents, etc. The main effect of the fixed surface charges and the difference in the work functions is to cause a voltage shift corresponding to the flat-band voltage V_{FB}. This in turn will cause a change in the threshold voltage V_T, e.g., in the linear region voltage V_T should be modified to

$$V_T = V_{FB} + 2\psi_B + \frac{\sqrt{2\varepsilon_s q N_A (2\psi_B)}}{C_i}$$

$$= \left(\phi_{ms} + \frac{Q_{fs}}{C_i}\right) + 2\psi_B + \frac{\sqrt{4\varepsilon_s q N_A \psi_B}}{C_i}. \qquad (33)$$

To consider the effect due to the diffusion current component, we first present in Fig. 12 the three-dimensional band diagram of an n-channel IGFET.[19] Figure 12(b) shows the flat-band zero-bias ($V_D = V_G = 0$) equilibrium condition. The equilibrium condition under a gate bias which causes a surface inversion is shown in Fig. 12(c). The nonequilibrium condition under both drain and gate biases is shown in Fig. 12(d) where we note the separation of the imrefs of electrons and holes. The drain current density including both drift and diffusion components is given by

$$J_D(x, y) = q\mu_n n \mathscr{E}_y + q D_n \nabla n$$

$$= -q D_n n(x, y) \nabla \psi_{Fn} \qquad (34)$$

where ψ_{Fn} is the electron imref measured from the bulk Fermi level. The total drain current is then[19]

$$I_D = \int_0^{x_i} J_D(x, y) Z \, dx$$

$$= \frac{1}{L} \int_0^L D_n q Z \left(\frac{\partial \psi_{Fn}}{\partial y}\right) \int_0^{x_i} n(x, y) \, dx \, dy$$

$$= \frac{Z}{L} \frac{\varepsilon_s \mu_n}{L_D} \int_0^{V_D} \int_{\psi_B}^{\psi_s} \frac{e^{(\beta\psi - \beta V)}}{F\left(\beta\psi, V, \frac{n_{po}}{p_{po}}\right)} \, d\psi \, dV \qquad (34a)$$

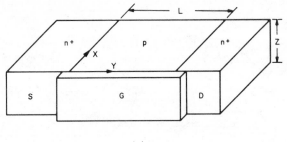

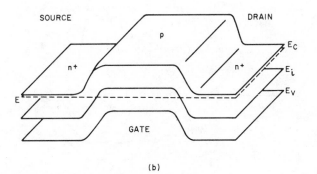

Fig. 12 Three-dimensional band diagram of an *n*-channel IGFET.
(a) Device configuration.
(b) Flat-band zero-bias ($V_D = V_G = 0$) equilibrium condition.

where x_i is the intrinsic point beyond which the electron concentration is negligible. The above result is still based on the gradual-channel approach, i.e., the longitudinal field $\mathcal{E}_y$ is smaller than the transverse field $\mathcal{E}_x$. The validity of this approach will be discussed later.

The gate voltage V_G is related to the surface potential ψ_s by

$$V_G' = V_G - V_{FB} = -\frac{Q_s}{C_i} + \psi_s$$

$$= \frac{2\varepsilon_s kT}{C_i q L_D} F\left(\beta\psi_s, V, \frac{n_{po}}{p_{po}}\right) + \psi_s. \tag{35}$$

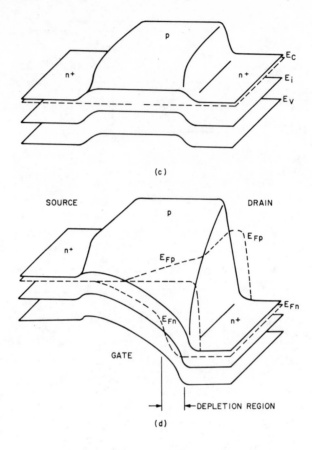

Fig. 12
(c) Equilibrium condition under a gate bias.
(d) Nonequilibrium condition under both drain and gate biases.
(After Pao and Sah, Ref. 19.)

For a particular device, Eq. (34a) can be calculated numerically. The input data are: the bulk impurity concentration N_A, the physical dimensions of the device, and the constant effective carrier mobility as discussed in Section 3. A typical result of the drain characteristics is shown in Fig. 13(a). It demonstrates the current saturation phenomena very well. The result is valid for the entire range of drain voltage from the linear region to the saturation region.

The relative importance of the two components can be easily illustrated from the ratio

$$\frac{D(dn/dy)}{\mu_n n\mathscr{E}_y} \simeq \frac{(N_A - p_s)}{n_s} \tag{36}$$

where p_s and n_s are the carrier concentrations at the semiconductor surface. Near the source junction in the channel $n_s \gg N_A$ and the diffusion current component can be neglected, while in the depletion region $n_s \ll N_A$ so that the diffusion component must be taken into account. As discussed in Section 4, in the short-channel devices (< 10 μm), the saturation current does not level off as illustrated in Fig. 13(a). This is due to the space-charge-region widening effect, which shortens the effective length of the channel. As shown in Fig. 7(c), this distance between the point Y and the drain contact edge can be given by the conventional formula for the depletion region of an n^+p junction:

$$L - L' = \sqrt{2\varepsilon_s(V_D - V_{D\,\mathrm{sat}})/qN_A}. \tag{37}$$

Now L is effectively reduced to L'. From Eqs. (31) and (37) the current now is given by

$$I_D(V_D > V_{D\,\mathrm{sat}}) = \frac{Z}{2L'}\,\mu_n C_i (V_G - V_T')^2$$

$$\approx \left\{\frac{L}{L - [2\varepsilon_s(V_D - V_{D\,\mathrm{sat}})/qN_A]^{1/2}}\right\} I_{D\,\mathrm{sat}}. \tag{38}$$

Because of the decrease of the effective channel length with increasing V_D, the channel conductance g_D in the saturation region[20] has a finite value and can be obtained from Eqs. (29) and (38):

$$g_D \equiv \frac{\partial I_D}{\partial V_D} = \frac{1}{r_D}$$

$$= \frac{Z\mu_n C_i \sqrt{2\varepsilon_s/qN_A}\, V_{\mathrm{sum}}^2}{12[L\sqrt{V_D - V_{D\,\mathrm{sat}}} - \sqrt{2\varepsilon_s/qN_A}\,(V_D - V_{D\,\mathrm{sat}})]} \tag{38a}$$

where V_{sum}^2 corresponds to the terms in the [ ] in Eq. (29a). Under the condition that we have a flat-band voltage shift, V_{FB}, the value of V_G should be replaced by $V_G' \equiv V_G - V_{FB}$. Plots of $V_{D\,\mathrm{sat}}$ versus V_G' for different bulk doping are shown in Fig. 13(b), and it is seen that, for a given V_G', $V_{D\,\mathrm{sat}}$ is smaller for larger bulk doping. In the limiting cases of an intrinsic semiconductor, $V_{D\,\mathrm{sat}} = V_G'$. For finite doping, the $V_{D\,\mathrm{sat}}$ versus V_G' curve has an offset along the V_G' axis; the physical meaning of this is that, below this gate voltage, there is no channel formation and hence no transistor action.

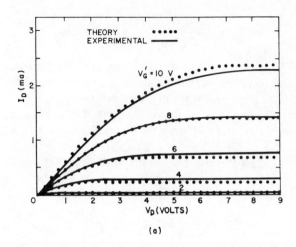

(a)

Fig. 13

(a) Theoretical (dots) and experimental (solid lines) output characteristics of a p-channel IGFET having $d = 2000$ Å, $N_D = 4.6 \times 10^{14}$ cm^{-3}, $\mu_p = 256$ cm^2/volt-sec, and an area of 8.4×10^{-4} cm^2.

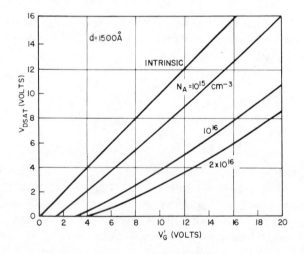

Fig. 13

(b) $V_{D\,\text{sat}}$ versus $V_G{}'$ for various bulk doping of silicon IGFET.

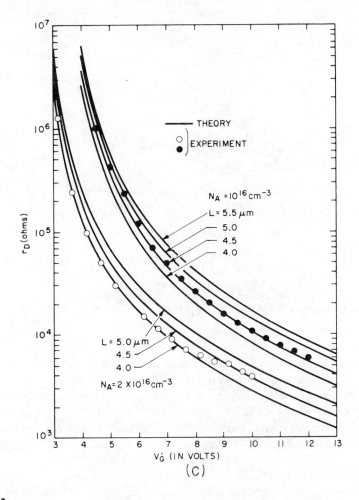

Fig. 13

(c) Drain resistance in saturation versus V_G' at $V_D = 10$ volts for different channel lengths and bulk dopings.

Plots of r_D versus V_G' at $V_D = 10$ volts are shown in Fig. 13(c) for different channel lengths and bulk dopings.[20] For the theoretical curves the following values have been assumed: gate oxide thickness = 1500 Å, dielectric constant of $SiO_2 = 3.7$, $Z = 0.084$ cm, and $\mu_n = 200$ cm^2/volt-sec. It is noticed from the curves that r_D decreases with decreasing channel length L and bulk doping

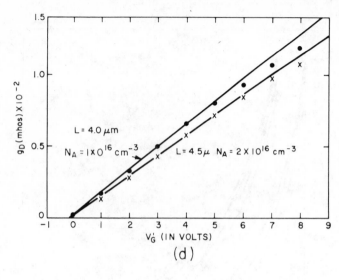

Fig. 13
(d) Channel conductance in the linear region versus gate voltage.
(After Sah et al., Refs. 19 and 20.)

for a given V_D and V_G'. It also decreases with increasing V_G'. Experimental points obtained with the devices having bulk dopings $N_A = 10^{16}$ and 2×10^{16} are also shown. Channel lengths of $L = 4$ and 4.5 μm show the best match for devices with dopings of 10^{16} and 2×10^{16} respectively. These lengths are in good agreement with the estimated values from the diffusion processes.

As discussed in Section 3, the mobility in the channel of a Si-SiO$_2$ system is approximately constant up to a surface field of about 10^5 V/cm. Based on the constant mobility assumption, the drain conductance at low drain voltage should then linearly increase with the gate voltage. This has indeed been found to be the case over a wide range of gate voltages as shown in Fig. 13(d).

(2) Device Parameters

The general expression as given in Eq. (34a) can be used to calculate other device parameters over the complete range of drain voltages. The short-circuit gate capacitance, C_{GS}, is given by

$$C_{GS} \equiv \frac{\partial Q_M}{\partial V_G}\bigg|_{V_D = \text{const}} \tag{39}$$

where Q_M is the total charge on the metal gate electrode for a given drain bias condition and is given by[19]

$$Q_M = \frac{\varepsilon_s L d}{I_D C_i Z L_D \mu_n} \left(\frac{q}{kT}\right) \int_0^{V_D} F\left(\beta\psi, V, \frac{n_{po}}{p_{po}}\right) \int_{\psi_B}^{\psi_s} \frac{e^{\beta(\psi-V)}}{F\left(\beta\psi, V, \frac{n_{po}}{p_{po}}\right)} \, d\psi \, dV.$$

(40)

The calculated values of C_{GS} from Eqs. (39) and (40) are shown in Fig. 14(a). We note that the results not only show all the detailed multiextrema structure of the capacitance but also agree remarkably well with the experimental data (solid lines).

The transconductance, g_m, may also be readily derived from Eqs. (34a) and (35). A typical example is shown in Fig. 14(b). The rounding off at the saturation points is clearly displayed. For very high gate voltages the observed discrepancy between the calculated values and the experimental results is believed to be mostly due to the field-dependent carrier mobility as discussed in Section 3. Figure 14(c) shows the theoretical and experimental drain conductance, $g_D \equiv (\partial I_D/\partial V_D)_{V_G}$. The drain (or channel) conductance is evaluated at the drain junction where the diffusion current becomes important at saturation. The residual drain conductance is due mainly to the reduction of the channel length, Eq. (38), which provides a finite drain conductance beyond saturation.

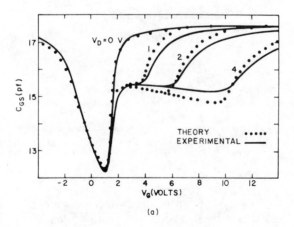

(a)

Fig. 14

(a) Theoretical and experimental short-circuit gate capacitance characteristics for an n-channel IGFET having $d = 1700$ Å, $N_A = 1.6 \times 10^{16}$ cm^{-3}, and an area of 8.4×10^{-4} cm^2.

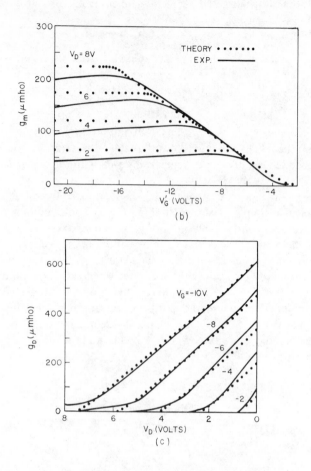

Fig. 14

(b) Theoretical and experimental transconductance characteristics of a p-channel IGFET having $d = 6200$ Å, $N_D = 6.4 \times 10^{14}$ cm^{-3}, $\mu_p = 326$ cm^2/volt-sec, and an area of 8.4×10^{-4} cm^2.

(c) Theoretical and experimental channel conductance of a p-channel IGFET having $d = 2100$ Å, $N_D = 1.6 \times 10^{15}$ cm^{-3}, $\mu_p = 343$ cm^2/volt-sec, and an area of 8.4×10^{-14} cm^2. (After Pao and Sah, Ref. 19.)

Figure 15 shows the variation of the surface potential (ψ_s), the electron imref (E_{Fn}), and the surface electron conconcentration along the channel with the IGFET operated just in saturation. We note in Fig. 15(a) that the surface potential increases steadily from source to drain, with the fastest change near the drain end. The drain voltage is mainly responsible for the

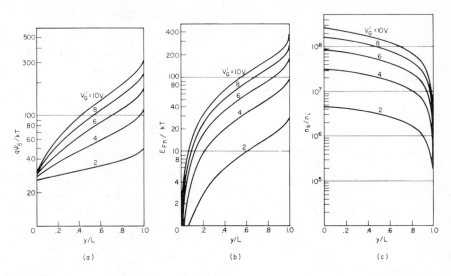

Fig. 15

(a) Surface potential along the channel with the IGFET operating just in saturation ($d = 2000$ Å, $N_D = 4.6 \times 10^{14}$ cm^{-3}, $L = 70\mu$m, $Z = 1200\mu$m).

(b) Imref along the channel with the IGFET operating just in saturation.

(c) Surface electron concentration along the channel with the IGFET operating just in saturation.

(Ref. 19.)

large change in ψ_s along the channel. In Fig. 15(b) a similar plot is given for the electron imref. It can be seen that for a given gate voltage, the difference between ψ_s and E_{Fn}/q is almost constant (~ 30 kT/q) until y/L approaches 1 at which value the carriers rapidly decrease toward the intrinsic concentration or E_{Fn} approaches E_i. Another way of showing this behavior is the plot of the surface electron concentration, in Fig. 15(c), where the concentration is given by

$$n_S = n_i \exp\left[\frac{q(\psi_s - \psi_B)}{kT} - \frac{E_{Fn}}{kT}\right]$$

$$= n_{po} \exp\left(\frac{q\psi_s - E_{Fn}}{kT}\right). \tag{41}$$

The variation of the transverse electric field $\mathscr{E}_{x\,\text{sat}}$ (at saturation) along the channel is plotted in Fig. 16(a). It is interesting to note that the field decreases as it goes toward the drain. If we examine the energy band diagram of

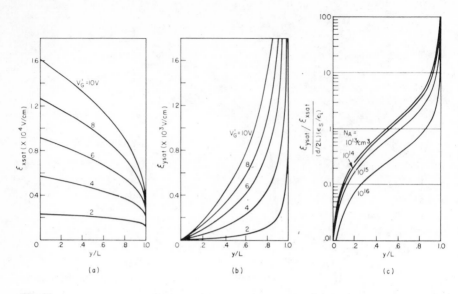

Fig. 16
(a) Transverse electrical field at the surface along the channel.
(b) Longitudinal electrical field at the surface along the channel.
(c) Ratio of the electric fields along the channel with the IGFET (same as Fig. 15) operating just in saturation.
(Ref. 19.)

Fig. 12(d), it becomes evident that for a cross section near the source the band-bending is narrow and very sharp whereas, for one near the drain, the band-bending extends further into the semiconductor and results in a lower electric field. The longitudinal electric field, $\mathscr{E}_{y\,sat}$, is shown in Fig. 16(b). As expected, it increases as it approaches the drain. The ratio of the longitudinal field to the transverse field normalized with respect to $(d/2L)(\varepsilon_s/\varepsilon_i)$ is plotted in Fig. 16(c) for different bulk impurity concentrations. This points out that the higher the concentration, the harder it is to pinch off the channel. It also shows explicitly that for devices with small values of the ratio $(d/2L)$, the one-dimensional gradual-channel approximation is valid for most of the possible biasing conditions. For example, for an IGFET operated in saturation with $N_A = 10^{16}$ cm^{-3}, an oxide thickness d of 1000 Å and a channel length L of 10 μm, the ratio $(d/2L)$ is only 0.005. The ratio of the longitudinal field to the transverse field at the drain is about 0.1. For lower drain voltage $(V_D < V_{D\,sat})$ the ratio would be even smaller.

(3) Equivalent Circuit and Types of IGFET

The IGFET is basically an ideal transadmittance amplifier with an infinite input resistance and a current generator at the output. In practice, however, we have also some other circuit parameters. An equivalent circuit[6,21] is shown in Fig. 17 for the common-source connection. The differential transconductance, g_m, has been discussed previously. The input conductance, G_{in}, is due to the leakages through the thin gate insulator. For a thermally grown silicon dioxide layer, the leakage current between the gate and the channel is very small, of the order of 10^{-10} amp/cm^2; thus the input conductance is negligibly small.[22] In practical devices the insulator layer and the metal gate may extend somewhat above the source and drain regions. Consequently C_{in} will be larger than the value given in Eq. (39). This fringe effect will also be the most important contribution to the feedback capacitance, C_{fb}. The output conductance, G_{out}, is equal to the drain conductance. The output capacitance consists mostly of the two p-n junction capacitances connected in series through the semiconductor bulk. Because of the nonlinear nature of the device, all the above circuit elements are differential and their magnitude depends on the operation points.

The maximum operating frequency (neglecting stray components) can be defined as the frequency at which the current through C_{in} is equal to the current of the current generator $g_m V_G$:

$$f_m = \frac{\omega_m}{2\pi} = \frac{g_m}{2\pi C_{in}} \approx \frac{\mu V_D}{2\pi L^2}. \tag{42}$$

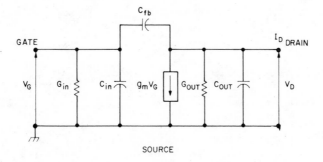

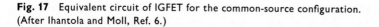

Fig. 17 Equivalent circuit of IGFET for the common-source configuration. (After Ihantola and Moll, Ref. 6.)

For low drain voltage the transit time of a carrier through the channel is given by

$$\tau = \frac{L}{\bar{v}} \approx \frac{L}{\mu \mathscr{E}_y} = \frac{L^2}{\mu V_D} = \frac{1}{\omega_m} \qquad (43)$$

where $\bar{v}$ is the average velocity. At saturation ($V_D = V_{D\text{sat}}$), however, it has been shown[6] that $\tau \simeq 2/\omega_m$.

There are basically four different types of IGFET depending on the types of inversion layer rather than the dopant type of the bulk semiconductor. If an n channel exists at zero gate bias, Fig. 18(a), we must apply a negative bias on the gate to deplete the carriers in the channel and thus reduce the channel conductance. This channel can be formed, for example, by shallow diffusion or by the built-in positive fixed charges; and this type is called the n-channel depletion IGFET. On the other hand, if the channel conductance is very low at zero gate bias, we must apply positive voltage on the gate in order to enhance the channel conductance, Fig. 18(b). The second type is called the

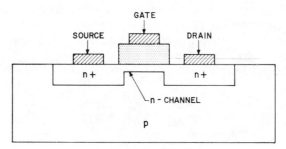

(a) n - CHANNEL DEPLETION

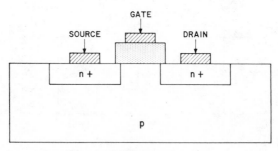

(b) n - CHANNEL ENHANCEMENT

Fig. 18 Basic types of IGFET. The four types are determined by the inversion layer rather than the dopant of the bulk semiconductor.

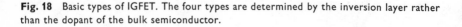

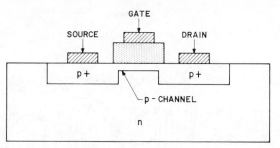

(c) p - CHANNEL DEPLETION

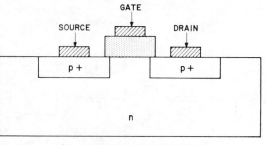

(d) p - CHANNEL ENHANCEMENT

Fig. 18 Basic types of IGFET. The four types are determined by the inversion layer rather than the dopant of the bulk semiconductor.

n-channel enhancement IGFET. Similarly we have the p-channel depletion and enhancement IGFET as shown in Fig. 18(c) and (d) respectively.

The electrical symbol, the transfer characteristics, and the output characteristics of the above four types are shown[23] in Fig. 19. We note that for the n-channel depletion type, substantial drain current flows at $V_G = 0$, and the current can be increased or decreased by varying the gate voltage. For the n-channel enhancement type, however, we must apply a positive gate bias larger than the threshold voltage, V_T, before a substantial drain current flows. The above discussion can be readily extended to p-channel devices by changing the polarities.

(4) Common-Gate Configuration[24]

In the previous discussion we have been concerned with the common-source electrode arrangement. In this section we shall derive the common-gate

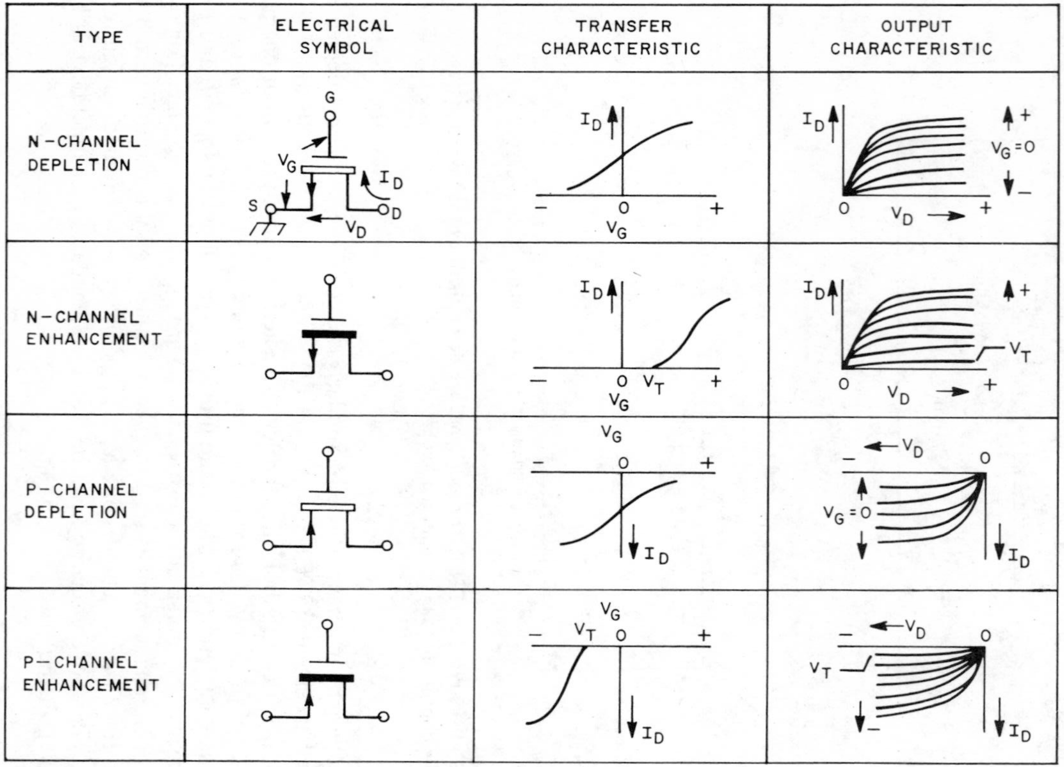

Fig. 19 Electrical symbol, transfer characteristics, and output characteristics of the four types. (After Gallagher and Corak, Ref. 23.)

538

characteristics based on the basic theory presented in Section 4. The two configurations of an n-channel IGFET are shown in Fig. 20. To define the voltages more clearly, two subscript letters will be used, e.g., V_{DS} means the voltage applied from the drain to the source. For the common-source configuration the drain current I_D is given by Eqs. (25) and (31):

$$I_D = \frac{Z\mu_n C_i}{L}\left[(V_{GS} - V_T)V_{DS} - \frac{V_{DS}^2}{2}\right] \quad \text{for} \quad V_{DS} < V_{GS} - V_T \quad (44a)$$

and

$$I_D = \frac{Z\mu_n C_i}{2L}(V_{GS} - V_T)^2 \quad \text{for} \quad V_{DS} \geq \overline{V}_{DS} \equiv V_{GS} - V_T. \quad (44b)$$

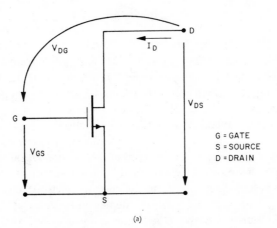

(a)

G = GATE
S = SOURCE
D = DRAIN

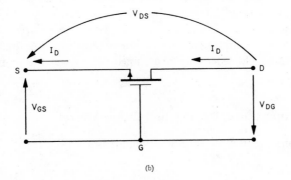

(b)

Fig. 20 Common-source and common-gate configurations of IGFET.

For the common-gate configuration the drain current I_D may be expressed by voltages V_{GS} and V_{DS}. From Fig. 20(b) we obtain

$$V_{DS} = V_{DG} + V_{GS}. \tag{45}$$

If we define $\overline{V}_{GS} \equiv V_{GS} - V_T$, the current equation, Eq. (44a), may be written as

$$I_D = \frac{Z\mu_n C_i}{L}\left[\overline{V}_{GS}(V_{DG} + \overline{V}_{GS} + V_T) - \frac{(V_{DG} + \overline{V}_{GS} + V_T)^2}{2}\right]$$

$$= \frac{Z\mu_n C_i}{2L}[(V_{GS} - V_T)^2 - (V_{DG} + V_T)^2]. \tag{46}$$

For the saturation region we obtain from Eq. (44b) a modified pinch-off condition for the common-gate arrangement:

$$\overline{V}_{DG} = \overline{V}_{DS} - V_{GS} = V_{GS} - V_T - V_{GS} = -V_T. \tag{47}$$

Substituting Eq. (47) into Eq. (46) we obtain an expression identical to Eq. (44b). It is evident that in the saturation region the drain current expression is independent of the configurations. The output characteristics (I_D versus V_{DG}) and the input characteristics (I_D versus V_{GS}) for the common gate configurations are shown in Fig. 21(a) and (b) respectively.

There is a remarkable difference between junction transistors and IGFET: junction transistor characteristics are commonly unsymmetrical when emitter and collector are interchanged in common-base configuration; field-effect transistors are theoretically symmetrical (but a possible asymmetry can be made in the actual devices, e.g., a partial-gate electrode covers only those portions near the source electrode in a depletion mode). By "symmetry" we mean that if electrodes S and D are interchanged in the common-gate configuration, the resulting properties are identical. In this case it is possible to conjugate the output characteristics from Fig. 21(a) and (b) in the single graphical presentation shown in Fig. 22. These generalized characteristics illustrate in one figure the whole region of the transistor operation where source and drain elctrodes may be assumed to be input and output electrodes respectively.

(5) Temperature, Doping, and Other Effects

In the linear region the threshold voltage is given by Eq. (33):

$$V_T = \phi_{ms} + \frac{Q_{fs}}{C_i} + 2\psi_B + \frac{\sqrt{4\varepsilon_s q N_A \psi_B}}{C_i}. \tag{48}$$

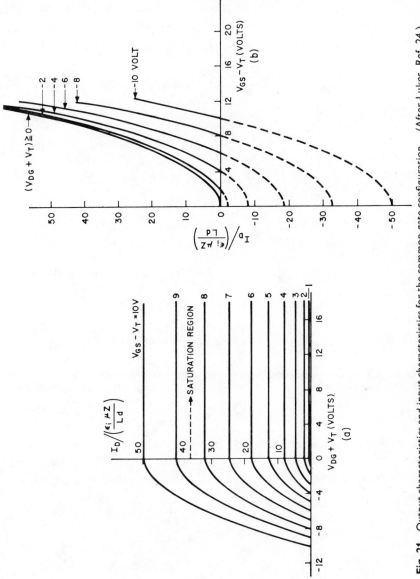

Fig. 21 Output characteristics and input characteristics for the common-gate configuration. (After Lukes, Ref. 24.)

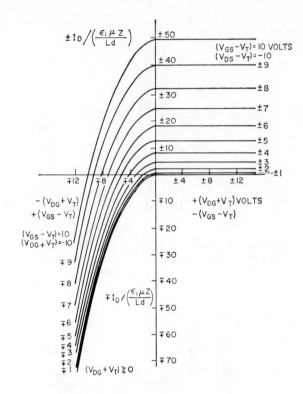

Fig. 22 Generalized characteristics for common-gate configuration.
(After Lukes, Ref. 24.)

If the work-function difference ϕ_{ms} and the surface charges are essentially independent of temperature, differentiation of the above equation with respect to temperature yields[25]

$$\frac{dV_T}{dT} = \frac{d\psi_B}{dT}\left[2 - \frac{1}{C_i}\sqrt{\frac{\varepsilon_s q N_A}{\psi_B}}\right] \tag{49}$$

where

$$\frac{d\psi_B}{dT} \simeq \pm\frac{1}{T}\left[\frac{E_g(T=0)}{2q} - |\psi_B(T)|\right]. \tag{49a}$$

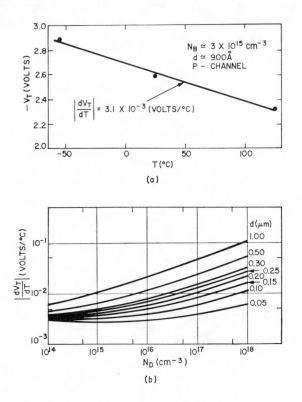

Fig. 23
(a) Experimental measurement of the threshold voltage as a function of temperature.
(b) $|dV_T/dT|$ of Si-SiO$_2$ system versus substrate impurity concentration with oxide thickness as a parameter.
(After Vadasz and Grove, Ref. 25.)

Typical experimental measurements of the threshold voltage near room temperature for Si-SiO$_2$ systems are shown in Fig. 23(a). The data can be represented by a straight line over this temperature range. Thus, a representative figure for device behavior can be obtained by evaluating Eq. (49) at room temperature. Results of such calculations are shown in Fig. 23(b) as a function of substrate dopings, for various values of the oxide thicknesses. The quantity (dV_T/dT) is positive for n-channel devices and negative for p-channel devices. We also note that for a given oxide thickness, the quantity dV_T/dT generally increases with increased doping.

At a given temperature (e.g., room temperature), the substrate impurity concentration also has profound effect on the threshold voltage, V_T. The relationship between V_T and the doping, N_A, is given by Eq. (48) where

$$\psi_B = \frac{kT}{q} \ln(N_A/n_i).$$

If an n-channel IGFET is of the depletion type (which means the device is normally on under the zero gate bias condition and that a negative bias is required on the gate to reduce the source-drain current), the threshold voltage V_T is less than zero. To make an n-channel enhancement-type IGFET, we can increase the substrate doping such that

$$\left[\frac{\sqrt{4\varepsilon_s N_A kT \ln(N_A/n_i)}}{C_i} + \frac{2kT}{q} \ln(N_A/n_i)\right] > \left| -\left(\phi_{ms} + \frac{Q_{fs}}{C_i}\right)\right| \quad (50)$$

and $V_T > 0$.

Figure 24 shows the variation[26] of the threshold voltage with substrate resistivity for a surface-charge density of 5.4×10^{11} cm^{-2}. The experimental points are in reasonable agreement with the calculated curve (solid line). It is clear that the threshold voltage can be varied by varying the temperature, the substrate doping (N_A), the gate metal (ϕ_{ms}), the surface charge (Q_{fs}), and the insulator thickness ($C_i = \varepsilon_i/d$).

In addition to the temperature and doping effects, the IGFET characteristics are also influenced by surface orientation and irradiation effects similar to those discussed in Chapter 9. In the Si-SiO$_2$ system, the fixed surface-charge density on the (100) surface is smaller than that on the (111) surface. It is thus expected that the threshold voltage of IGFET made on (100) substrate should be smaller. This is indeed verified experimentally to be the case.[27] When the MIS structure is under ionization radiation, as discussed in Chapter 9, a net positive space charge builds up near the semiconductor-insulator interface resulting in a shift of the flat-band voltage. This, in turn, causes a change in the threshold voltage. The expected behaviors have been generally observed experimentally.[28]

(6) Physical Limitations of IGFET

The first fundamental limitation of an IGFET is the maximum attainable drain voltage which is determined by substrate doping and the diffusion profile (which is related to the junction curvature effect);[29] above that value the drain p-n junction will break down.

The second limitation is due to the so-called "punch-through" effect. As the drain voltage increases, the effective channel length, L', decreases. Eventually, when L' is reduced to zero, the drain is punched through to the

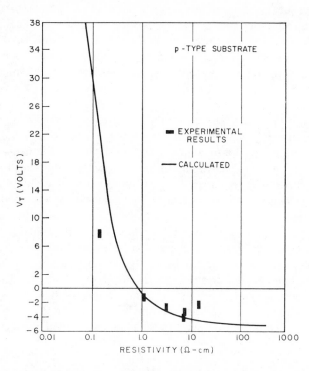

Fig. 24 Theoretical (solid line) and experimental results for the variation of the threshold voltage with substrate resistivity for a surface charge density of 5.4×10^{11} cm^{-2}. (After Brotherton, Ref. 26.)

source; and the drain current is no longer controlled. The drain voltage at punch-through can be obtained from Eq. (37) by letting $L' = 0$ and $V_D - V_{D\,\mathrm{sat}} = V_{PT}$ (punch-through):[5]

$$V_{PT} = \frac{L^2 q N_A}{2\varepsilon_s}. \tag{51}$$

For low substrate dopings and a long channel the voltage is very large, and generally the avalanche breakdown will take place first.

The transit time[6,30,31] across the channel is another limitation. Since the transit time is proportional to the square of the channel length, Eq. (43), it is desirable to make the channel as short as possible. For a 1-μm channel length the transit time is of the order of 10^{-11} second, corresponding to a maximum operating frequency of about 10 GHz.

The power limitation[9] of an IGFET is very similar to that of the junction transistor. The dissipation power is limited primarily by the thermal resistance of the transistor-package combination. For a channel length much smaller than the wafer thickness, the conduction of heat from the channel into the substrate is radial and leads to a logarithmic dependence of temperature on power dissipation.

The noise in IGFET imposes another limitation on the minimum signal levels to be detected or amplified. The three types of noise in IGFET are: thermal noise, generation-recombination noise, and $1/f$ noise—similar to those in a p-n junction. The thermal noise[32] arises mainly from the fluctuations of the current-charge carriers in the channel which modulates the channel conductance. This noise is important at high frequencies. The product of the equivalent input thermal noise resistance and the transconductance g_m of the IGFET equals 2/3 for devices with intrinsic substrate and increases with increasing substrate doping. The generation-recombination noise[33] is due to the fluctuation of the charge carriers at the recombination centers and the defect centers in the depletion region between the channel and the semiconductor substrate. This noise is important in intermediate frequency. The $1/f$ noise[34] is mainly due to the random fluctuation of the carriers in the surface states, and is the dominant noise at very low frequencies.

6 IGFET WITH SCHOTTKY BARRIER CONTACTS FOR SOURCE AND DRAIN[35]

In the previous discussions we were concerned with IGFET's having diffused source and drain contacts. Use of Schottky barrier contacts for the source and drain of an IGFET gives fabrication advantages. Elimination of the high-temperature diffusion steps promotes better quality in the oxides and better control of the geometry, particularly between source and gate. In addition the SB-IGFET can be made on semiconductors (such as CdS) where p-n junctions cannot be easily formed.

An SB-IGFET has been made using a PtSi-Si barrier for source and drain and SiO_2 as the gate insulator. The source barrier height is 0.85 eV on n type and 0.24 on p type; the lower barrier height applies in normal operation with a p-inversion layer over the n-Si. The basic device geometry and cross-sectional view are shown in Fig. 25. The platinum silicide films are formed at 650°C in the contact holes using the vacuum sintering technique.[36]

The output characteristics (drain current versus drain voltage) of the device are shown in Fig. 26 for two substrate temperatures. It is seen that at room

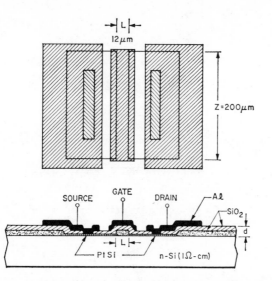

Fig. 25 Basic device geometry and cross-sectional view of SB-IGFET. The substrate is *n*-type, 1 ohm-cm, ⟨100⟩ oriented silicon. The first SiO₂ layer is thermally grown, and the second is deposited by thermal decomposition of ethylorothosilicate and trimethyl phosphate. (After Lepselter and Sze, Ref. 35.)

temperature, Fig. 26(a), the characteristics are similar to those of a conventional IGFET described previously. The turn-on voltage V_T is found to be about 2 volts, and the transconductance $g_m (\equiv \partial I_D / \partial V_G)$ in the saturation region is about 50 μmho for a gate voltage $V_G = -14$ volts.

At liquid nitrogen temperature, Fig. 26(b), the output characteristics are considerably different from those of a conventional IGFET. A more detailed result is shown in Fig. 27. At $V_G = 0$ the leakage current between the source and drain contacts is about 4×10^{-10} amperes. For an applied gate voltage and at low drain voltages the drain current I_D remains essentially constant and equal to the small leakage current. When the drain voltage reaches about one volt, I_D starts to increase rapidly and varies approximately as $(V_D)^m$ with m between 4 and 8. For a given gate voltage the current again saturates at the pinch-off point.

The rapid increase of current can be explained with the help of the simplified band diagrams shown in Fig. 28. At equilibrium with $V_G = V_D = 0$, the band diagram at the semiconductor surface corresponds to the two metal-semiconductor barriers with barrier height $\phi_{Bn} = 0.85$ V for the PtSi-Si

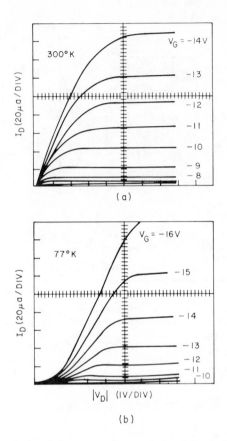

(a)

(b)

Fig. 26 Oscilloscope display of the output characteristics of the device at
(a) 300°K and
(b) 77°K, where I_D is the drain current, and V_D and V_G are the drain voltage and gate voltage
respectively.
(Ref. 35.)

system.[36] When $|V_G| > |V_T|$, the gate voltage is large enough to invert the surface from n-type to p-type, and the barrier height between source and inversion layer is now $\phi_{Bp} \simeq 0.25$ V for PtSi on p-type silicon (see Chapter 8 for more detailed discussion on Schottky barriers). Note that the source contact is reverse-biased under operating conditions. For a 0.25 V barrier the thermionic-type reverse saturation current density is of the order of 10^3 A/cm^2 at room temperature and about 10^{-11} A/cm^2 at liquid nitrogen temperature.[37]

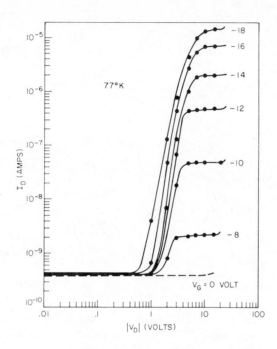

Fig. 27 Log-log plot of the output characteristics at 77°K.
(Ref. 35.)

With the present geometry and an inversion layer depth of the order of 1000 Å, the maximum thermionic current that can flow between the source and drain contacts is about 10^{-16} ampere at 77°K. Thus the thermionic current at 77°K is negligible. The dominant current is then the tunneling current which is proportional to the tunneling probability.[38]

$$J_t \sim \exp\left[-\frac{4}{3}\frac{\sqrt{2m^{*1/2}(q\phi_{Bp})^{3/2}}}{qh\mathscr{E}} \right]$$

where $\mathscr{E}$ is the electric field and m^* is the effective mass. If we assume that m^* equals the free electron mass, the ratio of current densities for fields at 3×10^5 V/cm and 10^5 V/cm is about 10^{10}. Thus a small increase in the drain voltage will cause a large increase in the drain current in agreement with the experimental results shown in Fig. 27.

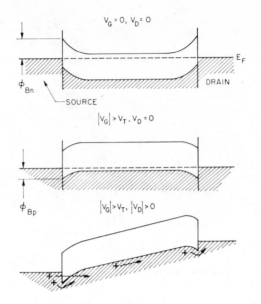

Fig. 28 Band diagrams at the semiconductor surface. Top: for equilibrium condition with no gate or drain voltage applied. Middle: for an inverted surface. Bottom: where a large drain voltage is applied such that tunneling current can flow. (Ref. 35.)

7 IGFET WITH A FLOATING GATE—A MEMORY DEVICE[39]

When the gate electrode of a conventional IGFET is modified to incorporate an additional metal-insulator sandwich (a floating gate), the new structure can serve as a memory device in which semipermanent charge storage is possible.

A schematic diagram of an IGFET with a floating gate is shown in Fig. 29 which is basically a p-channel enhancement-mode device. The structure of the gate electrode is layered like a sandwich: insulator $I(1)$, metal $M(1)$, insulator $I(2)$, and metal $M(2)$. The corresponding energy band diagram of the gate structure is shown in Fig. 30. If the thickness of $I(1)$ is thin enough that a field-controlled electron transport mechanism such as tunneling or internal field-emission is possible, a positive bias on $M(2)$ with respect to the semiconductor would cause electron accumulation in the floating gate $M(1)$, provided electron transport across $I(2)$ is small. These conditions can be met by

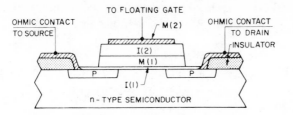

Fig. 29 Schematic diagram of an IGFET with a floating gate. $I(1)$ and $I(2)$ are insulators. $M(1)$ is the floating-gate metal-electrode, and $M(2)$ is the outer gate electrode. (After Kahng and Sze, Ref. 39.)

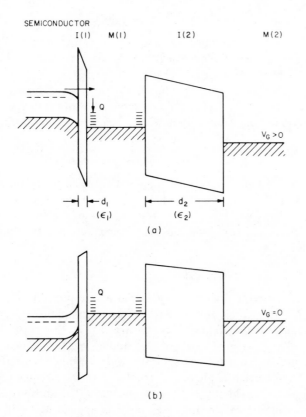

Fig. 30
(a) Energy band diagram along the gate structure when a positive bias is applied to the outer gate electrode.
(b) Energy band diagram along the gate structure after the bias voltage is removed.
(Ref. 39.)

choosing $I(1)$ and $I(2)$ such that the ratio of dielectric permittivity $\varepsilon_1/\varepsilon_2$ is small and/or the barrier height into $I(1)$ is smaller than that into $I(2)$. In addition, the $M(1)$ should be thick enough that emitted electrons are close to the Fermi level of $M(1)$ before reaching $I(2)$; and $I(2)$ should be thick enough that no carrier transport is allowed across it.

The stored charge Q, as a function of time when a step voltage function with amplitude V_G is applied across the sandwich, is given by

$$Q(t) = \int_0^t J\,dt' \qquad \text{coul/cm}^2. \tag{52}$$

When the emission is of the Fowler-Nordheim tunneling type,[38] then the current density, J, has the form

$$J = C_1 \mathscr{E}^2 \exp(-\mathscr{E}_0/\mathscr{E}), \tag{53}$$

where $\mathscr{E}$ is the electric field, and C_1 and $\mathscr{E}_0$ are constants in terms of effective mass and barrier height. (We have neglected the effects due to the image force lowering of the barrier, etc., but the essential feature is expected to be retained even after detailed corrections are made.) This type of current transport occurs in SiO_2 and Al_2O_3. When the field emission is of the internal Schottky or Frenkel-Poo le type, as occurs in Si_3N_4, the current density follows the form

$$J = C_2 \mathscr{E} \exp[-q(\phi_B - \sqrt{q\mathscr{E}/\pi\varepsilon_1})/kT] \tag{54}$$

where C_2 is a constant in terms of trapping density in the insulator, ϕ_B the barrier height in volts, and ε_1 the dynamic permittivity.

The electric field in $I(1)$ at all times is a function of the applied voltage V_G and $Q(t)$, and is obtainable from the displacement continuity requirement as

$$\mathscr{E} = \frac{V_G}{d_1 + d_2(\varepsilon_1/\varepsilon_2)} - \frac{Q}{\varepsilon_1 + \varepsilon_2(d_1/d_2)}, \tag{55}$$

where d_1 and d_2 are the thickness of $I(1)$ and $I(2)$ respectively.

Figure 31(a) shows the results of a theoretical computation using Eqs. (52), (53), and (55) with the following parameters: $d_1 = 50$ Å, $\varepsilon_1 = 3.8\,\varepsilon_0$ (for SiO_2), $d_2 = 1000$ Å, $\varepsilon_2 = 30\,\varepsilon_0$ (for ZrO_2), and $V_G = 50$ volts. One notes that initially the stored charge increases linearly with time and then saturates. For a short time the current is almost constant and then decreases rapidly. The field in $I(1)$ decreases slightly as the time increases. The above results can be explained as follows: When a voltage pulse is applied at $t = 0$, the initial charge Q is zero, and the initial electric field across $I(1)$ has its maximum value, $\mathscr{E}_{\max} = V_G/[d_1 + (\varepsilon_1/\varepsilon_2)d_2]$. When Q is sufficiently small that $\mathscr{E}$ remains essentially the same, the current will in turn remain the same, and Q will increase linearly with time. Eventually, when Q is large enough to reduce the

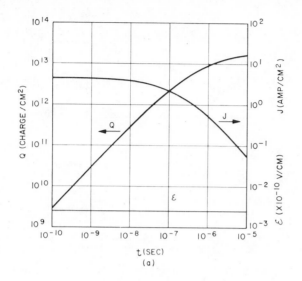

Fig. 31

(a) Theoretical results of the stored charge in $M(1)$, current through $I(1)$, and electric field in $I(1)$ as functions of time for $d_1 = 50$ Å, $d_2 = 1000$ Å, $\varepsilon_1 = 3.8\varepsilon_0$, $\varepsilon_2 = 30\varepsilon_0$, and $V_G = 50$ volts.

value of $\mathscr{E}$ substantially, the current will decrease rapidly with time and Q will increase slowly.

Figure 31(b) shows the stored charge as a function of time for the same ε_1 and ε_2 but different d_1, d_2, and V_G. To store a given amount of charge for a given structure, it is clear that one can increase either the applied voltage or the charging time (pulse width) or both. Figure 31(c) shows the calculated stored charge for the current transport described by Eq. (54). Here $I(1)$ is a 20-Å-thick Si_3N_4 film. There are marked decreases in the gate voltages required for a given charge compared to SiO_2. This is largely due to the much lower barrier height (1.3 volts) compared to SiO_2 (≈ 4.0 volts).

It is noted that the field in $I(1)$ for appreciable charge storage is in the 10^7 V/cm range. When the outer gate voltage is removed, the field in $I(1)$ due to the stored charge on the inner gate is only 10^6 V/cm or so, corresponding to 5×10^{12} charges/cm^2, a charge large enough to be detected easily. Since the transport across $I(1)$ is highly sensitive to the field, no charges flow back. The charge loss is actually controlled by the dielectric relaxation time of the sandwich structure, which is very long. When it is desired to discharge the floating gate quickly, it is necessary to apply to the outer gate a voltage about

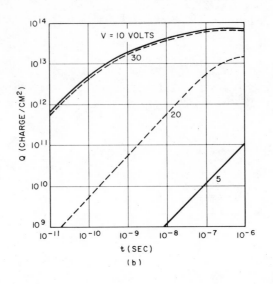

Fig. 31

(b) Theoretical results of Q, the stored charge as a function of time with the same ε_1 and
ε_2 as in (a), and $d_1 = 10$ Å, $d_2 = 100$ Å (solid lines), $d_1 = 30$Å, $d_2 = 300$ Å (dotted lines).

equal in magnitude but opposite in polarity to the voltage which was used
for charging. It is evident that net positive charges (loss of electrons) can also
be stored in the floating gate if the discharging gate voltages are appropriately
chosen for magnitude and duration.

The experimental results are shown in Fig. 32. The substrate of the device
is an n-type silicon, 1 ohm-cm, and $\langle 111 \rangle$ oriented. $I(1)$ is a 50 Å SiO_2
thermally grown in a dry oxygen furnace. $M(1)$ and $I(2)$ are Zr (1000 Å) and
ZrO_2 (1000 Å) respectively. $M(2)$ and the ohmic contact metals are aluminum
deposited in a vacuum system. Because of the relatively thick insulator layers,
a large voltage (~ 50 V) and a long pulse width ($\sim 0.5 \, \mu$s) must be applied in
order to store the required charge ($\sim 5 \times 10^{12}$ charges/cm^2). A positive pulse
of 50 volts is first applied to the gate electrode, and 60 ms later a negative pulse
of 50 volts is applied, Fig. 32(a). Then the pulsing cycle repeats. One notes
that, as shown in Fig. 32(b), when the positive pulse is applied, a sufficient
amount of charge is stored in the floating gate so that the silicon surface is
inverted; a conducting channel is thus formed, and the channel current is on.
It can be seen that the channel current decreases only slightly at the end of

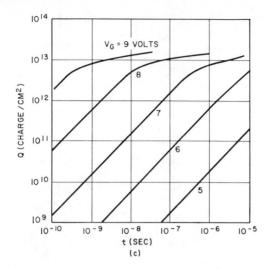

Fig. 31
(c) Theoretical results of the stored charge density as a function of time with $d_1 = 20$ Å, $\varepsilon_1 = 6\varepsilon_0$ (Si$_3$N$_4$), $d_2 = 200$ Å, $\varepsilon_2 = 30\varepsilon_0$ (ZrO$_2$), and various applied voltages. (Ref. 39.)

60 ms. When the negative pulse is applied, the stored charge is eliminated, and also the channel. The channel current reduces to its off state. Figure 32(c) shows results for pulses with the same widths but smaller amplitude (40 V). Since the stored charge is a strong function of the pulse amplitude, only a very small amount of charge is stored, one too small to cause inversion.

It is clear that the controlled field emission to the floating gate may be capacitively induced by pulsing the outer gate electrode. The IGFET with a floating gate can therefore be used as a memory device with holding time as long as the dielectric relaxation time of the gate structure (for read-in) and with continuous nondestructive read-out capability (from source-drain electrodes). The read-in read-out can also be performed in a very short time, e.g., in the nanosecond range or even shorter.

8 SURFACE FIELD EFFECTS ON p-n JUNCTIONS AND METAL-SEMICONDUCTOR DEVICES

(1) Capacitance and Reverse Current[15]

Figure 33(a) shows a gate-controlled diode in which the p-n junction characteristics near the corner are modulated by the applied surface field on

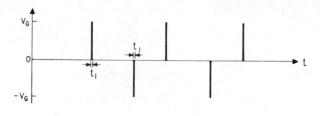

(a)

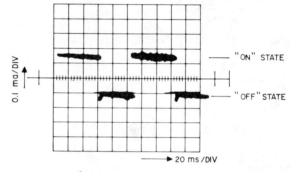

(b) $V_G = 50$ V, $t_1 = 0.5 \mu$s

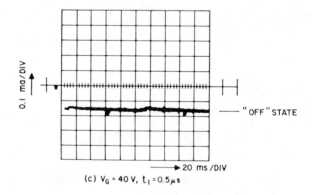

(c) $V_G = 40$ V, $t_1 = 0.5 \mu$s

Fig. 32 Experimental results.
(a) The applied gate pulse voltage.
(b) The source-drain current for $V_G = 50$ volts.
(c) The source-drain current for $V_G = 40$ volts.
(Ref. 39.)

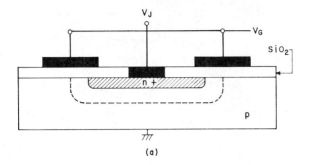

(a)

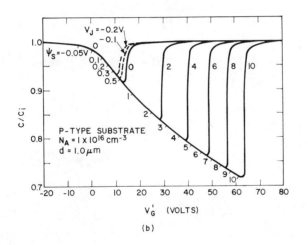

(b)

Fig. 33
(a) Gate-controlled diode.
(b) Theoretical low-frequency gate-to-substrate capacitance in the presence of an applied junction bias.
(After Grove and Fitzgerald, Ref. 15.)

the MIS annular structure. We shall now consider the capacitance-voltage characteristics of the MIS structure when a junction bias V_J is applied, i.e., under the nonequilibrium condition. The relationship between the gate voltage and the surface potential is given by

$$V_G - V_{FB} = -Q_s/C_i + \psi_s \tag{56}$$

where V_{FB} is the flat-band voltage shift due to work-function difference and surface-charge effect. In the depletion approximation, we have

$$\psi_s \simeq \frac{qN_A W^2}{2\varepsilon_s} \tag{57}$$

and

$$Q_s \simeq -qN_A W \tag{58}$$

where W is the depletion-layer width. From Eqs. (56), (57), and (58) we have

$$V_G - V_{FB} = \frac{qN_A W^2}{2\varepsilon_s} + \frac{qN_A W}{C_i}. \tag{59}$$

At the onset of strong inversion the surface potential is given by $\psi_s = 2\psi_B + V_J$, and the maximum depletion-layer width is given by

$$W_m \simeq \sqrt{\frac{2\varepsilon_s(V_J + 2\psi_B)}{qN_A}}. \tag{60}$$

Substitution of Eqs. (57), (58), and (60) into Eq. (56) yields the gate voltage required for strong inversion of the surface in the presence of an applied junction bias V_J:

$$V_G(V_J) - V_{FB} = V_J + 2\psi_B + \frac{1}{C_i}\sqrt{2\varepsilon_s qN_A(V_J + 2\psi_B)}. \tag{61}$$

Equation (61) can be solved to give the width of the surface depletion region as a function of the gate voltage,

$$W = \frac{\varepsilon_s}{C_i}\left\{\left[1 + \frac{2(V_G - V_{FB})C_i^2}{qN_A\varepsilon_s}\right]^{1/2} - 1\right\} \quad \text{for} \quad W \le W_m. \tag{62}$$

Thus, in the region where the semiconductor surface is merely depleted, the gate-to-substrate capacitance is given by

$$C = \frac{C_i C_D}{C_i + C_D} = \frac{C_i(\varepsilon_s/W)}{C_i + (\varepsilon_s/W)} = C_i\left(1 + \frac{C_i W}{\varepsilon_s}\right)^{-1} \tag{63}$$

or from Eq. (62)

$$\frac{C}{C_i} = \left[1 + \frac{2C_i^2(V_G - V_{FB})}{qN_A\varepsilon_s}\right]^{-1/2}. \tag{64}$$

The calculated results of the low-frequency gate-to-substrate capacitance in the presence of the applied junction bias are shown in Fig. 33(b) for a p-type substrate doped with 10^{16} acceptor atoms/cm^3. We note the delay in the capacitance rise corresponding to the onset of the inversion as a result of the applied reverse junction bias, and the earlier inversion as a result of an applied forward bias, compared to the $V_J = 0$ case.

The surface field across the MIS annular structure as shown in Fig. 33(a) also has profound effect on the reverse current of the *p-n* junction. Figure 34(a) illustrates this effect under three different surface conditions at a fixed junction voltage. Since the room-temperature reverse current of silicon *p-n* junctions is due to the generation processes through the generation-recombination centers in the depletion region, the magnitude of the reverse current depends on the total number of such centers included within the junction depletion region. When the surface under the gate electrode is accumulated, only those centers within the depletion region of the metallurgical *p-n* junction contribute to the generation current (I_1). When the surface under the gate is inverted, centers within the depletion region of the field-induced junction between the inversion layer and underlying substrate also contribute to the total generation current which is therefore larger than in the accumulation case ($I_1 + I_2$). When the surface is depleted, centers at the insulator-semiconductor interface provide yet another contribution to the total generation current resulting in a peak in the reverse-current versus gate-voltage characteristics ($I_1 + I_2 + I_3$). Figure 34(b) shows the measured reverse current as a

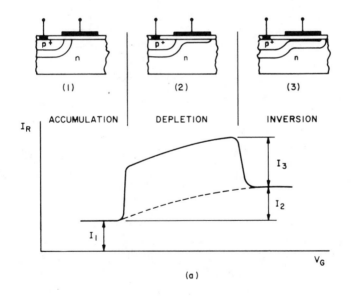

(a)

Fig. 34
(a) Effects of surface field on the *p-n* junction reverse saturation current.

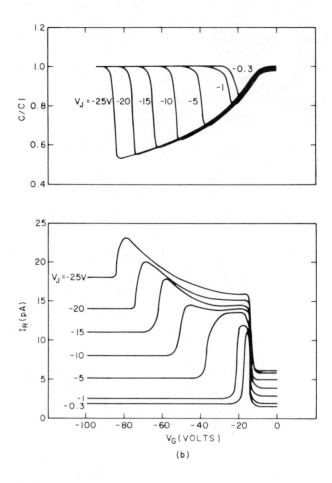

Fig. 34

(b) Measured reverse current and gate-to-substrate capacitance as a function of the gate
   voltage for various junction biases.

(After Grove and Fitzgerald, Ref. 15.)

function of the gate voltage with junction voltage V_J as a parameter. Also
shown are the corresponding gate-to-substrate capacitance measurements. It
is evident from this figure that at a given junction voltage a large increase in
the junction current occurs when the surface under the gate becomes depleted,
and that the current decreases when the surface becomes inverted. This condition is shown in a very clear manner by the I_R versus V_G characteristics. The
increase in current takes place at the same value of gate voltage independent

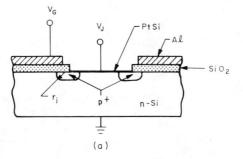

(a)

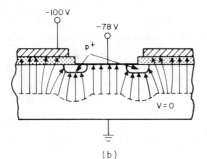

(b)

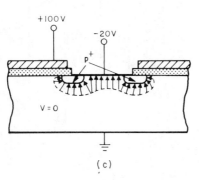

(c)

Fig. 35
(a) Metal-semiconductor diode with a diffused *p-n* junction guard ring and a gate electrode on the insulator.
(b) The breakdown condition of the guard ring when a negative gate bias is applied.
(c) The breakdown condition of the guard ring when a positive gate bias is applied.
(After Lepselter and Sze, Ref. 36.)

of the reverse bias, but the decrease in current occurs at the value of V_G corresponding to the onset of strong inversion as indicated by the non-equilibrium MIS capacitance curves.

(2)   Breakdown Voltage[36,40]

Figure 35(a) shows a metal-semiconductor diode which has a diffused p-n junction guard ring and a separate metal electrode on the insulator (the gate electrode). As discussed previously, if the junction radius near the edges of the diffused p^+ region is sufficiently small, the p^+n junction guard ring will break down first owing to the junction curvature effect. When a negative gate bias is applied, the surface field tends to smooth out the field concentration near the junction edge, Fig. 35(b). Thus the radius of curvature, r_j, is effectively increased. This in turn increases the breakdown voltage. As a positive gate bias is applied, the field profile near the junction is shown in Fig. 35(c) where the radius of curvature, r_j, is effectively reduced and results in a decreased breakdown voltage. Figure 36 shows the measured reverse I-V characteristics as a function of the gate voltage for a Schottky diode with a guard ring of

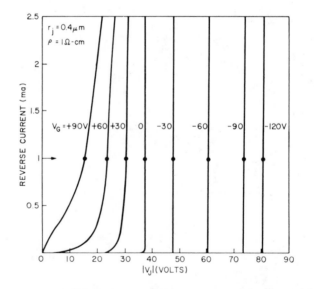

Fig. 36   Measured reverse I-V characteristics as a function of the gate bias for the device shown in Fig. 35 with junction depth of 0.4 μm. The breakdown voltage is defined as the voltage at which the reverse current reaches 1 ma. (Ref. 36.)

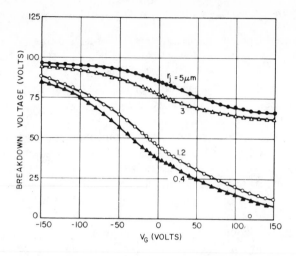

Fig. 37 Measured breakdown voltage versus gate voltage for four different junction depths. (Ref. 36.)

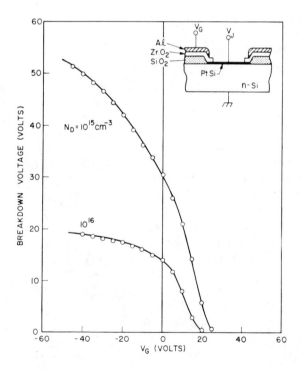

Fig. 38 Effect of surface field on the breakdown voltage of a metal-semiconductor diode. (Ref. 36.)

0.4 μm junction depth. As expected, the gate voltage does have a profound effect on the junction breakdown voltage. Figure 37 shows the measured breakdown voltage versus gate voltage for four different junction depths. One notes that the breakdown voltages all approach the theoretical value (~ 100 V) as $-V_G$ increases. Also at zero bias, the breakdown voltage decreases as junction depth r_j decreases.

Similar effects are observed on a metal-semiconductor diode that has no *p-n* junction guard ring but has a second MIS overlay as shown on the insert of Fig. 38, where ZrO_2 is formed near the periphery of the metal, and separate voltages are applied to the diode and the MIS guard ring electrodes. As the gate voltage increases in the negative direction, the edge field is gradually reduced. This results in an increased breakdown voltage (here defined as voltage drawing 1-ma current). When V_G increases in the positive direction, however, the enhanced edge field causes drastic lowering of the junction breakdown voltage as shown in Fig. 38.

REFERENCES

1. J. E. Lilienfeld, U.S. Patent No. 1,745,175 (1930).

2. O. Heil, British Patent No. 439,457 (1935).

3. W. Shockley and G. L. Pearson, "Modulation of Conductance of Thin Films of Semi-conductors by Surface Charges," Phys. Rev., *74*, 232 (1948).

4. D. Kahng and M. M. Atalla, "Silicon-Silicon Dioxide Field Induced Surface Devices," IRE Solid-State Device Research Conference, Carnegie Inst. of Tech., Pittsburgh, Penn. (1960).

5. H. K. J. Ihantola, "Design Theory of a Surface Field-Effect Transistor," Stanford Electronics Laboratories Technical Report No. 1661-1 (1961).

6. H. K. J. Ihantola and J. L. Moll, "Design Theory of a Surface Field-Effect Transistor," Solid State Electron., *7*, 423 (1964).

7. C. T. Sah, "Characteristics of the Metal-Oxide-Semiconductor Transistors," IEEE Trans. Electron Devices, *ED-11*, 324 (1964).

8. S. R. Hofstein and F. P. Heiman, "The Silicon Insulated-Gate Field-Effect Transistor," Proc. IEEE, *51*, 1190 (1963).

9. J. T. Wallmark and H. Johnson, *Field Effect Transistors, Physics, Technology, and Applications*, Prentice-Hall (1966).

10. W. Shockley, "A Unipolar Field-Effect Transistor," Proc. IRE, *40*, 1365 (1952).

11. R. K. Weimer, "An Evaporated Thin Film Triode," IRE-AIEE Solid-State Device Research Conference, Stanford U., Stanford, California (June 1961).

12. L. L. Chang and H. N. Yu, "The Germanium Insulated-Gate Field-Effect Transistor (FET)," Proc. IEEE, *53*, 316 (1965).

13. M. H. White and J. R. Cricchi, "Complementary MOS Transistors," Solid State Electron., *9*, 991 (1966).

13a. T. T. Kamins and R. S. Muller, "Statistical Considerations in MOSFET Calculations," Solid State Electron., *10*, 423 (1967).

13b. F. P. Heiman and H. S. Miller, "Temperature Dependence of *n*-Type MOS Transistors," IEEE Trans. Electron Devices, *ED-12*, 142 (1965).

14. H. Becke, R. Hall, and J. White, "Gallium Arsenide MOS Transistors," Solid State Electron., *8*, 813 (1965).

15. A. S. Grove and D. J. Fitzgerald, "Surface Effects on *p-n* Junctions: Characteristics of Surface Space-Charge Regions Under Nonequilibrium Conditions," Solid State Electron., *9*, 783 (1966).

16. O. Leistiko, A. S. Grove, and C. T. Sah, "Electron and Hole Mobilities in Inversion Layers on Thermally Oxidized Silicon Surfaces," IEEE Trans. Electron Devices, *ED-12*, 248 (1965).

17. J. R. Schrieffer, "Effective Carrier Mobility in Surface-Space Charge Layers," Phys. Rev., *97*, 641 (1955).

18. D. R. Frankl, *Electrical Properties of Semiconductor Surfaces*, Chap. 4, pp. 93-141, Pergamon Press (1967).

19. H. C. Pao and C. T. Sah, "Effects of Diffusion Current on Characteristics of Metal-Oxide (Insulator)-Semiconductor Transistors," Solid State Electron., *10*, 927 (1966).

20. V. G. K. Reddi and C. T. Sah, "Source to Drain Resistance Beyond Pinch-Off in Metal-Oxide-Semiconductor Transistors (MOST)," IEEE Trans. on Electron Devices, *ED-12*, 139 (1965).

21. W. Fischer, "Equivalent Circuit and Gain of MOS Field-Effect Transistors," Solid State Electron., *9*, 71 (1966).

22. F. J. Kennedy, "Gate Leakage Current in MOS Field-Effect Transistors," Proc. IEEE, *54*, 1098 (1966).

23. R. C. Gallagher and W. S. Corak, "A Metal-Oxide-Semiconductor (MOS) Hall Element," Solid State Electron., *9*, 571 (1966).

24. Z. Lukes, "Characteristics of the Metal-Oxide-Semiconductor Transistor in the Common-Gate Electrode Arrangement," Solid State Electron., *9*, 21 (1966).

25. L. Vadasz and A. S. Grove, "Temperature Dependence of MOS Transistor Characteristics Below Saturation," IEEE Trans. Electron Devices, *ED-13*, 863 (1966).

26. S. D. Brotherton, "Dependence of MOS Transistor Threshold Voltage on Substrate Resistivity," Solid State Electron., *10*, 611 (1967).

27. F. Leuenberger, "Dependence of Threshold Voltage of Silicon p-Channel MOS FET's in Crystal Orientation," Proc. IEEE, *54*, 1985 (1966).

28. A. G. Stanley, "Effects of Electron Irradiation of Metal-Nitride-Semiconductor Insulated-Gate Field-Effect Transistors," Proc. IEEE, *54*, 784 (1966).

29. S. M. Sze and G. Gibbons, "Effect of Junction Curvature on Breakdown Voltages in Semiconductors," Solid State Electron., *9*, 831 (1966).

30. G. F. Newmark, "Theory of the Influence of Hot Electron Effects on Insulated-Gate Field-Effect Transistors," Solid State Electron., *10*, 169 (1967).

31. S. R. Hofstein and G. Warfield, "Carrier Mobility and Current Saturation in the MOS Transistor," IEEE Trans. Electron Devices, *ED-12*, 129 (1965).

32. M. Shoji, "Analysis of High-Frequency Thermal Noise of Enhancement Mode MOS Field-Effect Transistors," IEEE Trans. Electron Devices, *ED-13*, 520 (1966).

33. S. Y. Wu, "Theory of Generation-Recombination Noise in MOS Transistors," Solid State Electron., *11*, 25 (1968).

34. I. Flinn, G. Bew, and F. Berg, "Low Frequency Noise in MOS Field-Effect Transistors," Solid State Electron., *10*, 833 (1967).

35. M. P. Lepselter and S. M. Sze, "SB-IGFET: An Insulated-Gate Field-Effect Transistor Using Schottky Barrier Contacts as Source and Drain," Proc. IEEE, *56* (August 1968).

36. M. P. Lepselter and S. M. Sze, "Silicon Schottky Barrier Diode With Near-Ideal *I-V* Characteristics," Bell. Sys. Tech. J., *47*, 195 (1968).

37. C. R. Crowell and S. M. Sze, "Current Transport in Metal-Semiconductor Barriers," Solid State Electron., *9*, 1035 (1966).

38. R. H. Fowler and L. Nordheim, "Electron Emission in Intense Electric Fields," Proc. Roy. Soc. (London), *119*, 173 (1928).

39. D. Kahng and S. M. Sze, "A Floating Gate and Its Application to Memory Devices," Bell Syst. Tech. J., *46*, 1283 (1967).

40. A. S. Grove, O. Leistiko, and W. W. Hooper, "Effect of Surface Field on the Breakdown Voltage of Planar Silicon *p-n* Junctions," IEEE Trans. Electron Devices, *ED-14*, 157 (1967).

■ INTRODUCTION

■ INSULATED-GATE
  THIN-FILM TRANSISTORS (TFT)

■ HOT-ELECTRON TRANSISTORS

■ METAL-INSULATOR-METAL STRUCTURE

11

Thin-Film Devices

1 INTRODUCTION

Thin-film devices are structures consisting of one or more thin layers of metal, semiconductor, or insulator. The insulated-gate thin-film transistor (TFT) was proposed by Weimer[1] in 1961 and used evaporated semiconductor, metal, and insulator layers to form a device which functions essentially as an IGFET. The same author has also recently reviewed the basic technology and applications of TFTs in integrated circuits.[2] Since the semiconductor layer is formed by deposition, more defects and crystalline imperfections in the layer than in the corresponding single-crystal semiconductor are expected. This results in more complicated transport processes in the TFT. We shall in the next section consider the basic characteristics, the effects due to traps and surface states, and the power limitations of a TFT.

In the past decade many attempts have been made to invent or discover solid-state phenomena capable of displacing the junction transistor in one or another circuit application. Among the most interesting candidates are the hot-electron transistors. The first hot-electron device was proposed by Mead[3] in 1960. Many others have been proposed in subsequent years. In Section 3 we shall compare, on a uniform basis, some important hot-electron transistors with the junction transistor. It will be shown that none of the proposed hot-electron transistors can compete with the junction transistor in ultimate high-frequency performance. The hot-electron transistors, however, are considered to be useful devices in the sense that they can be employed to study fundamental physical parameters and processes such as hot-electron lifetimes in metal films and transport mechanisms in insulators.

In the last section we shall consider a related thin-film device, the metal-insulator-metal structure. It should be pointed out that there are many other thin-film devices which are not included here, e.g., the thin-fim transducers[4] which utilize the piezoelectric and piezoresistive properties of semiconductor films, and thin-film optical detectors such as the lead-salt detectors.[5] Thin-film transducers have recently been reviewed by Foster.[4] Optical detectors will be considered in the next chapter on optoelectronic devices.

2 INSULATED-GATE THIN-FILM TRANSISTORS (TFT)

The insulated-gate thin-film transistor can be constructed in a variety of forms. Figure 1 shows cross-sectional diagrams of typical TFT structures. In the staggered-electrode structure of Fig. 1(a), the metal source and drain electrodes are deposited first on the insulating substrate separated by a narrow space corresponding to the channel length. A semiconductor layer is then deposited. Finally, the insulating layer and then the gate electrode are put down in registry with the channel. For the staggered-electrode structure of Fig. 1(b), the process sequence is exactly inverted. The coplanar-electrode

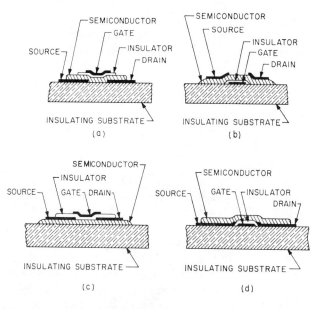

Fig. 1 Insulated-gate thin-film transistors (TFT).
(a) and (b) staggered-electrode structures,
(c) and (d) coplanar-electrode structures.
(After Weimer, Ref. 1.)

structures shown in Fig. 1(c) and 1(d) are somewhat simpler to fabricate than the staggered-electrode structures because all the evaporation-requiring precision masking can be carried out in one operation. Typically, glass or sapphire is used as the insulating substrate, cadmium sulfide or cadmium selenide as the semiconductor, gold or aluminum as the metal electrodes, and silicon oxide or aluminum oxide as the gate insulator.

(1) Basic Characteristics

The basic current-voltage characteristic of a TFT is virtually identical to that of an IGFET as discussed in the previous chapter. We shall make the following assumptions:[6] (1) the mobility of the carriers in the channel is a constant, (2) the gate capacitance is a constant independent of gate voltage, (3) the source and drain metal electrodes are ohmic contacts to the semiconductor, and (4) the initial charge density in the semiconductor is n_o charges/cm^3, which is positive for a depletion-type TFT having an initial excess of donor-type states and is negative for an enhancement-type TFT having an initial excess of unfilled traps or acceptor-type states. The main difference in the above assumptions as compared to those used by Ihantola and Moll[6a] for the IGFET is that the initial charge density in the semiconductor is taken into account.

The analysis is given for the simplified structure as shown in Fig. 2. All the symbols are identical to those used in the previous chapter: L for the channel length, Z for the channel width, and d for the insulator thickness. The thickness h of the semiconductor film is an additional parameter. The charge density $n(y)$, induced in the channel region by application of a gate voltage V_G, is given by

$$q\Delta n(y) = \frac{C_i}{h}\,[V_G - V(y)] \tag{1}$$

where C_i is the gate capacitance per unit area (ε_i/d), and $V(y)$ is the applied drain voltage at a distance y from the source. The total drain current I_D may be expressed as

$$I_D = (hZ)[\sigma_o + \Delta\sigma(y)]\mathscr{E}_y$$
$$= (hZ)q\mu_n[n_o + \Delta n(y)]\mathscr{E}_y \tag{2}$$

where σ_o and $\Delta\sigma(y)$ are the initial conductivity and the incremental conductivity due to $\Delta n(y)$. Combining Eqs. (1) and (2) gives

$$I_D = Z\mu_n C_i\left[\frac{qhn_o}{C_i} + V_G - V(y)\right]\frac{dV(y)}{dy} \tag{3}$$

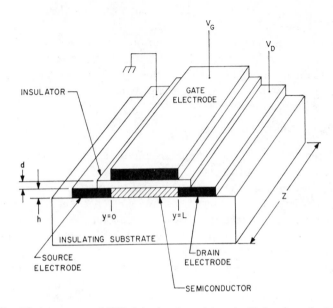

Fig. 2 Simplified structure of TFT. L is the channel length, Z the channel width, d the insulator thickness, and h the semiconductor thickness.

$$I_D \int_0^L dy = Z\mu_n C_i \int_0^{V_D} \left[\frac{qhn_o}{C_i} + V_G - V(y) \right] dV(y) \qquad (4)$$

which yields

$$I_D = \frac{Z\mu_n C_i}{L} \left[(V_G - V_T)V_D - \frac{V_D^2}{2} \right] \qquad (5)$$

where

$$V_T \equiv -qhn_o/C_i.$$

We note that the expression of Eq. (5) is identical to that derived for the IGFET. The threshold voltage, V_T, now depends on the initial charge density n_o. Equation (5) is valid for $0 \leq V_D \leq (V_G - V_T)$, up to the knee of the I_D versus V_D characteristic. The resulting characteristics calculated from Eq. (5) are shown by the heavy lines of Fig. 3. Beyond the knee the current is assumed to be substantially constant as in the IGFET.

Experimental results of TFTs using various semiconductors[7-7f] in both the coplanar and staggered-electrode structures have been in good agreement with the characteristics predicted by Eq. (5). Some typical I_D versus V_D plots are shown in Fig. 4 for the depletion-type and enhancement-type TFTs.[2]

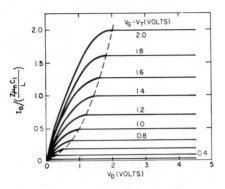

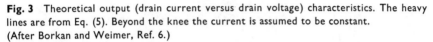

Fig. 3 Theoretical output (drain current versus drain voltage) characteristics. The heavy lines are from Eq. (5). Beyond the knee the current is assumed to be constant. (After Borkan and Weimer, Ref. 6.)

In the region below the knee the drain conductance, g_D, and the transconductance, g_m, are given by

$$g_D \equiv \left.\frac{\partial I_D}{\partial V_D}\right|_{V_D \to 0} = \frac{Z\mu_n C_i}{L}(V_G - V_T) \tag{6}$$

$$g_m \equiv \frac{\partial I_D}{\partial V_G} = \frac{Z\mu_n C_i}{L} V_D. \tag{7}$$

The drain conductance is thus linear with V_G, and the transconductance increases with V_D.

The saturation of the drain current occurs at a value of drain voltage given by $(V_G - V_T)$. This value gives the locus of knees shown in Fig. 3 and is obtained from the condition $\partial I_D/\partial V_D = 0$. The saturation mechanism is similar to that of the IGFET and is a geometry-dependent effect which results from the pinch-off of the conducting channel in the neighborhood of the drain. The saturated drain current (at the knee) can be obtained from Eq. (5):

$$I_{D\,\text{sat}} = \frac{Z\mu_n C_i}{2L}(V_G - V_T)^2. \tag{8}$$

This is the square-law relationship of the saturated drain current to the effective gate voltage $(V_G - V_T)$. The transconductance in the saturation region is given by

$$g_m = \frac{Z\mu_n C_i}{L}(V_G - V_T) = \left(\frac{2Z\mu_n C_i}{L} I_{D\,\text{sat}}\right)^{1/2}. \tag{9}$$

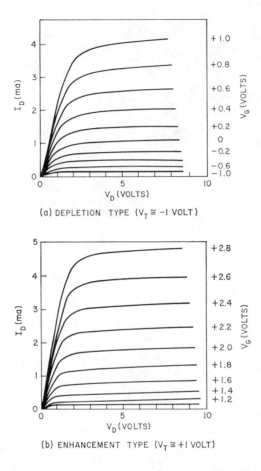

Fig. 4 Experimental output characteristics for depletion-type ($V_T < 0$) and enhancement-type ($V_T > 0$) TFT.
(After Weimer, Ref. 1.)

In a depletion-type TFT the saturated drain current at zero gate bias is

$$I_{DO} = \frac{Z\mu_n C_i V_T^2}{2L} = \frac{Z\mu_n}{2LC_i}(qhn_o)^2. \tag{10}$$

It is noted that I_{DO} can be small even for very high carrier density n_o, provided that the semiconductor thickness h is small and the insulator capacitance is large.

A figure of merit which characterizes the high-frequency performance of TFT is the gain-bandwidth product which is equivalent to the maximum operating frequency defined for IGFET:

$$f_m \equiv \frac{g_m}{2\pi C_i LZ}. \tag{11}$$

We obtain from Eqs. (6) and (7)

$$f_m \simeq \frac{\mu_n V_D}{2\pi L^2} \tag{11a}$$

for $V_D < (V_G - V_T)$, and

$$f_m \simeq \mu_n(V_G - V_T)/2\pi L^2 \tag{11b}$$

at or above the knee.

(2) Effects of Traps and Surface States

From the discussions of the last section it is clear that the electrical characteristics of a TFT are basically identical to those of an IGFET. The detailed transport processes in a TFT, however, are more complicated. The major complication arises from the fact that the semiconductor layer is formed by vacuum deposition, and a deposited layer, in general, contains many more defects and crystalline imperfections than the corresponding single-crystal semiconductor. Thus the trapping centers in the semiconductor layer will have profound effect on the device characteristics. In addition, the trapping centers in the deposited insulator layer, the surface states and surface charges near or at the semiconductor-insulator interface, and the metal contacts to the semiconductor will also influence the device performance.

Unlike the diffused source and drain contacts in an IGFET, the metal electrode of a TFT may give rise to nonohmic contacts resulting from the formation of Schottky-type barriers (see Chapter 8). The failure to achieve a good ohmic contact at the source electrode will give a low transconductance (and in some cases the "crowded" characteristics[2]) which, instead of continuing to increase with increasing gate bias, levels off and decreases toward zero as the family of I_D versus V_D curves crowd together at a maximum value of I_D.

The existence of large densities of traps and states can cause a reduction of the channel mobility and can affect the reliability, reproducibility, and performance of the TFT. In this section we shall consider the influences of these traps and states on the device characteristics.

The effect of the trapping centers in the semiconductor layer can be studied by a combination of optical and field-effect techniques.[8] The experimental result of a CdSe thin-film transistor is shown in Fig. 5(a). When illuminated, the device shows a significantly higher saturation current and transconductance than the same device measured in the dark. The transconductance is particularly affected by photons with energy slightly less than that necessary for band-to-band transitions. The energy band model for the illumination effect

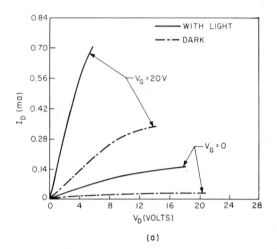

(a)

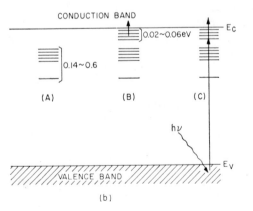

(b)

Fig. 5

(a) Experimental result of a CdSe TFT.

(b) Energy band model for the illumination effect where (A) shows the deep levels, (B) shows both the deep and the shallow levels in dark specimen, and (C) shows the enhancement of field effect by illumination.

(After Poehler and Abraham, Ref. 8.)

is shown in Fig. 5(b) where (A) shows the deep levels, (B) shows both the deep and the shallow levels in the dark specimen, and (C) shows the enhancement of the field effect by illumination. The incoming photon excites additional electrons to shallow levels, from which they are subsequently field-excited to the conduction band. The field, $\mathscr{E}$, required for excitation from such levels, can be found from the tunneling probability derived previously in Chapter 4.

$$T_t \simeq \exp\left[-\frac{4}{3}\frac{\sqrt{2m^*}(\Delta E)^{3/2}}{q\hbar\mathscr{E}}\right]. \qquad (12)$$

The calculated probability for carriers excited from a single level ($\Delta E = 0.04$ eV) below the conduction band edge is shown in Fig. 6 (dotted line); also shown is the experimental result (solid line) of the drain current as a function of applied gate voltage. Since the number of carriers and hence the current should be proportional to this probability, the agreement between the probability-field curve and the current-voltage curve is indicative of the importance of the traps in the semiconductor layer.

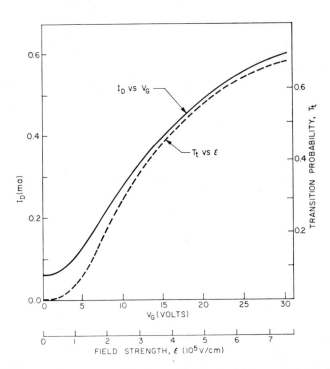

Fig. 6   Calculated tunneling probability for carriers excited from a single level $\Delta E = 0.04$ eV and the experimental drain current of the CdSe TFT versus the applied gate voltage. (After Poehler and Abraham, Ref. 8.)

We next consider the effect of the surface traps which are located at the semiconductor-insulator interface. The traps are assumed to have effective density N_t, a capture cross section σ_n, and an ionization energy E_t measured from conduction band edge. If the average velocity of mobile carriers is $\bar{v}$, then in one second a carrier traces out a volume $\bar{v}\sigma_n$ of possible recombination. If the density of filled traps is n_t, then the density of empty traps is $(N_t - n_t)$. The rate at which conduction carriers recombine with traps is $\bar{v}\sigma_n(N_t - n_t)n_c$ where n_c is the density of conduction carriers. The interaction of trapping centers and crystal lattice releases carriers in the traps at a rate proportional to their number. The rate of release is $\bar{v}\sigma_n n_t n_1$, where n_1 is the available density of sites to be refilled by the released electrons; in other words n_1 is the density of carriers which would be present in the semiconductor if the Fermi level were at the trap level, and is given by

$$n_1 = N_C \exp(-E_t/kT) \tag{13}$$

where N_C is the effective density of states in the conduction band. The difference in these rates produces the net rate of change of trapped carriers,[9]

$$\frac{\partial n_t}{\partial t} = \bar{v}\sigma_n(N_t - n_t)n_c - \bar{v}\sigma_n n_t n_1. \tag{14}$$

At equilibrium the rate of change of trapped carriers must be zero and therefore

$$n_c(N_t - n_t) = n_1 n_t. \tag{15}$$

The total charge per unit area induced which results from the applied gate voltage is given by

$$q(\Delta n_c + \Delta n_t) = q[(n_c + n_t) - (n_{co} + n_{to})] = \frac{C_i}{h} V_G \tag{16}$$

or

$$q(n_c + n_t) = \frac{C_i}{h}(V_G - V_T) \tag{17}$$

where h is assumed to be equal to the effective inversion depth and V_T is the threshold voltage:

$$V_T = -\frac{qh}{C_i}(n_{co} + n_{to}), \tag{18}$$

and n_{co} and n_{to} are the initial densities of conduction carriers and trapped carriers respectively. Substituting n_t from Eq. (15) into Eq. (17), we obtain[10]

$$V_c \equiv \frac{qn_c h}{C_i} = \tfrac{1}{2}[(V_G - V_T) - (V_t + V_1)]$$

$$+ \tfrac{1}{2}\{[(V_G - V_T) - (V_t + V_1)]^2 + 4V_1(V_G - V_T)\}^{1/2} \qquad (19)$$

where

$$V_t \equiv qhN_t/C_i \quad \text{and} \quad V_1 \equiv qhn_1/C_i.$$

The source-drain conductance at zero source-drain voltage is given by

$$g_D = \frac{ZC_i\mu_n}{L} V_c. \qquad (20)$$

The above expression reduces to the single result derived previously, Eq. (6), when V_1 is small and $(V_G - V_T) \gg (V_1 + V_t)$, i.e., in the case of small trap densities and large trap ionization energy. Figure 7 shows the theoretical fitting of the experimental conductance characteristics of an evaporated silicon TFT.[10] From the result, we obtained the following device parameters: $N_t = 4.6 \times 10^{17}$ cm^{-3}, $n_{co} = 9 \times 10^{16}$ cm^{-3}, and $n_1 = 1.5 \times 10^{15}$ cm^{-3}.

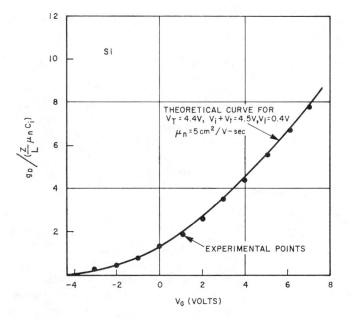

Fig. 7   Experimental conductance characteristic of an evaporated Si TFT. (After Salama and Young, Ref. 10.)

From Eq. (14) we can calculate the variation of the total charge density, n ($= n_c + n_t$ = total charge density), resulting from the dc and ac parts of the applied gate voltage.[11] We seek solutions of Eq. (14) of the form

$$n_c = n_{cD} + n_{cA} \exp(j\omega t) \tag{21a}$$

$$n_t = n_{tD} + n_{tA} \exp(j\omega t). \tag{21b}$$

If the ac parts of the quantities are treated as small signals, we find[11]

$$\frac{dn_c}{dn} = \frac{\omega_1 + j\omega}{\omega_1 + \omega_2 + j\omega} \tag{22a}$$

$$\frac{dn_t}{dn} = \frac{\omega_2}{\omega_1 + \omega_2 + j\omega} \tag{22b}$$

where ω_1 and ω_2 are characteristic of the traps and of the existing dc bias conditions:

$$\omega_1 \equiv N_C \exp(-E_t/kT)\sigma_n \bar{v} \left/ \left(1 - \frac{n_{tD}}{N_t}\right)\right. \tag{23a}$$

$$\omega_2 \equiv N_t \sigma_n \bar{v} \left(1 - \frac{n_{tD}}{N_t}\right). \tag{23b}$$

If we assume that a change in gate voltage results in changes of both the carrier concentration and carrier mobility, the conductivity in Eq. (2) should be amended to read

$$\Delta\sigma = q\Delta n_c \mu_n + q n_c \Delta\mu_n \tag{24}$$

where n_c is the density of conduction electrons.

The transconductance in the saturation region is then given by

$$g_m = \frac{Z\mu_n C_i(V_G - V_T)}{L} \left(\frac{dn_c}{dn} - \beta \frac{dn_t}{dn}\right) \tag{25}$$

where

$$\beta \equiv \frac{n_c}{\mu_n} \left(\frac{d\mu_n}{dn_t}\right),$$

n_t the density of trapped electrons, and n the density of total electrons. The first term in Eq. (25) is the direct result of electron density changes in the channel. For $dn_c/dn = 1$ (no trapping) this result is identical with Eq. (9). The second term in Eq. (25) expresses the effect of mobility variation on the transconductance.

The small-signal ac transconductance resulting from Eqs. (22) and (25) is now given by[11]

$$g_m(\omega) = g_0 \left\{ \frac{(\omega_1 + \beta\omega_2)^2 + \omega^2}{(\omega_1 + \omega_2)^2 + \omega^2} \right\}^{1/2} \exp(j\theta) \tag{26}$$

where

$$g_0 \equiv Z\mu_n(n_{tD})C_i/L \tag{26a}$$

$$\tan\theta = \frac{\omega\omega_2(1 - \beta)}{(\omega_1 + \omega_2)(\omega_1 + \beta\omega_2) + \omega^2}. \tag{26b}$$

From the above equations, we obtain for the two limiting cases:

$$g_m(\omega \to 0) = g_0 \left(\frac{\omega_1 + \beta\omega_2}{\omega_1 + \omega_2} \right)$$

and

$$g_m(\omega \to \infty) = g_o. \tag{27}$$

If the carrier mobility is constant, $\beta = 0$, then $|g_m(\omega)|$ is always an increasing function of frequency and $\tan\theta > 0$. When $\beta > 1$, $|g_m(\omega)|$ is a decreasing function of frequency and $\tan\theta < 0$. Typical curves are shown in Fig. 8(a) for $\beta = 0$, 1, and 2. Useful information concerning the trap or impurity level may thus be derived from the frequency dependence discussed above. It is convenient to introduce the peak frequency, ω_m and to rewrite Eq. (26b) as follows

$$\frac{\tan\theta}{\tan\theta_m} = \frac{2\left(\dfrac{\omega}{\omega_m} \right)}{\left[1 + \left(\dfrac{\omega}{\omega_m} \right)^2 \right]}. \tag{28}$$

Equation (28) is plotted in Fig. 8(b) (solid curve). The peak frequency ω_m will shift with dc bias and with temperature. For $\beta = 0$, we have

$$\omega_m \simeq \sqrt{\omega_1(\omega_1 + \omega_2)} \tag{29}$$

so that ω_m will increase with increasing gate voltage, since n_{tD} decreases with increasing gate voltage. From Eqs. (23) and (29) the activation energy derived from a plot of ω_m versus $1/T$ should vary from E_t (the trap ionization energy) at high temperature to $E_t/2$ at lower temperature. This is observed experimentally on CdS TFTs. The experimental phase data obtained at different temperatures are also plotted in Fig. 8(b). The agreement with theory is

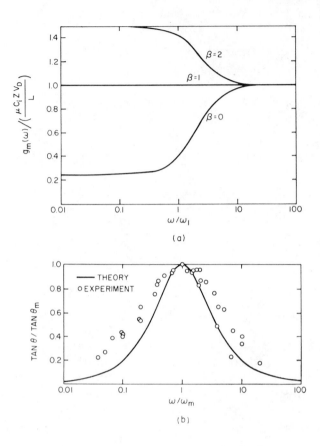

Fig. 8

(a) Small signal dc transconductance versus frequency. $\beta \equiv \dfrac{n_c}{\mu_n}\left(\dfrac{d\mu_n}{dn_t}\right)$ where n_c is the density of the conduction electrons, n_t the density of the trapped electrons, and μ_n is the electron mobility. $\beta = 0$ for constant mobility.
(After Haering, Ref. 11.)
(b) Experimental phase data versus frequency of a CdS TFT.
(After Miksic et al., Ref. 12.)

reasonably good.[12] The variation of transconductance with temperature is shown in Fig. 9 for a CdS TFT. The behavior is the result of traps as well as of mobility variations. It is obvious that more than one trap level plays a role in these data.

The trapping centers in the insulator can cause additional effects of drift and instability in TFT. The kinetics of carrier transfer are still given by the

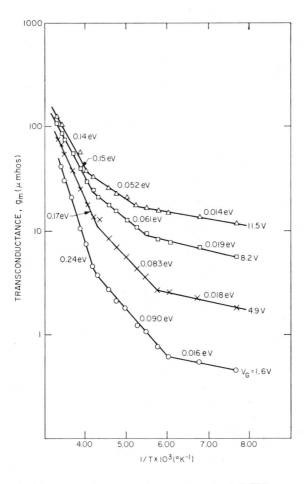

Fig. 9 Variation of transconductance with temperature of a CdS TFT. (After Miksic et al., Ref. 12.)

same expression as in Eq. (14) where N_t and n_t are the total density of traps and filled traps in the insulator. The density n_1 is now given by[13]

$$n_1 = N_C \exp\left[-\frac{E_{cs} - E(x)}{kT}\right] \tag{30}$$

where E_{sc}, as shown in Fig. 10(a), is the energy difference between the conduction band and the Fermi level at the interface, and $E(x)$ is the energy

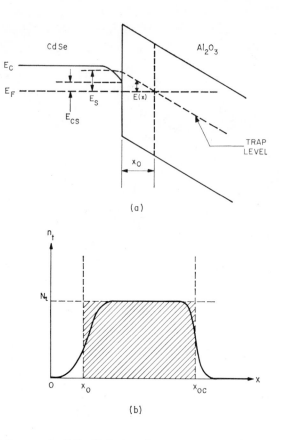

Fig. 10
(a) Energy band diagram of a CdSe-Al₂O₃ interface.
(b) The trap distribution in the insulator as a function of distance.
(After Koelmans and DeGraaff, Ref. 13.)

difference between the trap level and the Fermi level in the semiconductor:

$$E(x) = E_s[(x_o - x)/x_o] \qquad (31)$$

where E_s is the energy difference at $x = 0$. $E(x)$ is negative for $x > x_o$. It is predominantly the traps at $x > x_o$ which will be filled during the drift process. If the transport mechanism of carriers to the traps is due to tunneling, then the capture cross section σ_n has the following form:

$$\sigma_n(x) = \sigma_n(0)\exp(-ax). \qquad (32)$$

$\sigma_n(0)$ is the capture cross section at the interface. The constant "a" depends on the height of the barrier and the effective mass of the carriers. Substitution of Eqs. (30), (31), and (32) into Eq. (14) and integration with respect to t yield the concentration of filled traps at time t and distance x:[13]

$$n_t(x, t) = \frac{n_c N_t}{n_c + n_1}\{1 - \exp[-t\sigma_n(0)\bar{v}(n_c + n_1)\exp(-ax)]\}$$

$$= \frac{n_c N_t}{n_c + n_1}\{1 - \exp[-\exp(ax_{oc} - ax)]\}$$

where

$$x_{oc} \equiv \frac{1}{a}\ln[\sigma_n(0)\bar{v}(n_c + n_1)t].$$

The distribution represented by the above equation is sketched in Fig. 10(b). At $x = x_o$ the concentration of filled traps quickly rises to N_t because of the x dependence of n_1. At $x = x_{oc}$, n_t quickly drops to zero. We can now approximate the actual distribution of Eq. (33) by the rectangular one shaded in Fig. 10(b), and obtain for the total number, N_e, of carriers per cm^3 transferred into the insulator at time t:

$$N_e \simeq N_t(x_{co} - x_o)$$

$$\simeq N_t\left(\frac{1}{a}\right)\ln[\sigma_n(0)\bar{v}n_c t] - N_t x_o. \tag{34}$$

When N_e is measured between the time t_o and t, we obtain

$$N_e(t_o \to t) \simeq N_t\frac{1}{a}\ln(t/t_o). \tag{35}$$

As the variations of the conductivity, $\Delta\sigma$, are relatively small, $\Delta\sigma$ is about proportional to N_e. It is thus expected that, at a given V_G and V_D, the drain current should decay, and the time dependence of this decay is logarithmic.

The experimental results[13] of CdSe TFT are in good agreement with the above model in which tunneling of carriers from the channel to traps in the insulator is assumed. In the experiment it is found that $1/a \simeq 3$ Å, and $x_o < 60$ Å which are reasonable for the tunneling model.

From the above discussion it is clear that because of the large densities of traps which exist in the semiconductor, in the insulator, and at the interface, the detailed transport processes in a TFT are very complicated. It is necessary to reduce the trap densities in order to improve device reliability, reproducibility, and performance.

(3) Power Limitations[14]

Power dissipation constitutes one of the important device limitations. In order to obtain a more realistic expression for the gain-bandwidth product, power dissipation will be introduced into the basic theoretical characteristics. The power dissipation per unit area at the knee of the I_D versus V_D plot is

$$P = \frac{I_{D\,sat}\, V_{D\,sat}}{ZL} \qquad (36)$$

where

$$I_{D\,sat} = \frac{Z\mu_n \varepsilon_i}{2Ld}\, (V_{D\,sat})^2 \qquad (37a)$$

and

$$V_{D\,sat} = V_G - V_T. \qquad (37b)$$

Combination of Eqs. (36) and (37) yields

$$V_{D\,sat} = \left(\frac{2LdP}{\mu_n \varepsilon_i}\right)^{1/3}. \qquad (38)$$

The gain-bandwidth product at the knee is then [from Eq. (11a)]

$$f_m \approx \frac{\mu_n V_{D\,sat}}{2\pi L^2} = \frac{1}{2\pi}\left(\frac{2d}{\varepsilon_i}\right)^{1/3}\left(\frac{\mu_n}{L^2}\right)^{2/3} P^{1/3}. \qquad (39)$$

Thus the gain-bandwidth product increases as the cube root of the permitted power dissipation per unit area and as the factor $(\mu_n/L^2)^{2/3}$. This is in contrast to the factor (μ_n/L^2) when power dissipation is neglected.

The heat flow pattern in a TFT is shown in Fig. 11(a). Heat is generated in the active region of the device which is in the form of a long narrow strip on the surface of the substrate. The heat can be removed by conduction through the substrate, conduction through the electrodes, and by radiation. The heat flow through an element distance r from the center of the heated region is given by

$$H_1 = -\pi r \kappa Z \frac{dT}{dr} \qquad (40)$$

where κ is the thermal conductivity of the insulating substrate. Integration of this expression yields the heat flow:

$$H_1 = \frac{\pi \kappa Z(T_1 - T_2)}{\ln(b/a)}. \qquad (41)$$

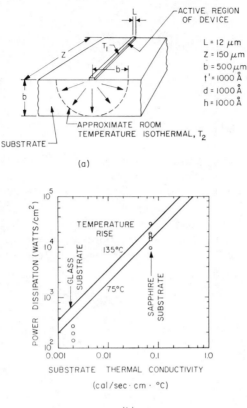

(a)

(b)

Fig. 11
(a) Heat flow pattern in a TFT.
(b) Theoretical and experimental power dissipation for a TFT with device geometry given in (a).
(After Page, Ref. 14.)

Heat will also be conducted away from the active region through the source and drain electrodes and is given by

$$H_2 = \frac{2t'Z\kappa'(T_1 - T_2)}{b} \qquad (42)$$

where t' is the electrode thickness and κ' the electrode thermal conductivity. The heat loss by radiation is given by

$$H_3 = 2.74 \times 10^{-12}E_1(T_1^4 - T_2^4)ZL \qquad (43)$$

where E_1 = emissivity $\simeq$ 5 to 70 depending on metal and surface. The total power dissipation is then given by

$$P(\text{max}) = (H_1 + H_2 + H_3)/ZL. \tag{44}$$

For practical values of device geometrics and solid state materials the most important heat loss is due to the substrate conduction. The theoretical power dissipation (solid lines) for a TFT with device geometry given in Fig. 11(a) is shown in Fig. 11(b) as a function of the substrate thermal conductivity. Also shown are the experimental points for glass substrate and sapphire substrate. We note that, for a temperature rise of 75°C, the maximum power dissipation is about 300 W/cm^2 on glass substrate and 1×10^4 W/cm^2 on sapphire substrate. The theoretical characteristics of a TFT with geometry outlined in Fig. 11(a) is shown in Fig. 12. The electron mobility is assumed to be 300 cm^2/V-sec. Superimposed on this figure are the locus of $V_{D\,\text{sat}}$ and the constant power-dissipation lines. For the device to operate in the saturation region and not exceed the power dissipation limitation, operation must be confined to the right of the locus and below the power curve, e.g., the angle ABC of Fig. 12 for a device on glass.

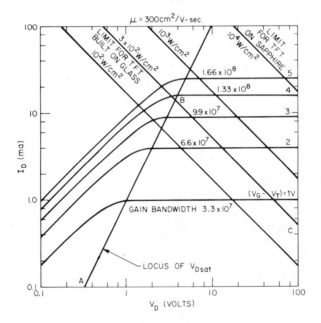

Fig. 12 Theoretical output characteristics at the constant-power dissipation lines. (After Page, Ref. 14.)

3 HOT-ELECTRON TRANSISTORS

(1) Hot Electrons in Metals

As defined previously, a useful solid-state device is one which can be used in electronic applications or can be used to study the fundamental physical parameters. All the hot electron transistors, at the present time, belong to the latter category. We shall consider first the hot-electron lifetime in metals. By hot electron we mean an electron having an energy more than a few kT above the Fermi energy, where k and T are Boltzmann's constant and lattice temperature respectively, and thus the electron is not in thermal equilibrium with the lattice.

The simplest approach which gives the dependence of the hot-electron lifetime on energy is illustrated in Fig. 13. Because of the Pauli Exclusion Principle, the excited electrons with energy $(E - E_F)$ above the Fermi level can only interact with the conduction electrons in the energy range $-(E - E_F)$ relative to the Fermi energy (the shaded area). The lifetime, τ, of a hot electron in its excited state is inversely proportional to the number of conduction electrons it can excite, $\sim N(E_F)(E - E_F)$, and also is inversely proportional to the number of states available for it to go to after excitation, likewise

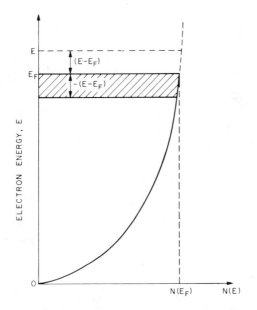

Fig. 13 Electron energy distribution based on a free electron model.

$\sim N(E_F)(E - E_F)$. The functional form of the lifetime for low-energy excitation is thus given by

$$\tau \sim (E - E_F)^{-2} \tag{45}$$

which means that the larger the excitation energy above the Fermi energy, the shorter the hot-electron lifetime.

More detailed calculation based on the free electron model using self-consistent dielectric constant approach[15] gives the following expression for the lifetime of hot electrons with initial energy close to the Fermi energy:

$$\frac{1}{\tau} = \frac{1}{32\pi} \frac{q^2}{a_B \hbar} \frac{1}{(\gamma r_s/4)^{3/2}} \left[\tan^{-1} \frac{2}{(\gamma r_s)^{1/2}} + \frac{2(\gamma r_s)^{1/2}}{4 + \gamma r_s} \right] \frac{(E - E_F)^2}{E_F^{3/2} E^{1/2}} \tag{46}$$

$$\tau \sim r_s^{3/2}(E - E_F)^{-2} E_F^{3/2} E^{1/2} \tag{47}$$

where a_B is the Bohr radius (0.53 Å), r_s the electron spacing measured in the Bohr radius such that $\frac{4}{3}\pi(r_s a_B)^3 = 1/n$, n the electron density, and $\gamma \equiv 0.656$. The values of r_s and conductivity data for fifteen common metals are listed in Table 11.1, where l_p is the electron-phonon mean free path, and v_F is the electron velocity at the Fermi level.[16]

Figure 14 shows the lifetime as a function of r_s and the excitation energy from the bottom of the conduction band to twice the Fermi energy.[15,17] For $E < E_F$, the lifetime is that for hot holes. For a hot hole whose initial energy is very close to the Fermi energy, the lifetime is identical to that for the hot electron given by Eq. (46). For $|(E - E_F)/E_F| \approx 1$, i.e., for holes close to the bottom of the conduction band or for electrons at twice the Fermi energy, the lifetime has been obtained by numerical computation. The electron-electron mean free path is given by the product of the lifetime and the electron velocity $(v \sim \sqrt{E})$:

$$l_e \equiv \tau v \sim (r_s E_F)^{3/2}(E - E_F)^{-2} E. \tag{48}$$

For higher electron energies such that $E > (E_F + \hbar\omega_p)$, we must consider the creation of plasmons which are the collective modes of vibration of the electron gas, where ω_p is the plasma frequency defined by

$$\omega_p = \sqrt{nq^2/m\varepsilon_o}. \tag{49}$$

The lifetime for a plasmon is given by[15]

$$\tau = \frac{a_B \sqrt{2mE}}{\hbar\omega_p} \left[\ln\left(\frac{\sqrt{E_F + \omega_p} - \sqrt{E_F}}{\sqrt{E} - \sqrt{E - \hbar\omega_p}} \right) \right]^{-1}. \tag{50}$$

TABLE 11.1

VALUES OF r_s AND CONDUCTIVITY DATA

Metal	Free Electrons per cm^{-3} $n \times 10^{-22}$	Electrons per Atom	E_F (eV)	r_o (Å)	$r_s \equiv r_o/a_B$	l_p @ 300°K (Å)	$v_F \times 10^{-8}$ cm/sec
Cs	0.85	1	1.5	3.05	5.8	160	0.75
Rb	1.13	1	1.8	2.8	5.2	220	0.80
K	1.37	1	2.1	2.6	4.9	370	0.85
Fe	1.7	0.2	2.8	2.2	4.2	220	0.91
Na	2.6	1	3.1	2.1	4.0	350	1.07
Ba	3.4	1	3.8	1.9	3.6	—	1.16
Pd	3.8	0.55	4.1	1.85	3.5	110	1.2
Pt	4.0	0.6	4.3	1.8	3.4	110	1.23
Li	4.8	1	4.7	1.72	3.25	110	1.30
Ni	5.4	0.6	5.2	1.7	3.2	133	1.37
Ag	5.8	1	5.5	1.61	3.0	570	1.39
Au	5.9	1	5.5	1.6	3.0	406	1.40
Co	6.2	0.7	5.7	1.5	2.9	130	1.42
Cu	8.7	1	7.1	1.4	2.6	420	1.58
Al	18.4	3	11.7	1.1	2.0	—	2.04

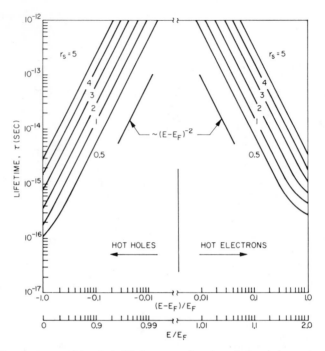

Fig. 14 Hot-electron and hot-hole lifetimes as a function of r_s and the excitation energy where r_s is the electron spacing measured in Bohr radius such that exactly one electron is contained in a sphere with radius r_s. (After Matizuki and Sparks, Ref. 17.)

The hot-electron lifetimes in gold ($r_s = 3.0$) and aluminum ($r_s = 2.0$) are shown in Fig. 15(a) and 15(b) respectively. We note that the lifetime is a monotonically decreasing function of energy, and an abrupt decrease is expected at the plasma frequency. The values of the plasma frequencies are listed in Table 11.2. Since the plasmon created must take up the momentum k_p as well as the energy, the modified frequencies $\hbar\omega_p(k_p)$, which are also listed in Table 11.2, are about 15% to 20% larger than the corresponding $\hbar\omega_p$.

We conclude that, from the above approaches based on the free-electron model, the lifetime is strongly dependent on the initial energy and is expected to decrease with increasing excitation energy.

The experimental results of the hot-electron lifetime and electron-electron mean free path are obtained by the use of hot-electron transistors and other related thin film devices to be considered next. It will be shown later that there is general agreement between the experimental results and the theoretical predictions.

TABLE 11.2

PLASMON ENERGIES (FREE-ELECTRON MODEL)

Metal	r_s	ω_p 10^{16} rad/sec	$\hbar\omega_p$ (eV)	$\hbar\omega_p(k_p)$ (eV)
Cs	5.8	0.54	3.52	4.2
Rb	5.2	0.59	3.92	4.6
K	4.9	0.66	4.36	5.2
Fe	4.2	0.83	5.4	6.4
Na	4.0	0.88	5.8	6.8
Ba	3.6	1.03	6.8	8.0
Pd	3.5	1.1	7.2	8.5
Pt	3.4	1.13	7.4	8.7
Li	3.25	1.21	8.0	9.4
Ni	3.2	1.30	8.5	10.0
Ag	3.0	1.37	9.0	10.5
Au	3.0	1.38	9.0	10.5
Co	2.9	1.40	9.1	10.7
Cu	2.6	1.64	10.8	12.7
Al	2.0	2.4	15.8	18.7

(2) Comparison of Hot-Electron Transistors

In the past decade many transistor-like three-terminal structures have been proposed which basically consist of alternating layers of metal and insulator or semiconductor. The first of these structures, a metal-insulator-metal-insulator-metal structure (MIMIM), in which current flow through the insulator layer occurs by tunneling, Fig. 16, was made by Mead[3] in 1960. Spratt et al.[18] pointed out that the current gain of such structures could be greatly improved by replacing the collector insulator by a Schottky barrier semiconductor layer, Fig. 17(b). Rose,[19] Attala and Kahng,[20] and Geppert[21] continued the process of development by suggesting that the tunnel emitter be replaced by a Schottky barrier emitter, Fig. 17(c). Wright[22] in 1962 suggested

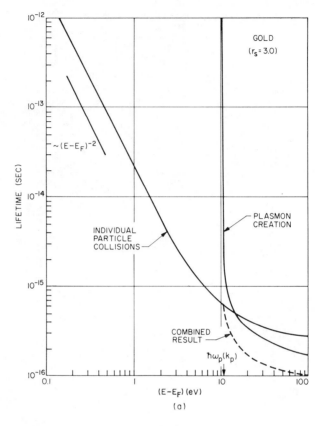

Fig. 15
(a) Hot-electron lifetimes in gold ($r_s = 3.0$).

a transistor structure using a space-charge-limited emitter, Fig. 17(d). The hot-electron transport and electron tunneling in various thin-film structures have been reviewed recently by Crowell and Sze.[23]

In this section we shall compare these transistors with the junction transistor based on their high-frequency performance. In the next two sections we shall consider the basic physical parameters which can be obtained from these transistor structures.

The main difference in these transistors shown in Fig. 17 is the method by which the electrons are injected into the base.[24] For the tunnel transistor the electrons are injected by tunneling through the thin insulator layer. For the space-charge-limited transistor (SCLT) the electrons travel from the emitter

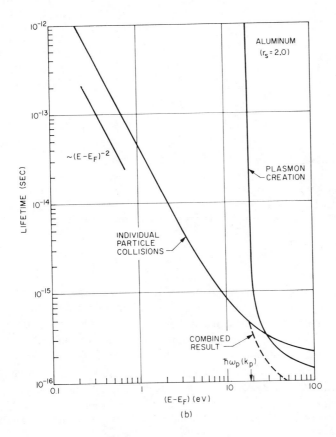

Fig. 15
(b) Hot-electron lifetimes in aluminum ($r_s = 2.0$).
(After Quinn, Ref. 15.)

by a space-charge-limited process. For the semiconductor-metal-semiconductor transistor (SMST) the electrons are emitted by the Schottky-type thermionic emission process. Once the electrons are injected into the metal base, the transit time through the thin metal film is very short (e.g., for a 100 Å film, the time is about 10^{-14} sec). Thus during this time interval the electrons are not in thermal equilibrium with the lattice—hence the name hot-electron transistor.

(A)   Emitter Conductance and Emitter Charging Time.   Since all the hot-electron transistors differ only in their emitter structure, we shall

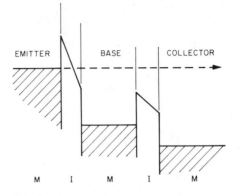

Fig. 16 Band diagram of the first proposed hot-electron device—a metal-insulator-metal structure.
(After Mead, Ref. 3.)

compare the emitter characteristics such as the current-voltage relationship, the capacitance-voltage relationship, the emitter conductance, and the emitter charging time. The results are listed in Table 11.3 for the four transistors shown in Fig. 17. The assumptions and notations used are as follows:[24,25]

(a) For the *n-p-n* junction transistor the emitter junction is a step n^+p junction of unity injection efficiency. The base width is W_B and concentration N_B. A unity base-transport factor is assumed. V_D is the diffusion potential, and V_{EB} the applied emitter-base voltage. D_n is the diffusion coefficient of electrons in the base. We shall use $N_B = 10^{17}$ cm^{-3} and $W_B = 1$ μm as the parameters for a typical high-frequency transistor.

(b) For the semiconductor-metal-semiconductor transistor, the emitter efficiency is unity (i.e., no minority carriers are injected). The emitter is uniformly doped to a concentration N_E, the emitter barrier height is ϕ_B, and A^{**} is the effective Richardson constant (defined in Chapter 8). Typical parameters of an SMST are $\phi_B = 0.8$ V, $N_E = 10^{16}$ cm^{-3}, and $W_B = 100$ Å.

(c) For the tunnel transistor the current flow is assumed to obey the Fowler-Nordheim relation of field emission similar to that derived for tunnel diodes. The thickness and the permittivity of the insulator are W_E and ε_e respectively, the metal-insulator barrier height is ϕ_B. Typical parameters to be considered are $\phi_B = 1$ V, $W_E = 20$ Å, and $\varepsilon_e/\varepsilon_0 = 4$.

(d) For the space-charge-limited transistor, it is assumed that the emitter region W_E is free of fixed charges or traps with only one type of carrier present, and that throughout the region the carrier velocity equals $\mu\mathscr{E}$ where μ and $\mathscr{E}$ are the carrier mobility and electric field respectively.

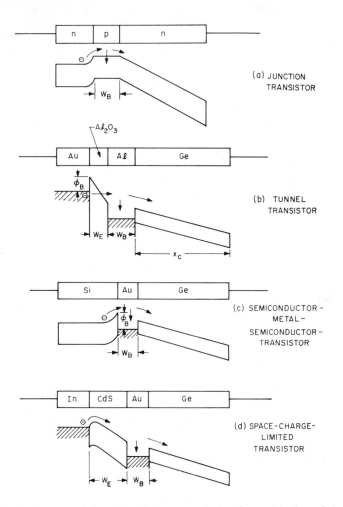

Fig. 17 Band diagrams of three hot-electron transistors along with that of the bipolar transistor.
(After Moll, Ref. 24.)

Based on the relations listed in Table 11.3, the emitter conductance $g_e \equiv \partial J_E / \partial V_{EB}$ is calculated and is shown in Fig. 18(a). It is seen that the bipolar transistor and SMST have the highest emitter conductances due to their strong exponential dependence of emitter current on voltage. The space-charge-limited transistor has the lowest conductance due to its weak current-voltage dependence ($J_E \sim V_{EB}^2$).

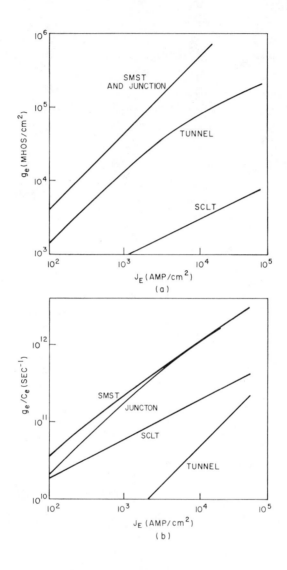

Fig. 18
(a) Theoretical emitter conductances as a function of emitter current density for the four transistors as shown in Fig. 17.
(b) Theoretical emitter figures of merit as a function of emitter current density, for the four transistors.
(After Atalla and Soshea, Ref. 25.)

Figure 18(b) shows the calculated emitter figure of merit,[25] g_e/C_e, which is the reciprocal of the emitter charging time τ_e. It is to be noted that here the SMST and the junction transistor have essentially the same emitter performance, both having the highest figures of merit. They are followed by the SCLT and finally the tunnel transistor which exhibits the most serious emitter limitation.

(B) Maximum Oscillation Frequency (f_{max}). To compare the frequency performance of the hot-electron transistors, we shall use the maximum-oscillation frequency as a figure of merit. For simplicity we shall use a stripe geometry shown in Fig. 19, with an emitter stripe width S, spaced $S/2$ from the base stripes.[26] The base resistance due to the metal film is

$$r_b \simeq \frac{2}{3}\left(\frac{S}{L}\right)\left(\frac{\rho_m}{W_B}\right),$$

and the collector capacitance is $C_c = 2SL\,\varepsilon_c/x_c$ where L is the stripe length, ε_c the collector semiconductor permittivity, x_c the collector depletion-layer width, and ρ_m is the resistivity of the metal base.

The maximum oscillation frequency (f_{max}) is given by[27]

$$f_{max} = \frac{1}{4\pi}\left[\frac{\alpha}{r_b\,C_c\,A_c\,\tau_{ec}}\right]^{1/2} \tag{51}$$

where α is the common-base current gain, and τ_{ec} is the emitter-to-collector signal delay time. It will be shown in the next section that the common-base current gain, α, in a hot-electron transistor can be expressed as

$$\alpha = \alpha^* \exp(-W_B/L_B) \tag{52}$$

where α^* is the value of α extrapolated to zero film thickness, and L_B is the ballistic mean free path, i.e., the mean distance an electron can travel in the

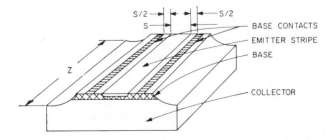

Fig. 19 Transistor geometry with an emitter stripe width S spaced $S/2$ from the base stripes. (After Early, Ref. 26.)

metal base without undergoing any scattering event. For the hot-electron transistors considered, the base transit time is small and can be neglected, hence

$$\tau_{ec} = C_e/g_e + \frac{x_c}{2v_{sl}} \tag{53}$$

where x_c is the collector depletion-layer width and v_{sl} is the carrier scattering-limited velocity.

When we substitute Eqs. (52) and (53) into (51), and maximize the f_{max} with respect to the metal film thickness, we obtain

$$f_{max} = \frac{1}{4\pi S} \left[\frac{\alpha^*(3L_B/2e\rho_m \varepsilon_c)}{1/v_{sl} + 2C_e/g_e x_c} \right]^{1/2}. \tag{54}$$

A comparison of f_{max} versus emitter current density for the hot-electron transistors is shown in Fig. 20. It is clear that the highest-frequency performance should be obtainable by the SMST, followed by SCLT, and finally by the tunnel transistor; and that the only hot-electron transistor which has the potential to give high-frequency performance superior to that of the junction transistor is the SMST. In the next section we shall consider in more detail the limitations and the usefulness of the SMST.

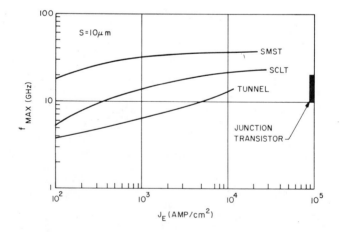

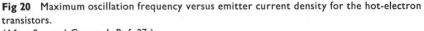

Fig 20 Maximum oscillation frequency versus emitter current density for the hot-electron transistors.
(After Sze and Gummel, Ref. 27.)

(3) Semiconductor-Metal-Semiconductor Transistor

Although the SMST has the potential to give a superior high-frequency performance, the small current gain α of an SMST severely limits its application in practical microwave circuits. The basic reasons that give rise to the low current gain are as follows: (1) the base transport loss due to electron-electron and electron-phonon scatterings, (2) the electron-phonon interaction in the emitter and collector semiconductors, and (3) the quantum-mechanical reflection at the metal-collector-semiconductor interface. The base transport factor is given by $\exp(-W_B/L_B)$ as in Eq. (52). The ballistic mean free path, L_B, is given by[28]

$$1/L_B = 1/l_p + 1/l_e + 1/l_i \tag{55}$$

where l_p is the electron mean free path for phonon scattering, l_e the electron-electron mean free path, and l_i the electron mean free path for impurity and defect scattering. To improve the base transport, we must use a metal with a long ballistic mean free path. In gold film it has been found that $L_B \simeq 220$ Å for 1-eV electrons at room temperature. The collection efficiency α_c, due to electron-phonon interaction in the collector semiconductor can be given, to a first approximation, as the probability that an electron can reach the potential energy maximum x_m (refer to Chapter 8, x_m is measured from the metal-semiconductor interface) without being scattered by optical phonons with mean free path λ; and α_c is given by[29]

$$\alpha_c = \exp(-x_m/\lambda). \tag{56}$$

From the dependence of x_m on electric field $\mathscr{E}$ due to the Schottky effect and the dependence of λ on temperature due to emission and absorption of phonons, we can write α_c as follows:

$$\alpha_c = \exp\{-\sqrt{q/(16\pi\varepsilon_s \mathscr{E})}/[\lambda_o \tanh(E_p/2kT)]\}. \tag{57}$$

It is clear from this equation that in order to improve the collection efficiency, we must apply a large field, and use a semiconductor with a long phonon mean free path (λ_o) and a large phonon energy (E_p). The third limitation, quantum-mechanical reflection at the collector, comes about because of the sudden change in potential as an electron travels from the metal to the semiconductor.[30]

The above effects are summarized[27] in Fig. 21. Curve (1) is for the quantum-mechanical transmission (QMT) which increases with increasing incoming

TABLE 11.3

EMITTER CHARACTERISTICS OF JUNCTION AND HOT-ELECTRON TRANSISTORS

Type of Transistor	Ideal I-V Characteristics	Emitter Capacitance, C_e, (f/cm²)	Emitter Conductance, $g_e = \partial J_E/\partial V_{EB}$ @ 1000 amp/cm²	Emitter Charging Time, C_e/g_e @ 1000 amp/cm² (sec)
Junction*	$J_E = J_s(e^{qV_{EB}/kT} - 1)$ $J_s \simeq \dfrac{qn_{po}D_n}{W_B}$	$\sqrt{\dfrac{q\varepsilon_s N_B}{2(V_{bi} - V_{EB})}}$	4×10^4	5.5×10^{-12}
Semiconductor-Metal-Semiconductor†	$J_E = J_s(e^{qV_{EB}/kT} - 1)$ $J_s = A^{**}T^2 e^{-q\phi_B/kT}$	$\sqrt{\dfrac{q\varepsilon_s N_E}{2(V_{bi} - V_{EB})}}$	4×10^4	4.5×10^{-12}
Tunnel‡	$J_E = qvn\exp\left\{-\dfrac{4\sqrt{2m^*}}{3}\dfrac{(q\phi_B)^{3/2}}{q\hbar}\dfrac{W_E}{V_{EB}}\right\}$	$\dfrac{\varepsilon_e}{W_E}$	1.4×10^4	2×10^{-10}
Space-Charge-Limited§	$J_E = \dfrac{9}{8}\mu\varepsilon_e V_{EB}^2/W_E^3$	$\dfrac{3}{2}\dfrac{\varepsilon_e}{W_E}$	10^3	1.7×10^{-11}

* $N_D = 10^{17}$ cm⁻³, $W_B = 1\mu m$

‡ $\phi_B = 1$V, $W_E = 20$Å, $\varepsilon_e/\varepsilon_0 = 4$

† $\phi_B = 0.8$ V, $N_E = 10^{16}$ cm⁻³

§ $\varepsilon_e/\varepsilon_0 = 10$, $\mu = 200$ cm²/V-sec, $W_E = 1\ \mu m$

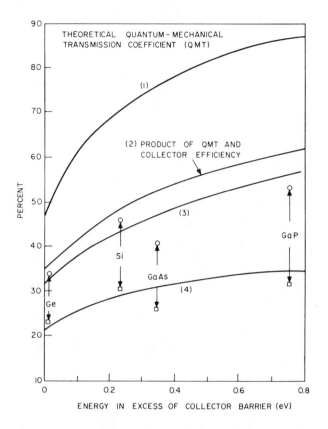

Fig. 21   Curve (1) is for quantum-mechanical transmission (QMT). Curve (2) combines QMT and the collector efficiency due to phonon effects. Curve (3) further incorporates the effect due to electron-phonon scattering in the emitter semiconductor. Curve (4) is the final result of the common-base current transfer ratio.
(After Sze and Gummel, Ref. 27.)

electron energy. The second curve is the combined result due to QMT and the collector efficiency. Curve (3) further incorporates the effect due to the electron-phonon scattering in the emitter semiconductor. The final result for the common-base current gain is given in the bottom curve, which includes the base transport loss through a 100 Å gold film. Also shown in Fig. 21 are some experimental results which are in reasonable agreement with the theoretical prediction. We note that the current gain is only of the order of 0.3. Because of this low value of α, the SMST shows current loss and can

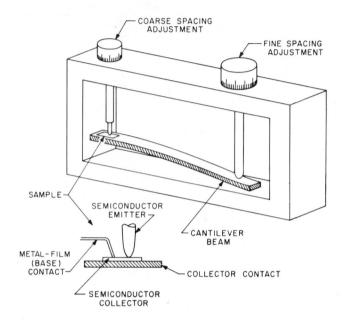

Fig. 22 Schematic diagram for the experimental setup of a semiconductor-metal-semi-conductor transistor.
(After Sze et al., Ref. 31.)

attain power gain only by means of impedance gain. The SMST with low α is also expected to have large partition noise. For $\alpha = 0.3$, the minimum noise figure is found[27] to be 7.9 dB [refer to Eq. (56) of Ch. 6]. In addition to the above drawbacks, we have the unsolved technological problem of deposition of device quality semiconductor materials on thin metal films.

It is thus unlikely that the junction transistor can be replaced by any of the proposed hot-electron transistors. The hot-electron transistors are nevertheless useful devices in the sense that they can be used to study fundamental physical parameters such as the hot-electron lifetime and the transport properties through thin films.

A schematic diagram[31] is shown in Fig. 22 for the experimental setup of SMST. The cantilever beam is used as a stable micromanipulator capable of incremental adjustments of the order of 10 Å to make a nondestructive contact between a smoothly pointed and freshly cleaned semiconductor needle and a thin metal film evaporated on polished semiconductor substrates. The substrate served as the collector semiconductor, the metal film as the base, and the needle as the emitter semiconductor.

The collector current versus voltage is then measured with the emitter current as a parameter.[31] A typical collector characteristic is shown in Fig. 23. The emitter-collector incremental current gains ($\alpha = \partial I_C / \partial I_E$) are then obtained. Figure 24 shows the current gain as a function of gold film thickness for a GaP emitter point on Au-Ge, -Si, -GaAs, -CdSe, and -CdS collector barriers. As expected from Eq. (52), there is a linear relationship between $\ln\alpha$ and W_B. The slope can be used to determine the ballistic mean free path, L_B. We note that even when $W_B = 0$, $\alpha < 1$, and the slopes are essentially independent of the collector semiconductors. These results are in agreement with the previous discussions; because of the phonon scattering and quantum-mechanical reflection at the collector barrier, the overall collection efficiency is reduced. Among the five collector semiconductors, Ge gives the highest α at any given metal thickness. This is because Ge has a very large optical-phonon mean free path (~ 65 Å), and the energy difference between the emitter (Au-GaP) and collector (Au-Ge) barrier height is large enough (~ 0.7 eV) that the quantum-mechanical reflection is small.

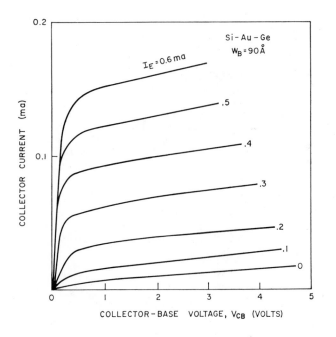

Fig. 23 Collector characteristics of a Si-Au-Ge SMST with 90 Å Au film. (Ref. 31.)

Figure 25 shows similar results obtained at 105°K and 298°K for a Si point emitter in contact with Au-Si and Au-Ge Schottky diodes with varying metal film thicknesses.[28] The measured L_B is 300 Å $\pm$ 10 Å at 298°K and 370 Å $\pm$ 30 Å at 105°K. The temperature dependence of L_B is mainly due to l_p, the electron mean free path of phonon scattering. If we assume that l_p has its bulk values (406 Å at 298°K and 1150 Å at 105°K), the predicted L_B at 105°K is 360 Å $\pm$ 20 Å and is in good agreement with the experimental result. Another interesting result in Fig. 25 is that the values of α^* (at $W_B = 0$) increase with decreasing temperature. This is mainly due to the increase of the optical phonon mean free paths in the emitter and collector semiconductors at lower temperatures.

To study the energy dependence of L_B, we shall briefly discuss the electron-phonon and electron-impurity scattering processes. For free electrons an electron-phonon mean free path proportional to the square of the electron

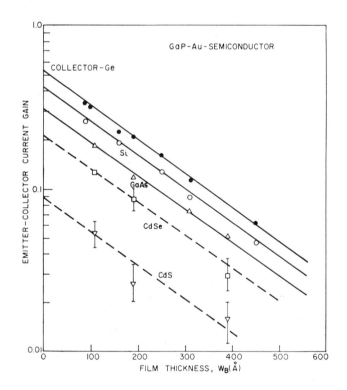

Fig. 24 Current gain versus the gold film thickness for a GaP emitter point on Au-Ge, —Si, —GaAs, —CdSe, —CdS collector barriers.

kinetic energy has been predicted by Wilson.[32] In most of the metal films there are, however, additional scattering mechanisms such as the electron-defect and electron-impurity interactions which are found by Mott[33] to have an energy dependence similar to that of the electron-phonon interaction. Thus when carriers in a single parabolic band are involved, one can combine Wilson's and Mott's results to relate the hot-electron mean free path for lattice-related scattering mechanisms, l_c, to the electron mean free path for electrical conductivity, l_σ. Then

$$l_c = l_\sigma (E/E_F)^2 \tag{58}$$

where the electron energy E is measured from the bottom of the conduction band. It can be shown from Eqs. (48), (55), and (58) that the following relationship between the ballistic mean free path L_B and the electron energy is expected[23]

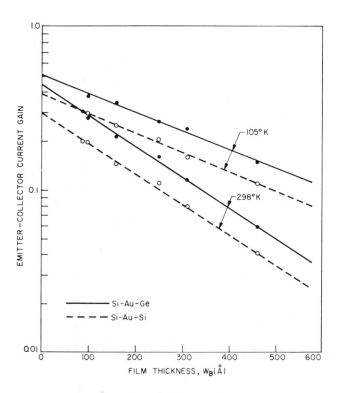

Fig. 25   Current gains versus the gold film thickness for Si-Au-Ge and Si-Au-Si structures. (After Crowell and Sze, Ref. 28.)

$$1/L_B \equiv 1/l_c + 1/l_e = (E_F/E)^2/l_\sigma + \frac{1}{\xi}(E - E_F)^2(E_F/E) \qquad (59)$$

or

$$(E/E_F)^2/L_B = 1/l_\sigma + \frac{1}{\xi}(E - E_F)^2(E/E_F) \qquad (60)$$

where ξ is a constant of proportionality which includes parameters such as r_s.

From Eq. (60) a plot of $(E/E_F)^2/L_B$ versus $(E - E_F)^2/(E/E_F)$ should yield a straight line. Figure 26 shows the mean free paths reported for GaP, GaAs,

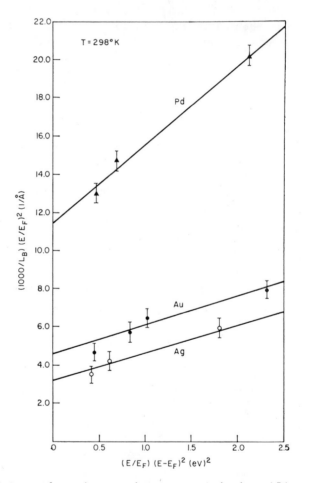

Fig 26   Ballistic mean free paths versus electron energy in Au, Ag, and Pd. (After Crowell and Sze, Ref. 23.)

Si, and Ge point emitters on Au films and GaP, Si, and Ge point emitters on Ag and Pd films.[31] These results are in reasonable agreement with Eq. (60). The L_B values and the lifetimes deduced from the parameters of best fit for electrons 1 eV above the Fermi level are listed in Table 11.4. The quantity l_σ which is obtained from Fig. 26 for $(E - E_F) = 0$ are comparable to the mean free path determined from the electrical conductivity measurement. The measured hot-electron lifetimes are also in reasonable agreement with the predicted values.

It is clear from the above discussions that the SMST is a useful device for studying the hot-electron transport in thin metal films. In addition it can give qualitative results concerning the electron-phonon interactions in the semiconductors and the quantum-mechanical reflection at the metal-semiconductor interface.

TABLE 11.4

ELECTRON MEAN FREE PATHS AND LIFETIMES

Metal	$(E - E_F) = 1$ eV		$E = E_F$	
	L_B(Å)	$\tau(10^{-14}$ sec)	l_σ(Å)	l_p(Å)
Ag	265	6.8	310	570
Au	220	7.0	220	406
Pd	87	3.9	88	110

(4) Tunnel Emission Devices

If the collector insulator-metal assembly as shown in Fig. 16 is replaced by a vacuum, we have a simplified metal-insulator-metal tunnel-emission diode structure. Its band diagram is shown in Fig. 27(a). This structure has been used extensively to study the transport mechanisms of hot electrons in both the thin metal films and the thin insulating films. A schematic experimental setup is shown in Fig. 27(b). A metal (e.g., Al) is first deposited on a glass substrate. Next, an insulating film (e.g., Al_2O_3) of the order of 50 Å is formed or deposited on the metal. The " base " metal overlayer film (e.g., Au) with a thickness of the order of 100 Å is then deposited onto the insulator. When a voltage, V_a, larger than the metal-vacuum work function (ϕ_M) is applied between the metals, electron emission into the vacuum can occur, Fig. 27(a).

Figure 28 shows a typical experimental result of Al—Al_2O_3—Au structures.[34] The current ratio $I_1/(I_1 + I_2)$ is plotted against the Au film thickness for an applied bias of 6 volts where I_1 is the emitted current which is collected

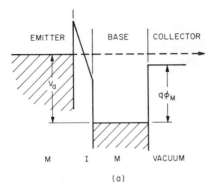

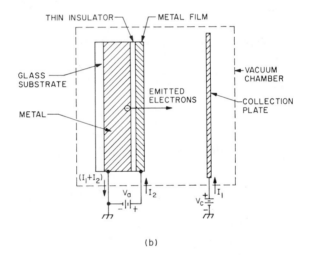

Fig. 27
(a) Energy band diagram of a metal-insulator-metal tunnel emission structure.
(b) Schematic experimental setup.

by the collection plate and I_2 is the diode current. We note that typically less than 1 % of the electrons which reach the metal base escape into the vacuum. We also note the strong increase in emission current when a monolayer of barium is deposited on the gold surface to lower the surface work function from 4.8 volts to about 2.7 volts. From Fig. 28 the hot-electron ballistic mean free path is obtained to be about 50 Å for electrons in gold with an energy of 6 eV.

To study the effect of the work function and the transport losses in the insulator and the metal films, we shall use a more detailed schematic energy band diagram under emitting condition as shown in Fig. 29. Consider first the energy spectrum of the electrons injected into the conduction band of the insulator. The applied voltage necessary to produce emission over the surface potential barriers (ϕ_m) of the gold film is larger than the internal barriers ϕ_1 and ϕ_2 by an amount sufficient to permit use of the simple Fowler-Nordheim equation for the tunneling process.[34a] Using this approximation, the half-width $\delta(\frac{1}{2})$ of the total energy distribution $N_1(E)$ at $0°K$ is given by[35]

$$\delta(\tfrac{1}{2}) = \frac{10^{-8}\mathscr{E}\ln 2}{\sqrt{q\phi_1}} \quad (eV) \tag{61}$$

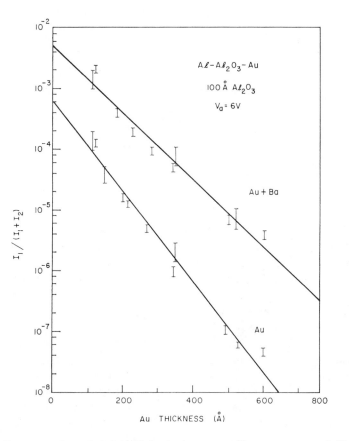

Fig. 28 Experimental results of Al-Al$_2$O$_3$-Au structures. The current ratio $I_1(I_1 + I_2)$ is plotted against the Au film thickness.
(After Kanter, Ref. 34.)

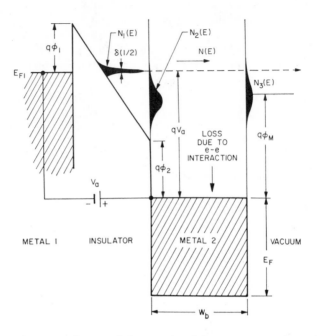

Fig. 29 Schematic energy diagram of the tunnel emission structure under emitting condition. Superimposed on the diagram are the electron energy distributions. (After Handy, Ref. 35.)

where $\mathscr{E}$ is the field in V/cm, and ϕ_1 is the barrier height. A typical operating field is in the range of 10^6 to 10^7 V/cm, which gives for $\phi_1 = 2$V, a half-width range of $\delta(\frac{1}{2}) = 0.005$ to 0.08 eV. At higher temperatures the half-width will be widened, e.g., at 10^7 V/cm and 300°K, $\delta(\frac{1}{2}) \simeq 0.15$ eV. Thus after tunneling into the conduction band of the insulator, the electron energy distribution is very narrow and can be effectively described by a delta function. This distribution is shown schematically in Fig. 29.

Once electrons have tunneled to the conduction band in the insulator, the generation of optical phonons occurs as the electrons gain kinetic energy from the electric field. As shown in Fig. 29, the electron energy distribution spreads as electrons traverse the insulator, and becomes $N_2(E)$ at the insulator-metal interface. Since the hot-electron transport in the second metal film is essentially ballistic, i.e., any scattering event will remove the electron permanently, the electron distribution $N_3(E)$ at the metal-vacuum interface is given by

$$N_3(E) \simeq N_2(E)\exp(-W_B/L_B). \tag{62}$$

The electrons that can be emitted into the vacuum are those with energies above the metal work function (ϕ_m):

$$I_1 \sim \int_{q\phi_m}^{\infty} N_2(E)\exp(-W_B/L_B)\,dE = \exp(-W_B/L_B)\int_{q\phi_m}^{\infty} N_2(E)\,dE.$$

(63)

From Eqs. (62) and (63) we note that the measured energy distribution of the emitter electrons actually corresponds to the distribution at the insulator-metal interface apart from a constant factor, $\exp(-W_B/L_B)$. The experimental results of the normal energy distributions for Al—Al$_2$O$_3$—Au structures are shown in Fig. 30 for four different oxide thicknesses.[36] It is obvious that the peak energy of the energy distribution falls well below the peak energy of the injected distributions, and that the half-widths are larger than $\delta(\tfrac{1}{2})$, Eq. (61), by roughly an order of magnitude. This indicates the strong scattering effect in the insulator. The dependence of the peak energy on the insulator thickness is plotted in Fig. 31 for similar gold film thicknesses and electric fields. The slope is about 0.03 eV/Å in Al$_2$O$_3$. Since the Debye temperature of aluminum oxide[37] is 1200 to 1500°K, the optical-phonon energy should be $\sim$0.1 eV.

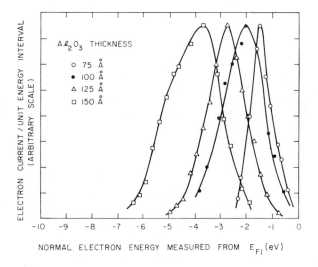

Fig. 30 Experimental results of the normal energy distributions for Al-Al$_2$O$_3$-Au structures for four different oxide thicknesses.
(After Kanter and Feibelman, Ref. 36.)

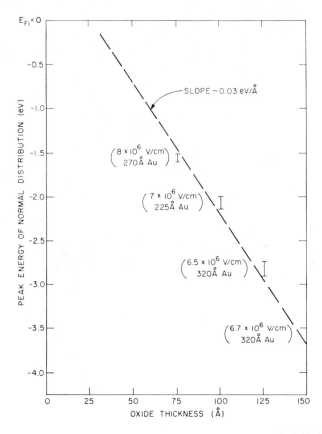

Fig. 31 Dependence of the peak energy on the insulator thickness of Al-Al$_2$O$_3$-Au structures with similar gold film thicknesses and electric fields.
(After Handy, Ref. 35.)

Consequently the hot-electron mean free path in Al$_2$O$_3$ is about 3 Å [0.1 eV/(0.03 eV/Å)]. This value is in reasonable agreement with the result obtained from a transistor structure of Al—Al$_2$O$_3$—Al—Al$_2$O$_3$—Al films[38] (Fig. 16).

The scatterings in the insulator thus cause a rapid degradation of energy in the electron distribution and tend to randomize the electron momentum distribution before the electrons are injected into the metal base. This makes it very difficult for electron emission to occur with appreciable efficiency. And this is the main reason that the emitted electrons from such structures are typically 1 % of the electrons which reach the metal base.

A comparison[23] of the theoretical hot-electron lifetime in gold and the experimental results is shown in Fig. 32. The experimental data are obtained from the transport measurements in SMST, tunnel emission structures, and Schottky barriers.[39–39c] The results shown in Fig. 32 indicate that there is general agreement between the lifetimes deduced from the measurements of various hot-electron structures. Over the energy range 0.1 to 10 eV the measurements are also in general agreement with the theoretical expectations, but the theoretical results cannot be expected to reflect the details of the band structure far from the Fermi level. When the d-band shielding effect[40] and the exchange effect[41] due to the antisymmetry of the wave functions are included, the lifetime is expected to be longer. This will bring closer the theoretical predictions and the experimental results.

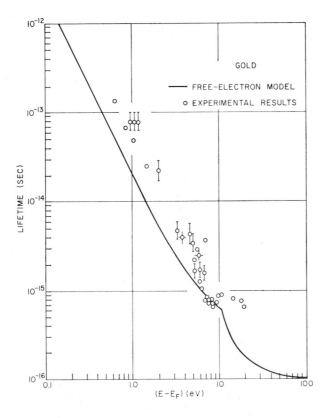

Fig. 32 Comparison of theoretical (solid line) and experimental lifetimes in gold. (After Crowell and Sze, Ref. 23.)

4 METAL-INSULATOR-METAL STRUCTURE

Another related thin-film device is the metal-insulator-metal (MIM) structure in which the electrons from the first metal can tunnel[42] and/or thermionically emit[43,44] into the insulator film and can be collected by the second metal. The MIM structure is different from the tunnel emission devices discussed previously in that the second metal of the MIM is very thick and we are not concerned with the hot-electron transport in the second metal.

In this section we shall be mainly concerned with the tunnel effect. The basic energy band diagrams of MIM with similar metal electrodes are shown in Fig. 33 under three biasing conditions. If the barrier height ϕ_0 is large enough and the insulator thickness d is small enough, the dominant mechanism of current transport is tunneling. The tunneling probability is given by the WKB approximation:

$$T_t(E_x) = \exp\left\{-\frac{4\pi}{h}\int_{d_1}^{d_2}\sqrt{2m[qV(x) - E_x]}\,dx\right\} \tag{64}$$

where $E_x \equiv mv_x^2/2$ and is the energy component of the incident electron in the normal direction. The number N_1 of electrons tunneling from electrode 1 to electrode 2 is given by[42]

$$N_1 = \frac{4\pi m^2}{h^3}\int_0^{E_m} T_t(E_x)\,dE_x \int_0^\infty F(E)\,dE_r \tag{65}$$

where E_m is the maximum energy of the electrons in the electrode, $F(E)$ is the Fermi-Dirac function, and E_r is the energy associated with the transverse velocities or $E_r \equiv m(v_y^2 + v_z^2)/2$. The number N_2 of electrons tunneling from electrode 2 to electrode 1 is given by an expression similar to Eq. (65) except that the function $F(E)$ is replaced by $F(E + qV)$ where V is the applied voltage. The net flow of current through the barrier is then

$$J \sim qN = q(N_1 - N_2) = q\int_0^{E_m} T_t(E_x)\,dE_x\left\{\frac{4\pi m^2}{h^3}\int_0^\infty [F(E) - F(E + qV)]\,dE_r\right\}. \tag{66}$$

At $0°K$, for an arbitrary barrier, the above equation is simplified to[42]

$$J = J_0[\bar{\phi}\exp(-A\sqrt{\bar{\phi}}) - (\bar{\phi} + V)\exp(-A\sqrt{\bar{\phi} + V})] \tag{66a}$$

where

$$J_0 \equiv q^2/[2\pi h(\Delta d)^2]$$

$$A \equiv 4\pi(\Delta d)\sqrt{2mq}/h$$

and $\bar{\phi}$ is the mean barrier height above the Fermi level as shown in Fig. 34(a). Equation (66a) can be interpreted as a current density $J_0 \bar{\phi} \exp(-A\sqrt{\bar{\phi}})$ flowing from electrode 1 to electrode 2 and a current density $J_0(\bar{\phi} + V)\exp(-A\sqrt{\bar{\phi} + V})$ flowing from electrode 2 to electrode 1.

We now apply Eq. (66a) to the ideal symmetrical MIM structure as shown in Fig. 33. By ideal we mean that the temperature effect, the image-force

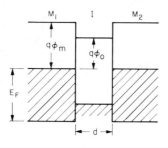

(a) V = 0

(b) V < ϕ_0

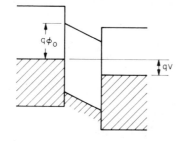

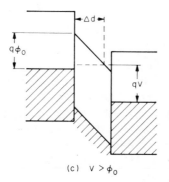

(c) V > ϕ_0

Fig. 33 Energy band diagrams of metal-insulator-metal structures with similar metal electrodes.

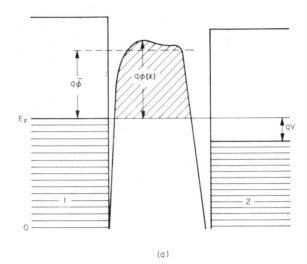

(a)

Fig. 34
(a) General barrier of a metal-insulator-metal structure (MIM).

effect, and the field-penetration effect in metal electrodes are neglected. For $0 \leq V \leq \phi_0$, $\Delta d = d$, and $\bar{\phi} = \phi_0 - V/2$, the current density is given by

$$J = J_0\{(\phi_0 - V/2)\exp(-A\sqrt{\phi_0 - V/2}) - (\phi_0 + V/2)\exp(-A\sqrt{\phi_0 + V/2})\}$$

(67)

For larger voltage, $V > \phi_0$, we have $\Delta d = d\phi_0/V$, and $\bar{\phi} = \phi_0/2$. The current density is then[42]

$$J = \frac{q^2 \mathscr{E}^2}{4\pi h \phi_0} \{\exp(-\mathscr{E}_0/\mathscr{E}) - (1 + 2V/\phi_0)\exp(-\mathscr{E}_0\sqrt{1 + 2V/\phi_0}/\mathscr{E})\}$$

(68)

where $\mathscr{E} = V/d$ is the field in the insulator and $\mathscr{E}_0 \equiv \frac{8}{3}\sqrt{\pi}q(\phi_0)^{3/2}$. For very high voltage such that $V > (\phi_0 + E_F/q)$, the second term in Eq. (68) can be neglected, and we have the well-known Fowler-Nordheim equation. The computed results for the tunnel resistance (J/V) are shown in Fig. 34(b) where various barrier heights and insulator thicknesses are used. We note the tunnel resistance decreases rapidly with increasing applied voltage.

For an ideal asymmetrical MIM structure (shown in Fig. 35) in the low-voltage range $0 < V < \phi_1$, the quantities $\Delta d = d$, and $\bar{\phi} = (\phi_1 + \phi_2 - V)/2$ are independent of the polarities. Thus the J-V characteristics are also independent of the polarity. At higher voltages, $V > \phi_2$, the energy diagram

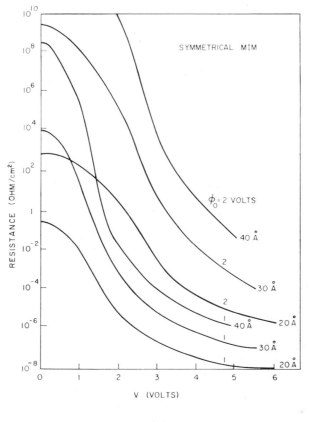

Fig. 34
(b) Computed results for the tunnel resistance of a symmetrical MIM structure.
(After Simmons, Ref. 42.)

for the "reverse-biased" condition is shown in 35(a) and $\bar{\phi} = \phi_1/2$, $\Delta d = d\phi_1/(V - \phi_1 - \phi_2)$. For the "forward-biased" condition, $\bar{\phi} = \phi_2/2$ and $\Delta d = d\phi_2/(V - \phi_1 - \phi_2)$. Thus the currents for different polarities are different. It follows then that the current shows rectifying characteristics in the range $V > \phi_2$. Figure 35(c) illustrates the tunnel resistance as a function of V for $d = 20$, 30, and 40 Å, $\phi_2 = 2$ volts, and $\phi_1 = 1$ volt. The reverse and forward directions are depicted by the full and chain-dotted curves respectively. Initially the resistance is smaller in the forward direction; at higher

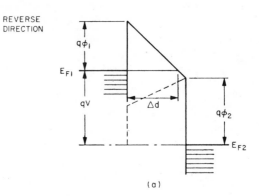

(a)

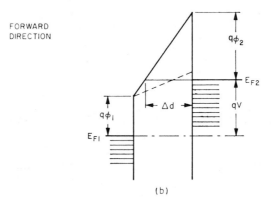

(b)

Fig. 35
(a) Energy diagram of a reverse-biased MIM.
(b) Energy diagram of a forward-biased MIM.

voltages, however, the forward and reverse characteristics cross over. A comparison of the theory with experimental results is shown in Fig. 36. The functional dependence of J on V shows excellent agreement over nine decades of current.[45] The effective area used for the experimental data in Fig. 35 is approximately 1% of the electrode area.

This small tunneling area can be explained by the statistical nature in the formation of the insulating film.[46] Only the thinnest portion in the film is responsible for the tunneling current. Because of this statistical fluctuation of the thickness, the capacitance of the MIM structure is always larger than that calculated based on an average thickness of the insulator film.[46,47] Another

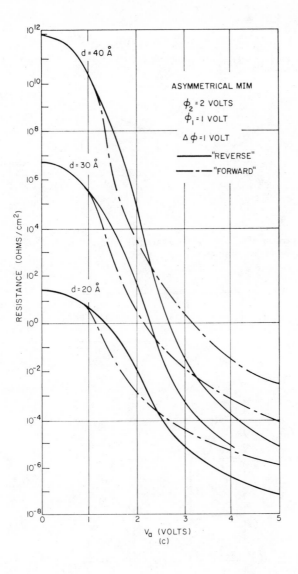

Fig. 35
(c) Theoretical tunnel resistance as a function of the applied voltage for an asymmetrical
   MIM structure.
(After Simmons, Ref. 42.)

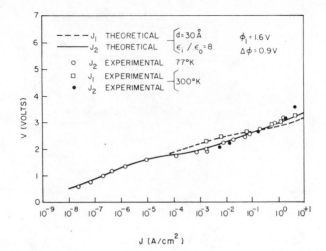

Fig. 36 Comparison of theory and experimental results of Al-Al₂O₃-Al structures. (After Pollack and Morris, Ref. 45.)

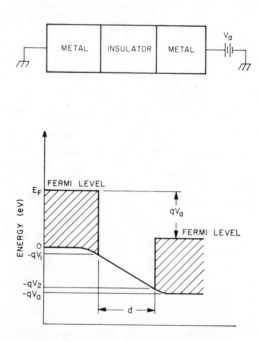

Fig. 37 Penetration of electric field into the metal in an MIM structure. (After Ku and Ullman, Ref. 48.)

factor which can also affect the capacitance value is the contribution from the potential distribution in the metal electrodes. Because of the penetration of the electric field into the metal as shown in Fig. 37, the total capacitance is effectively equal to two capacitances in series:[48]

$$1/C = d/\varepsilon_i + 2.3\lambda_m/\varepsilon_m \tag{69}$$

where ε_i and ε_m are the permittivities of the insulator and the metal respectively and λ_m is the characteristic penetration length in the metal (~ 0.5 Å for the noble metals). From Eq. (69) we note that the capacitance due to the second term is of the order of 10^6 pf/cm^2. The total capacitance per unit area will be much smaller than the value ε_i/d when d approaches a value of the order of 5 Å.

REFERENCES

1. P. K. Weimer, "An Evaporated Thin-Film Triode," IRE-AIEE Solid State Device Research Conference, Stanford, California (July 1961), and also "The TFT—A New Thin-Film Transistor," Proc. IRE, *50*, 1462 (1962).

2. P. K. Weimer, "The Insulated-Gate Thin-Film Transistor," a chapter of *Physics of Thin Films, Vol. 2* edited by G. Hass and R. E. Thun, Academic Press (1964).

3. C. A. Mead, "Tunnel-Emission Amplifiers," Proc. IRE, *48*, 359 (1960).

4. N. F. Foster, "Piezoelectric and Piezoresistive Properties of Films," a chapter of *Handbook of Thin Film Technology*, McGraw-Hill Co., New York (1968).

5. P. E. Bode, "Lead Salt Detectors," a chapter of *Physics of Thin Films*, Vol. 3, edited by G. Hass and R. E. Thun, Academic Press (1966).

6. H. Borkan and P. K. Weimer, "An Analysis of the Characteristics of Insulated-Gate Thin-Film Transistors," RCA Review, *24*, 153 (1963).

6a. H. K. J. Ihantola and J. L. Moll, "Design Theory of a Surface Field-Effect Transistor," Solid State Electron., *7*, 423 (1964).

7. J. F. Skalski, "A PbTe Single-Crystal Thin-Film Transistor," Proc. IEEE, *53*, 1792 (1965).

7a. P. K. Weimer, "A *p*-Type Tellurium Thin-Film Transistor," Proc. IEEE, *52*, 608 (1964).

7b. T. P. Brody and H. E. Kunig, "A High-Gain InAs Thin-Film Transistor," Appl. Phys. Letters, *9*, 259 (1966).

7c. W. B. Pennebaker, "PbS Thin-Film Transistors," Solid State Electron, *8*, 509 (1965).

7d. F. V. Shallcross, "Cadmium Selenide Thin-Film Transistors," Proc. IEEE, *51*, 851 (1963).

7e. D. Darmagna and J. Reynand, "A GaAs Thin-Film Transistor," Proc. IEEE, *54*, 2020 (1966).

7f. V. L. Frantz, "Indium Antimonide Thin-Film Transistors," Proc. IEEE, *53*, 760 (1965).

8. T. O. Poehler and D. Abraham, "Electric Field Excitation of Electrons from Shallow Traps in CdSe Thin-Film Triodes," J. Appl. Phys., *35*, 2452 (1964).

9. T. J. O'Reilly, "Effect of Surface Traps on Characteristics of Insulated-Gate Field-Effect Transistors," Solid State Electron., *8*, 267 (1965).

10. C. A. T. Salama and L. Young, "Evaporated Silicon Thin-Film Transistors," Solid State Electron., *10*, 473 (1967).

11. R. R. Haering, "Theory of Thin-Film Transistor Operation," Solid State Electron., *7*, 31 (1964).

12. M. G. Miksic, E. S. Schlig, and R. R. Haering, "Behavior of CdS Thin-Film Transistors," Solid State Electron., *7*, 39 (1964).

13. H. Koelmans and H. C. DeGraaff, "Drift Phenomena in CdSe Thin Film FET's," Solid State Electron., *10*, 997 (1967).

14. D. J. Page, "Thermal Limitations of the Thin Film Transistor," Solid State Electron., *11*, 87 (1968).

15. J. J. Quinn, "Range of Excited Electrons in Metals," Phys. Rev., *126*, 1453 (1962).

16. N. F. Mott and H. Jones, *The Theory of the Properties of Metals and Alloys*, Dover Publication, New York (1958).

17. K. Matizuki and M. Sparks, "Range of Excited Electrons and Holes in Metals and Semiconductors," J. Phys. Society of Japan, *19*, 486 (1964).

18. J. P. Spratt, R. F. Schwartz, and W. M. Kane, "Hot Electrons in Metal Films: Injection and Collection," Phys. Rev. Letters, *6*, 341 (1961).

19. A. Rose, "Interim Rept.," No. 6A RCA (June 1960); Govt. Contract Rept. No. bsr 77523. Supported in part by the U.S. Navy, September 1960.

20. M. M. Atalla and D. Kahng, "A New Hot Electron Triode Structure with Semiconductor-Metal Emitter," IRE-AIEE Solid State Device Research Conference, University of New Hampshire (July 1962).

21. D. V. Geppert, "A Metal-Base Transistor," Proc. IRE, *50*, 1527 (1962).

22. G. T. Wright, "The Space-Charge-Limited Dielectric Triode," Solid State Electron., *5*, 117 (1962).

23. C. R. Crowell and S. M. Sze, "Hot Electron Transport and Electron Tunneling in Thin-Film Structures," a chapter of *Physics of Thin Films*, Vol. 4, edited by G. Hass and R. E. Thun, Academic Press (1967).

24. J. L. Moll, "Comparison of Hot Electrons and Related Amplifiers," IEEE Trans. Electron Devices, *10*, 299 (1963).

25. M. M. Atalla and R. W. Soshea, "Hot-Carrier Triodes with Thin-Film Metal Base," Solid State Electron., *6*, 245 (1963).

26. J. M. Early, "Structure Determined Gain-Band Product of Junction Triode Transistors," Proc. IRE, *46*, 1924 (1958).

27. S. M. Sze and H. K. Gummel, "Appraisal of Semiconductor-Metal-Semiconductor Transistor," Solid State Electron., *9*, 751 (1966).

28. C. R. Crowell and S. M. Sze, "Ballistic Mean Free Path Measurements of Hot Electrons in Au Films," Phys. Rev. Letters, *15*, 659 (1965).

29. C. R. Crowell and S. M. Sze, "Electron-Optical-Phonon Scattering in the Emitter and Collector Barriers of Semiconductor-Metal-Semiconductor Structures," Solid State Electron., *8*, 979 (1965).

30. C. R. Crowell and S. M. Sze, "Quantum-Mechanical Reflection at Metal-Semiconductor Barrier," J. Appl. Phys., *37*, 2683 (1966).

31. S. M. Sze, C. R. Crowell, G. P. Carey, and E. E. Labate, "Hot-Electron Transport in Semiconductor-Metal-Semiconductor Structures," J. Appl. Phys., *37*, 2690 (1966).

32. A. H. Wilson, *Theory of Metals*, p. 264, Cambridge Univ. Press, London and New York (1955).

33. N. F. Mott, Proc. Cambridge Phil. Soc., *32*, 281 (1936). Also N. F. Mott and H. Jones, Ref. 16.

34. H. Kanter, "Slow Electron Transfer Through Evaporated Au Films," J. Appl. Phys. *34*, 3629 (1963).

34a. R. H. Fowler and L. Nordheim, "Electron Emission in Intense Electric Fields," Proc. Roy. Soc., *119*, 173 (1928).

35. R. M. Handy, "Hot Electron Energy Loss in Tunnel Cathode Structures," J. Appl. Phys., *37*, 4260 (1966).

36. H. Kanter and W. A. Feibelman, "Electron Emission for Thin Al-Al_2O_3-Au Structures," J. Appl. Phys., *33*, 3580 (1962).

37. A. Goldsmith, T. Waterman, and H. Hirshorn, *Handbook of Thermophysical Properties of Solid Materials*, Vol. III, p. 35, Macmillan Co., New York (1961).

38. O. L. Nelson and D. E. Anderson, "Hot-Electron Transfer Through Thin-Film Al-Al_2O_3 Triodes," J. Appl. Phys., *37*, 66 (1965).

39. R. W. Soshea and R. C. Lucas, "Attenuation Length of Hot Electrons in Gold," Phys. Rev., *138*, A1182 (1965).

39a. R. Stuart, F. Wooten, and W. E. Spicer, "Monte Carlo Calculations Pertaining to the Transport of Hot Electrons in Metals," Phys. Rev., *135*, A495 (1964).

39b. C. R. Crowell, W. G. Spitzer, L. E. Howarth, and E. E. Labate, "Attenuation Length Measurements of Hot Electrons in Metal Films," Phys. Rev., *137*, 2006 (1962).

39c. S. M. Sze, J. L. Moll, and T. Sugano, "Range-Energy Relation of Hot Electrons in Gold," Solid State Electron., *7*, 509 (1964).

40. J. J. Quinn, "The Range of Hot Electrons and Holes in Metals," Appl. Phys. Letters, *2*, 167 (1963).

41. R. H. Ritchie and J. C. Ashley, "The Interaction of Hot Electrons with a Free Electron Gas," J. Phys. Chem. Solids, *26*, 1689 (1965).

42. J. G. Simmons, "Generalized Formula for the Electric Tunnel Effect between Similar Electrodes Separated by a Thin Insulating Film," J. Appl. Phys., *34*, 1793 (1963).

43. J. G. Simmons, "Potential Barriers and Emission-Limited Current Flow between Closely Spaced Parallel Metal Electrodes," J. Appl. Phys., *35*, 2472 (1964).

44. S. M. Sze, "Current Transport and Maximum Dielectric Strength of Silicon Nitride Films," J. Appl. Phys., *38*, 2951 (1967).

45. S. R. Pollack and C. E. Morris, "Tunneling through Gaseous Oxidized Films of Al_2O_3," Trans. AIME, *233*, 497 (1965).

46. Z. Hurych, "Influence of Nonuniform Thickness of Dielectric Layers on Capacitance and Tunnel Currents," Solid State Electron., *9*, 967 (1966).

47. J. Pochobradsky, "On the Capacitance of Metal-Insulator-Metal Structures with Nonuniform Thickness," Solid State Electron., *10*, 973 (1967).

48. H. Y. Ku and F. G. Ullman, "Capacitance of Thin Dielectric Structures," J. Appl. Phys., *35*, 265 (1964).

PART IV

OPTOELECTRONIC DEVICES

- Optoelectronic Devices
- Semiconductor Lasers

■ INTRODUCTION

■ ELECTROLUMINESCENT DEVICES

■ SOLAR CELL

■ PHOTODETECTORS

12

Optoelectronic Devices

1 INTRODUCTION

Optoelectronic devices include those which convert electrical energy into optical radiation, or vice versa, and those which detect optical signals through electronic processes. One of the most important optoelectronic devices is the semiconductor laser. Because of its importance and its involved lasing phenomenon, we shall devote Chapter 13 entirely to the physics of lasers.

In this chapter we shall first consider electroluminescent devices which emit incoherent optical radiation when a current or voltage is applied.The emitted light is due to spontaneous emission with a wide spectral linewidth (typically ~100 Å) in contrast to the narrow linewidth of a laser output (~0.1 Å). These devices can be used as tools to study impurity or other energy levels in semiconductor band gaps. However their practical applications are mainly in optical coupling, optical display, and illumination.

We next consider photovoltaic devices which convert optical radiation to electrical energy. The most important photovoltaic device is the solar cell. At the present time the solar-cell power plant furnishes the most important long-duration power supply for satellites and space vehicles. In Section 3 we shall consider the characteristics of solar cells, especially the radiation effect due to electron and proton bombardments.

In Section 4 we shall discuss three classes of photodetectors: photoconductors which are mainly used for infrared detection, depletion-layer photodiodes which are useful for high-speed coherent and incoherent detection, and avalanche photodiodes which are promising solid-state detectors because of their internal current gain.

2 ELECTROLUMINESCENT DEVICES

Luminescence is the emission of optical radiation (ultraviolet, visible, or infrared) as a result of electronic excitation of a material, excluding any radiation which is the result purely of the temperature of the material (incandescence). Figure 1 is a chart of the electromagnetic spectrum.[1] Although radiation of different wavelengths must be excited by different methods all are alike as far as fundamentals are concerned. The visual range of the human eye extends only from about 0.4 μm to 0.7 μm. The infrared region extends from 0.7 μm to about 200 μm; and the ultraviolet region includes wavelengths from 0.4 μm to about 0.002 μm (20 Å). In this and the subsequent chapter, we are primarily interested in the wavelength range from near infrared (~ 1 μm) to near ultraviolet (~ 0.1 μm).

Types of luminescence may be distinguished in accordance with the source of the input energy:[2a] (1) photoluminescence involving excitation by optical radiation, (2) cathodoluminescence by electron beams or cathode ray, (3) radioluminescence by other fast particles or high-energy radiation, and (4) electroluminescence by electric field or current.

We shall in this section be mainly concerned with the physics of electroluminescence and of semiconductor electroluminescent devices such as the GaAs infrared source, useful for optical coupling; and the GaP diode lamp, useful for applications where the human eye is the detector.

(1) Radiative Transitions

The electronic transitions which follow the excitation and which result in luminescent emission are generally the same for the various types of excitations. Figure 2 shows a schematic diagram of the basic transitions in a semiconductor or insulator. These may be classified as follows:[3]

A. Transitions involving chemical impurities or physical defects (lattice vacancies, etc.): (a) conduction band to acceptor, (b) donor to valence band, and (c) donor to acceptor (pair emission).

B. Interband transitions: (a) intrinsic or edge emission corresponding very closely in energy to the band gap, where phonons and/or excitons may be involved, and (b) higher energy emission involving energetic or hot carriers, sometimes related to avalanche emission.

C. Intraband transitions involving hot carriers, sometimes called deceleration emission.

It should be pointed out that not all the transitions can occur in the same material or under the same conditions. Nor are all electronic transitions radiative. An efficient luminescent material is one in which radiative transi-

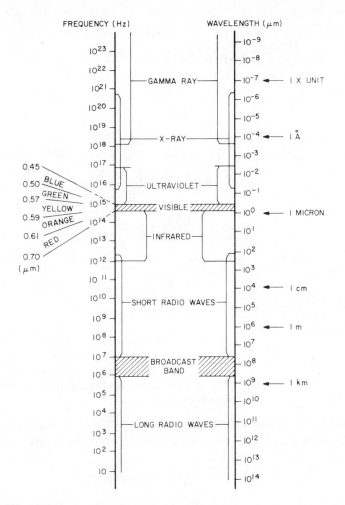

FREQUENCY (Hz)   WAVELENGTH (μm)

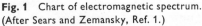

Fig. 1 Chart of electromagnetic spectrum.
(After Sears and Zemansky, Ref. 1.)

tions predominate over nonradiative ones (such as Auger nonradiative recombination).[2b]

(2) Emission Spectra

When electron-hole pairs are generated by external excitations, radiative transitions resulting from the hole-electron recombination may occur. The

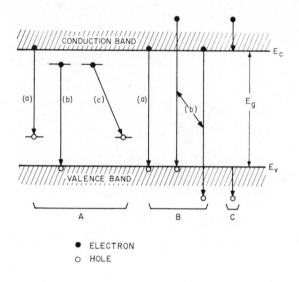

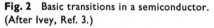

Fig. 2 Basic transitions in a semiconductor.
(After Ivey, Ref. 3.)

radiative transitions in which the sum of electron and photon wave vectors is conserved are called direct transitions as against indirect transitions which involve other scattering agents such as phonons.

A. Intrinsic Transitions (Band-to-Band). The emission spectrum for intrinsic emission by direct transition of hole-electron recombination is given by[4]

$$I(hv) \sim v^2 \langle M \rangle^2 F_C(E) F_V(E) \rho(hv) \tag{1}$$

where $\langle M \rangle$ is the transition matrix element, $F_C(E)$ and $F_V(E)$ are, respectively, the electron and hole Fermi-Dirac distribution functions, and $\rho(hv)$ is the density of states per unit range of transition energy hv. For energy bands with constant effective masses $\rho(hv) \sim (hv - E_g)^{1/2}$. If $\langle M \rangle^2$ is approximately a constant, and the distribution functions approximated by classical (Boltzmann) distribution, we get

$$I(hv) \sim v^2 (hv - E_g)^{1/2} \exp\left[-\frac{(hv - E_g)}{kT} \right]. \tag{2}$$

From Eq. (2) it can be shown that the peak intensity occurs near E_g and the width of the spectrum (at half value of peak intensity) is proportional to kT.

If the electron distribution is highly degenerate and the hole distribution non-degenerate as in most of the experiments, we get for the intrinsic emission by direct transition:[4]

$$I(hv) \sim v^2(hv - E_g)^{1/2} \exp\left[-\frac{m_e^*}{m_e^* + m_h^*}\left(\frac{hv - E_g}{kT}\right)\right]$$

$$\times \left\{\exp\left(\left[\frac{m_h^*}{m_e^* + m_h^*}(hv - E_g) - E_{Fn}\right]/kT\right) + 1\right\}^{-1} \quad (3)$$

where m_e^* and m_h^* are, respectively, the electron and hole effective masses, and E_{Fn} is the quasi Fermi level of electrons or the electron imref.

The emission spectrum for intrinsic emission due to indirect transition [corresponding to the conditions for Eq. (2)] is given by

$$I(hv) \sim v^2(hv - E_g)^2 \exp\left[-\frac{(hv - E_g)}{kT}\right]. \quad (4)$$

B. Extrinsic Transitions. Transitions of carriers from one energy band to impurity levels near the opposite band give rise to emission with hv smaller than E_g. The emission spectrum for electron transitions from the conduction band to acceptor levels near the valence band is given by

$$I(hv) \sim v^2(hv - E_g + E_a)^{1/2}\left\{\exp\left[\frac{(hv - E_g + E_a - E_{Fn})}{kT}\right] + 1\right\}^{-1} \quad (5)$$

where E_a is the impurity ionization energy. This expression resembles Eq. (3) with a shift of hv by E_a. In this case the peak intensity occurs near $(E_g - E_a)$, and the width of the spectrum is proportional to kT.

Figure 3(a) shows the luminescent emission spectrum[4] for an intrinsic (band-to-band) transition in an n-type InSb sample at 4.2°K. We note there is good agreement between the experimental result and the theoretical calculation.[4] Figure 3(b) shows the emission spectrum of a similar sample. The peak at 0.234 eV corresponds to the intrinsic emission. The peak at 0.228 eV is the main impurity emission from the conduction band to the Zn acceptor levels corresponding to an impurity ionization energy of 0.006 eV. The peak at 0.212 eV is due to a photon-assisted band-to-band transition.

(3) Luminescent Efficiency

For a given input excitation energy, the radiative recombination process is in direct competition with the nonradiative processes. Luminescent efficiency is defined as the ratio of the energy associated with the radiative process to the total input energy. To derive the efficiency we shall use a simple model[5,6] as

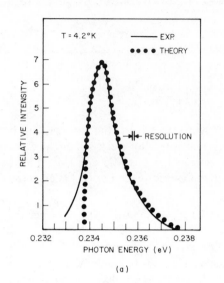

(a)

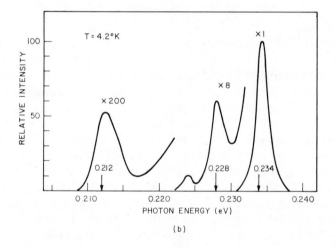

(b)

Fig. 3

(a) Luminescent emission spectrum for a band-to-band transition in a pure *n*-type InSb
($n = 5 \times 10^{13}$ cm^{-3}). Experimental result (solid line), theory (points).

(b) Emission spectrum of the same sample. The peak at 0.234 eV is the band-to-band emission.
The peak at 0.228 eV (multiplied by 8) is the main impurity emission. The peak at 0.212
eV (multiplied by 200) is due to phonon-assisted band-to-band transitions.

(After Mooradian and Fan, Ref. 4.)

shown in Fig. 4 where we assume that there is one trapping level with energy E_t below the bottom of conduction band with a trapping density N_t and a captured electron density n_t. We shall also assume that there is one luminescent level with energy E_l above the valence band edge with density N_l. In Fig. 4, G is the hole-electron generation rate and the other symbols represent the time rates of one kind or another associated with the following processes:

β: radiative recombination through capture of an electron by an empty luminescent center

α_0: nonradiative recombination through capture of a hole from the valence band by the electron trap

α_1: electron capture by the trap

α_2: production of empty luminescent center by capture of a hole (or release of an electron)

γ_1: thermal release of trapped electron

γ_2: thermal filling of an empty luminescent center by capture of a valence electron (or release of a captured hole).

The rate equations for the densities of the free electrons (n) and captured electrons (n_t) under steady-state condition are

$$\frac{dn}{dt} = 0 = G - \alpha_1 n(N_t - n_t) - \beta n p_l + \gamma_1 n_t, \tag{6a}$$

$$\frac{dn_t}{dt} = 0 = \alpha_1 n(N_t - n_t) - \alpha_0 n_t p - \gamma_1 n_t \tag{6b}$$

where p_l is the captured hole density in the luminescent level. Similar

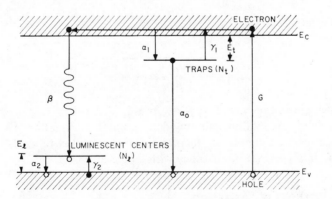

Fig. 4 Simple band model for derivation of luminescent efficiency. (After Ivey, Ref. 5.)

expressions can be written for the rate equations of the densities of free holes (p) and captured holes (p_l). A combination of Eqs. (6a) and (6b) yields

$$G = \beta n p_l + \alpha_0 n_t p. \tag{7}$$

The first term on the right-hand side, $\beta n p_l$, is associated with the radiative recombination process, while the second term is for the nonradiative process. Therefore the luminescence efficiency is given by

$$\eta = \frac{\beta n p_l}{G} = \frac{1}{\left(1 + \dfrac{\alpha_0 n_t p}{\beta n p_l}\right)}. \tag{8}$$

For low levels of excitation it can be assumed that the condition of thermal equilibrium prevails between the electron traps and the conduction band, and between the luminescent centers and the valence band. Then

$$\gamma_1 n_t \simeq \alpha_1 n(N_t - n_t)$$
$$\gamma_2 p_l \simeq \alpha_2 p(N_l - p_l). \tag{9}$$

In addition, the Fermi level is assumed to be located between E_t and E_l; thus

$$N_t - n_t \simeq N_t \tag{10a}$$

$$N_l - p_l \simeq N_l \tag{10b}$$

and

$$\frac{n_t}{p_l} \simeq \left(\frac{N_t}{N_l}\right) \exp\left[-\frac{(E_t - E_l)}{kT}\right]. \tag{10c}$$

The efficiency becomes

$$\eta = \frac{1}{\left[1 + \dfrac{N_t}{N_l}\dfrac{\alpha_0}{\beta}\dfrac{p}{n}\exp\left(-\dfrac{E_t - E_l}{kT}\right)\right]}. \tag{11}$$

The efficiency will be high if $N_l \gg N_t$, and $E_t - E_l \gg kT$, i.e., the luminescent centers should be shallow and abundant in number. Equation (11) also indicates that as the temperature increases, the exponential term also increases resulting in a decrease of efficiency with temperature.

(4) Methods of Excitation

Electroluminescence may be excited in a variety of ways including intrinsic, injection, avalanche, and tunneling processes which we now briefly discuss.

A. Intrinsic. When a powder of a semiconductor (e.g., ZnS) is embedded in a dielectric (plastic or glass) and submitted to an alternating electric field, usually at frequencies in the audio range, electroluminescence may occur. Generally the efficiency is low ($\sim 1\%$) and such materials are used only in display devices. The mechanism is mainly due to impact ionization by accelerated electrons and/or field emission of electrons from trapping centers.[3]

B. Injection. Under forward-bias conditions the injection of minority carriers in a *p-n* junction can give rise to radiative recombination. The energy band diagram for a Cd-doped GaP *p-n* junction is shown[7] in Fig. 5(a). Several

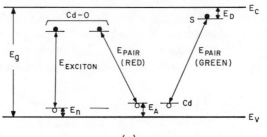

(a)

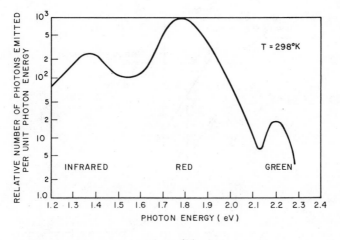

(b)

Fig. 5
(a) Energy band diagram for a Cd-doped GaP *p-n* junction Cd-O is the cadmium-oxygen complex. Transitions between the exciton level of the Cd-O complex to the acceptor level of Cd give rise to the red light emission. Transitions between the donor level (S) and acceptor level (Cd) give rise to the green light emission.
(After Henry et al., Ref. 7.)
(b) Measured emission spectrum from a GaP diode.
(After Gershenzon, Ref. 7a.)

different transitions for electron-hole recombination are indicated. The transition between the exciton energy level of the cadmium-oxygen complex and the Cd acceptor level is mainly responsible for the red band (~ 1.8 eV) as shown[7a] in Fig. 5(b). The green band (~ 2.2 eV) is due to the transition between the donor level (sulfur) and the Cd acceptor level. Similar results have been obtained for Zn-doped GaP *p-n* junctions. The relative intensity of the red and the green bands can be varied by varying the impurity concentrations. The brightness of the red-light or the green-light emission from the GaP *p-n* junction at room temperature is sufficiently high to merit electroluminescent applications.

C. Avalanche. When a *p-n* junction or a metal-semiconductor contact is reverse-biased into avalanche breakdown,[8] Fig. 6, the hole-electron pairs generated by impact ionization may result in emission of either interband (avalanche emission) or intraband (deceleration emission) transitions respectively.

D. Tunneling. Electroluminescence can result from tunneling into both forward-biased and reverse-biased junctions. In addition, light emission[9] can

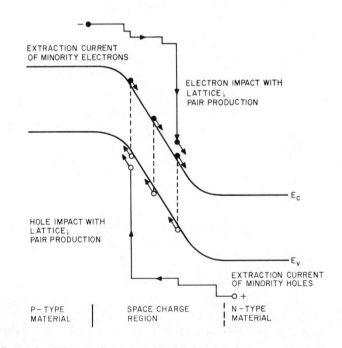

Fig. 6   Energy band diagram for reverse-biased *p-n* junction. (After Henisch, Ref. 8.)

occur in a reverse-biased metal-semiconductor contact (p-type degenerate) as shown in Fig. 7. Figure 7(a) corresponds to the equilibrium condition. When a sufficiently large reverse bias is applied, Fig. 7(b), holes at the metal Fermi level can tunnel into the valence band and subsequently recombine with the electrons which have tunneled from the valence band to the conduction band.

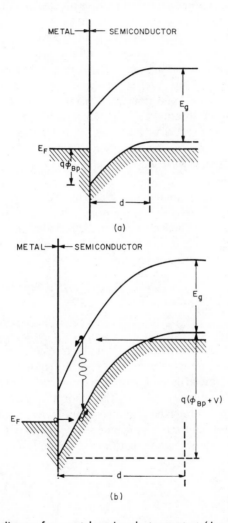

Fig. 7 Energy band diagram for a metal-semiconductor contact (degenerate p type):
(a) equilibrium condition,
(b) under reverse bias.
(After Eastman et al., Ref. 9.)

(5) GaAs Infrared Source

At the present time, the highest electroluminescent efficiency has been obtained experimentally in forward-biased GaAs diodes.[10] This is expected because (1) the forward-bias injection is a very efficient method, since electric energy can be converted directly into photons; (2) GaAs is a direct-gap semiconductor, thus the radiative recombination process is a first-order transition process (no phonons involved);[11] and (3) GaAs has the most advanced material technology of all the direct-gap semiconductors.

The cross-sectional views of three EL diodes are shown in Fig. 8 for (1) rectangular, (2) hemispherical,[12] and (3) parabolic geometries.[13] For a fixed internal efficiency, η_0, the external efficiency, η, depends upon two main factors (1) total internal reflection and (2) internal absorption. The critical angle θ_c at which total internal reflection occurs, is $\theta_c = \sin^{-1}(1/\bar{n})$ where $\bar{n}$ is the refractive index. For GaAs with $\bar{n} = 3.6$, the critical angle is about $16°$. The internal absorption can be expressed as $\exp[-\alpha(\lambda)x]$ where $\alpha(\lambda)$ is the absorption coefficient at wavelength λ and x is the distance from the radiative recombination center.

For the rectangular geometry, Fig. 8(a), ray A originating from the recombination center is attenuated by bulk absorption and is only partially reflected at the air interface. Other rays (B) which strike the semiconductor-air interface at an angle $\theta \geq \theta_c$ are internally reflected. The total efficiency for electrical-to-optical conversion is given by[12]

$$\eta_R = \frac{q}{P} \frac{4\bar{n}}{(\bar{n}+1)^2} (1 - \cos \theta_c) \frac{\int \Phi(\lambda)(1 + R_1 e^{-2\alpha_1(\lambda)x_1}) e^{-\alpha_2(\lambda)x_2} \, d\lambda}{\int \Phi(\lambda) \, d\lambda} \quad (12)$$

where P is the power input, $4\bar{n}/(\bar{n}+1)^2$ is the transmission coefficient of light from the bulk semiconductor to air, $(1 - \cos \theta_c)$ is the solid cone, $\Phi(\lambda)$ is the photon generation rate in units of photon/sec-cm^2, R_1 is the reflection coefficient at the back contact, and $\alpha(\lambda)$ and x are the absorption coefficient and thickness of the respective p- and n-type regions of the device. Similar expressions of efficiency can be written for the hemispherical (η_H) and parabolic (η_P) sources. The main difference, however, is that the solid cone is unity. Thus the ratio is given by

$$\frac{\eta_H}{\eta_R} \quad \text{or} \quad \frac{\eta_P}{\eta_R} \approx \frac{1}{1 - \cos \theta_c} = \frac{1}{1 - \sqrt{1 - 1/\bar{n}^2}} \simeq 2\bar{n}^2 \quad \text{for} \quad \bar{n} \gg 1.$$
$$(13)$$

This means that for GaAs with $\bar{n} = 3.6$, an increase in efficiency by an order of magnitude is expected for the hemispherical or parabolic source.

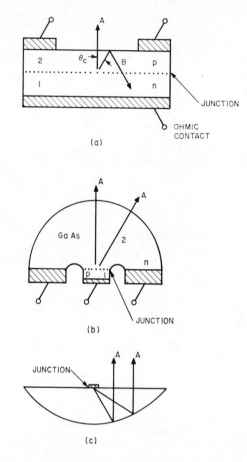

Fig. 8 Cross-sectional views of three electroluminescent (EL) diodes:
(a) rectangular,
(b) hemispherical.
(After Carr and Pittman, Ref. 12.)
(c) parabolic geometry.
(After Galginaitis, Ref. 13.)

The typical radiation patterns[13] for the geometries shown in Fig. 8 are illustrated in Fig. 9(a). It is apparent that one can design the geometry of the device to give a desired radiation pattern. The cross-sectional view of a typical arrangement[10] used for efficiency measurement is shown in the insert of Fig. 9(b). The emission is measured directly as the short-circuit current of the solar cell. The measured external efficiency for the hemispherical source is

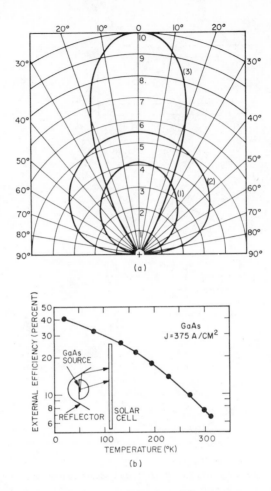

(a)

(b)

Fig. 9
(a) Radiation patterns of the EL diodes shown in Fig. 8, (1) rectangular, (2) hemispherical, and (3) parabolic geometry.
(After Galginaitis, Ref. 13.)
(b) External efficiency as a function of temperature for a GaAs infrared source. Insert: cross-sectional view of arrangement used for efficiency measurement.
(After Carr, Ref. 10.)

shown in Fig. 9(b) as a function of temperature.[10] The efficiency, as expected, decreases with increasing temperature, and has a value of 40, 32, or 7% at 20, 77, or 300°K respectively.

Figure 10(a) shows the emission spectra[10] observed at 77°K and 300°K. The peak photon energy decreases with increasing temperature mainly

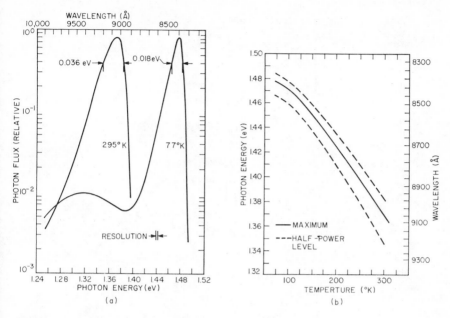

Fig. 10
(a) GaAs diode emission spectra observed at 300°K and 77°K.
(b) Dependence of the emission peak and half-power width as a function of temperature.
(After Carr, Ref. 10.)

because of the decrease of band gap with temperature. A more detailed plot is shown in Fig. 10(b) for the peak photon energy and the half-power points from the diode emission spectrum as a function of temperature. The slight increase of the width of the half-power points with temperature is expected from the discussion of the emission spectra, Eqs. (2) through (5). The optical response time of the light sources, i.e., the time required for light output after a current pulse is applied, is determined by the spontaneous radiative lifetime required for recombination of the injected carriers. In most direct-gap semiconductors, because of their short lifetime, the response time is of the order of 1 ns or less.

The GaAs infrared sources as well as other electroluminescent (EL) diodes have major application in areas where decoupling between the input signal, or control signal, and the output is of special importance. Figure 11 shows two typical photocoupled isolators using a GaAs diode as the light source and a *p-i-n* diode as the photodetector. When an input electrical signal is applied to the GaAs diode, an infrared light is generated and subsequently detected by

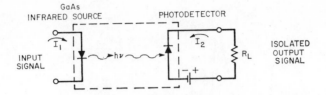

(a)   PHOTOCOUPLED ISOLATOR

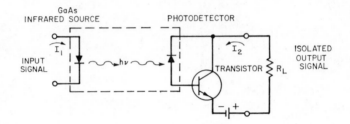

(b) HIGH-GAIN PHOTOCOUPLED ISOLATOR

Fig. 11

(a) Photocoupled isolator using a GaAs diode as infrared source and a *p-i-n* diode as the photodetector.

(b) High-gain photocoupled isolator where the signal is amplified by a junction transistor.

the photodetector. The light is then converted back to electrical signal as a current which can flow through a load resistor as shown in Fig. 11(a) with a current ratio, I_2/I_1, of the order of 10^{-3}; or be amplified as in Fig. 11(b) with a current ratio of the order of 0.1. They are called photocoupled isolators because there is no feedback from the output to the input and the signal is transmitted at the speed of light.

In addition to a single electroluminescent diode unit, integrated arrays of electroluminescent diodes can be made using the planar and beam-lead technology.[14] The diode array can be used to optically trigger matching arrays of photosensitive devices, such as the Si *p-n-p-n*, and in memory applications.

3 SOLAR CELL

The reciprocal effect to electroluminescence which converts electrical energy to radiation is the photovoltaic effect. The most important photovoltaic device is the solar cell which is basically a *p-n* junction with a large surface area and which can convert solar radiation directly into electrical energy with high overall efficiency (of the order of 10% or more).

The high-efficiency solar cell was first developed[15] by Chapin, Fuller, and Pearson in 1954 using a diffused silicon *p-n* junction. A schematic diagram of a typical solar cell is shown in Fig. 12(a). To date, solar cells have been made in many other semiconductors such as the diffused GaAs cell,[16] the Cu_2 S-CdS cell,[17] the Se-CdS cell,[18] and thin film GaAs and CdS cells.[19,20] Solar cells now are the most important long-duration power supply for satellites and space vehicles.

The photovoltaic power conversion may best be defined by considering the case of an ideal *p-n* junction with a constant-current source in parallel with the junction, Fig. 12(b). The constant-current source results from the excitation

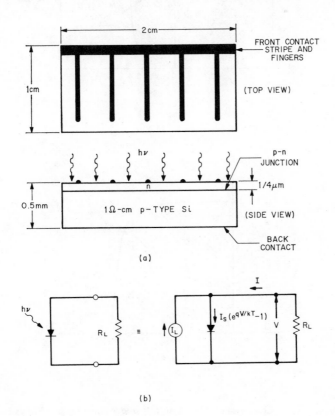

(a)

(b)

Fig. 12
(a) Typical schematic representation of a solar cell.
(b) The idealized equivalent circuit of a solar cell.

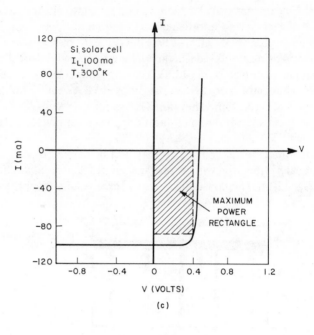

(c)

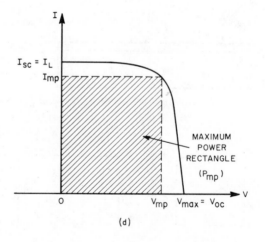

(d)

Fig. 12
(c) Current-voltage characteristics of a solar cell under illumination. The photocurrent is 100 mA.
(d) Inversion of (a) about the voltage axis.
(After Prince, Ref. 21.)

of excess carriers by solar radiation. Actually the nature of the source need not concern us in the subsequent analysis; it may be a photon source (solar energy, γ radiation, incandescent lamp, x-ray, etc.), a high-energy particle source (electron gun, β-radioactive elements, α particles, protons, neutrons, etc.), or any other means of creating electron-hole pairs without changing the properties of the ideal junction appreciably. The following discussion is concerned only with the solar cell. However, the results are also applicable (with minor modifications) to other particle detectors.

(1) Basic Characteristics

The solar cell as shown in Fig. 12(a) consists of a shallow *p-n* junction formed on the surface (e.g., by diffusion), a front ohmic contact stripe and fingers, and a back ohmic contact which covers the entire back surface. The simplest equivalent circuit of the solar cell under radiation is shown in Fig. 12(b) where I_L is the strength of constant current source due to the incident light, and I_s is the saturation current.

The *I-V* characteristics of such a device are given by

$$I = I_s(e^{qV/kT} - 1) - I_L \tag{14}$$

and

$$I_s = qn_i^2 \left[\frac{1}{N_A} \left(\frac{D_n}{\tau_n} \right)^{1/2} + \frac{1}{N_D} \left(\frac{D_p}{\tau_p} \right)^{1/2} \right]. \tag{14a}$$

A plot[21] of Eq. (14) is given in Fig. 12(c) for $I_L = 0.1$ amp, $I_s = 10^{-9}$ amp, and $T = 300°K$. It is seen that the curve passes through the fourth quadrant and therefore that power can be extracted from the device. By properly choosing a load, it is possible to extract close to 80% of the product $I_{SC} \times V_{OC}$ where I_{SC} is the short-circuit current and V_{OC} is the open-circuit voltage of the cell, as indicated by the maximum power rectangle. The *I-V* curve is more generally represented by Fig. 12(d) which is simply an inversion of Fig. 12(c) about the voltage axis. We also define in Fig. 12(d) the quantities I_{mp} and V_{mp} which correspond, respectively, to the current and voltage for the maximum power output $P_{mp}(\equiv I_{mp} \times V_{mp})$.

From Eq. (14) we obtain for the open-circuit voltage

$$V_{OC} \equiv V_{max} = \frac{1}{\beta} \ln \left(\frac{I_L}{I_s} + 1 \right) \tag{15}$$

where $\beta \equiv q/kT$. The output power is given by

$$P = IV = I_s V(e^{\beta V} - 1) - I_L V. \tag{16}$$

The condition for maximum power can be obtained when $\partial P/\partial V = 0$, or

$$(1 + \beta V_{mp})\exp(\beta V_{mp}) = \left(1 + \frac{I_L}{I_s}\right), \tag{17}$$

$$I_{mp} = |I_s(e^{\beta V_{mp}} - 1) - I_L| = I_s \beta V_{mp} e^{(\beta V_{mp})}. \tag{18}$$

The efficiency of solar energy conversion is given then by

$$\eta \equiv \frac{\text{maximum power output}}{\text{power input}} = \frac{I_{mp} V_{mp}/\text{cm}^2}{P_{in}/\text{cm}^2}$$

$$= \frac{I_L \beta V_{mp}^2}{(1 + \beta V_{mp})A}\left(1 + \frac{I_s}{I_L}\right)\frac{1}{P_{in}/\text{cm}^2} \tag{19}$$

where A is the exposed front area of the solar cell and P_{in}/cm^2 is the solar power density outside the atmosphere. The predicted efficiency for various semiconductors is plotted in Fig. 13 as a function of energy band gap and

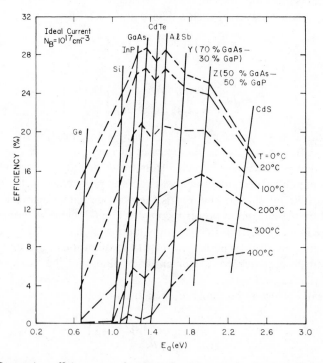

Fig. 13 Conversion efficiency as a function of energy gap for ideal current-voltage characteristics.
(After Wysocki and Rappaport, Ref. 22.)

for different lattice temperatures.[22] The generated current I_L is determined by solar conditions outside the atmosphere where the solar power density is 135 mW/cm². The number of photons effective in creating electron-hole pairs is taken from the published values. The collection efficiency is assumed to be unity, and losses due to reflection, leakage conductance, and series resistance are neglected. Under the above idealized conditions and for the idealized current-voltage characteristics, Eq. (14), it is predicted that the material with optimum efficiency at room temperature is GaAs. The optimum shifts to higher band-gap materials as the temperature is increased. The efficiency has also been calculated using the method of detailed balance limit,[23] and the result is essentially the same as the "semiempirical approach" shown in Fig. 13.

For a practical Si of GaAs solar cell at room temperature, the efficiency is approximately a factor of two smaller than the predicted value (10 to 15% instead of 22 to 28%). This is mainly due to the reflection loss, the leakage conductance loss, and the effect of series resistance to be discussed later.

(2) Spectral Response

The spectral response of a solar cell is defined as the short-circuit current as a function of the wavelength of the incident light. To derive the spectral response, we shall use the simple one-dimensional geometry shown in Fig. 14(a) where d is the junction depth, and L_n and L_p are the minority diffusion lengths in the p side and n side respectively. We shall assume that the depletion width of the junction is much smaller than L_n or L_p. For incident photons with energies in excess of the energy gap ($hv > E_g$), the density of photons in the semiconductor varies as $\Phi = \Phi_0 \exp(-\alpha x)$ where Φ is in the units of photon/sec-cm² and α is the absorption coefficient which is a function of wavelength. The hole-electron generation rate by photons is given by

$$G(x) = \Phi_0 \alpha e^{-\alpha x}. \tag{20}$$

In the n side the minority carriers (holes) created at a distance x will have a fraction proportional to $\exp[-(d-x)/L_p]$ diffuse to the junction. The total number of minority carriers reaching the junction due to creation of hole-electron pairs in the n side is given by

$$N \sim \int_0^d \Phi_0 \alpha \exp(-\alpha x)\exp\left[-\frac{|d-x|}{L_p}\right] dx$$

$$= \frac{\Phi_0 \alpha}{\alpha - \dfrac{1}{L_p}} [\exp(-d/L_p) - \exp(-\alpha d)]. \tag{21}$$

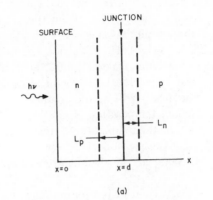

(a)

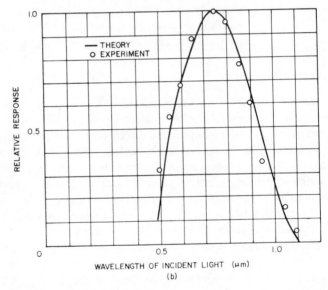

(b)

Fig. 14

(a) One-dimensional geometry of a solar cell with junction depth d and minority carrier diffusion lengths L_n and L_p for the p side and n side, respectively.

(b) Typical comparison between measured and calculated response curves for p-on-n solar cell with etched surface. Data points are measured results for $d = 2.0\ \mu$m. Solid curve is calculated response for $d = 2.0\ \mu$m, $L_n = 0.5\ \mu$m and $L_p = 10.0\ \mu$m.

(After Terman, Ref. 24.)

Similarly, the number of minority carriers (electrons) reaching the junction as a result of the creation of hole-electron pairs in the p-region is the integral of Eq. (21) from $x = d$ to $x = \infty$ with L_p replaced by L_n. Thus the total number of carriers crossing the p-n junction is[24]

$$N_T \sim \frac{\Phi_0 \alpha}{\alpha - \frac{1}{L_p}} [\exp(-d/L_p) - \exp(-\alpha d)] + \frac{\Phi_0 \alpha \exp(-\alpha d)}{\left(\alpha + \frac{1}{L_n}\right)}. \quad (22)$$

In the steady state the current through the cell is a constant, hence it is proportional to N_T. For an equal-energy spectrum, the photon density is proportional to the wavelength of the incident light, since the energy of an individual photon is proportional to frequency. Therefore Φ_0 itself is proportional to wavelength for an equal-energy spectrum. The short-circuit current per unit wavelength is then given from Eq. (22):

$$\frac{dI_{SC}(\lambda)}{d\lambda} \sim \alpha\lambda \left\{ \frac{L_p}{1 - \alpha L_p} [\exp(-\alpha d) - \exp(-d/L_p)] + \frac{L_n \exp(-\alpha d)}{1 + \alpha L_n} \right\}.$$

$$(23)$$

Under the conditions that αL_n, $\alpha L_p \ll 1$, and $d/L_p \gg 1$, this equation reduces to

$$\frac{dI_{SC}(\lambda)}{d\lambda} \sim \alpha\lambda(L_n + L_p)\exp(-\alpha d). \quad (23a)$$

The above equations are for n-on-p solar cells. For p-on-n solar cells we have only to replace the quantity L_p with L_n. Figure 14(b) indicates the good agreement that is obtained between the measured and calculated response curves.[24] The points are the measured values of a silicon solar cell with an etched surface and 2-μm p-type diffusion, and the solid line is the theoretical curve calculated from Eq. (23) for $d = 2.0$ μm, $L_n = 0.5$ μm, $L_p = 10$ μm, and using published values of $\alpha(\lambda)$. From Eq. (23) it can be shown that in order to increase the short-wavelength response the junction should be made closer to the surface since $1/\alpha$ is small for short wavelength; while in order to increase the long-wavelength response the junction must be made comparatively deep below the surface. If the surface recombination velocity is high (such as a lapped surface), the lifetime near the surface is reduced, thus reducing the response to short wavelengths of incident light.

(3) Recombination Current and Series Resistance

If the forward current is dominated by the recombination current within the depletion region, the efficiency is generally reduced in comparison with that of an ideal diode discussed previously. For single-level centers, the recombination current can be expressed as[25]

$$I_{recom} = I_s' \left[\exp\left(\frac{qV}{2kT}\right) - 1 \right] \quad (24)$$

with

$$I_s' = \frac{q n_i W}{\sqrt{\tau_p \tau_n}}$$

where W is the depletion width and τ_p and τ_n are the lifetimes of holes and electrons respectively. The energy conversion equation could again be put into closed form yielding equations similar to Eqs. (17) through (19) with the exception that I_s is replaced by I_s' and the exponential factor is divided by 2. Figure 15 shows the efficiency for the case of recombination current.[22] Although the optimum band gap is roughly the same as for the ideal current case, the efficiency is much less for the present case.

The results of some experimental measurements[22] are plotted in Fig. 16 together with theoretical curves for the ideal case and the case with the recombination current at a doping level of 10^{17} cm^{-3}. Figure 16(a) is a plot of V_{max} versus T in silicon. The experimental points fall between the theoretical

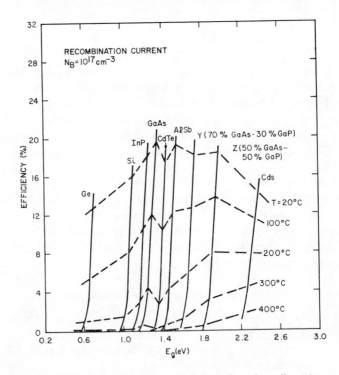

Fig. 15 Conversion efficiency as a function of band gap for solar cells with recombination currents.
(After Wysocki and Rappaport, Ref. 22.)

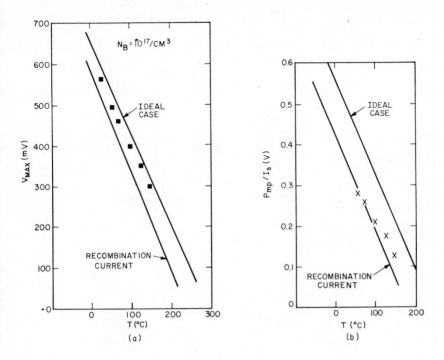

Fig. 16
(a) Measured open-circuit voltage. V_{max} versus temperature and
(b) measured ratio of maximum power output to saturation current, P_{mp}/I_s, versus temperature for a silicon solar cell. The solid lines are theoretical calculations for an ideal-current case and a recombination-current case.
(After Wysocki and Rappaport, Ref. 22.)

curves, and the rate of decrease agrees well with the theoretical prediction. Similar agreement is found in Fig. 16(b) where (P_{mp}/I_s) is plotted versus T.

For solar cells containing many defects, the forward current may show an exponential dependence[26] on the forward voltage as $\exp(qV/nkT)$ with $n \sim 3$. These defects cause a further reduction of the efficiency.

The series resistance imposes another limitation on the maximum attainable efficiency of solar energy conversion. The equivalent circuit containing both a shunt and a series resistance is shown in the insert of Fig. 17. It can be shown that the I-V characteristic is given by[21]

$$\ln\left(\frac{I + I_L}{I_s} - \frac{V - IR_S}{I_s R_{sh}} + 1\right) = \frac{q}{kT}(V - IR_S). \tag{25}$$

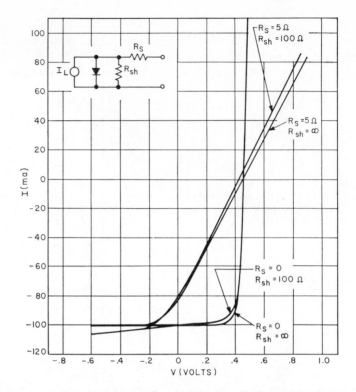

Fig. 17 Theoretical I-V characteristics for various solar cells that include series and shunt resistances. The insert shows the equivalent circuit. The parameters are identical to that used in Fig. 12(c).
(After Prince, Ref. 21.)

Plots of this equation, with combinations of $R_S = 0$, 5 ohms and $R_{sh} = \infty$, 100 ohms, are given in Fig. 17 with the same parameters I_s, I_L, and T as in Fig. 12(c). It can be seen that a shunt resistance even as low as 100 ohms does not appreciably change the power output of a unit, whereas a series resistance of only 5 ohms reduces the available power to less than 30% of the optimum power with $R_S = 0$. We can thus neglect the effect of R_{sh}. The output current and output power are:

$$I = I_s \left\{ \exp\left[\frac{q(V - IR_S)}{kT} \right] - 1 \right\} - I_L \tag{26}$$

$$P = |IV| = I \left[\frac{kT}{q} \ln\left(\frac{I + I_L}{I_s} + 1 \right) + IR_S \right]. \tag{27}$$

The relative maximum available power is 1, 0.77, 0.57, 0.27, or 0.14 for R_S of 0, 1, 2, 5, or 10 ohms respectively. The series resistance of a solar cell depends on the junction depth, impurity concentrations of p-type and n-type regions, and the arrangement of the front surface ohmic contacts. For a typical Si solar cell with the geometry shown in Fig. 12(a), the series resistance is about 0.7 ohm for n-on-p cells and about 0.4 ohm for p-on-n cells.[27] The difference in resistance is mainly the result of lower resistivity in n-type substrates.

(4) Radiation Effect

The most important application of solar cells is in satellite and space vehicles. The high-energy particle radiation in outer space produces defects in semiconductors which cause a reduction in solar cell power output. It is important to assess the expected useful life of the solar cell power plant. Its lifetime is the length of time the power plant is capable of delivering the electrical power necessary for successful operation of the satellite.

From the expression of short-circuit current, Eq. (23), we note that the current will decrease with decreasing diffusion length L_n and/or L_p. If the lifetime τ of the excess minority carriers at any point in the bombardment is given by[28]

$$\frac{1}{\tau} = \frac{1}{\tau_o} + K'\Phi \tag{28}$$

where τ_o is the initial lifetime, K' is a constant, and Φ is the bombardment flux. The above expression states that the recombination rate of the minority carriers is proportional to the initial number of recombination centers present plus the number introduced during bombardment, the latter being proportional to the flux. Since the diffusion length is equal to $\sqrt{D\tau}$ and D is a slowly varying function with bombardment (or doping concentration), Eq. (28) can be expressed as

$$\frac{1}{L^2} = \frac{1}{L_o^2} + K\Phi \tag{29}$$

where L_o is the initial diffusion length, and $K = K'/D$. Figure 18 shows the measured substrate diffusion length as a function of a 1-MeV electron flux for three different silicon solar cells. The blue-sensitive n-on-p cell is one with n-type diffusion and with antireflection coating. The diffusion depth is adjusted to give a large spectral response near blue light (0.45 to 0.5 μm) which corresponds to the maximum of the solar energy distribution. The blue-sensitive p-on-n cell is similar to the n-on-p cell except that the roles of p-type and n-type are interchanged. We see that the experimental results can indeed be reasonably fitted by Eq. (29). The curve passing through the points for the

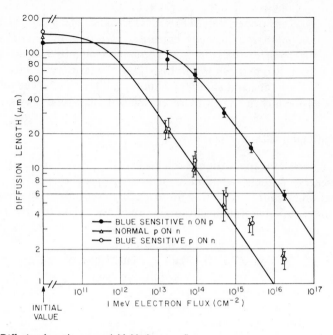

Fig. 18 Diffusion length versus 1-MeV electron flux. (After Rosenzweig et al., Ref. 28.)

n-on-p cell is computed using Eq. (29) with $L_o = 119$ μm and $K = 1.7 \times 10^{-10}$ The data points for the blue-sensitive p-on-n and normal p-on-n cells can be approximately fitted by Eq. (29) with $L_o = 146$ μm and $K = 1.22 \times 10^{-8}$. It is apparent from Fig. 18 that the radiation resistance of n-on-p cells is higher than that of p-on-n cells.

The short-circuit current per unit wavelength for an n-on-p cell at a given wavelength λ, Eq. (23), can be written as

$$\frac{dI_{SC}(\lambda)}{d\lambda} \sim \lambda \left[1 - \exp(-\alpha d) + \frac{\exp(-\alpha d)}{1 + \frac{1}{\alpha}\sqrt{\frac{1}{L_o^2} + K\Phi}} \right] \qquad (30)$$

where the diffusion length L_n has been replaced by the expression of Eq. (29). In Eq. (30) it is also assumed that $d/L_p \ll 1$ and $\alpha L_p \gg 1$. Figure 19 shows the measured contributions (data points) to the total short-circuit current for outer space sunlight falling into the indicated wavelength intervals as a function of bombardment flux.[28] The solid curves are the theoretical calculations based on Eq. (30) and the values of L_o and K obtained from Fig. 18. The

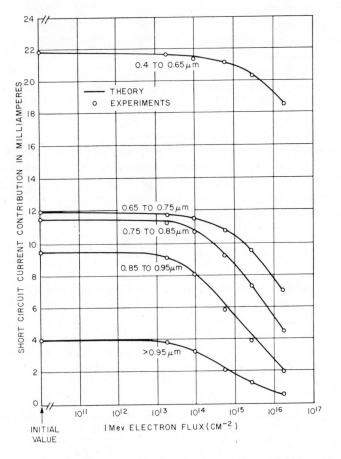

Fig. 19 Contribution to outer space short-circuit current by sunlight in various wave-length intervals for Si *n-on-p* solar cells as a function of 1-MeV electron flux. (After Rosenzweig et al., Ref. 28.)

agreement between the theory and experiment points is noted to be good. Similar effects have been observed on solar cells treated with proton bombardment.[29,30]

4 PHOTODETECTORS

The extension of coherent and incoherent light sources into the far infrared region on one hand and the ultraviolet region on the other has increased the need for high-speed as well as sensitive photodetectors.[31] For a general photo-detector there are basically three processes: (1) carrier generation by incident light (2) carrier transport and/or multiplication by whatever current-gain mechanism may be present, and (3) interaction of current with external circuitry to provide the output signal.

To describe the performance of a photodetector under specified operating conditions, certain figures of merit are defined.[32,32a] One of the important figures of merit is the signal-to-noise ratio. Since all photodetectors are ultimately limited by noise, it is important to have a large signal-to-noise ratio. Another commonly used figure of merit is the noise equivalent power (NEP) which is defined as the root mean square value of the sinusoidally modulated radiant power falling upon a photodetector which will give rise to an rms signal voltage equal to the rms noise voltage from the detector. For infrared detectors the most used figure of merit is the detectivity D^* which is defined as

$$D^* \equiv \frac{A^{1/2} B^{1/2}}{\text{NEP}} \tag{31}$$

where A is the area of the photodetector, and B is the bandwidth. In order to remove any ambiguity in D^*, one must state whether the radiation is from a black-body source or a monochromatic source and at what modulation frequency. It is recommended that D^* be expressed as $D^*(\lambda, f, 1)$ or $D^*(T, f, 1)$ where λ is the wavelength in μm, f is the frequency of modulation in Hz, T is the black-body temperature in °K, and the reference bandwidth is always 1 Hz.

In this section we shall consider three important classes of photodetectors: the photoconductor, the depletion-layer photodiode, and the avalanche photodiode.

(1) Photoconductor

A photoconductor is a device consisting simply of a slab of semiconductor (in bulk or thin-film form) with ohmic contacts affixed to opposite ends, Fig. 20. When incident light falls on the surface of the photoconductor, carriers are generated either by band-to-band transitions (intrinsic) or by transitions involving forbidden-gap energy levels (extrinsic), resulting in an increase of the conductivity. For the intrinsic photoconductor, the conductivity is given by $\sigma = q(\mu_n n + \mu_p p)$, and the increase of conductivity under illumination is mainly due to the increase in the number of carriers. The long-wavelength limit for this case is given by

$$\lambda_c = \frac{hc}{E_g} = \frac{1.24}{E_g(\text{eV})} \qquad (\mu\text{m}) \tag{32}$$

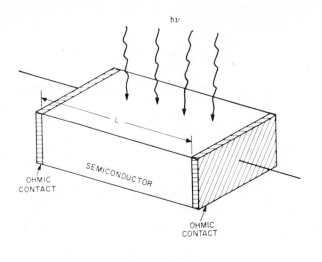

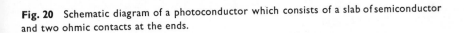

Fig. 20   Schematic diagram of a photoconductor which consists of a slab of semiconductor and two ohmic contacts at the ends.

where λ_c is the wavelength corresponding to the semiconductor band gap E_g. For wavelengths shorter than λ_c, the incident radiation is absorbed by the semiconductor, and hole-electron pairs are generated. For the extrinsic case, photoexcitation may occur between a band edge and an energy level in the energy gap. Photoconductivity can take place by the absorption of photons of energy equal to or greater than the energy separation of the band-gap levels and the conduction or valence band.

The performance of a photodetector in general and a photoconductor in particular is measured in terms of two parameters: the photoconductivity gain and the response time of the detector.[33] The photoconductivity gain is defined as the number of charge carriers which pass between the contact electrodes per second for each photon absorbed per second.

$$\text{Gain} \equiv \frac{\Delta I}{q G_{\text{pair}}} \tag{33}$$

where ΔI is the photocurrent in amperes and G_{pair} is the total number of electron-hole pairs created in the photoconductor per second by the absorption of light. The gain may also be expressed as the ratio of the lifetime of a free carrier to the transit time for that carrier, i.e., the time required for the carrier to move between the electrodes. For a semiconductor in which one-carrier conductivity dominates,

$$\text{Gain} = \frac{\tau}{t_r} = \frac{\tau \mu V}{L^2} \tag{33a}$$

where τ is the lifetime of a free carrier, t_r the transit time for this carrier, μ the mobility, V the applied voltage, and L the electrode spacing as shown in Fig. 20. The gain is thus dependent upon the lifetime of the free carriers as a critical parameter. For long-lifetime sample with short electrode spacing the gain can be substantially greater than unity. Some typical values[33] of gain and response time are given in Table 12.1.

TABLE 12.1

Typical Values of Gain and Response Time

Photodetector	Gain	Response Time (sec)
Photoconductor	10^5	10^{-3}
p-n Junction	1	10^{-11}
p-i-n Junction	1	10^{-10}
Junction Transistor	10^2	10^{-8}
Avalanche Photodiode	10^4	10^{-10}
Metal-Semiconductor Diode	1	10^{-11}
Field-Effect Transistor	10^2	10^{-7}

For sensitive infrared detection, the photoconductor must be cooled in order to reduce thermal effects (which causes thermal ionization and depletes the energy levels) and to increase the gain and detection efficiency. However, carrier lifetimes are correspondingly increased with a resulting long response time. For a background-limited photoconductor, the ideal D^* for unit quantum efficiency is given by[32a]

$$D^*(\lambda, f, 1) = \frac{c \exp(\zeta)}{2\sqrt{\pi h k T}\, v^2 (1 + 2\zeta + 2\zeta^2)^{1/2}} \qquad \text{cm(Hz)}^{1/2}/\text{watt} \tag{34}$$

where c is the velocity of light, T the background temperature in °K, and $\zeta \equiv h v/kT$. The ideal D^* is plotted in Fig. 21 (dotted curve) for a background temperature of 300°K. Also shown are some typical D^* values for available photoconductors such as CdS, PbS, gold-doped p-type germanium (Ge:Au), and mercury-doped germanium (Ge:Hg) detectors[34] (refer to Fig. 12 of Chapter 2 for impurity energy levels). We note that, for example, in order to detect blue-green light near 0.5 μm, a CdS photoconductor gives the highest sensitivity, while at 2 μm, a PbS photoconductor is the best choice.

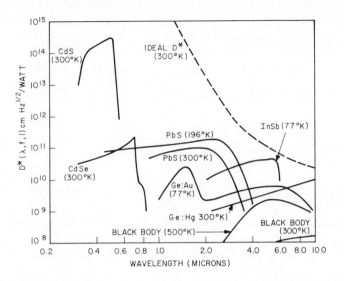

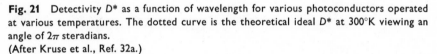

Fig. 21 Detectivity $D*$ as a function of wavelength for various photoconductors operated at various temperatures. The dotted curve is the theoretical ideal $D*$ at 300°K viewing an angle of 2π steradians.
(After Kruse et al., Ref. 32a.)

For high-speed photodetection, especially for detection of coherent light waves, we must consider the available power and the signal-to-noise performance. The generation of dc photocurrent, when a photon flux density Φ_0 in units of photons/sec-cm^2 is incident on the device (assuming total uniform absorption), is given by[35]

$$I_o = \eta q \Phi_0 A \frac{\tau}{t_r} \qquad (35)$$

where η is the quantum efficiency, A is the area, and τ and t_r are defined in Eq. (33a).

For ac detection we assume that the incident photon flux density Φ is time dependent and can be expressed as

$$\Phi = \Phi_0 + \Phi_1 e^{j\omega t} \qquad (36)$$

where $\omega/2\pi$ is the frequency (generally in the microwave region). Equation (36) applies directly when the incident light is intensity modulated, in which case $m \equiv \Phi_1/\Phi_0$ is the modulation index. If the incident light is composed instead of two optical signals at frequencies $\omega_m/2\pi$ and $\omega_n/2\pi$ whose difference is

$\omega/2\pi$, we then have the combined photon intensity (for the photomixing process):[36]

$$(\sqrt{\Phi_m}\, e^{j\omega_m t} + \sqrt{\Phi_n}\, e^{j\omega_n t})^2$$
$$= \tfrac{1}{2}(\Phi_m + \Phi_n) + \sqrt{\Phi_m \Phi_n}\, e^{j\omega t} + \sqrt{\Phi_m \Phi_n}\, e^{j(\omega_n + \omega_n)t} + \dots \quad (37)$$

By ignoring any frequencies beyond $\omega/2\pi$, we obtain the ratio between the amplitude of the instantaneous difference frequency flux Φ_1 and the dc flux Φ_0 as

$$\frac{\Phi_1}{\Phi_0} = \frac{2\sqrt{\Phi_m \Phi_n}}{\Phi_m + \Phi_n} \quad (37a)$$

where Φ_m and Φ_n are the absorbed photon flux densities of the two optical signals.

The equivalent circuits for the high-frequency photoconductor and its noise equivalent circuit are shown in Fig. 22(a) and (b) respectively. In Fig. 22(a) the quantity G is the *rf* conductance, and $I_{pc}(\omega)$ is the ac photoconductor current which is given by

$$I_{pc}(\omega) = \eta q \Phi_1 A \frac{\tau}{t_r} \frac{1}{(1 + j\omega\tau)} \quad (38)$$

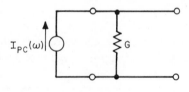

(a)

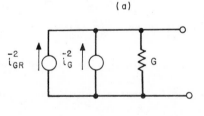

(b)

Fig. 22
(a) Equivalent circuit and
(b) noise equivalent circuit of a photoconductor where G is the conductance.
(After DiDomenico and Svelto, Ref. 35.)

where uniform absorption is assumed. The available power output (for operating into a match load) can be obtained from Fig. 22(a):

$$P_{av} = \frac{1}{8} |I_{pc}(\omega)|^2 \frac{1}{G} \simeq \frac{1}{8} (\eta q \Phi_1 A)^2 \frac{1}{G(\omega t_r)^2} \Big|_{\omega\tau \gg 1}. \tag{39}$$

The signal-to-noise ratio can be obtained from Fig. 22(b) when I_{GR} denotes the generation-recombination noise source and I_G gives the thermal noise in the shunt conductance G. The mean-square current fluctuation due to generation-recombination noise in bandwidth B is[37]

$$\overline{i_{GR}^2} = 4q \frac{\tau}{t_r} \frac{I_s}{(1 + \omega^2 \tau^2)} B \tag{40}$$

where I_s is the reverse-biased saturation current and B is the bandwidth. Since

$$\overline{i_G^2} = 4kTBG \tag{41}$$

one finds from Eqs. (38), (40), and (41) that the S/N can be expressed as

$$(S/N)_{\text{power}} = \frac{\eta}{8B} m^2 \Phi_0 A \left[1 + \frac{kT}{q} \frac{t_r}{\tau} (1 + \omega^2 \tau^2) \frac{G}{I_s} \right]^{-1}. \tag{42}$$

The above formula will be used later to compare the photoconductor performance with that of the photodiode. It will be shown that a photoconductor is an extremely broad-band device and can give comparable performance for high-level (strong light intensity) detection. For low-level detection at microwave frequencies, however, a photodiode will give considerably more signal power and considerably higher signal-to-noise ratio. Thus photoconductors have found limited use as high-frequency optical demodulators such as in high-level optical mixing. They have been, however, extensively used for infrared detections especially beyond a few microns, where, in spite of intensive work, no really satisfactory alternative detection techniques exist as yet.

(2) Depletion-Layer Photodiode

A. General Consideration. Depletion-layer photodiodes consist, in essence, of a reverse-biased semiconductor diode whose reverse current is modulated by the electron-hole pairs produced in or near the depletion layer by the absorption of light. For the photodiodes, the applied reverse voltage is not large enough to cause avalanche breakdown. This is in contrast to the avalanche photodiodes to be discussed in the next subsection in which an internal current gain is obtained as a result of the impact ionization under avalanche breakdown conditions. The depletion-layer photodiode family includes the *p-i-n* diode, the *p-n* junction, metal-semiconductor diode (Schottky barrier), the heterojunction, and the point contact diode.

We shall now briefly consider the characteristics of a general depletion-layer photodiode—its wavelength response, modulation-frequency response, available power, and signal-to-noise ratio.

The wavelength response is the wavelength range in which appreciable photocurrent can be generated. One of the key factors that determines the wavelength response is the absorption coefficient α. Figure 23 shows the measured intrinsic absorption coefficients for Ge and Si as a function of wavelength.[38] Also shown are the emission wavelengths of some important lasers.

Since α is a strong function of the wavlength, for a given semiconductor the wavelength range in which appreciable photocurrent can be generated is limited. The long-wavelength cutoff λ_c is established by the energy gap of the semiconductor, Eq. (32), e.g., about 1.7 μm for Ge and 1.1 μm for Si. For wavelengths longer than λ_c, the values of α are too small to give appreciable absorption. The short-wavelength cutoff of the photoresponse comes about because the values of α for short wavelengths are very large ($\sim 10^5$ cm^{-1}), and the radiation is absorbed very near the surface where the recombination time is short. The photocarriers thus can recombine before they are swept out. Figure 24 shows typical plots of quantum efficiency versus wavelength for both silicon and germanium high-speed photodiodes.[39]

The modulation-frequency response is limited by a combination of three factors: diffusion of carriers, drift time in the depletion region, and capacitance of the depletion region. Carriers generated outside the depletion region must diffuse to the junction resulting in considerable time delay. To minimize the diffusion effect, the junction should be formed very close to the surface. Most light will be absorbed when the depletion region is sufficiently wide (of the order of $1/\alpha$); with sufficient reverse bias the carriers will drift at their scattering-limited velocities. The depletion layer must not be too wide, however, or transit-time effects will limit the frequency response. Neither should be too thin, or excessive capacitance C will result in a large RC time constant where R is the series resistance. The optimum compromise occurs when the depletion layer is chosen such that the transit time is of the order of one half the modulation period. For example, for a modulation frequency of 10 GHz the optimum depletion layer thickness in Si (with a velocity of 10^7 cm/sec) is about 5 μm.

To consider the available power, we shall use the equivalent circuit shown[35] in Fig. 25(a) where C is the depletion capacitance and R is the series resistance. The current $I_{PD}(\omega)$ is given by the expression $\eta q \Phi_1 A$ which is essentially the same as Eq. (35) except that Φ_0 is replaced by Φ_1 for the ac case and τ equals t_r, since for a photodiode under sufficient reverse bias the carriers are swept across the depletion region before they have a chance to recombine. The

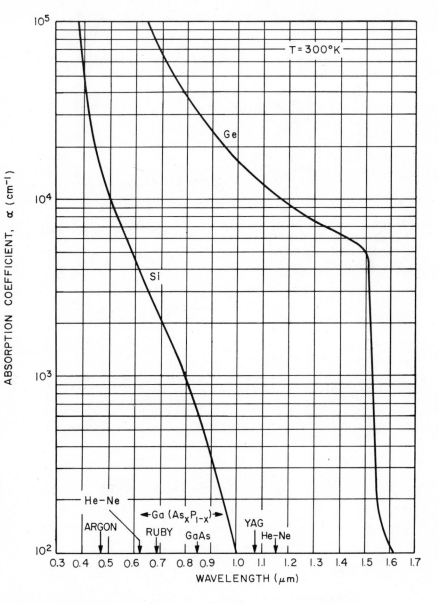

Fig. 23 Absorption coefficient versus wavelength for Ge and Si at 300°K
(After Dash and Newman, Ref. 38.)
Some laser emission wavelengths are indicated.

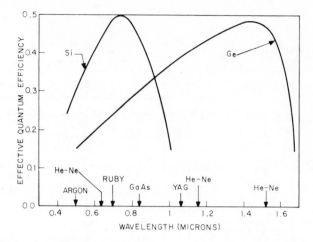

Fig. 24 Effective quantum efficiency (hole-electron pairs/photon) versus wavelength for Ge and Si photodetectors.
(After Melchior and Lynch, Ref. 39.)

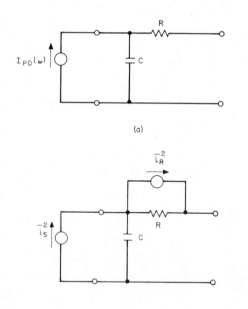

Fig. 25
(a) Equivalent circuit and
(b) noise equivalent circuit of a photodiode, where R is the series resistance and C is the junction capacitance.
(After DiDomenico and Svelto, Ref. 35.)

available power for the photodiode is then

$$P_{av} = \frac{1}{8}|I_{PD}(\omega)|^2 \frac{1}{\omega^2 C^2 R} = \frac{1}{8}(\eta q \Phi_1 A)^2 \frac{R}{(\omega CR)^2}. \qquad (43)$$

It is interesting to compare Eq. (43) with Eq. (39) for the photoconductor. For a typical photodiode with area 10^{-4} cm^2, $C = 2$ pf, and $R = 2$ ohms, and a photoconductor with the same area, $t_r = 10^{-7}$ sec, and $G = 10^{-3}$ mhos, the available power from the photodiode is about a million times larger than that from the photoconductor.

The signal-to-noise performance of a photodiode can be found from the equivalent noise circuit shown in Fig. 25(b). Here i_s^{-2} represents the thermal noise source due to the series resistance, and i_R^{-2} represents the shot noise source. The signal-to-noise ratio (SNR) in a bandwidth B is then given by

$$(S/N)_{power} = \frac{\eta}{4B} m^2 \Phi_0 A \left[1 + \frac{2kT}{q} \frac{(\omega RC)^2}{RI_s}\right]^{-1}. \qquad (44)$$

Comparing Eq. (44) with Eq. (42) for the photoconductor, we conclude that at high-level detection where the terms involving kT can be neglected, the SNR is comparable; at low-level detection where thermal noise predominates, however, the SNR of the photodiode is considerably higher.

B. The p-i-n Photodiode. The p-i-n photodiode is the most common depletion-layer photodetector. This is because the depletion region thickness (the intrinsic layer) can be tailored to optimize the sensitivity range and the frequency response. A typical structure of a p-i-n photodiode[40] is shown in Fig. 26(a). Absorption of light in the semiconductor produces hole-electron pairs. Pairs produced in the depletion region or within a diffusion length of it will eventually be separated by the electric field, leading to current flow in the external circuit as carriers drift across the depletion layer.

Under steady-state conditions the total current density through the reverse-biased depletion layer is given by[41]

$$J_{tot} = J_{dr} + J_{diff} \qquad (45)$$

where J_{dr} is the drift current due to carriers generated inside the depletion region and J_{diff} is the diffusion current density due to carriers generated outside the depletion layer in the bulk of the semiconductor and diffusing into the reverse-biased junction. We shall now derive the total current under the assumptions that the thermal generation current can be neglected and that the surface n layer is much thinner than $1/\alpha$. Referring to Fig. 26(b), the hole-electron generation rate is given by

$$G(x) = \Phi_0 \alpha e^{-\alpha x} \qquad (46)$$

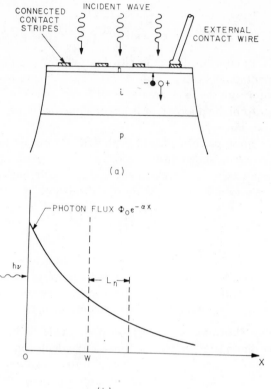

Fig. 26
(a) *p-i-n* photodiode. The holes and electrons produced in the intrinsic region will drift
   towards the *p* side and *n* side, respectively, under the applied reverse bias.
(After Rietz, Ref. 40.)
(b) Light intensity versus distance where *W* is the width of the depletion region.

where Φ_0 is the total incident photon flux. J_{dr} is thus given by

$$J_{dr} = -q \int_0^W G(x)\, dx = q\Phi_0(1 - e^{-\alpha W}) \qquad (47)$$

where *W* is the boundary of the depletion layer. For $x > W$, the minority
carrier density (electrons) in the bulk semiconductor is determined by the one-
dimensional diffusion equation:

$$D_n \frac{\partial^2 n_p}{\partial x^2} - \frac{n_p - n_{po}}{\tau_n} + G(x) = 0 \qquad (48)$$

where D_n is the diffusion coefficient for electrons, τ_n the lifetime of excess carriers, and n_{po} the equilibrium electron density. The solution of Eq. (48) under the boundary conditions $n_p = n_{po}$ for $x = \infty$ and $n_p = 0$ for $x = W$ is given by

$$n_p = n_{po} - (n_{po} + C_1 e^{-\alpha W})e^{(W-x)/L_n} + C_1 e^{-\alpha x} \tag{49}$$

with $L_n = \sqrt{D_n \tau_n}$ and

$$C_1 \equiv \left(\frac{\Phi_0}{D_n}\right)\frac{\alpha L_n^2}{(1 - \alpha^2 L_n^2)}. \tag{50}$$

The diffusion current density is given by $J_{\text{diff}} = -qD_n (\partial n_p/\partial x)|_{x=W}$:

$$J_{\text{diff}} = q\Phi_0 \frac{\alpha L_n}{(1 + \alpha L_n)} e^{-\alpha W} + qn_{po}\frac{D_n}{L_n}. \tag{51}$$

And the total current density is obtained as

$$J_{\text{tot}} = q\Phi_0 \left[1 - \frac{e^{-\alpha W}}{(1 + \alpha L_n)}\right] + qn_{po}\frac{D_n}{L_n}. \tag{52}$$

Under normal operating conditions the term involving n_{po} is much smaller so that the total photocurrent is proportional to the photon density flux. In addition, for a large response from Eq. (52) it is desirable to have $\alpha W \gg 1$ and $\alpha L_n \gg 1$. However, for $W \gg 1/\alpha$ there may be considerable transit time delay. We shall consider the transit time effect next.

Since the carriers require a finite time to traverse the depletion layer, a phase difference between the photon flux and the photovoltage and current will appear when the incident light intensity is modulated rapidly. To obtain a quantitative result for this effect the simplest case is shown in Fig. 27(a). The applied voltage is assumed to be high enough to deplete the intrinsic region and to ensure carrier scattering-limited velocity, v_{sl}. For a photon flux density given by $\Phi_1 e^{j\omega t}$ (photons/sec-cm^2), the conduction current density J_{cond} at point x is found to be

$$J_{\text{cond}}(x) = q\Phi_1 e^{j\omega(t - x/v_{sl})}. \tag{53}$$

Since $\nabla \cdot J_{\text{tot}} = 0$, we can write

$$J_{\text{tot}} = \frac{1}{W}\int_0^W \left(J_{\text{cond}} + \varepsilon_s \frac{\partial \mathscr{E}}{\partial t}\right) dx \tag{54}$$

where the second term in the parentheses is the displacement current density and ε_s and $\mathscr{E}$ are the permittivity and electric field respectively. Substitution of Eq. (53) into Eq. (54) yields

$$J_{\text{tot}} = \left[\frac{j\omega\varepsilon_s V}{W} + q\Phi_1 \frac{(1 - e^{-j\omega t_r})}{j\omega t_r}\right]e^{j\omega t} \tag{55}$$

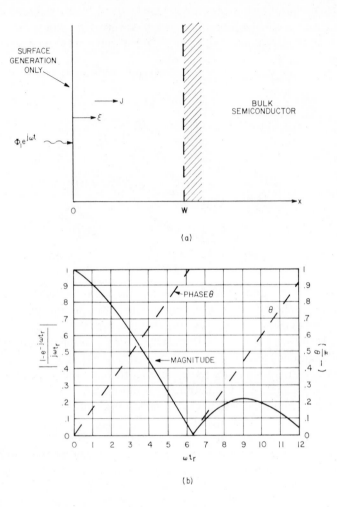

Fig. 27

(a) Geometry assumed for the analysis of transit-time effect.

(b) Photoresponse (normalized current or voltage) versus normalized frequency of incident photon flux where $\theta \equiv \omega t_r / 2$.

(After Gartner, Ref. 41.)

where V is the applied voltage and $t_r \equiv W/v_{sl}$ is the transit time of carriers through the depletion region. From Eq. (55) the short-circuit current density $(V = 0)$ is given by

$$J_{sc} = \frac{q\Phi_1(1 - e^{j\omega t_r})}{j\omega t_r} e^{j\omega t} \qquad (56)$$

Figure 27(b) shows the transit time effects at high frequencies where the amplitude and phase angle of the normalized current are plotted as a function of the normalized modulation frequency.[41] We note that the amplitude of the ac photocurrent decreases rapidly with frequency when ωt_r exceeds unity. At $\omega t_r = 2.4$, the amplitude is reduced by $\sqrt{2}$, and is accompanied by a phase shift of $2\pi/5$. The response time of the photodetector is thus limited by the carrier transit time through the depletion layer. A reasonable compromise[31] between high-frequency response and high quantum efficiency is obtained for an absorption region of thickness $1/\alpha$.

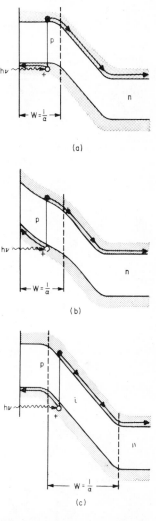

(a)

(b)

(c)

For the p-i-n diode the thickness of the i region is assumed equal to $1/\alpha$. The carrier transit time is the time required for carriers to drift through the i region. From Eq. (56) the 3-dB frequency, with respect to the low-frequency level, is given by

$$f_{\mathrm{drift}} \geq \frac{2.4}{2\pi t_r} \simeq 0.4 \frac{v_{sl}}{W} \approx 0.4\alpha v_{sl}. \quad (57)$$

C. p-n Junction Photodiodes. A p-n junction photodiode may be defined as a photodiode in which most of the absorption takes place in either the n or the p region, rather than in the depletion region. The energy band diagram of a conventional p-n junction[42] under reverse bias is shown in Fig. 28(a). If one varies the doping profile in such a way as to create an enhanced electric field[43] near the illuminated surface, the diode is a graded p-n junction with a built-in field as shown in Fig. 28(b). For comparison, Fig. 28(c) shows the band diagram of a p-i-n diode discussed in the previous subsection.

For the structure shown in Fig. 28(a), the depletion region is assumed to be sufficiently thin that the drift transit time through that region is smaller than the diffusion time

Fig 28 Energy band diagrams for
(a) a p-n junction,
(b) a graded p-n junction with a built-in field, and
(c) a p-i-n photodiode.
(After Lucovsky et al., Ref. 36.)

through the p region. Here 3-dB frequency for those carriers collected by diffusion through the p-region is

$$f_{\text{diff}} = \frac{2.4D_n}{2\pi W^2} \cong 0.4\alpha^2 D_n \tag{58}$$

where W is the depletion-layer width (which is assumed equal to $1/\alpha$) and D_n is the diffusion length of electrons. For light incident from the n side, D_p should be used for the diffusion length. Since generally $D_n > D_p$, the structure shown in Fig. 28(a) is preferred.

For the graded structure with built-in field the carrier transport involves both drift and diffusion processes and the 3-dB frequency is restricted to the range:[43,36]

$$0.4\alpha v_{sl} > f_{\text{drift-diff}} > 0.4\alpha^2 D_n. \tag{59}$$

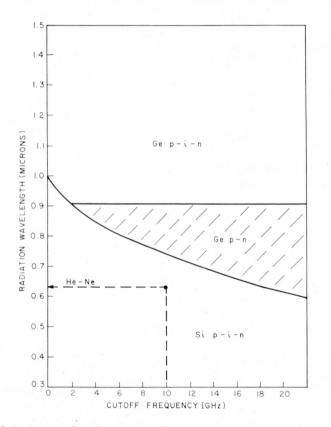

Fig. 29 Radiation wavelength versus cutoff frequency operating region for Ge and Si photodiodes.
(After Lynch, Ref. 44.)

For a given modulation frequency and for a given radiation wavelength, one must consider both the wavelength response and the modulation frequency response of various photodiodes in order to make the optimum choice. Figure 29 represents the areas of operation of Si and Ge photodiodes as a function of the 3-dB frequency desired and the radiation wavelength.[44] In this figure it is assumed that collection efficiencies are to be greater than 60% independent of losses due to contact areas or reflection. This plot is obtained by consideration of the dependence of α on λ, Fig. 23, and the cutoff frequency relationships, Eqs. (57) through (59). For example, in order to detect the output of a He-Ne laser light (~ 6300 Å) at a modulation frequency of 10 GHz, a Si p-i-n photodiode is the best choice. In cases where either Si or Ge may be used, Si photodiodes are preferred because of their lower series resistance (higher impurity solubility in the diffused layer) and lower capacitance (lower dielectric constant). Where neither Si nor Ge p-i-n photodiodes suffice, Ge p-n structures are chosen.

D. Metal-Semiconductor Photodiode. A metal-semiconductor diode can also be used as a high-efficiency photodetector.[45,46] The energy band diagram and current transport in a metal-semiconductor diode have been considered in Chapter 8. A typical configuration is illustrated in Fig. 30. To avoid large reflection and absorption losses when the diode is illuminated through the metal contact, the metal film must be very thin (~ 100 Å) and an antireflection coating must be used.

The equilibrium energy band diagram of a Schottky diode is shown in Fig. 30(b) where ϕ_{Bn} represents the Schottky barrier height. The diode can be operated in various modes depending on the photon energies and the biasing conditions:

(a) For $E_g > h\nu > q\phi_{Bn}$ and for $V \ll V_B$, Fig. 31(a), where V_B is the avalanche breakdown voltage, the photo-excited electrons in the metal can surmount the barrier and be collected by the semiconductor. This process has been extensively used to determine the Schottky barrier height, and to study the hot-electron transport in metal films.[47]

(b) For $h\nu > E_g$ and $V \ll V_B$, Fig. 31(b), the radiation produces hole-electron pairs in the semiconductor, and the general characteristics of the diode are very similar to those of a p-i-n photodiode.

(c) For $h\nu > E_g$ and $V \simeq V_B$ (high reverse-bias voltage), Fig. 31(c), the diode can be operated as an avalanche photodiode[48] (to be discussed in the next subsection).

In addition it can be used as an efficient ultraviolet detector.[49,50] In the ultraviolet region the absorption coefficients α in most of the common semiconductors are very high, of the order of 10^5 cm^{-1} or more, corresponding to an effective absorption length of $1/\alpha \simeq 0.1$ μm or less. It is possible to choose a

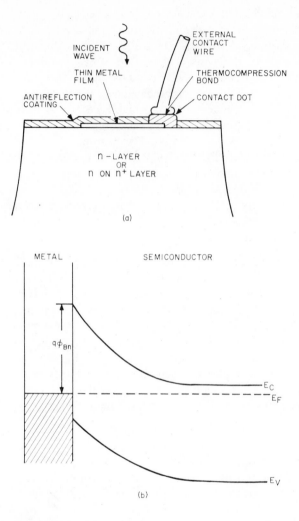

Fig 30.
(a) Schematic metal-semiconductor photodiode with antireflection coating.
(b) Energy band diagram of a metal-semiconductor barrier.
(After Schneider, Ref. 46.)

proper metal and a proper antireflection coating such that a large fraction of the incident radiation will be absorbed near the surface of the semiconductor where the maximum multiplication occurs.

An interesting example is shown in Fig. 32 to demonstrate that the transmission with low loss into the semiconductor substrate is feasible.[46] Figure

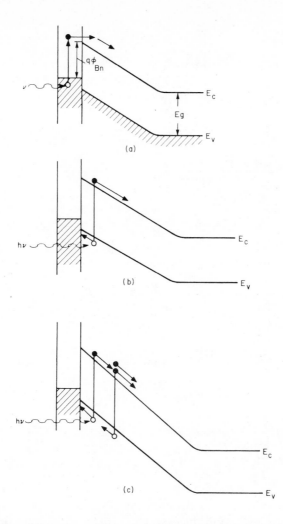

Fig. 31
(a) Photoelectric emission of excited electrons from metal to semiconductor $(E_g > h\nu > q\phi_{Bn})$.
(b) Band-to-band excitation of hole-electron pairs $(h\nu > E_g)$.
(c) Hole-electron pair generation and avalanche multiplication under large reverse bias $(h\nu > E_g$ and $V \approx V_B$ where V_B is the breakdown voltage).

32(a) shows a particular choice of antireflection coating of 500 Å ZnS with a refractive index of 2.30. The gold film has a complex refractive index $\bar{n} = (0.28 - j0.301)$ at $\lambda = 6328$ Å (He-Ne laser wavelength), while at this wavelength the silicon substrate has a refractive index of $(3.75 - j0.018)$. Figure 32(b) shows the calculated transmittance, reflectance, and loss in the

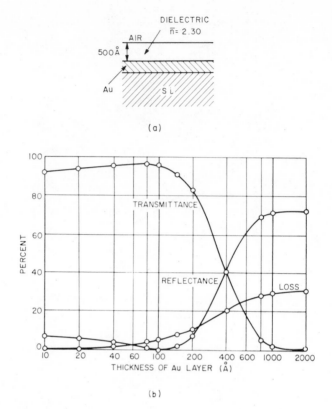

Fig. 32   Reflectance, transmittance, and loss from Au-Si surface barrier with 500 Å thick ZnS antireflection coating.
(After Schneider, Ref. 46.)

Au films as a function of the Au layer thickness. It is apparent that at about 100 Å more than 95 % of the incident light will be transmitted into the silicon substrate. Experimental results with photodiodes similar to that shown in Fig. 30(a) show a net quantum efficiency of 70 % at $\lambda = 6328$ Å, and a very fast response of the order of 0.1 ns (1 nanosecond $= 10^{-9}$ sec) pulse rise time.[46]

E.   Heterojunction and Point Contact Photodiodes.   A depletion-layer photodiode can also be realized in a heterojunction in which the junction is formed from two semiconductors of different band gaps (refer to Section 8 of Chapter 3). For a heterojunction photodiode, the frequency response does not depend critically on the distance of the junction from the surface but depends rather on the relative absorption of the incident radiation in the two materials making the heterojunction. Figure 33(a) shows schematically a photodiode containing an *n*-GaAs and *p*-Ge junction, which is designed for high-speed response to the 8450 Å radiation emitted from a GaAs diode infrared source.[51] At this wavelength the absorption coefficient is 10 cm^{-1} in GaAs and 3.2×10^4 cm^{-1} in Ge. Thus most of the light is absorbed in Ge within 1 μm. Since the carriers are produced near the junction, not near the surface, these diodes should have a high-frequency response, and are relatively insensitive to surface conditions.

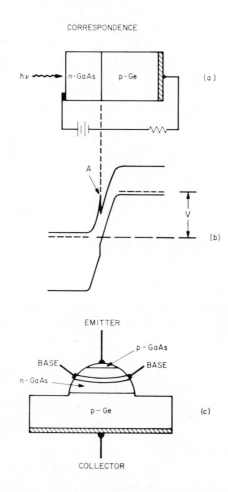

Fig. 33
(a) Schematic diagram ef heterojunction photodiode with applied reverse bias.
(b) Energy band diagram of the diode where *A* is due to the discontinuity in conduction band.
(c) Schematic diagram of a beam-of-light transistor.
(After Rediker et al., Ref. 51.)

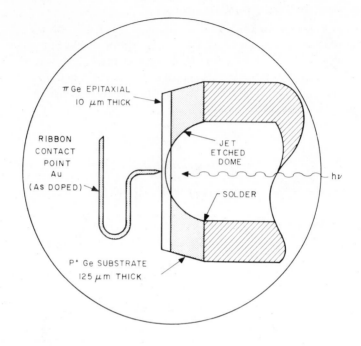

Fig. 34   Point-contact photodiode.
(After Sharpless, Ref. 52)

The band diagram of the heterodiode is shown in Fig. 33(b). Point A is due to the discontinuity in the conduction band. The barrier at A can be penetrated by tunneling or be surmounted by carriers with sufficiently high energy. As an extension of the heterodiode it is possible to fabricate a transistor with an infrared-emitting GaAs p-n junction as the emitter, and an n-GaAs p-Ge heterjunction as a collector. Such a structure is shown in Fig. 33(c) and is called the beam-of-light transistor.[51] The basic advantages of this device are that the transport from emitter to collector is at the speed of light in the material, and that essentially all the radiation incident on the collector junction produces carriers within one micron of this junction.

Photodiodes can also be made using a point contact geometry[52,53] as shown in Fig. 34. The active volume is very small, and as a result both the drift time and capacitance are extremely small. It is thus suitable for very high modulation frequencies. The device is limited, however, to applications where the radiation can be focused on a spot a few microns in diameter.

(3) Avalanche Photodiode

Avalanche photodiodes,[48] which are operated at high reverse bias voltages where avalanche multiplication takes place, are the most promising solid-state photodetectors with internal current gain. The current gain-bandwidth product of the multiplication mechanism can be as high as 100 GHz, so that substantial gain can be achieved at microwave frequencies.[31]

The basic requirement for the production of useful current gain in such a device is the elimination of microplasmas, i.e., small areas in which the breakdown voltage is less than that of the junction as a whole. To obtain spatially uniform multiplication, one must also eliminate the edge breakdown that results from the junction curvature effect.[55] Two typical experimental structures are shown in Fig. 35(a) and (b) for, respectively, the planar diode with diffused guard ring[54] and the mesa diode with diffused guard ring.[39] The guard ring has a lower impurity gradient and a sufficiently large radius of curvature that the central active region breaks down first. The probability of microplasmas occurring in the active area itself can be minimized by use of substrate material of uniform doping concentration, low dislocation density, and by designing the active area to be no larger than necessary to accommodate the incident light beam (generally from 10 μm to 500 μm in diameter).

The photomultiplication factor M_{ph} which is defined as the multiplied photocurrent I_{ph} divided by the photocurrent I_{pho} at low voltages where no carrier multiplication takes place, can be described by

$$M_{ph} \equiv \frac{I_{ph}}{I_{pho}} = \frac{1}{1 - \left(\dfrac{V}{V_B}\right)^n} \tag{60}$$

where V_B is the breakdown voltage, V the applied voltage, and n a constant depending on the semiconductor material, doping profile, and radiation wavelength.[39] A typical set of static current-voltage characteristics is shown in Fig. 36 for various primary currents I_p which are the sum of the photocurrent I_{pho} and the dark current I_{do}. The multiplication factor, defined as the ratio of the total output current I to the primary current, is given by

$$M = \frac{I}{I_p} = \frac{I_{ph} + I_d}{I_{pho} + I_{do}} = \frac{1}{1 - \left(\dfrac{V - IR}{V_B}\right)^n} \tag{61}$$

where I_d is the multiplied dark current and R is the sum of the series resistance and the space-charge resistance. Under the conditions that $I_{pho} \gg I_{do}$, and $IR \ll V_B$, the maximum value of the photomultiplication for a chopped light beam is given by[39]

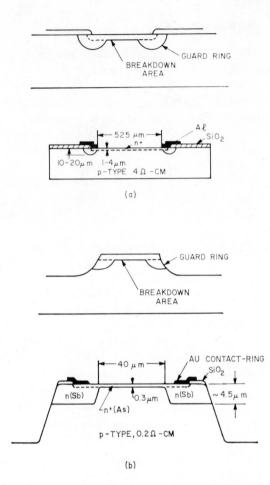

Fig. 35
(a) Cross-section view of planar photodiode with diffused guard ring.
(After Anderson et al., Ref. 54; and Baertsch, Ref. 56.)
(b) Cross-section view of mesa photodiode with diffused guard ring.
(After Melchior and Lynch, Ref. 39.)

$$(M_{ph})_{\text{opt}} \equiv \frac{I - I_d}{I_p - I_{do}} \simeq \left. \frac{I}{I_p} \right|_{\text{opt}} = \left. \frac{1}{1 - \left(\dfrac{V - IR}{V_B} \right)^n} \right|_{V \to V_B} \simeq \frac{1}{\dfrac{nIR}{V_B}}, \quad (62a)$$

or

$$(M_{ph})_{\text{opt}} \simeq \sqrt{\frac{V_B}{nI_p R}}. \quad (62b)$$

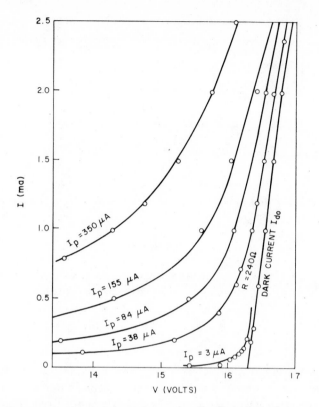

Fig. 36 Static current-voltage characteristics of a Ge photodiode for various primary currents.
(After Melchior and Lynch, Ref. 39.)

The above equation shows that the maximum photomultiplication depends inversely upon the square root of I_p. This indicates why it is important to keep the dark current I_{do} as low as possible in order to achieve high carrier multiplication and sensitivity.

The equivalent circuit[39] of an avalanche photodiode is shown in Fig. 37 where $I_s(\omega)$ and $i_n^{-2}(\omega)$ are the signal and noise current generators respectively, M is the multiplication factor, C is the junction capacitance, and R_S and R_L are the series and load resistances respectively. Under normal operating conditions, the available power is given by

$$P_{av} \simeq \frac{I_s^2(\omega)M^2 R_L}{1 + \omega^2 C^2 (R_S + R_L)^2}.$$

(63)

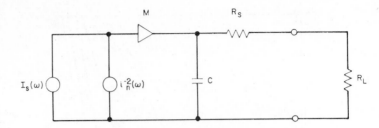

Fig. 37 Equivalent circuit of an avalanche photodiode.
(Ref. 39.)

The noise generated by the multiplication process in the white noise part of the spectrum is given by[57]

$$i_n^{-2} = 2qI_p BM^3 \tag{64}$$

for $\alpha_n = \alpha_p$ where α_n and α_p are the ionization rates of electrons and holes respectively. If $\alpha_n \neq \alpha_p$ the noise can be expressed approximately as

$$i_n^{-2} = 2qI_p BM^p \qquad \text{with} \qquad 2 \le p \le 4. \tag{65}$$

The theoretical noise spectral density as a function of the multiplication factor is shown in Fig. 38(a) where $k \equiv (\alpha_p/\alpha_n)$. Also shown in Fig. 38(a) are some experimental results obtained from silicon avalanche photodiodes. The upper values (0) represent the noise due to hole primary photocurrent resulting from radiation of short wavelength as shown in Fig. 38(b). The lower values (Δ) represent the noise due to electron primary photocurrent. The agreement between theory and experiments is very good, since the ratio α_p/α_n in Si is about 0.1 to 0.2. From Eq. (65) and the equivalent circuit, Fig. 37, the noise power is given by[39]

$$P_N = \frac{2qI_p BM^p R_L}{1 + \omega^2 C^2 (R_S + R_L)^2}. \tag{66}$$

Figure 39 shows the results of the signal and noise output of a Ge avalanche photodiode for a modulation frequency of 1.5 GHz and an average primary current of 40 μamp. We note that between $M = 1$ and 30, the signal power increases as M^2 and the noise power as M^3. This is in good agreement with the theoretical prediction, since for Ge the ionization rates are almost equal so that $p = 3$, Eq. (64). We also note that in Fig. 39 the highest signal-to-noise ratio (~ 20 dB) is obtained at $M \simeq 4$, i.e., where the noise contribution from

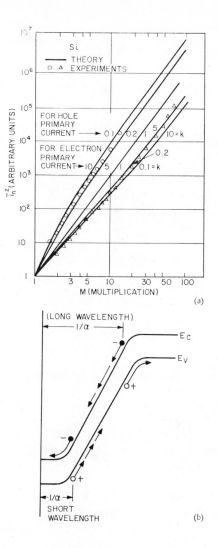

Fig. 38

(a) Equivalent saturated noise diode current at 600 kHz for a Si photodiode with 0.1 μA primary current. Solid curves are theoretical calculations.

(After McIntyre, Ref. 57.)

Data points are experimental results.

(After Baertsch, Ref. 56.)

(b) Energy band diagram of an avalanche photodiode with electron or hole primary current depending on the wavelength of the incident light.

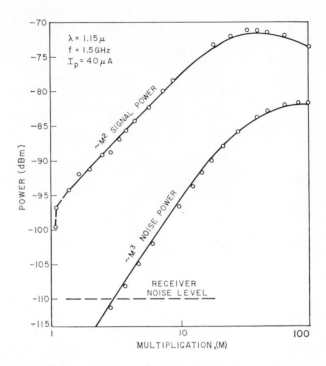

Fig. 39 Signal and noise power output of a Ge avalanche photodiode. Both are measured with respect to 1 mW power level.
(After Melchior and Lynch, Ref. 39.)

the diode is about equal to the receiver noise. At higher values of M the S/N ratio decreases because avalanche noise increases faster than multiplied signal.

The characteristics of Schottky-barrier avalanche photodiodes are similar to those of diffused photodiodes.[50] Schottky-barrier photodiodes have been made on a 0.5-Ω-cm n-type silicon substrate with a thin PtSi film (~ 100 Å) and a diffused guard ring[58] as shown in Fig. 40. Since the edge leakage current is eliminated by the guard ring, an ideal reverse saturation current density of 2×10^{-7} amp/cm^2 can be obtained. For the Schottky-barrier photodiode, avalanche multiplication can amplify the peak value of fast photocurrent pulses of 0.8 ns duration by factors of up to 35 as shown in Fig. 41. The difference between the photocurrent I_{ph} (at low voltage without multiplication) and the multiplied photocurrent shows how much current gain is possible at a certain level for pulses of certain duration (0.8 ns for the present case).

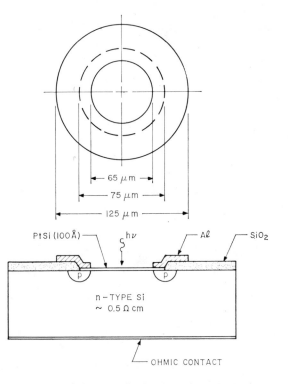

Fig. 40   Cross-sectional view of platinum silicide-silicon Schottky-barrier photodiode with guard ring.
(After Melchior et al., Ref. 50.)

The average light power from the 6328 Å laser is 0.8 mW and the peak power of the pulses corresponding to zero dB in Fig. 41 is approximately 7 mW. It is expected that higher current gain than that shown in Fig. 41 can be obtained by improvement of the guard ring structure and substrate material.

Noise measurements for avalanche multiplication in PtSi-Si diodes are shown in Fig. 42 which shows the noise spectral density in a narrow frequency range ($B = 0.6$ MHz) centered around 30 MHz. The noise of the multiplied photocurrent is found to increase approximately with the third power of the multiplication ($\sim M^3$) for excitation with light in the visible range. As the wavelength decreases, the multiplication noise increases less rapidly with M. The above results are in good agreement with McIntyre's theory[57] as shown in Fig. 38.

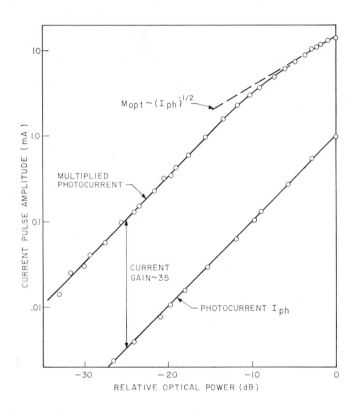

Fig. 41 Amplitudes of current pulses in Schottky-barrier photodiode as a function of intensity of 0.8 *ns* light pulse at 6328 Å (He-Ne laser pulse). Lower curve shows photocurrent at low reverse bias voltages. Upper curve shows highest pulse amplitudes of multiplied photocurrent reached at the breakdown voltage.
(Ref. 50.)

Schottky-barrier avalanche photodiodes with *n*-type silicon substrates promise to be particularly useful as high-speed photodetectors for UV light. UV light which is transmitted through the thin metal electrodes is absorbed within the first 100 Å of the silicon. The carrier multiplication is then mainly initiated by electrons; this results in a low noise and high gain-bandwidth product. Avalanche multiplication of photocurrent increases the sensitivity of detection systems consisting of Schottky-barrier photodiodes and a receiver. Amplification of high-speed photocurrent pulses is also possible.

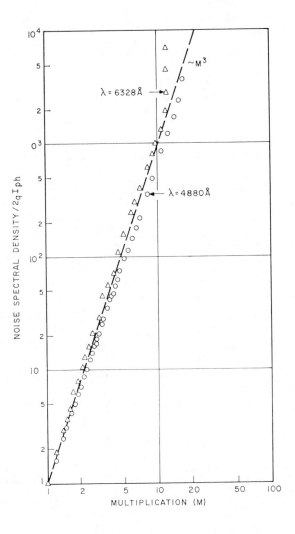

Fig. 42 Spectral density of avalanche multiplication noise at 30 MHz normalized to shot noise of photocurrent versus multiplication of photocurrent. Measurements for two different wavelengths of light are given.
(Ref. 50.)

REFERENCES

1. F. W. Sears and M. W. Zemansky, *College Physics*, 3rd Ed., Addison-Wesley Book Co. (1960).

2a. P. Goldberg, ed., *Luminescence of Inorganic Solids*, Academic Press, New York and London (1966).

2b. P. J. Dean, "A Survey of Radiative and Nonradiative Recombination Mechanisms in the III-V Compound Semiconductors," Trans. Metallurgical Society of AIME, *242*, 384 (1968).

3. H. F. Ivey, "Electroluminescence and Semiconductor Lasers," IEEE J. Quantum Electronics, *QE-2*, 713 (1966).

4. A. Mooradian and H. Y. Fan, "Recombination Emission in InSb," Phys. Rev, *148*, 873 (1966).

5. H. F. Ivey, *Electroluminescence and Related Effects*, Suppl. 1, to *Advances in Electronics and Electron Physics*, p. 205, Academic Press, New York and London (1963).

6. S. Wang, *Solid State Electronics*, McGraw-Hill Book Co., New York (1966).

7. C. H. Henry, P. J. Dean, and J. D. Cuthbert, "New Red Pair Luminescence from GaP," Phys. Rev., *166*, 754 (1968).

7a. M. Gershenzon, "State of the Art of GaP Electroluminescent Junction," Bell Syst. Tech. J., *45*, 1599 (1966).

8. H. K. Henisch, *Electroluminescence*, Macmillan Book Co., New York (1962).

9. P. C. Eastman, R. R. Haering, and P. A. Barnes, "Injection Electroluminescence in Metal-Semiconductor Tunnel Diodes," Solid State Electron., *7*, 879 (1964).

10. W. N. Carr, "Characteristics of a GaAs Spontaneous Infrared Source with 40 Percent Efficiency," IEEE Trans. Electron Devices, *ED-12*, 531 (1965).

11. D. E. Hill, "Internal Quantum Efficiency of GaAs Electroluminescent Diodes," J. Appl. Phys., *36*, 3405 (1965).

12. W. N. Carr and G. E. Pittman, "One Watt GaAs *p-n* Junction Infrared Source," Appl. Phys. Letters, *3*, 173 (1963).

13. S. V. Galginaitis, "Improving the External Efficiency of Electroluminescent Diodes," J. Appl. Phys., *36*, 460 (1965).

14. W. T. Lynch and R. A. Furnanage, "Planar Beam-Lead Gallium Arsenide Electroluminescent Arrays," IEEE Trans. Electron Devices, *ED-14*, 705 (1967).

15. D. M. Chapin, C. S. Fuller, and G. L. Pearson, "A New Silicon *p-n* Junction Photocell for Converting Solar Radiation into Electrical Power," J. Appl. Phys., *25*, 676 (1954).

16. J. J. Wysocki, P. Rappaport, E. Davison, and J. J. Loferski, "Low-Energy Proton Bombardment of GaAs and Si Solar Cells," IEEE Trans. Electron Devices, *ED-13*, 420 (1966).

17. E. R. Hill and B. G. Keramidas, "A Model for the CdS Solar Cell," IEEE Trans. Electron Devices, *ED-14*, 22 (1967).

18. A. Kunioka and Y. Sakai, "Optical and Electrical Properties of Se-CdS Photovoltaic Cells," Solid State Electron., *8*, 961 (1965).

19. P. Vohl, D. M. Perkins, S. G. Ellis, R. R. Addiss, W. Hui, and G. Noel, "GaAs Thin-Film Solar Cells," IEEE Trans. Electron Devices, *ED-14*, 26 (1967).

20. A. E. Spakowski, "Some Problems of the Thin-Film CdS Solar Cells," IEEE Trans. Electron Devices, *ED-14*, 18 (1967).

21. M. B. Prince, "Silicon Solar Energy Converters," J. Appl. Phys., *26*, 534 (1955).

22. J. J. Wysocki and P. Rappaport, "Effect of Temperature on Photovoltaic Solar Energy Conversion," J. Appl. Phys., *31*, 571 (1960).

23. W. Shockley and H. J. Queisser, "Detailed Balance Limit of Efficiency of *pn* Junction Solar Cells," J. Appl. Phys., *32*, 510 (1961).

24. L. M. Terman, "Spectral Response of Solar-Cell Structures," Solid State Electron., *2*, 1 (1961).

25. R. N. Hall, "Electron-Hole Recombination in Germanium," Phys. Rev., *87*, 387 (1952); W. Shockley and W. T. Read, "Statistics of the Recombination of Holes and Electrons," Phys. Rev., *87*, 835 (1952).

26. H. J. Queisser, "Forward Characteristics and Efficiencies of Silicon Solar Cells," Solid State Electron., *5*, 1 (1962).

27. R. J. Handy, "Theoretical Analysis of the Series Resistance of a Solar Cell," Solid State Electron., *10*, 765 (1967).

28. W. Rosenzweig, H. K. Gummel, and F. M. Smits, "Solar Cell Degradation Under 1-MeV Electron Bombardment," Bell Syst. Tech. J., *42*, 399 (1963).

29. R. L. Statler, "An Evaluation by Solar Stimulation of Radiation Damage in Silicon Solar Cells," IEEE Trans. Electron Devices, *ED-14*, 31 (1967).

30. R. V. Tanke and B. J. Faraday, "Proton-Irradiation Study of Pulled and Float-Zone Silicon Solar Cells," Proc. IEEE, *55*, 234 (1967).

31. For an excellent review on high-speed photodiodes see L. K. Anderson and B. J. McMurtry, "High Speed Photodetectors," Proc. IEEE, *54*, 1335 (1966).

32. For a review, see, for example, M. Ross, *Laser Receivers—Devices, Techniques, Systems*, John Wiley & Sons, Inc., New York (1966).

32a. P. W. Kruse, L. D. McGlauchlin, and R. B. McQuistan, *Elements of Infrared Technology*, John Wiley & Sons, Inc. (1962).

33. For a lucid review, see R. H. Bube, "Comparison of Solid State Photoelectronic Radiation Detectors," AIME Trans., *239*, 291 (1967).

34. H. Levinstein, "Extrinsic Detectors," Appl. Opt., *4*, 639 (1965).

35. M. DiDomenico, Jr., and O. Svelto, "Solid State Photodetection Comparison between Photodiodes and Photoconductors," Proc. IEEE, *52*, 136 (1964).

36. G. Lucovsky, M. E. Lasser, and R. B. Emmons, "Coherent Light Detection in Solid-State Photodiodes," Proc. IEEE, *51*, 166 (1963).

37. A. Van der Ziel, *Fluctuation Phenomena in Semiconductors*, Academic Press Inc., New York, Ch. 6. (1959).

38. W. C. Dash and R. Newman, "Intrinsic Optical Absorption in Single Cyrstal Germanium and Silicon at $77°K$ and $300°K$," Phys. Rev., *99*, 1151 (1955).

39. H. Melchior and W. T. Lynch, "Signal and Noise Response of High Speed Germanium Avalanche Photodiodes," IEEE Trans. Electron Devices, *ED-13*, 829 (1966).

39a. R. B. Emmons and G. Lucovsky, "The Frequency Response of Avalanching Photodiodes," IEEE Trans. Electron Devices, *ED-13* (1966).

40. R. P. Rietz, "High Speed Semiconductor Photodiodes," Rev. Sci. Int., *33*, 994 (1962).

41. W. W. Gartner, "Depletion-Layer Photoeffects in Semiconductors," Phys. Rev., *116*, 84 (1959).

42. D. E. Sawyer and R. H. Rediker, "Narrow Base Germanium Photodiodes," Proc. IRE, *46*, 1122 (1958).

43. A. G. Jordan and A. G. Milnes, "Photoeffect on Diffused *p-n* Junctions with Integral Field Gradients," IRE Trans. Electron Devices, *ED-7*, 242 (1960).

44. W. T. Lynch, unpublished results.

45. E. Ahlstrom and W. W. Gartner, "Silicon Surface-Barrier Photocells," J. Appl. Phys., *33*, 2602 (1962).

46. M. V. Schneider, "Schottky Barrier Photodiodes with Antireflection Coating," Bell Syst. Tech. J., *45*, 1611 (1966).

47. For a review on hot electron transport, see, for example, C. R. Crowell and S. M. Sze, "Hot Electron Transport and Electron Tunneling in Thin Film Structures," a chapter of *Physics of Thin Films*, Vol. 4, p. 325-371, ed. R. E. Thun, Academic Press, New York (1967).

48. K. M. Johnson, "High-Speed Photodiode Signal Enhancement at Avalanche Breakdown Voltage," IEEE Trans. Electron Devices, *ED-12*, 55 (1965).

49. R. B. Campbell and H. C. Chang, "Detection of Ultraviolet Radiation Using Silicon Carbide *p-n*Junctions," Solid State Electron., *10*, 949 (1967).

50. H. Melchior, M. P. Lepselter, and S. M. Sze, "Metal-Semiconductor Avalanche Photodiode," paper presented at IEEE Solid-State Device Research Conf., Boulder, Colorado, (June, 17-19 1968).

51. R. H. Rediker, T. M. Quist, and B. Lax, "High Speed Heterojunction Photodiodes and Beam-of-Light Transistors," Proc. IEEE, *51*, 218 (1963).

52. W. M. Sharpless, "Cartridge-Type Point Contact Photodiode," Proc. IEEE, *52*, 207 (1964).

53. L. U. Kibler, "A High-Speed Point Contact Photodiode," Proc. IEEE, *50*, 1834 (1962).

54. L. K. Anderson, P. G. McMullin, L. A. D'Asaro, and A. Goetzberger, "Microwave Photodiodes Exhibiting Microplasma—Free Carrier Multiplication," Appl. Phys. Letters, *6*, 62 (1965).

55. S. M. Sze and G. Gibbons, "Effect of Junction Curvature on Breakdown Voltage in Semiconductors," Solid State Electron., *9*, 831 (1966).

56. R. D. Baertsch, "Noise and Ionization Rate Measurements in Silicon Photodiodes," IEEE Trans. Electron Devices, *ED-13*, 987 (1966).

57. R. J. McIntyre, "Noise Theory for Read Type Avalanche Diode," IEEE Trans. Electron Devices, *ED-13*, 164 (1966).

58. M. P. Lepselter and S. M. Sze, "Silicon Schottky Barrier Diode with Near-Ideal I-V Characteristics," Bell Syst. Tech. J., *47*, 195 (1968).

- INTRODUCTION
- SEMICONDUCTOR LASER PHYSICS
- JUNCTION LASERS
- HETEROSTRUCTURE AND CONTINUOUS ROOM-TEMPERATURE OPERATION
- OTHER PUMPING METHODS AND LASER MATERIALS

13

Semiconductor Lasers

1 INTRODUCTION

A powerful new coherent source of energy became available in 1954 when Townes[1] and his collaborators operated the first maser (*M*icrowave *A*mplication by *S*timulated *E*mission of *R*adiation). Four years later Schawlow and Townes[2] proposed an extension of the microwave theory into the visible and infrared region of the spectrum. Their detailed theory formed the basis for the construction of the first optical maser or laser (*L* replacing *M* in maser and standing for Light)—a pulsed solid-state ruby laser by Maiman[3] in 1960. The three-level ruby laser system used the energy levels of Cr^{+3} imbedded in Al_2O_3. This system was soon followed by that of Sorokin and Stevenson[4] who made a four-level laser using U^{+3} in CaF_2, and by that of Javan et al.[5] which used a gas discharge in a He-Ne mixture.

During the years 1957 to 1961, Nishizawa and Watanabe,[6] Basov et al.,[7] and Aigrain[8] independently proposed the *p-n* junction semiconductor laser. In 1962 Dumke[9] showed that laser action was indeed possible in direct band-gap semiconductors and set forth important criteria for such action. Using available absorption data, he showed the possibility of using interband transitions in direct band-gap semiconductors such as gallium arsenide (GaAs). The theoretical calculations of Bernard and Duraffourg[10] in 1961 set forth the necessary conditions for lasing, using the concept of quasi-Fermi levels.

With many groups actively investigating lasing possibilities in semiconductors, it was not surprising that almost simultaneously three groups, headed by Hall,[11] Nathan,[12] and Quist and Rediker[13] announced achievement of lasing in late 1962. The pulsed radiation at 8400 Å was obtained from a liquid-nitrogen-cooled, forward-biased GaAs *p-n* junction. Shortly thereafter, Holonyak and Bevacqua[14] announced laser action in the mixed crystal of a

Ga(As$_{1-x}$P$_x$) junction at 7100 Å. This discovery demonstrated that the laser wavelength could be varied by varying the relative amounts of As and P. Since the time of these initial discoveries, many new laser materials have been found. The wavelength of coherent radiation has been extended through the visible into the ultraviolet and out to the mid-infrared.[15]

Recently continuous operation of junction lasers at room temperature has been achieved by Hayashi, Panish et al.[49,50] by the use of double hetero-junctions. This structure was first proposed by Kroemer[51] and Alferov.[52] By the use of a stripe geometry, continuous operating temperature higher than 80°C has been reported by Ripper et al.[53]

Semiconductor lasers are similar to other lasers (such as the conventional solid-state ruby laser and He-Ne gas laser) in that the emitted radiation has spatial and temporal coherence. This means that laser radiation is highly monochromatic (of small bandwidth) and that it produces highly directional beams of light.[16] However, semiconductor lasers differ from other lasers in several important respects:

(1) In conventional lasers the quantum transitions occur between discrete energy levels while in semiconductor lasers the transitions are associated with the band properties of materials.

(2) A semiconductor laser is very compact in size (of the order of 0.1 mm long). In addition, because the active region is very narrow (of the order of 1 μm thick) the divergence of the laser beam is considerably larger than in a conventional laser.

(3) The spatial and spectral characteristics of a semiconductor laser are strongly influenced by the properties of the junction medium (such as doping and band tailing).

(4) For the p-n junction laser, the laser action is produced by simply passing a current through the diode itself. Thus the pumping energy is used directly to populate the conduction band. The result is a very efficient overall system which can be modulated easily by modulating the current. Since semiconductor lasers have very short stimulated-emission lifetimes, modulation at high frequencies can be achieved.

In the next section we shall consider the basic principles of semiconductor lasers. The characteristics of p-n junction lasers will be discussed in Section 3 which will be primarily concerned with GaAs lasers. In Section 4 we shall consider the heterostructures and the room-temperature operation of junction lasers. Other pumping schemes, as well as other semiconductor materials for lasers, will be discussed in Section 5.

2 SEMICONDUCTOR LASER PHYSICS

(1) Transition Processes

There are three basic transition processes that relate to laser operation, namely: absorption, spontaneous emission, and stimulated emission. We shall use a simple system to demonstrate these processes.[17] Consider two energy levels E_1 and E_2 in an atom, as shown in Fig. 1, where E_1 is the ground state and E_2 is an excited state. Any transition between these states involves, according to Planck's law, the emission or absorption of a photon with frequency v_{12} given by $hv_{12} = E_2 - E_1$, where h is Planck's constant. At ordinary temperatures most of the atoms are in the ground state. This situation is disturbed when a photon of energy exactly equal to hv_{12} impinges on the system. An atom in state E_1 will absorb the photon and thereby go to the excited state E_2. This is the absorption process, Fig. 1(a). The excited state of the atom is

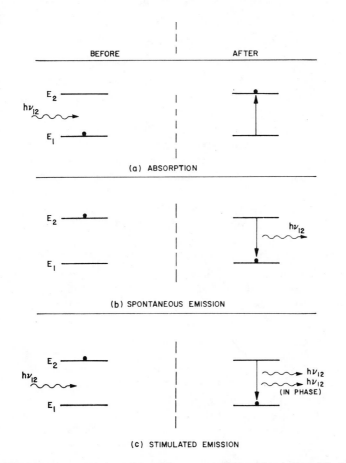

Fig. 1 The three basic transition processes between two energy levels E_1 and E_2. The black dots indicate the state of the atom. The initial state is at the left; the final state, after the process has occurred, is at the right.
(After Levine, Ref. 17,)

unstable and after a short time it will, without any external stimulus, make a transition to the ground state giving off a photon of energy hv_{12}. This process is called spontaneous emission, Fig. 1(b). The lifetime for spontaneous emission, i.e., the average time of the excited state, varies considerably ranging typically from 10^{-9} to 10^{-3} second depending on various semiconductor parameters such as band gap (direct or indirect) and density of recombination centers. An important and interesting event occurs when a photon of energy hv_{12} impinges on an atom while it is still in the excited state. In this case the atom is immediately stimulated to make its transition to the ground state and gives off a photon of energy hv_{12} which is in phase with the incident radiation. This process is called stimulated emission, Fig. 1(c).

(2)   Direct and Indirect Band-Gap Semiconductors

As discussed in Chapter 2, the energy band structures of semiconductors can be classified into two types: direct band-gap semiconductors such as GaAs in which the lowest conduction band minimum and the highest valence band maximum are at the same wave vector in the Brillouin zone; and indirect band-gap semiconductors such as Si in which the extrema are at different wave vectors. Since the optical transition in a direct band-gap semiconductor is a first-order process (i.e., the momentum is automatically conserved), the optical gain is expected to be high. On the other hand, the optical transition [Process (c) of Fig. 26 in Chapter 2] in an indirect band-gap semiconductor is a second-order process (i.e., it involves phonons or other scattering agents in order to conserve momentum and energy), thus the radiative transitions are weaker, and the available gain is smaller.

The above ideas apply only to band-to-band or exciton transitions in fairly pure semiconductors. For heavy doping and impurity transitions, the momentum selection rule does not necessarily hold. Nevertheless, at present all the well substantiated lasing semiconductors have direct-band structures.

(3)   Population Inversion

A population is said to be inverted when there are more atoms in the excited state than in the ground state. If photons of energy hv_{12} are incident on a system where the population of level E_2 is inverted with respect to E_1 (refer to Fig. 1), stimulated emission will exceed absorption and more photons of energy hv_{12} will leave the system than entered it. Such a phenomenon is called quantum amplification.

To consider the inversion condition for a semiconductor laser, we refer to Fig. 2 which shows the energy versus density of states in a direct band-gap semiconductor. Figure 2(a) shows the equilibrium condition at $T = 0°K$ for an

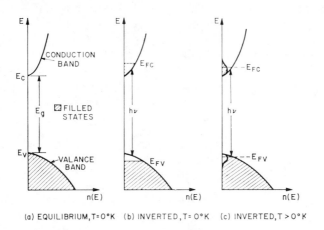

(a) EQUILIBRIUM, T=0°K (b) INVERTED, T=0°K (c) INVERTED, T>0°K

Fig. 2 Energy versus density of states in a semiconductor. (After Nathan, Ref. 15b.)

intrinsic semiconductor in which the shaded area represents the filled states. Figure 2(b) shows the situation for an inverted population at 0°K. This inversion can be achieved, for example, by photoexcitation with photon energy greater than the band gap E_g. The valence band is empty of electrons down to an energy E_{FV}, and the conduction band filled up to E_{FC}. Now photons with energy hv such that $E_g < hv < (E_{FC} - E_{FV})$ will cause downward transition and hence stimulated emission.

At finite temperatures, the carrier distributions will be "smeared" out in energy as shown in Fig. 2(c). Although overall thermal equilibrium does not exist, the carriers in a given energy band will be in thermal equilibrium with each other. The occupation probability of a state in the conduction band is given by the Fermi-Dirac distribution

$$F_C(E) = \frac{1}{1 + \exp\left(\dfrac{E - E_{FC}}{kT}\right)} \tag{1}$$

where E_{FC} is the imref (quasi-Fermi level) for electrons in the conduction band. A similar expression holds for the valence band.

Consider the rate of photon emission at v due to a transition from a group of upper states near E in the conduction band to lower states at $(E - hv)$ in the valence band. The rate for this emission is proportional to the product of the density of occupied upper states $n_C(E)F_C(E)$ and the density of unoccupied lower states $n_V(E - hv)[1 - F_V(E - hv)]$. The total emission rate is obtained by integrating over all energies:

$$W_{\text{emission}} \sim \int n_C(E)F_C(E)n_V(E - h\nu)[1 - F_V(E - h\nu)] \, dE. \tag{2}$$

In a similar manner we can write for the absorption rate

$$W_{\text{absorption}} \sim \int n_V(E - h\nu)F_V(E - h\nu)n_C(E)[1 - F_C(E)] \, dE \tag{3}$$

with the same proportionality constant which includes the square of the transition probability matrix element. For a net amplification we require that $W_{\text{emission}} > W_{\text{absorption}}$. From Eqs. (1), (2), and (3) with the appropriate imrefs E_{FV} and E_{FC}, we obtain[10]

$$E_{FC} - E_{FV} > h\nu. \tag{4}$$

This is a necessary condition for stimulated emission to be dominant over absorption for conduction-to-valence-band transitions in an intrinsic semiconductor. If the semiconductor contains impurities, and the impurity energy states are either the initial or final state, it is necessary only to use the imrefs for the impurities with the proper degeneracy for the states.

(4) Gain

A. Discrete Two-Level System. In a semiconductor laser the gain (g), i.e., the incremental energy flux per unit length, depends on the energy-band structure and is a complicated function of doping levels, current density, temperature, and frequency. In order to obtain some basic ideas of the gain function we shall first consider the simple two-level energy system shown in Fig. 1. The result for this system will be used later to give a qualitative expression for the threshold current to be discussed in Section 3(2).

The gain for the two-level system can be obtained by the following arguments originally discussed by Einstein.[18] Consider the interaction of an atom with a blackbody radiation field at temperature T whose energy density per unit frequency is given by

$$\rho(\nu) = \frac{8\pi h\nu^3}{c^3} \frac{1}{e^{h\nu/kT} - 1} \tag{5}$$

where c is the velocity of light in a vacuum. The total transition rate from level 2 to level 1 in the presence of the field is given by [Fig. 1(b) and 1(c)]:

$$W_{21} = B_{21}\rho(\nu) + 1/\tau_{\text{spont}} \tag{6}$$

and the transition rate from level 1 to 2, Fig. 1(a), is

$$W_{12} = B_{12}\rho(\nu) \tag{7}$$

where τ_{spont} is the lifetime of the spontaneous emission, and B_{21} and B_{12} are constants to be determined. At thermal equilibrium the rate of $2 \rightarrow 1$ transitions is equal to that of the $1 \rightarrow 2$ transitions, hence

$$N_2 W_{21} = N_1 W_{12}. \tag{8}$$

The population ratio N_2/N_1 is given by the Boltzmann factor

$$\frac{N_2}{N_1} = \frac{g_2}{g_1} e^{-h\nu/kT} \tag{9}$$

where g_2, g_1 are the degeneracies of levels 2 and 1 respectively. Combination of Eqs. (6) through (9) yields

$$\rho(\nu) = \frac{\dfrac{1}{\tau_{\text{spont}}} (g_2/g_1) e^{-h\nu/kT}}{B_{12} - B_{21}(g_2/g_1) e^{-h\nu/kT}}. \tag{10}$$

Since the whole system (atoms and field) is at thermal equilibrium, $\rho(\nu)$ as given by Eq. (10) is the same as that in Eq. (5). To satisfy the above equality, we obtain

$$B_{12} = B_{21}(g_2/g_1) \tag{11a}$$

and

$$\frac{1}{B_{21}} = \frac{8\pi h\nu^3 \tau_{\text{spont}}}{c^3}. \tag{11b}$$

From Eqs. (6) and (11) the stimulated transition rate due to interaction with the field is

$$(W_{21})_{\text{stim}} = \frac{c^3}{8\pi h\nu^3 \tau_{\text{spont}}} \rho(\nu). \tag{12}$$

The more general form of Eq. (12) is given by

$$(W_{21})_{\text{stim}} = \int_{-\infty}^{\infty} \frac{c^3 \rho(\nu')}{8\pi h\nu'^3 \tau_{\text{spont}}} A(\nu') \, d\nu' \tag{13}$$

where $A(\nu')$ is a normalized lineshape function for the transition, or $\int A(\nu') \, d\nu' = 1$. It is clear that, if $\rho(\nu')/\nu'^3$ is essentially constant over the absorption band of frequencies, Eq. (13) reduces to Eq. (12).

In a monochromatic radiation field of frequency ν and radiation density ρ_ν, the energy density per unit frequency is $\rho_\nu \delta(\nu' - \nu)$ which from Eq. (13) gives

$$W_{21} = \frac{c\lambda^2 \rho_\nu}{2\pi h\nu \tau_{\text{spont}}} A(\nu). \tag{14}$$

If the monochromatic wave propagates in a vacuum (in direction z) with level populations N_2 and N_1, the variation of the energy flux $I_\nu \equiv c\rho_\nu$ due to stimulated transition in the y-direction is

$$\frac{1}{I_\nu} \frac{dI_\nu}{dz} = (N_2 W_{21} - N_1 W_{12})h\nu$$

$$= -\frac{c^2[N_1(g_2/g_1) - N_2]A(\nu)}{8\pi\nu^2\tau_{\text{spont}}}. \tag{15}$$

The intensity will vary consequently as

$$I_\nu(z) = I_\nu(0)e^{-g(\nu)z} \tag{15a}$$

where

$$g(\nu) = -\frac{dI_\nu/dz}{I_\nu} = \frac{c^2[N_2 - N_1(g_2/g_1)]}{8\pi\nu^2\tau_{\text{spont}}} A(\nu). \tag{16}$$

The actual lineshapes encountered in experimental situations can often be approximated by either a Gaussian or a Lorentzian curve. The normalized Gaussian curve is

$$A(\nu) = \frac{2}{\Delta\nu} \sqrt{\frac{\ln 2}{\pi}}\, e^{-4\ln 2 \cdot (\nu - \nu_o)^2/\Delta\nu} \tag{17}$$

where ν_o is the central frequency and $\Delta\nu$ is the full width at half intensity. The normalized Lorentzian curve is

$$A(\nu) = \frac{\Delta\nu}{2\pi[(\nu - \nu_o)^2 + (\Delta\nu/2)^2]}. \tag{18}$$

To obtain $g(\nu)$ from Eq. (16), one can replace $A(\nu)$ by $A(\nu_o) \simeq 1/\Delta\nu$ (for a Lorentzian or a Gaussian curve one would use $2/\pi\Delta\nu$ or $2\sqrt{\ln 2/\pi}/\Delta\nu$ respectively).

B. Semiconductor Band System.   In a semiconductor laser the gain function is more complicated than the two-level system discussed above. As the excitation rate is increased, the distribution functions $F_C(E)$ and $F_V(E)$ change, i.e., E_{FC} increases and E_{FV} decreases. The shape of the gain curve versus photon energy also changes. If the matrix element is the same for all initial and final states, the spontaneous and the net stimulated emission functions can be written as[19]

$$W_{\text{spont}}(\nu) = B \int n_C(E)n_V(E - h\nu)F_C(E)[1 - F_V(E - h\nu)]\, dE \tag{19a}$$

$$W_{\text{stim}}(\nu) = B \int n_C(E)n_V(E - h\nu)[F_C(E) - F_V(E - h\nu)]\, dE. \tag{19b}$$

The coefficient B is given by

$$B = (4\bar{n}q^2 h\nu/m^2\hbar^2 c^3)|\langle M\rangle|^2\text{Vol} \tag{20}$$

where $\langle M \rangle$ is the matrix elements, Vol is the volume of the crystal, and $\bar{n}$ is the refractive index of of the semiconductor.

The total emission rate W_T is particularly simple when there is no selection rule, since one can integrate over conduction and valence bands independently to find

$$W_T = Bnp \tag{21}$$

where n and p are the concentrations of electrons and holes. W_T gives the rate at which photons are spontaneously emitted in unit volume, thus the coefficient B has dimensions cm^3/sec. From these relations and the Fermi distribution functions defined previously, the stimulated and spontaneous functions are related to each other by

$$W_{\text{stim}}(v) = W_{\text{spont}}(v)\left\{1 - \exp\left[\frac{E - (E_{FC} - E_{FV})}{kT}\right]\right\}. \tag{22}$$

The gain is related to the stimulated emission function by[19]

$$g(v) = \frac{\pi^2 c^2 \hbar^3}{\bar{n}^2 (hv)^2} W_{\text{stim}}(v)$$

$$= \frac{\pi^2 c^2 \hbar^3}{\bar{n}^2 (hv)^2} W_{\text{spont}}(v)\left\{1 - \exp\left[\frac{hv - (E_{FC} - E_{FV})}{kT}\right]\right\} cm^{-1}. \tag{23}$$

We see that the gain is positive corresponding to amplification when $hv < (E_{FC} - E_{FV})$. This is the inversion condition discussed in Section 2(3). On the other hand when $hv > (E_{FC} - E_{FV})$, the gain is negative, corresponding to the absorption of light.

We shall now calculate $g(v)$ for a special distribution of density of states as shown in Fig. 3 where the conduction band has a band tail as given by Kane[20] (discussed in Chapter 4) and the valence band has a high density of impurity states which merge with the main valence band. This distribution is important, since it is applicable to most of the GaAs injection lasers. Once the distribution of density of states is specified, for a given difference $\Delta E \equiv (E_{FC} - E_{FV})$ and a given temperature, the spontaneous function can be calculated from Eq. (19a), and the gain can be obtained from Eq. (23) as a function of ΔE. In presenting the results, one can also give $g(v)$ as a function of a nominal current density defined as[21]

$$J_{\text{nom}} \equiv qBnp(d_{\text{nom}})$$

$$= 1.6 \times 10^{-23} W_T \tag{24}$$

which is the current that must flow to maintain the rate W_T per cm^2-sec in a layer 1 μm $(=d_{\text{nom}})$ thick if the internal quantum efficiency is 100%.

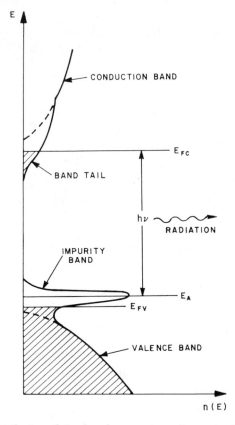

Fig. 3 A special distribution of density of states where the conduction band has a band tail and the valence band has a high density of impurity states which merges with the main valence band.
(After Popov, Ref. 48.)

Figure 4 shows the dependence of gain[21] on nominal current density for GaAs with donor and acceptor concentrations of 1×10^{18} and 4×10^{18} cm^{-3} respectively, for five temperatures between 0°K and 300°K. We note that the theoretical gain varies nearly linearly with current (unity slope) up to about 160°K. This result will be shown to be in good agreement with experiments. Figure 5 shows the theoretical temperature dependence of the nominal current density required to reach a gain of 100 cm^{-1} for four compositions with $N_A - N_D = 3 \times 10^{18}$ cm^{-3}. We note the striking variation of the temperature dependence with donor content. At lower temperatures the nominal current density increases with increasing doping. At higher temperatures and for a

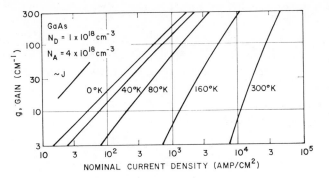

Fig. 4 Theoretical dependence of gain on nominal current density for GaAs lasers with donor and acceptor concentration of 1×10^{18} and 4×10^{18} cm^{-3}, respectively, for diode temperatures between 0°K and 300°K. (After Stern, Ref. 21.)

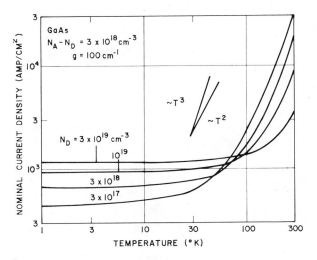

Fig. 5 Theoretical temperature dependence of the nominal current density required to reach a gain of 100 cm^{-1} in GaAs lasers. (After Stern, Ref. 21.)

given doping the current varies approximately as T^{γ} where γ lies between 2 and 3.

(5) Axial Modes

To discuss the most important concept of mode selection we shall employ the one-dimensional (z-direction) schematic laser structure, as shown in

Fig. 6, which has an active region sandwiched between two inactive absorbing regions. This simple structure will be considerably modified, however, when we investigate mode confinement and spectral distribution in a junction laser with three-dimensional electromagnetic wave propagations.

The basic mode selection for the z-direction (axial direction) in Fig. 6 arises from the requirement that only an integral number "q" of half wavelengths fits between the reflection planes (note here that "q" is not the electronic change but the axial mode number). Thus

$$q\left(\frac{\lambda}{2\bar{n}}\right) = L \qquad (25)$$

or

$$q\lambda = 2L\bar{n} \qquad (25a)$$

where $\bar{n}$ is the refractive index in the medium corresponding to the wavelength λ, and L is the length of the semiconductor. The spacing, $\Delta\lambda$, between these allowed modes in z-direction is the difference in the wavelengths corresponding to q and $q + 1$. Differentiating Eq. (25a) with respect to λ we obtain for large q,

$$\frac{L\Delta\lambda}{\lambda^2\Delta q} \simeq \frac{-1}{2\bar{n}\left(1 - \dfrac{\lambda}{\bar{n}}\dfrac{d\bar{n}}{d\lambda}\right)}. \qquad (26)$$

In the parentheses in Eq. (26) is the term which arises from dispersion. For gas lasers, $\bar{n}$ is nearly independent of λ and the dispersion term is only a small

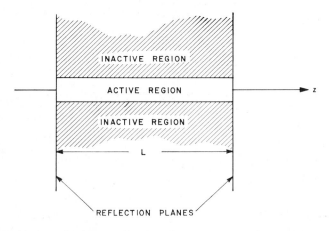

Fig. 6 One dimensional schematic laser structure with an active region sandwiched between two inactive regions.

correction. The wavelength separation between modes in the z direction is inversely proportional to the length L. Because of the short length of a semiconductor laser the separation $\Delta\lambda$ is much larger than that in a gas laser.

By the use of Eq. (25) we can study the temperature dependence of the modes, since $\bar{n}$ is also a function of temperature. Differentiating Eq. (25a) with respect to temperature we obtain (for a given value of q)

$$\frac{1}{\lambda}\frac{d\lambda}{dT} = \frac{1}{\bar{n}}\left(\frac{\partial\bar{n}}{\partial\lambda}\right)_T \frac{d\lambda}{dT} + \frac{1}{\bar{n}}\left(\frac{\partial\bar{n}}{\partial T}\right)_\lambda + \frac{1}{L}\frac{dL}{dT}. \tag{27}$$

If the linear expansion term (which is of the order of 10^{-6}) is negligible, the temperature dependence of the modes is determined by the temperature and wavelength dependence of the index of refraction only, or

$$\frac{1}{\lambda}\frac{d\lambda}{dT} = \frac{1}{\bar{n}}\left(\frac{\partial\bar{n}}{\partial T}\right)_\lambda \frac{1}{\left[1 - \dfrac{\lambda}{\bar{n}}\left(\dfrac{\partial\bar{n}}{\partial\lambda}\right)_T\right]}. \tag{28}$$

We can combine Eqs. (26) and (28) to give the temperature dependence of the index of refraction,

$$\left(\frac{\partial\bar{n}}{\partial T}\right)_\lambda = \frac{1}{2}\left(-\frac{\lambda^2\Delta q}{L\Delta\lambda}\right)\left(\frac{1}{\lambda}\frac{d\lambda}{dT}\right)$$

$$= \frac{1}{2}\left(\frac{\lambda^2}{L\Delta\lambda}\right)\left(\frac{1}{\lambda}\frac{d\lambda}{dT}\right) \qquad \text{for} \qquad \Delta q = -1. \tag{29}$$

Thus if λ, $\partial\lambda/\partial T$, and $\Delta\lambda$ are known, we can deduce $(\partial\bar{n}/\partial T)$ from the above equation.

3 JUNCTION LASERS

To achieve laser action there are three basic requirements: (1) a method of excitation or pumping of electrons from a lower level to a higher level, (2) a large number of electrons which are inverted to give sufficient stimulated emission to overcome the losses, and (3) a resonant cavity to provide positive feedback and quantum amplification.

The most common pumping method in semiconductor lasers is electron-hole injection in a *p-n* junction. The semiconductor which has been most extensively studied is GaAs. The most widely used resonant cavity is the Fabry-Perot cavity[2,22] with two opposite sides cleaved or polished and the other two sawed or roughened.

(1)   Device Fabrication

The basic structure of a *p-n* junction laser is shown in Fig. 7. To achieve the inversion condition, both *p* and *n* sides of the junction are heavily doped (similar to a tunnel diode but generally not as heavily doped as a tunnel diode) such that the imrefs are within the conduction and valence band edges respectively. A pair of parallel planes are cleaved or polished perpendicular to the plane of the junction. The two remaining sides of the diode are roughened to eliminate lasing in directions other than the main one. This structure is called a Fabry-Perot cavity.

Other junction laser structures such as the cylindrical,[23] triangular,[24] and rectangular[24a] (the so-called nondirectional cavity with four sides polished) structures shown in Fig. 8, are less extensively studied because they are generally more difficult to construct and have inferior lasing properties.

Most *p-n* junction lasers are made by the diffusion method.[25,26] Typical GaAs laser junctions are fabricated by diffusion of acceptor atoms such as Zn into an *n*-type substrate having a carrier concentration in the range from 10^{17} to 10^{19} cm^{-3}. Junction depths can vary from one μm to 100 μm depending on the diffusion time and temperature. An interesting laser structure as reported by Dyment and D'Asaro[27] is shown in Fig. 9. The device is a GaAs junction laser with stripe geometry contacts, very thin *p*-type diffusion layer ($\sim$2 μm), and mounted on a type II diamond heat sink. This laser can be

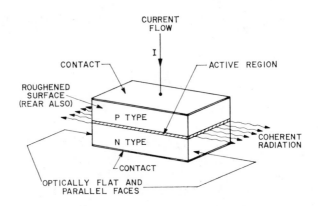

FABRY-PEROT CAVITY

Fig. 7 Basic structure of a junction laser in the form of a Fabry-Perot cavity, i.e., a pair of parallel planes are cleaved or polished and two remaining sides roughed.

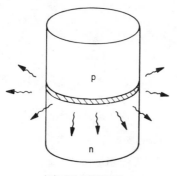

(a) CYLINDRICAL

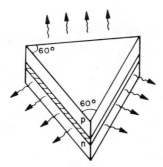

(b) TRIANGULAR

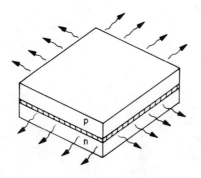

(c) RECTANGULAR

Fig. 8   Three junction laser structures other than the Fabry-Perot cavity
(a) cylindrical,
(b) triangular, and
(c) rectangular or nondirectional with four sides polished.

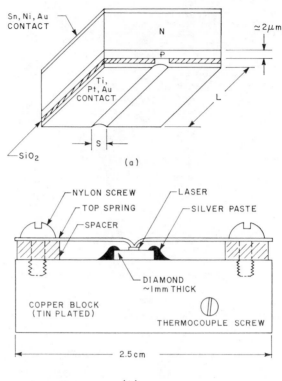

Fig. 9 GaAs laser with stripe geometry contact. The laser is mounted with p^+ side down on a metallized type II diamond heat sink.
(After Dyment and D'Asaro, Ref. 27.)

continuously operated at a heat sink temperature of 200°K, the highest temperature yet reported for continuous operation. The diffusion is accomplished in an inert gas atmosphere using a 2% solution of Zn in Ga saturated with GaAs. A 2.5-μm p layer is diffused in 4 hours at 800°C. The n-type substrate is doped with Te at 3 to 4×10^{18} cm^{-3}. Since the diffusion produces a high-enough sheet resistance (about 150 ohms/square), the stripe contact limits the spreading of current transversely in the p layer and thus confines the laser action to a narrow region. The use of a diamond heat sink significantly reduces the junction temperature rise at a given temperature as compared with a copper heat sink. Further properties of the stripe-geometry laser will be discussed in Sections 3(3) and 3(4).

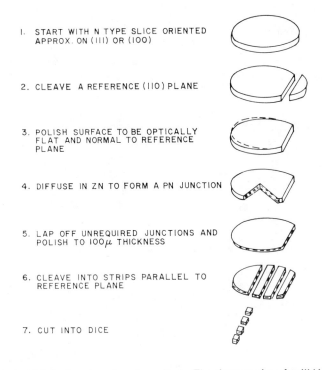

1. START WITH N TYPE SLICE ORIENTED APPROX. ON (III) OR (IOO)

2. CLEAVE A REFERENCE (IIO) PLANE

3. POLISH SURFACE TO BE OPTICALLY FLAT AND NORMAL TO REFERENCE PLANE

4. DIFFUSE IN ZN TO FORM A PN JUNCTION

5. LAP OFF UNREQUIRED JUNCTIONS AND POLISH TO IOOμ THICKNESS

6. CLEAVE INTO STRIPS PARALLEL TO REFERENCE PLANE

7. CUT INTO DICE

Fig. 10 Typical fabrication steps for a laser diode. The cleavage plane for III-V compounds is in ⟨110⟩ direction.
(After Burrell, Ref. 30.)

Laser junctions can also be formed by the solution regrowth method[28] and the vapor regrowth method.[29] For the solution regrowth method a solution of GaAs plus dopants dissolved in Ga or Sn near its saturation temperature is allowed to wet an oppositely doped GaAs substrate. Upon cooling, an epitaxial growth is formed on the wafer and a *p-n* junction at the interface. Since some of the substrate is first dissolved, exposing a fresh surface where the junction is formed, a good plane junction can be obtained. For the vapor regrowth method a vapor containing the desired dopant is passed through the substrate, and epitaxial regrowth can be achieved under proper temperature conditions.

Figure 10 shows some of the steps[30] required in the fabrication of laser diodes, using any of the above doping techniques. The diode fabricated is the conventional Fabry-Perot type.

(2)   Threshold Current Density

Figure 11 shows an energy band diagram of a junction laser. It is similar to that of a tunnel diode, however, the doping is lower, such that generally the forward I-V characteristic of a laser diode has no or a negligibly small negative-resistance region. We note that at equilibrium, Fig. 11(a), the Fermi level lies in the valence band on the p side and in the conduction band on the n side. This is because both sides are heavily doped to become degenerate. When a forward bias is applied, Fig. 11(b), electrons can flow over the top of the barrier to the p side where they can make transitions to empty states in the valence band and emit photons with energy approximately equal to E_g. It is also possible that the transitions occur between the conduction band edge and the acceptor impurity levels (or band) close to the top of the valence band. In this case the photon energy is less than E_g. In addition, holes may flow to

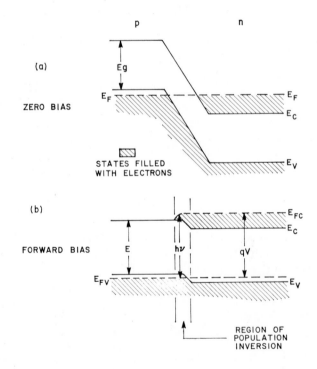

Fig. 11   Energy band diagram of a junction laser (note that it is similar to that of a tunnel diode).

the n side where they recombine with electrons and emit photons. The predominant process is determined by the relative impurity densities, the mobilities, and the carrier lifetimes. In any case, for high-enough applied voltage, there can be an active region in the vicinity of the junction where the population is inverted. Because the active region is generally quite thin (of the order of 1 μm), the maximum gain over a reasonable distance will be in the plane of the junction. This is the reason that the Fabry-Perot cavity is extensively used, so that the feedback path is directly in the plane of maximum gain.

At low current there is spontaneous emission in all directions. As the current is increased, the gain increases until the threshold for lasing is reached, i.e., the gain will satisfy the condition that a light wave make a complete traversal of the cavity without attenuation:

$$R \exp \left[(g - \alpha)L \right] = 1 \tag{30}$$

or

$$g(\text{threshold gain}) = \alpha + \frac{1}{L} \ln \left(\frac{1}{R} \right) \tag{30a}$$

where g is the gain per unit length at threshold, α the loss per unit length which is mainly due to free-carrier absorption and defect-center scattering, L the length of the cavity, and R the reflectivity of the ends of the cavity (if the reflectivities of the ends are different, $R = \sqrt{R_1 R_2}$). At and above lasing threshold, the light output is increased and coherent beams are emitted from the reflecting ends. With the help of the continuity equation we can write the current density J as[15(b)]

$$\frac{J}{d} \simeq \frac{qn}{\tau} \tag{31}$$

where d is the thickness of the active region and τ is the lifetime due to both radiative and nonradiative processes. The combination of Eqs. (16), (30), and (31) with $N_2 \simeq n$ and $N_1 \simeq 0$ yields the threshold current at 0°K,

$$J_t = \frac{8\pi q \bar{n}^2 v^2 \, d\Delta v}{\eta c^2} \left[\alpha + \frac{1}{L} \ln \left(\frac{1}{R} \right) \right] \tag{32}$$

$$= \frac{1}{\beta} \left[\alpha + \frac{1}{L} \ln \left(\frac{1}{R} \right) \right] = \frac{g(\text{threshold})}{\beta} \tag{32a}$$

where $1/\Delta v$ is used to replace $A(v)$, $\bar{n}$ is the refractive index of the semiconductor [$\bar{n} = 1$ for a vacuum as used in Eq. (16)—in a semiconductor medium the velocity of light is given by $c/\bar{n}$], and η is the quantum efficiency (defined as τ/τ_{spont}) which is the fraction of exciting particles that produces radiative recombination in the spectral range of interest. This equation is valid only at 0°K, since the lower band has been assumed to be unoccupied by electrons ($N_1 \simeq 0$). Although this equation is derived based on a simple two-level system, the functional form is expected to be valid. From Eq. (32) we can see at once the requirements for a good laser with a low threshold current density are narrow linewidth, high quantum efficiency, long cavity length, and low loss. From Eq. (32a) the gain at threshold is proportional to the threshold current density, or $g = \beta J_t$ where β as defined in Eq. (32a) and is called the gain factor. This relationship is consistent with the computed result shown in Fig. 4.

To compare the theoretical expression with the experimental threshold current, one can study the dependence of J_t on the laser cavity length. Equation (32a) predicts a linear dependence. Figure 12 shows experimental results[31] at four different temperatures for a set of laser diodes made from the same substrate but with different cavity lengths (7.5×10^{-3} cm to 10^{-1} cm). The substrate material is n-type Sn-doped GaAs with 5×10^{18} cm^{-3}, and the p-n junction is formed by Zn diffusion at 850°C for 3 hours. As can be seen, the linear relation holds very well at all temperatures up to room temperature. (To avoid sample heating at high current densities, short current pulses of 27 ns duration and a repetition rate of 10 to 50 pps are used.) This linear dependence indicates (1) that the functional form of Eq. (32a) is also valid for higher temperatures and (2) that the gain is linearly dependent on the current density up to room temperature.

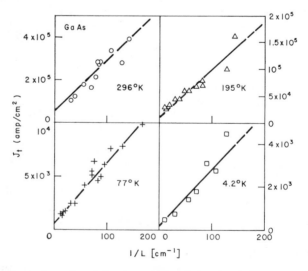

Fig. 12 Experimental results of the dependence of the threshold current density J_t on the cavity length L for a set of GaAs laser diodes. (After Pilkuhn et al., Ref. 31.)

We note also from Fig. 12 that, for a given cavity length, e.g., 10^{-2} cm, the threshold current density increases from 3×10^3 amp/cm^2 at 4.2°K to about 3×10^5 amp/cm^2 at 296°K, a 100-fold increase. The corresponding gain factor β decreases from 5.1×10^{-2} to 4.9×10^{-4} cm/amp, and the loss factor α increases from 13 to 30 cm^{-1}. Figure 13 shows more clearly the temperature and doping dependence of the threshold current density.[32] At lower temperatures, J_t increases with doping density; at high temperatures, J_t varies approximately as T^3. These results are in good agreement with the theoretical calculations,[21] Fig. 5, based on the band-tail model. To convert the nominal current density used in Fig. 5 to the threshold current density, we can write

$$J_t = \left(\frac{d}{\eta \Gamma}\right) J_{nom} \qquad (33)$$

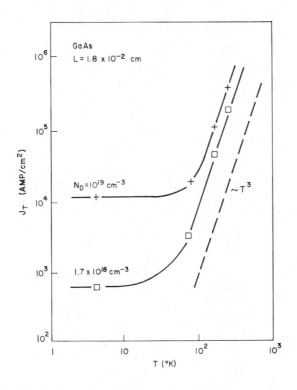

Fig. 13 Measured doping and temperature dependence of the threshold current density in GaAs junction lasers.
(After Pilkuhn and Rupprecht, Ref. 32.)

where d is the thickness of the active layer measured in μm, η is the quantum efficiency, and Γ takes into account the incomplete confinement of the mode propagating along the active layer to the layer itself. In GaAs the value of η is found to be close to 1 at low temperatures, and is of the order of 0.5 at room temperature. In a diffused diode the active layer thickness d is of the order of 1 μm, and the coefficient Γ is about 0.9. Thus the threshold current density at a given doping and temperature is approximately equal to the corresponding nominal current density shown in Fig. 5 (for GaAs).

(3) Spectral Distribution

In this section we shall first discuss the gross features of the frequency spectrum below and above threshold. We shall then present the detailed theory and experimental results concerning the high-resolution spectral distribution (intensity versus frequency or wavelength) under lasing conditions. Figure 14 shows spectra from a typical GaAs laser diode.[33] Below lasing threshold the spectrum is due to the spontaneous emission process and is quite broad. When the current increases above the threshold, the emission line is considerably narrowed. This is because the light intensity depends exponentially on the gain function, Eq. (15a). Thus above threshold a particular frequency close to the central frequency v_o, Eqs. (17) or (18), will be amplified the most. [The frequency corresponding to the maximum gain is not exactly v_o because of the term v^3 in Eq. (13).] At 77°K and $\lambda = 8400$ Å, the dispersion of the refractive index $\bar{n}$ with respect to λ is found to contribute about 50% to the value of the right-hand side of Eq. (26), i.e., $(\lambda/\bar{n})((dn/d\lambda) \simeq 0.5$. Figure 15 shows the temperature dependence of the spectral distributions.[34] As temperature increases, the stimulated emission line shifts toward longer wavelengths (or lower photon energy) in agreement with the change of energy band gap with temperature. In other words, $d\lambda/dT$ varies in accordance with the variation of band gap. The variation of refractive index $\bar{n}$ with temperature can be determined from the measurement of $\Delta\lambda$ and $d\lambda/dT$. For GaAs at 77°K and 8400 Å, one obtains $d\bar{n}/dT = 2.9 \times 10^{-4}/°$K from Eq. (29). The pressure dependence of the refractive index can be similarly determined from Eq. (29) with pressure P replacing T, and one obtains $d\bar{n}/dP \simeq 10^{-6}$ atm^{-1}.

To understand the detailed spectral distribution under lasing conditions, we refer first to Fig. 16 which defines an (x, y, z) Cartesian coordinate system relative to the junction laser. The planes $z = 0$, and $z = -L$ coincide with the two reflecting ends or the laser " mirrors " which are either cleaved or polished surfaces. The front mirror can radiate into the half-space $z > 0$. The transverse coordinates x and y coincide with the directions perpendicular and parallel to the junction plane respectively.

It has been observed experimentally that the electromagnetic energy is

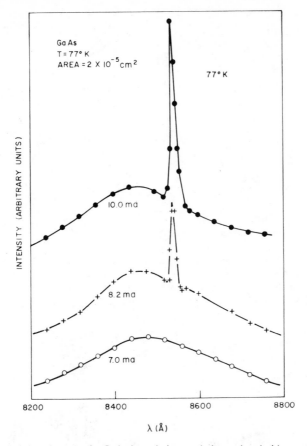

Fig. 14 Spectral distribution of a GaAs laser below and above threshold. (After Pilkuhn et al., Ref. 33.)

confined in a narrow region near the junction plane. This confinement (or focusing of radiation) will be shown to be mainly due to the variation of the refractive index in the transverse planes z = constant. To account for observed focusing of radiation in the two transverse directions, a dielectric constant profile $\varepsilon(x, y)$ is assumed which has a maximum value along a line coinciding with the z-axis and decreasing away from this axis with increasing values of $|x|$ and $|y|$. An arbitrary $\varepsilon(x, y)$, independent of the propagation direction z, can be represented by a general power series expansion about the z-axis. Because of symmetry the dielectric constant is an even function of x and y. Retaining terms up to the second order only, the refractive index

$$\bar{n}(x, y) = [\varepsilon(x, y)/\varepsilon_o]^{1/2} \tag{34}$$

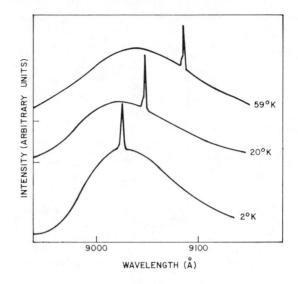

Fig. 15 Temperature dependence of the spectral distribution of an InP laser diode output. The stimulated emission line shifts toward longer wavelength with increasing temperature in agreement with the change of band gap with temperature. (After Burns et al., Ref. 34.)

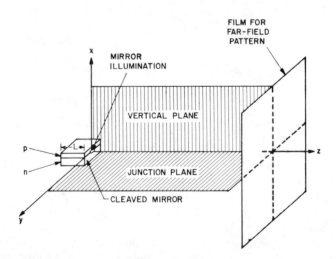

Fig. 16 Cartesian coordinate system relative to the junction laser. The planes $z = 0$, and $z = -L$ coincide with the two reflecting ends of the laser cavity. The field intensity distribution over the laser mirror is called "mirror illumination." This term is used in preference to the commonly known "near-field pattern," which often appears in the literature. We avoid the use of "near field" since it is not the field distribution on the laser mirror but refers to the region of points in the immediate neighborhood of the aperture.

takes the form

$$\bar{n}(x, y) = \bar{n}_o \left[1 - \left(\frac{x}{x_o}\right)^2 - \left(\frac{y}{y_o}\right)^2 \right]^{1/2} \tag{35}$$

where $\bar{n}_o$ is the maximum value of $\bar{n}(x, y)$ which coincides with the z-axis, and x_o and y_o are constants which measure the rate of decrease of $\bar{n}$ perpendicular and parallel, respectively, to the junction plane. This approximate represent-tion of an arbitrary refractive index is justified by the fact that, for low-order transverse modes, the electromagnetic fields are confined in the neighborhood of the z-axis so that $(x/x_o)^2$, $(y/y_o)^2 \ll 1$. This is the model proposed by Zachos and Ripper.[35] It gives results which are in good agreement with the observed spatial and spectral distributions.

Maxwell's equation for a laser medium characterized by a permittivity $\varepsilon_o \bar{n}^2$ and a permeability μ_o can be written as

$$\begin{aligned}
\mathscr{D} &= \varepsilon_o \bar{n}^2 \mathscr{E} \\
\mathscr{B} &= \mu_o \mathscr{H} \\
\nabla \times \mathscr{H} &= -j\omega \varepsilon_o \bar{n}^2 \mathscr{E} \\
\nabla \times \mathscr{E} &= j\omega \mu_o \mathscr{H}
\end{aligned} \tag{36}$$

where $\mathscr{E}$ and $\mathscr{D}$ are the electric field and electric displacement respectively, $\mathscr{B}$ and $\mathscr{H}$ are the magnetic induction and magnetic intensity respectively, $\omega = 2\pi\nu$ is the radian frequency of the radiation, and the time dependence of the radiation is assumed to be of the form $\exp(-j\omega t)$.

For the refractive index, Eq. (35), solutions to Maxwell's equations are hybrid modes characterized by mode numbers m, n, and q. The integers m and n indicate that the intensity of each transverse field component passes through m zeros in the x direction and n zeros in the y direction. The integer q which has been defined in Section 3(5) for axial modes specifies the number of half-wavelength variations within the medium in the z direction, between the laser mirrors. Under the assumption that $m\lambda/\pi x_o$, $n\lambda/\pi y_o \ll 1$, we obtain the simplified wave equation from Eq. (36)

$$[\nabla^2 + (\bar{n}k)^2]\mathscr{E} = 0 \tag{37}$$

where $k \equiv 2\pi/\lambda$ is the free-space propagation constant and λ is the free-space wavelength.

With $\bar{n}$ given by Eq. (35) and subject to the boundary conditions of vanishing electromagnetic field where x and y approach infinity, the solutions (by separation of variables) of Eq. (37) are given by

$$\mathscr{E}_{mnq}^{\pm}(x, y, z) = A_{mnq} X_m(x) Y_n(y) \exp(\pm jk\gamma_{mnq} z) \tag{38}$$

where A_{mnq} is a constant,

$$X_m(x) = H_m\left(\sqrt{\frac{2\pi\bar{n}_o}{\lambda x_o}}\, x\right)\exp\left[-\frac{x^2}{\left(\dfrac{\lambda x_o}{\pi\bar{n}_o}\right)}\right] \tag{39}$$

$$Y_n(y) = H_n\left(\sqrt{\frac{2\pi\bar{n}_o}{\lambda y_o}}\, y\right)\exp\left[-\frac{y^2}{\left(\dfrac{\lambda y_o}{\pi\bar{n}_o}\right)}\right] \tag{40}$$

$$\gamma_{mnq} = \bar{n}_o\left(1 - \frac{2m+1}{\bar{n}_o x_o k} - \frac{2n+1}{\bar{n}_o y_o k}\right)^{1/2}. \tag{41}$$

The function $H_m(\xi)$ is the Hermite polynomial of order of m, and both $X_m(x)$ and $Y_n(y)$ are Hermite-Gaussian functions. The $\pm$ signs associated with $\mathscr{E}_{mnq}^{\pm}$ are for positive-going and negative-going waves.

Oscillations in the resonant cavity are set up by the superposition of the two opposite-going traveling waves given by Eq. (38). Considering one of these $\mathscr{E}^+$, the definition of a mode requires that

$$\mathscr{E}_{mnq}^+(x, y, 0) = \pm\mathscr{E}_{mnq}^+(x, y, -L). \tag{42}$$

From Eq. (38) the above requirement is satisfied when

$$\gamma_{mnq} = \frac{q\pi}{kL} = \frac{q\lambda}{2L} \tag{43}$$

where q is an integer. Substitution of Eq. (43) into Eq. (41) yields the resonant frequencies for the laser cavity:

$$\nu_{mnq} = \frac{c}{4\pi\bar{n}_o}\left(\frac{2m+1}{x_o} + \frac{2n+1}{y_o}\right) + \frac{cq}{2L\bar{n}_o}\left\{1 + \left[\frac{L}{2\pi q}\left(\frac{2m+1}{x_o} + \frac{2n+1}{y_o}\right)\right]^2\right\}^{1/2}. \tag{44}$$

The superposition of waves traveling in the positive and negative z directions gives the standing wave pattern $\mathscr{E}_{mnq}(x, y, z, t)$ in the laser cavity which is now given by

$$\mathscr{E}_{mnq}(x, y, z, t) = 2A_{mnq}X_m(x)Y_n(y)\cos\left(\frac{q\pi}{L}z\right)\cos(2\pi\nu_{mnq}t). \tag{45}$$

Each resonant mode is thus characterized by a field distribution $\mathscr{E}_{mnq}$ given by Eq. (45) and a corresponding frequency of oscillation ν_{mnq} given by Eq. (44). These equations express the field and the frequency in terms of the mode number (m, n, q), the laser length L, and the focusing properties of the medium x_o, y_o.

The frequency separation $\Delta\lambda$ of the resonant modes can be obtained by differentiation of Eq. (44). For the cases of low-order transverse modes, such that

$$\frac{L}{2\pi q}\left(\frac{2m+1}{x_o}+\frac{2n+1}{y_o}\right) \ll 1 \tag{46}$$

we obtain

$$L\frac{\Delta\lambda}{\lambda^2} = -\frac{1}{2\bar{n}_e}\left[\frac{L}{\pi x_o}\Delta m + \frac{L}{\pi y_o}\Delta n + \Delta q\right] \tag{47}$$

where $L(\Delta\lambda/\lambda^2)$ can be considered as the normalized bandwidth, and

$$\bar{n}_e \equiv \bar{n}_o\left(1 - \frac{\lambda}{\bar{n}_o}\frac{d\bar{n}_o}{d\lambda}\right)$$

is an equivalent refractive index which accounts for the presence of a dispersive medium. Similar expression has already been derived in Eq. (26). When $\Delta m = \Delta n = 0$, Eq. (47) gives the separation of the axial modes which is identical to that given by Eq. (26):

$$\frac{\Delta\lambda}{\Delta q} \approx -\frac{\lambda^2}{2\bar{n}_e L}. \tag{48}$$

The wavelength separations between resonances whose mode numbers differ only in the x direction, perpendicular to the junction plane, are obtained by setting $\Delta n = \Delta q = 0$ in Eq. (47), yielding

$$\frac{\Delta\lambda}{\Delta m} \approx \frac{-\lambda^2}{2\pi\bar{n}_e x_o}. \tag{49}$$

Similarly along the junction plane when $\Delta m = \Delta q = 0$,

$$\frac{\Delta\lambda}{\Delta n} \approx \frac{-\lambda^2}{2\pi\bar{n}_e y_o}. \tag{50}$$

Hence, for any two resonances characterized by the mode numbers (m, n, q) and $(m, n, q + 1)$, the separation $(\Delta\lambda)_q$ depends on L. For modes (m, n, q) and $(m, n+1, q)$, $(\Delta\lambda)_n$ depends on y_o which is a measure of the focusing along the junction plane. A similar statement holds for the modes (m, n, q) and $(m+1, n, q)$.

The high-resolution spectral distribution[35] for a stripe geometry GaAs junction laser, Fig. 9, is shown in Fig. 17. The laser has a stripe width of 13 μm, a length of 380 μm, and is operated continuously at 77°K with a current density 13% above the threshold. The spectral distribution shows the multimode character of the laser radiation. The spectrum, which is centered approximately at $\lambda = 8383$ Å, consists of several "axial mode" groups. A

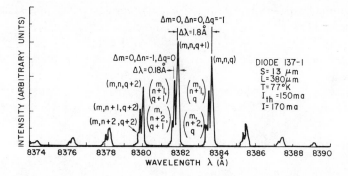

Fig. 17 Experimental high-resolution spectral distribution in which each "axial mode" contains regular satellite frequencies.
(After Zachos and Ripper, Ref. 35.)

number of closely spaced frequencies are found within each axial mode group. These clusters of lines are called "satellite" frequencies. The separation between a given mode (m, n, q) in a particular mode group and the corresponding mode $(m, n, q + 1)$ in an adjacent mode group is the "axial mode" separation $(\Delta\lambda)_q$ which is approximately 1.8 Å in the present case. Two adjacent satellite frequencies in a given axial mode group are separated by an amount $(\Delta\lambda)_n = 0.18$ Å. The reason for the association of separation with the y-axis will be discussed as follows.

We can now compare the experimental results with the theory.[35] The values of $\bar{n}_e$, x_o, and y_o at 77°K have been determined from independent measurements to be $\bar{n}_e \simeq \bar{n}_o \times 1.5 = 5.4$ (with $\bar{n}_o = 3.6$), $x_o = 17\mu m$, and $y_o \simeq 1400 \ \mu m$. For $\lambda = 8383$ Å, we calculate $\lambda/\pi x_o = 3.8 \times 10^{-2}$ and $\lambda/\pi y_o = 1.9 \times 10^{-4}$, thus confirming the assumption that $\lambda/\pi x_o$, $\lambda/\pi y_o \ll 1$. A typical value for the longitudinal mode number q may be obtained by noting that for low-order transverse mode numbers (m, n), Eq. (41) gives $\gamma_{mnq} \simeq \bar{n}_o$. Hence from Eq. (43), $q \simeq 2\bar{n}_o L/\lambda = 3260$ for $L = 380 \ \mu m$. Thus Eq. (46) is satisfied thereby justifying all the earlier assumptions. The frequency separation $(\Delta\lambda)_q$ between modes (m, n, q) and $(m, n, q + 1)$ is obtained by setting $\Delta q = -1$ in Eq. (48), and we obtain $(\Delta\lambda)_q = 1.8$ Å, in agreement with the experimental results. From Eq. (49) the frequency separation $(\Delta\lambda)_m$ between modes (m, n, q) and $(m + 1, n, q)$ is obtained to be 13 Å. Similarly the frequency separation between modes (m, n, q) and $(m, n + 1, q)$ is calculated to be $(\Delta\lambda)_n = 0.16$ Å. Since $(\Delta\lambda)_m \gg (\Delta\lambda)_n$, the satellite frequencies in a given axial mode are associated with the y-axis, $(\Delta\lambda)_n$, and the theoretical separation of 0.16 Å is in good agreement with the experimental separation of 0.18 Å.

The "clean" spectral distribution, shown in Fig. 17, can be obtained because the stripe metallic contact, Fig. 9, confines the light emission to a single region, i.e., under the contact. In a nonstripe geometry junction laser, Fig. 7, the light emission may occur at two or more unpredictable regions along the junction plane. For each region, several modes with different transverse field variations may oscillate simultaneously. One set of modes may be coupled with or uncoupled from an adjacent set. The various modes, whose oscillation cannot be controlled, interfere with each other and produce spectral distributions which generally differ from diode to diode and produce very complicated spectra.

(4) Spatial Distribution

We shall now consider the far-field pattern produced by the mirror illumination on the front laser mirror shown in Fig. 18 for a stripe-geometry junction laser. Let r_o be the distance from the origin of the (x, y, z) coordinate system to a far-field observation point $P(x, y, z)$. The distance from P to a point $(\xi, \zeta, 0)$ on the laser mirror is denoted by r. Under the conditions that $kr \gg 1$, ξ/r_o, and $\zeta/r_o \ll 1$, the far-field $\mathcal{E}_{mnq}(x, y, z)$ in the half-space $z > 0$ is given by

$$\mathcal{E}_{mnq}(x, y, z) = \frac{k}{2\pi j} \frac{z}{r_o} \frac{e^{jkr_o}}{r_o} \int\int_{-\infty}^{\infty} \mathcal{E}_{mnq}(\xi, \zeta, 0) e^{-jk[(x/r_o)\,\xi + (y/r_o)\,\zeta]} \, d\xi \, d\zeta. \quad (51)$$

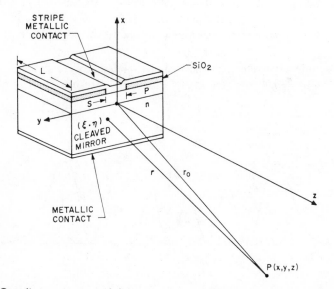

Fig. 18 Coordinate, system, and distance parameters for defining the far-field patterns. (After Zachos and Ripper, Ref. 35.)

Substituting Eq. (45) into Eq. (51) and omitting the time dependence, we find[35]

$$\mathcal{E}_{mnq}(x, y, z) = B_{mnq} \frac{z}{r_o} \frac{e^{jkr_o}}{r_o} H_m\left(\frac{\sqrt{2}x}{W_x}\right) H_n\left(\frac{\sqrt{2}y}{W_y}\right) e^{-[(x/W_x)^2 + (y/W_y)^2]}. \quad (52)$$

where B_{mnq} is a constant which includes the transmission coefficient of the GaAs mirror, and W_x and W_y are the scaling factors given by

$$W_x \equiv \sqrt{\frac{\lambda \bar{n}_o}{\pi x_o}} z \quad \text{and} \quad W_y \equiv \sqrt{\frac{\lambda \bar{n}_o}{\pi y_o}} z \quad (53)$$

which define the far-field beam widths of the longitudinal modes in the x and y directions respectively. It is thus expected that the far-field pattern should have a Hermite-Gaussian symmetry. For a fixed value of x,

$$\mathcal{E}(y) \sim H_n\left(\frac{\sqrt{2}y}{W_y}\right) \exp\left[-\left(\frac{y}{W_y}\right)^2\right].$$

A theoretical intensity for a mode with

$$n = 5, H_5\left(\frac{\sqrt{2}y}{W_y}\right) \exp\left[-\left(\frac{y}{W_y}\right)^2\right],$$

is shown in Fig. 19 in terms of the dimensionless parameter $\sqrt{2}y/W_y$. The intensity is proportional to the square of the Hermite-Gaussian function. This plot predicts maximum intensities and widths at the outside of the pattern with zeros at $\sqrt{2}y/W_y = 0, \pm 0.96, \pm 2.02$.

The Hermite-Gaussian patterns in junction lasers were first reported by Dyment[36] for stripe-geometry GaAs diodes. The results for a diode with

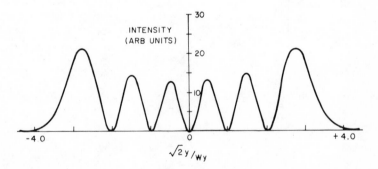

Fig. 19 Theoretical intensity for the Hermite-Gaussian function with mode number $n = 5$. (After Dyment, Ref. 36.)

$S = 50 \, \mu\text{m}$, $L = 380 \, \mu\text{m}$, and operated at $77°\text{K}$ are shown in Fig. 20 for (a) the mirror illumination and (b) the far-field pattern obtained at $z = 10$ cm away from the mirror. It is found that both (a) and (b) agree with the Hermite-Gaussian function within about 5%. The predicted pattern width at the mirror is 43 μm which compares favorably with the measured width of ~40 μm.

Figure 21 shows the correlation between spectral and spatial distributions. The laser is a GaAs stripe-geometry diode with a stripe width of 50 μm. In Fig. 21(a), the current is just slightly above the lasing threshold. We note from the frequency spectrum that the dominant mode values are $m = 0$ and $n = 3$. This spectrum should give a far-field pattern of Gaussian profile in the x-direction and Hermite-Gaussian profile with $n = 3$ in the y-direction. This is indeed observed as shown in Fig. 21(a). The additional minor lobes are due to the slight asymmetry in the refractive index $\bar{n}(x, y)$ along the x-axis. This asymmetry, which arises from different concentrations in the n and p sides of the junction, is not taken into account by the aforementioned theory. As the current increases, higher order spectrum lines appear, Fig. 21(b), and the far-field pattern is a superposition of the patterns of the Hermite-Gaussian profile with $n = 0$ to 5.

In general, the laser mirror illumination is asymmetric perpendicular to the junction plane and consists of a primary intensity maximum along with several weaker secondary peaks. If one places a slit in the image plane to suppress the secondary peaks thereby eliminating their effects in the diffraction field, an isolated primary maximum with a Gaussian distribution will be observed, and it will retain its Gaussian profile as it propagates into the far-field region.[37] This is demonstrated in Fig. 22. For the same mirror illumination, the far-field pattern is asymmetric along the x-axis, (a) Fig. 22, without the slit, and is symmetric with the slit. Recorder tracings of relative power intensity along lines A and B, (b) of Fig. 22 are shown in Fig. 23(a) and (b) respectively. The field has been scanned at a distance of 108.4 cm from the slit. The measurements of Fig. 23 show that the pattern has profiles along and perpendicular to the junction plane which may be approximated by the Gaussian functions shown by the dashed lines.

(5) Stress Effect, Magnetic-Field Effect, Time Delay, and Laser Degradation

The performance of a semiconductor laser can be affected by many factors such as applied stresses and magnetic field. Under pulsed operation there is usually a time delay between the coherent light output and the application of the current pulse. Also under pulsed or continuous operation the laser output

|← 50 μm →|

(a) MIRROR ILLUMINATION

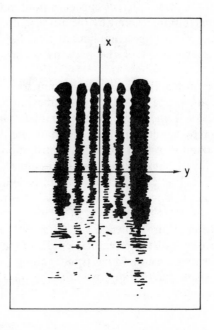

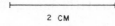

2 CM

(b) FAR FIELD PATTERN

Fig. 20 Experimental mirror illumination and far-field pattern of a stripe-geometry type GaAs laser with $n = 5$.
(After Dyment, Ref. 36.)

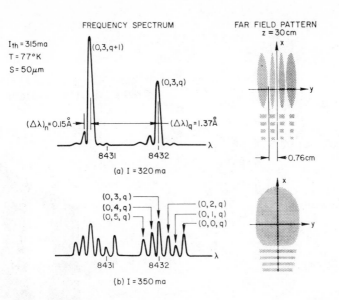

FREQUENCY SPECTRUM

FAR FIELD PATTERN
z = 30 cm

$I_{th} = 315\,ma$
$T = 77°K$
$S = 50\mu m$

$(0,3,q+1)$

$(0,3,q)$

$(\Delta\lambda)_n = 0.15\mathring{A}$

$(\Delta\lambda)_q = 1.37\mathring{A}$

8431   8432

(a) I = 320 ma

0.76 cm

$(0,3,q)$
$(0,4,q)$
$(0,5,q)$

$(0,2,q)$
$(0,1,q)$
$(0,0,q)$

8431   8432

(b) I = 350 ma

Fig. 21 Correlation between spectral and spatial distributions for a GaAs stripe-geometry diode laser with stripe width 50 μm operated at 77°K.
(a) current slightly above the threshold,
(b) current about 11% above the threshold.
(After Zachos and Ripper, Ref. 35.)

either gradually or suddenly decreases in intensity as a result of laser degradation. In this section we shall briefly discuss the above topics.

Uniaxial stress lowers the threshold current of a GaAs junction laser and shifts the emission peak to shorter wavelength, as shown in the typical plot[38] of Fig. 24. In explanation, since the uniaxial strain destroys the cubic symmetry, it is expected to cause anisotropy in the spatial distribution of spontaneously emitted photons. At threshold, there are sufficient photons in a particular mode to satisfy the gain requirement of Eq. (30a). If the strain increases the probability of spontaneous emission of photons in a direction parallel to the junction plane and decreases the probability in a perpendicular direction, then more photons would be emitted into the preferred modes and the threshold would be lowered.

It has been observed[39] that magnetic fields transverse to the current flow in a junction laser lower the threshold current for laser action and shift the peak of emission. Figure 25(a) shows the type of threshold current density lowering that occurs when a magnetic field is applied to an InSb laser diode. Figure 25(b) shows the shift of the peak of emission as a function of the magnetic field

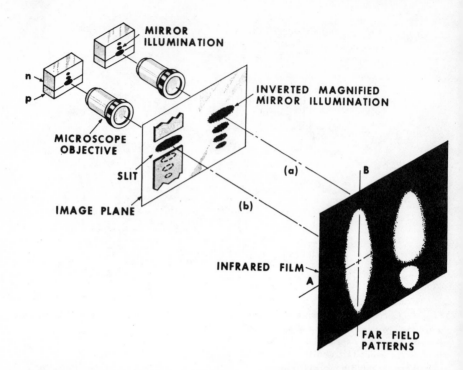

Fig. 22 Generation of Gaussian beams from GaAs lasers. The asymmetric intensity characteristic of the laser radiation (a) is transformed into a Gaussian beam by a microscope objective and a slit (b). For the purpose of illustration the figure is not drawn to scale. The magnification from the cleaved diode face to the image plane is about 35. The objective lens has a 0.6 numerical aperture.
(After Zachos, Ref. 37.)

applied. All measurements were made on a pulsed lasing basis. From these results, it is suggested that the transition probability increases with the magnetic field strength, resulting in a decrease of the threshold current density.

It has been observed experimentally that when lasers are operated under pulsed conditions, there is generally a delay time t_d between the application of the pulse and the onset of stimulated coherent emission. The delay depends strongly on the substrate material, the junction formation process, and the subsequent heat treatments.[40a] At 77°K, t_d is usually less than 2 ns and is essentially the time required to establish the population inversion. At room temperature the situation is radically different, and delays of several hundred nanoseconds have been observed. The delays have been suggested by Fenner[40b] as primarily due to absorption centers located in the vicinity

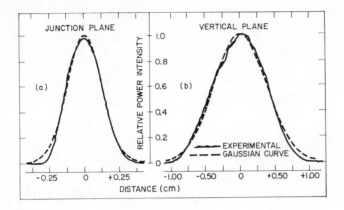

Fig. 23 Far-field relative power intensity measurements of the Gaussian beam
(a) along and
(b) perpendicular to the junction plane.
The scans are made along the lines A and B of (b) in Fig. 22. The dashed lines approximate
the data with mathematical Gaussian curves.
(After Zachos, Ref. 37.)

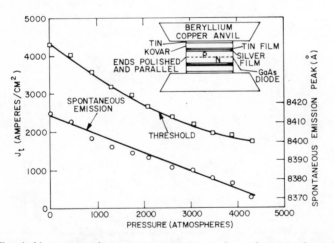

Fig. 24 Threshold current and spontaneous emission peak as a function of uniaxial stress.
(After Ryan and Miller, Ref, 38.)

of the junction. These centers cause optical losses which prevent lasing until
they are filled by electrons injected into the p^+ side of the junction.

Degradation of the optical output of semiconductor lasers can be classified
into two categories: (1) slow internal degradation and (2) catastrophic degra-

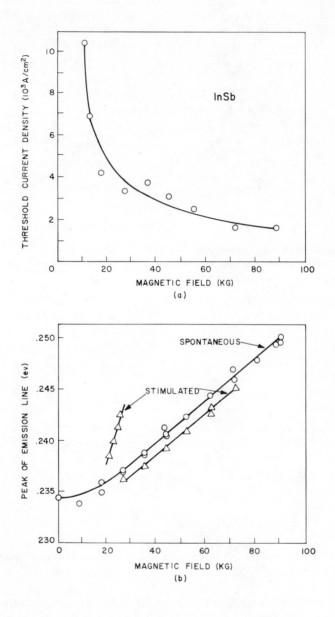

Fig. 25
(a) Threshold current density versus magnetic field for an InSb laser diode.
(b) Shift of the peak of emission as a function of the magnetic field.
(After Melngailis and Rediker, Ref. 39.)

dation. For the former, the output intensity decreases gradually and only occurs over many millions of pulses. The diode current-voltage characteristics also gradually change. The reason for this gradual deterioration in the GaAs laser is believed to be diffusion of zinc or copper under high-injection conditions. This is similar to a tunnel diode in which the peak-to-valley ratio decreases under large forward bias. For the catastrophic degradation,[41] the optical output under high power levels decreases after a few thousand pulses or even a single pulse. The laser mirror is permanently damaged due to the formation of pits or grooves on the mirror. In contrast to the slow internal degradation, at low temperatures the *I-V* characteristics of the diode remain essentially the same before and after the catastrophic degradation. To avoid this degradation, one can decrease the pulse width or decrease the optical flux density at the surface by the use of a tailored diffusion profile near the mirror.

4 HETEROSTRUCTURE AND CONTINUOUS ROOM-TEMPERATURE OPERATION

As considered in the previous section, the threshold current density, J_T, for a conventional *p-n* junction laser (also called homostructure laser) increases rapidly with increasing temperature. A typical value of J_T (obtained by pulse measurement) is about 10^5 A/cm^2 at room temperature. Such a large current density imposes serious difficulties to operate the laser continuously at 300°K.

To reduce the threshold current density, Kroemer[51] and Alferov[52] proposed the heterostructure laser as shown in Fig. 26. The new structure has two outstanding features: (1) it can confine the carriers in the active region by the potential barriers, and (2) it can confine the light intensity within the active region by the sudden reduction of the refractive index outside the active region. These confinements will enhance the population inversion and provide significant quantum amplification in the active region; therefore, the temperature dependence of the threshold, as well as the value of J_T, should be much reduced.

As discussed in Section 8 of Chapter 3, the selection of suitable materials to form the heterojunction is critical, since interface states will generally produce a high density of recombination centers. Panish and Sumski[54] found that a GaAs − Al$_x$Ga$_{1-x}$As alloy interface shows negligibly small numbers of states, and the band gap of Al$_x$Ga$_{1-x}$As can be made as large as 2 eV. Single heterostructures using the above alloy interface were made, and threshold current density as low as 6000 A/cm^2 at 300°K was obtained.[55,56] Using Fabry-Perot type cavity, a double heterostructure laser was successfully operated continuously up to 38°C in 1970. The threshold current density is

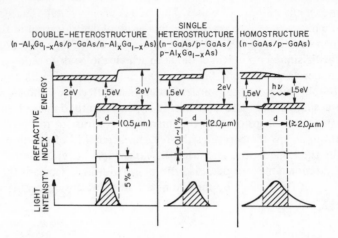

Fig. 26 Comparison of some characteristics of double heterostructure, single heterostructure, and homostructure lasers. The top row shows energy band diagrams under forward bias. The refractive index change for GaAs/Al$_x$Ga$_{1-x}$As is about 5%. The change across a homostructure is less than 1%. The confinement of light is shown in the bottom row (after Hayashi, Panish et al., Ref. 49,50).

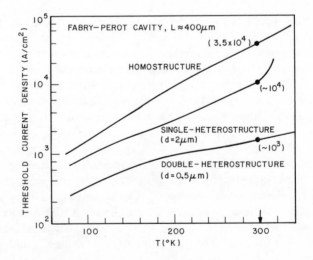

Fig. 27 Threshold current density versus temperature for three laser structures (Ref. 49,50).

about two orders of magnitude lower than a corresponding homostructure laser.

A comparison of J_T versus operating temperature is shown in Fig. 27 for a homostructure, a single heterostructure, and a double-heterostructure laser.[49,50] We note that the temperature dependence is the least for the double heterostructure laser.

5 OTHER PUMPING METHODS AND LASER MATERIALS

(1) Other Pumping Methods

Various methods are available for obtaining the population inversion as mentioned in Section 1. In the previous two sections we were concerned with the junction injection method that uses an electric current. Other means of pumping, such as optical pumping, electron-beam pumping, and avalanche breakdown will be briefly discussed in this section.

Semiconductor lasers have been optically pumped by Basov et al.[42] mainly for the purpose of understanding the interaction between a powerful radiation and a semiconductor. Phelan and Rediker[43] have optically pumped an n-type InSb semiconductor laser with a GaAs laser. The geometry of the experiment is shown in Fig. 28. Above a threshold input current (~ 14 amps) the output from the InSb laser varies linearly with the GaAs input current. Holonyak et al.[44] have reported successful laser operation of CdSe pumped with a Ga(AsP) laser diode. In both cases above, the laser pumping frequency was of a shorter wavelength than the pumped laser. An InAs laser has also been pumped by a GaAs laser by Melngailis.[45] One problem with optical pumping is that the pumping light is absorbed very close to the surface of the semiconductor unless the wavelength of the pump is close to or less than the wavelength corresponding to the energy gap. This makes the laser very sensitive to surface properties. Basov[42] has pumped GaAs material with a Raman shifted pulsed ruby laser (which closely matches the energy gap of GaAs) and has reported a peak power output of 3×10^4 watts, which is the highest for any semiconductor laser studied to date. An advantage for all optically pumped (and also electron beam pumped, see the following) semiconductor laser is that a large volume of materials can be excited. This is especially true for those in which p-n junction is difficult to form.

Another method for obtaining population inversion is by the bombardment of the semiconductor with a beam of energetic electrons. It was first reported by Guillaume and Debever.[46] The basic arrangement is similar to

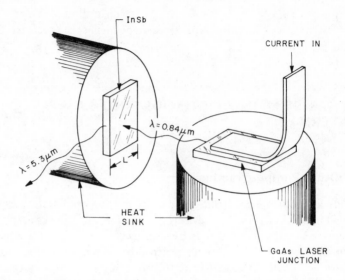

Fig. 28 Optical pumping scheme. An *n*-type InSb semiconductor laser (no junction) is optically pumped by a GaAs junction laser.
(After Phelan and Rediker, Ref. 43.)

that shown in Fig. 28 except that the pumping laser is replaced by the electron beam. As the electron beam penetrates the material (to a 5- or 10-μm depth) the electron hole pairs form degenerate populations as shown in Fig. 2. Hence spontaneous emission builds up until a threshold condition is reached and lasing occurs. Using a pulsed electron beam excitation on crystals of ZnS, Hurwitz[47] obtained ultraviolet lasing. Peak output powers of 1.7 watts were measured in the spectral range from 3245 to 3300 Å. The efficiency of electron-beam-excited lasers is lower than that of junction lasers. This is related to the inevitable loss of energy by hot electrons and holes created by the electron beam.

The avalanche breakdown pumping scheme has been used by Weiser and Woods[48] in a *p*-π-*p* GaAs structure. The avalanche breakdown is initiated by holes in the high field region of the *p*-π boundaries. When the field is high enough that a degenerate distribution is formed and the gain overcomes the losses in the cavity, laser action takes place perpendicular to the field direction.

(2)   Laser Materials

The list of semiconductor materials which have exhibited laser action has continued to grow. Table 13.1 shows the list as compiled by Nathan[15b] for the materials which have lased to date, together with the photon energy of oscillation, the corresponding wavelength, and the method of excitation. GaAs was the first material to lase, and it has had the most extensive study and development. The laser properties of other III-V compounds are all quite similar to GaAs. The stimulated emission usually occurs at an energy slightly less than the energy band gap of the material, which means that the radiative transition involves impurity states. In some III-V compounds laser action has also been observed at the energy gap from direct band-to-band recombination.

The IV-VI compounds, especially the lead salts such as PbS, PbTe, and PbSe also exhibit laser action. They are direct band-gap materials with their extrema located along the $\langle 111 \rangle$ directions in the Brillouin zone, in contrast to GaAs with extrema located at the zone center. The elemental semiconductor Te and several II-VI compounds have been made to lase. In addition, laser

TABLE 13.1
SEMICONDUCTOR LASER MATERIALS

Material	Photon Energy (eV)	Wavelength (microns)	Method of Excitation
ZnS	3.82	0.32	electron beam
ZnO	3.30	0.37	electron beam
CdS	2.50	0.49	electron beam, optical
GaSe	2.09	0.59	electron beam
CdS_xSe_{1-x}	1.80-2.50	0.49-0.69	electron beam
CdSe	1.82	0.68	electron beam
CdTe	1.58	0.78	electron beam
$Ga(As_xP_{1-x})$	1.41-1.95	0.88-0.63	p-n junction
GaAs	1.47	0.84	p-n junction, electron beam, optical, avalanche
InP	1.37	0.90	p-n junction
$In_xGa_{1-x}As$	1.5	0.82	p-n junction
GaSb	0.82	1.5	p-n junction, electron beam
InP_xAs_{1-x}	0.77	3.1	p-n junction, electron beam, optical
InSb	0.23	5.2	p-n junction, electron beam, optical
Te	0.34	3.64	electron beam
PbS	0.29	4.26	p-n junction, electron beam
PbTe	0.19	6.5	p-n junction, electron beam, optical
PbSe	0.145	8.5	p-n junction, electron beam
$Hg_xCd_{1-x}Te$	0.30-0.33	3.7-4.1	optical
$Pb_xSn_{1-x}Te$	0.09-0.19	6.5-13.5	optical

action has also been observed in alloys of several of the compounds such as $Ga(As_{1-x}P_x)$ and $Cd(S_x Se_{1-x})$.

As can be seen from Table 13.1, the wavelength has been extended from the ultraviolet to far-infrared. The semiconductor lasers have the advantages of small size ($\sim 10^{-4}$ cm^3), high efficiency ($\sim 40\%$ external quantum efficiency), and rugged construction (such as stripe-geometry lasers[53]); and the achievement of continuous room-temperature operation will lead to increasing applications in science and technology of semiconductor lasers.

REFERENCES

1. J. P. Gordon, H. J. Zeiger, and C. H. Townes, "Molecular Microwave Oscillator and New Hyperfine Structure in the Microwave Spectrum of NH$_3$," Phys. Rev., *95*, 282 (1954).

2. A. L. Schawlow and C. H. Townes, "Infrared and Optical Masers," Phys. Rev. *112*, 1940 (1958).

3. T. H. Maiman, "Stimulated Optical Radiation in Ruby Masers," Nature, *187*, 493 (1960).

4. P. P. Sorokin and M. J. Stevenson, "Stimulated Infrared Emission from Trivalent Uranium," Phys. Rev. Letters, *5*, 577 (1960).

5. A. Javan, W. B. Bennett, Jr., and D. R. Herriott, "Population Inversion and Conti n uous Optical Maser Oscillation in a Gas Discharge, Containing a He-Ne Mixture," Phys. Rev. Letters, *6*, 106 (1961).

6. J. I. Nishizawa and Y. Watanabe, Japanese Patent, April 1957. Also Electronics, p. 117 (December 11, 1967).

7. N. G. Basov, O. N. Krokhin, and Y. M. Popov, Soviet Physics Uspekhi, *3*, 7 (1961).

8. P. Aigrain, unpublished lecture at the "International Conference on Solid State Physics in Electronics and Telecommunications," Bruxelles (1958). The same idea was suggested independently by N. G. Basov, O. N. Krokhin, and Y. M. Popov, Second International Conference on Quantum Electronics.

9. W. P. Dumke, "Interband Transitions and Maser Action," Phys. Rev., *127*, 1559 (1962).

10. M. G. A. Bernard and G. Duraffourg, "Laser Conditions in Semiconductors," Physica Status Solidi, *1*, 699 (1961).

11. R. N. Hall, G. E. Fenner, J. D. Kingsley, T. J. Soltys, and R. O. Carlson, "Coherent Light Emission from GaAs Junctions," Phys. Rev. Letters, *9*, 366 (1962).

12. M. I. Nathan, W. P. Dumke, G. Burns, F. H. Dill, Jr., and G. J. Lasher, "Stimulated Emission of Radiation from GaAs *p-n* Junction," Appl. Phys. Letters, *1*, 62 (1962).

13. T. M. Quist, R. H. Rediker, R. J. Keyes, W. E. Krag, B. Lax, A. L. McWhorter, and H. J. Zeiger, "Semiconductor Maser of GaAs," Appl. Phys. Letters, *1*, 91 (1962).

14. N. Holonyak, Jr., and S. F. Bevacqua, "Coherent (Visible) Light Emission from Ga(As$_{1-x}$P$_x$) Junction," Appl. Phys. Letters, *1*, 82 (1962).

15. For a comprehensive review see (a) G. Burns and M. I. Nathan, "*P-N* Junction Lasers," Proc. IEEE, *52*, 770 (1964), and (b) M. I. Nathan, "Semiconductor Lasers," Proc. IEEE *54*, 1276 (1966).

16. For a general reference on various lasers see W. V. Smith and P. P. Sorokin, *The Laser*, McGraw-Hill Book Co., New York (1966).

17. A. K. Levine, "Lasers," American Scientist, *51*, 14 (1963).

18. For a lucid discussion, see A. Yariv, *Quantum Electronics*, John Wiley & Son, New York (1967).

19. G. Lasher, and F. Stern, "Spontaneous and Stimulated Recombination Radiation in Semiconductors," Phys. Rev., *133*, A553 (1964).

20. E. O. Kane, "Thomas-Fermi Approach in Impure Semiconductors," Phys. Rev., *131*, 79 (1963).

21. F. Stern, "Effect of Band Tails on Stimulated Emission of Light in Semiconductors," Phys. Rev., *148*, 186 (1966).

22. A. G. Fox and T. Li, "Resonant Modes in a Maser Interferometer," Bell Sys. Tech. J., *40*, 453 (1961).

23. K. M. Arnold and S. Mayburg, "Cylindrical GaAs Laser Diode," J. Appl. Phys., *34*, 3136 (1963).

24. I. Ladany, "Some Observation on Triangular GaAs Lasers," Proc. IEEE, *52*, 1353 (1964).

24a. R. A. Laff, W. P. Dumke, F. H. Dill, and G. Burns, "Directionality Effect of GaAs Light-Emitting Diodes: Part II," IBM J. Research and Development, *7*, 63 (1963).

25. J. C. Marinace, "Diffused Junctions in GaAs Injection Lasers," J. Electrochem. Soc., *110*, 1153 (1963).

26. M. H. Pilkuhn and H. S. Rupprecht, "Diffusion Problems Related to GaAs Injection Lasers," Trans. AIME, *230*, 296 (1964).

27. J. C. Dyment and L. A. D'Asaro, "Continuous Operation of GaAs Junction Lasers on Diamond Heat Sinks at 200°K," Appl. Phys. Letters, *11*, 292 (1967).

28. H. Nelson, "Epitaxial Growth from the Liquid State and Its Application to the Fabrication of Tunnel and Laser Diodes," RCA Rev., *24*, 603 (1963).

29. N. N. Winogradoff and H. K. Kessler, "Light Emission and Electrical Characteristics of Epitaxial GaAs Lasers and Tunnel Diodes," Solid State Commun., *2*, 119 (1964).

30. G. J. Burrell, "Semiconductor Injection Lasers and Lamps," Tech. Report No. 65095, Royal Aircraft Establishment (May 1965).

31. M. Pilkuhn, H. Rupprecht, and S. Blum, "Effect of Temperature on the Stimulated Emission from GaAs *p-n* Junctions," Solid State Electron., *7*, 905 (1964).

32. M. H. Pilkuhn and H. S. Rupprecht, "Influence of Temperature on Radiative Recombination in GaAs *p-n* Junction Lasers," 7th International Conf. on the Physics of Semiconductors, Radiative Recombination in Semiconductors, Paris (1964). Paris: Dunod pp. 195-199 (1965).

33. M. Pilkuhn, H. Rupprecht, and J. Woodall, "Continuous Stimulated Emission from GaAs Diodes at 77°K," Proc. IEEE, *51*, 1243 (1963).

34. G. Burns, R. S. Levitt, M. I. Nathan, and K. Weiser, "Some Properties of InP Lasers," Proc. IEEE, *51*, 1148 (1963).

35. T. H. Zachos and J. E. Ripper, "Resonant Modes of GaAs Junction Lasers," IEEE Trans. Quantum Electronics (1968).

36. J. C. Dyment, "Hermite-Gaussian Mode Patterns in GaAs Junction Lasers," Appl. Phys. Letters, *10*, 84 (1967).

37. T. H. Zachos, "Gaussian Beams from GaAs Junction Lasers," Appl. Phys. Letters (May 1968).

38. F. M. Ryan and R. C. Miller, "The Effect of Uniaxial Strain on the Threshold Current and Output of GaAs Lasers," Appl. Phys. Letters, *3*, 162 (1963).

39. I. Melngailis and R. H. Rediker, "Properties of InAs Lasers," J. Appl. Phys. *37*, 899 (1966).

40a. J. C. Dyment and J. E. Ripper, "Temperature Behavior of Stimulated Emission. Delays in GaAs Diode and a Proposed Trapping Model," IEEE Quantum Electronics, Special Semiconductor-Laser Issue (1968).

40b. G. E. Fenner, "Delay of the Stimulated Emission in GaAs Laser Diodes Near Room Temperature," Solid State Electron., *10*, 753 (1967).

41. H. Kressel and H. Mierop, "Catastrophic Degradation in GaAs Injection Lasers," J. Appl. Phys., *38*, 5419 (1967).

42. N. G. Basov, Fiz Tverd. Tela, *8*, 2816 (1966); Soviet Phys.—Solid State S. 2254 (1967).

43. R. J. Phelan and R. H. Rediker, "Optically Pumped Semiconductor Lasers," Appl. Phys. Letters, *6*, 70 (1965).

44. N. Holonyak, Jr., M. D. Sirkis, G. E. Stillman, and M. R. Johnson, "Laser Operation of CdSe Pumped With a Ga(AsP) Laser Diode," Proc. of the IEEE, *54*, 1068 (1966).

45. I. Melngailis, "Optically Pumped Indium Arsenide Laser," IEEE J. of Quantum Electronics, *QE-1* 104 (1965).

46. C. B. a la Guillaume and J. M. Debever, Solid State Comm., *2*, 145 (1965), also in Symposium on Radiative Recombination, p. 255, Dunod., Paris (1964).

47. C. E. Hurwitz, "Efficient Ultraviolet Laser Emission in Electron-Beam-Excited ZnS," Appl. Phys. Letters, *9*, 116 (Aug. 1966).

48. K. Weiser and J. F. Woods, Appl. Phys. Letters *7*, 225 (1965).

49. M. B. Panish, I. Hayashi, and S. Sumski, Appl. Phys. Letters *16*, 326 (1970).

50. I. Hayashi, M. B. Panish, P. W. Foy, and S. Sumski, Appl. Phys. Letters *17*, 109 (1970).

51. H. Kroemer, Proc. IEEE *51*, 1782 (1963).

52. Zh. I. Alferov and R. F. Kazarinov, Inventor's Certificate No. 181737 (1963); Zh. I. Alferov, Fiz. i Tekhnika Poluprovodnikov *1*, 436 (1967).

53. J. E. Ripper, J. C. Dyment, L. A. D'Asaro, and T. L. Paoli, Appl. Phys. Letters (Feb. 1971).

54. M. B. Panish and S. Sumski, J. Phys. Chem. Solids *30*, 129 (1969).

55. I. Hayashi, M. B. Panish and F. W. Foy, IEEE J. Quan. Elec. *QE-5*, 211 (1969).

56. H. Kressel and H. Nelson, RCA Review *30*, 106 (1969).

PART

BULK-EFFECT DEVICES

- Bulk Effect Devices
- Author Index
- Subject Index

■ INTRODUCTION

■ BULK DIFFERENTIAL NEGATIVE
RESISTANCE

■ RIDLEY-WATKINS-HILSUM
(RWH) MECHANISM

■ GUNN OSCILLATOR AND
VARIOUS MODES OF OPERATION

■ ASSOCIATED BULK-EFFECT DEVICES

14

Bulk-Effect Devices

I INTRODUCTION

Bulk-effect semiconductor devices are those which do not consist of any *p-n* junctions or any interfaces except ohmic contacts, and which utilize the semiconductor bulk properties to function under various external influences such as electric and magnetic fields. We have already considered a few bulk-effect devices in the previous chapters. For example, the photoconductor in Chapter 12 is a two-terminal optoelectronic device which utilizes the effect that under illumination the bulk resistivity decreases owing to band-to-band or impurity-level-to-band transitions. The current limiter in Chapter 7 is also a two-terminal bulk-effect device which utilizes the effect that in Ge or Si the drift velocity saturates at sufficiently high electric field.

In this chapter we shall consider a few other bulk-effect devices, in particular the microwave oscillations associated with the bulk voltage-controlled differential negative resistance. Gunn[1] discovered in 1963 that when the applied dc electric field across a randomly orientated, short, *n*-type sample of GaAs or InP exceeded a critical threshold value of several thousand volts per centimeter, coherent microwave output was generated. The frequency of oscillation was approximately equal to the reciprocal of the carrier transit time across the length of the sample. Later, Kroemer[2] pointed out that all the observed properties of the microwave oscillation were consistent with a theory of differential negative resistance independently proposed by Ridley and Watkins,[3] and Hilsum.[4] The mechanism responsible for the differential negative resistance is a field-induced transfer of conduction-band electrons from a low-energy, high-mobility valley to higher-energy, low-mobility satellite valleys. The GaAs pressure experiments of Hutson et al.[5] and the $GaAs_{1-x}P_x$ alloy experiments of Allen et al.,[6] which demonstrated that the threshold electric field decreases with decreasing energy separation between the valley minima, provided convincing evidence that the Ridley-Watkins-Hilsum (RWH) mechanism was indeed responsible for the Gunn oscillation.

In Section 2, we shall consider the general characteristics of devices with bulk differential negative resistance. In Section 3, we shall discuss the theory of intervalley carrier transport. The various modes of operation of Gunn-type devices utilizing the RWH mechanism are outlined in Section 4. We shall also consider a few other bulk-effect devices including thermistor and Hall devices in Section 5.

2 BULK DIFFERENTIAL NEGATIVE RESISTANCE

(1) Basic Characteristics[7]

A bulk semiconductor exhibiting differential negative resistance is inherently unstable. This is because a random fluctuation of carrier density at any point in the semiconductor produces a momentary space charge which grows exponentially in space and time. Bulk negative-resistance devices can be classified into two groups: voltage-controlled *differential negative resistance* (DNR) and current-controlled differential negative resistance. The general current-density versus electric field characteristics for these two groups are shown in Fig. 1(a) and (b) respectively. Also shown are the corresponding differential resistivity ($\partial \mathscr{E}/\partial J$) from zero bias to the onset of the negative-resistance region. In voltage-controlled DNR the electric field can be multivalued, and in current-controlled DNR it is the electric current which can be multivalued.

We have already seen similar *I-V* characteristics as shown in Fig. 1 in previous chapters which are associated with junction or contact phenomena. For example, the tunnel diode exhibits a voltage-controlled DNR, while the semiconductor-controlled rectifier exhibits a current-controlled DNR. In this chapter we shall consider differential negative resistance associated with the microscopic bulk semiconductor properties which are not affected by the sample geometry or boundary contact conditions. There are various physical causes which give rise to the bulk differential negative reistance. A famous example of voltage-controlled DNR is the Ridley-Watkins-Hilsum (RWH) mechanism (also referred as the Gunn effect) due to intervalley carrier transport. The devices associated with the RWH mechanism will be considered in the next section. For the bulk current-controlled DNR devices, an interest-

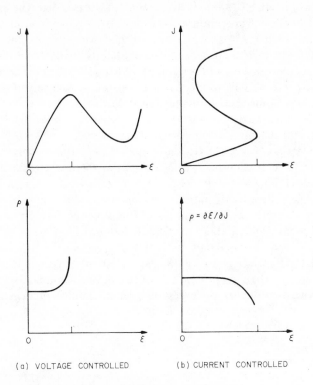

(a) VOLTAGE CONTROLLED   (b) CURRENT CONTROLLED

Fig. 1 Current density versus electric field characteristics for
(a) voltage-controlled differential negative resistance (DNR)
(b) current-controlled differential negative resistance.
Also shown are the corresponding differential resistivities prior to the onset of the DNR region.
(After Ridley, Ref. 7.)

ing example is the cryosar which involves impact ionization of shallow impurity levels in compensated semiconductors.[8]

In this section we shall consider the general behavior of bulk negative-resistance devices regardless of their physical causes underlying the appearance of a particular negative resistance.

Because of the differential negative-resistance, the semiconductor, initially homogeneous, becomes electrically heterogeneous in an attempt to reach stability. We shall now present a simple argument which shows that for voltage-controlled DNR devices, high-field domains (or an accumulation layer) will form, and for current-controlled DNR devices, high-current filaments will form.[7] The simple argument will be modified when we discuss the various modes of operation in later sections.

We note that in Fig. 1(a) the differential positive resistance increases with field or $dR/d\mathscr{E} > 0$. If there is a region of slightly higher field as shown in Fig. 2(a), the resistance in this region is larger. Thus less current will flow through this region. This results in an elongation of the region and a high-field domain is formed separating regions of low field. The interfaces separating low- and high-field domains lie along equipotentials so that they are in planes perpendicular to the current direction.

For current-controlled DNR the initial differential positive resistance decreases with increasing field or $dR/d\mathscr{E} < 0$. If there is a region of slightly higher field as shown in Fig. 2(b), the resistance in this region is smaller. Thus more current will flow into this region. This results in an elongation of this region along the current path, and finally in the formation of a high-current filament running along the field direction.

To consider in more detail the space-charge instability of a voltage-controlled DNR, we refer to Fig. 3. Part (a) shows a typical instantaneous $J\text{-}\mathscr{E}$ plot and part (b) shows the profile of this device. Assume that at point A in the device there exists an excess (or accumulation) of negative charge

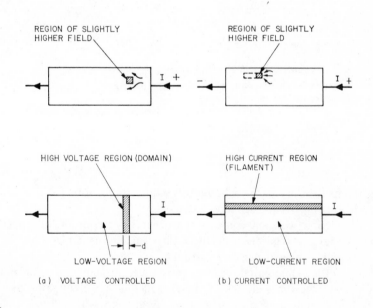

Fig. 2
(a) Formation of high-field domain in a voltage-controlled DNR.
(b) Formation of high-current filament in a current-controlled DNR.

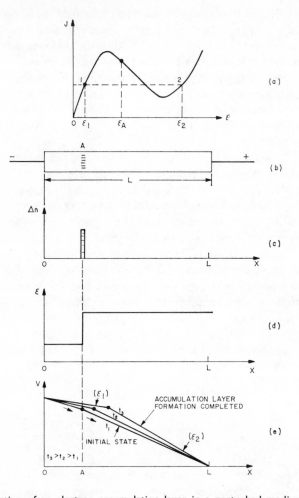

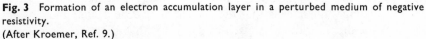

Fig. 3 Formation of an electron accumulation layer in a perturbed medium of negative resistivity.
(After Kroemer, Ref. 9.)

which could be due to a random noise fluctuation or possibly a permanent nonuniformity in doping. Integration once of Poisson's equation yields the electric field distribution as shown in Fig. 3(d) where the field to the left of point A is lower than that to the right of A. If the device is biased at point $\mathscr{E}_A$ on the J-$\mathscr{E}$ curve, this condition would imply that the carriers (or current) flowing into point A are greater than that flowing out of point A, thereby increasing the excess negative space charge at A. Now the field to the left of point A is even less than it was originally and the field to the right is greater than originally, resulting in an even greater space-charge accumulation. This process continues until the high and low fields both obtain values outside the DNR region and settle at points 1 and 2 in Fig. 3(a) where the currents in the two field regions are equal. As a result, a traveling space-charge accumulation is formed. This process, of course, is dependent on the condition that the number of electrons inside the crystal is large enough to allow the necessary amount of space charge to be built up during the transit time of the space-charge layer.[9]

The pure accumulation layer discussed above is the simplest form of space-charge instability. When there are positive and negative charges separated by a small distance, then one has a dipole formation (or domain) as shown in Fig. 4. The electric field inside the dipole would be greater than the fields on either side of it, Fig. 4(c). Because of the negative differential resistance, the current in the low-field region would be greater than that in the high-field region. The two field values will tend toward equilibrium levels outside the DNR region where the high and low currents are the same, Fig. 3(a). (Assumed here is a domain with zero wall thickness.) The dipole has now reached a stable configuration. The dipole layer moves through the crystal and disappears at the anode, at which time the field begins to rise uniformly across the sample through threshold, i.e., $\mathscr{E} > \mathscr{E}_T$ as shown in Fig. 5(a), thus forming a new dipole layer, and the process repeats itself. To estimate the dipole width d, we can use the following expression

$$V = \mathscr{E}_A L = \mathscr{E}_2 d + (L - d)\mathscr{E}_1 \tag{1}$$

or

$$d = L \frac{(\mathscr{E}_A - \mathscr{E}_1)}{(\mathscr{E}_2 - \mathscr{E}_1)} \tag{2}$$

where L is sample length. We shall now make the assumption that the most stable situation is obtained when the input energy is a minimum. For a constant voltage, this means that the current should be a minimum. From Fig. 5(a) the dipole width d is given by Eq. (2) with $\mathscr{E}_{2m}$ and $\mathscr{E}_{1m}$ replacing $\mathscr{E}_2$ and $\mathscr{E}_1$ respectively.

The case of current-controlled DNR can be treated similarly. Instead of a domain we consider a filament of cross-sectional area a. For a given current I, under steady-state condition, we obtain from Fig. 5(b)

$$I = J_A A = J_2 a + (A - a)J_1 \tag{3}$$

or

$$a = A \frac{(J_A - J_1)}{(J_2 - J_1)} \tag{4}$$

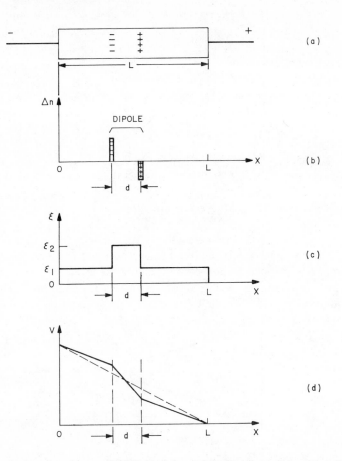

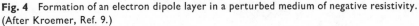

Fig. 4 Formation of an electron dipole layer in a perturbed medium of negative resistivity. (After Kroemer, Ref. 9.)

where A is the device cross-sectional area. Under the condition of minimum energy the voltage should be a minimum, and the area of the filament is given by Eq. (4) with J_{2m} and J_{1m} replacing J_2 and J_1 respectively.

It is important to point out that the J-$\mathscr{E}$ curves shown in Figs. 1 and 5 refer to the microscopic instantaneous characteristic of the bulk semiconductor and are not the dc characteristics. These curves should be distinguished from similar curves of differential negative resistance due to macroscopic effect such as the temperature effect in a thermistor (to be considered later). The latter current-voltage characteristics will be influenced by sample geometry and boundary conditions such as the heat sink.

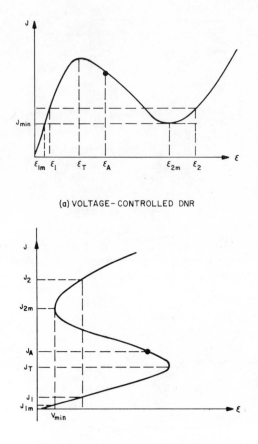

(a) VOLTAGE - CONTROLLED DNR

(b) CURRENT-CONTROLLED DNR

Fig. 5
(a) Minimum current density and corresponding electric field for a voltage-controlled DNR.
(b) Minimum electric field and corresponding current density for a current-controlled DNR.
(After Ridley, Ref. 7.)

(2) Cryosar

The cryosar is a bulk current-controlled differential negative-resistance device.[8] It has the potential to be used in high-speed switching and memory applications, and utilizes the low-temperature avalanche breakdown produced by impact ionization of impurities. The name of the device was derived from low-temperature (*cryo-*) *s*witching by *a*valanche and *r*ecombination.

At low temperatures (e.g., liquid helium temperature of 4.2°K) with low applied voltages, the bulk semiconductor may have a very high resistivity, since almost all the carriers are "frozen" to the impurity centers. The residual conductivity is due either to those few free carriers generated thermally and by stray radiation, or to a conduction process in the impurity levels themselves. As the applied voltage increases, it becomes possible for the free carriers to gain sufficient energy in the electric field to ionize the impurities upon impact. Finally at some critical electric field, the impact ionization rate exceeds the recombination rate, and a reversible nondestructive breakdown occurs similar to the avalanche multiplication due to band-to-band transitions. At the end of the avalanche process almost all the impurities are ionized, and the resistivity can be reduced by many orders of magnitude.

The basic equation governing the switching process for a p-type sample is the rate equation which relates the thermal excitation rate (G_T) and impact ionization rate (G_I) of the neutral acceptors by energetic holes to the corresponding inverse process of single-hole recombination rate R_T. If p, N_A, and N_D are the carrier concentration, acceptor, and donor concentrations respectively, the rate equation is given by[10]

$$\frac{dp}{dt} = G_T(N_A - N_D - p) + pG_I(N_A - N_D - p) - pR_T(N_D + p). \quad (5)$$

In steady state, holes are being ionized as fast as they recombine, i.e., $dp/dt = 0$, and we obtain for $p \ll N_D$, $(N_A - N_D)$:

$$p \cong \frac{G_T(N_A - N_D)}{R_T N_D - G_I(N_A - N_D)}. \quad (6)$$

Avalanche breakdown occurs when p approaches infinity, or

$$R_T N_D = G_I(N_A - N_D) \quad (7)$$

$$R_T/G_I = N_A/N_D - 1. \quad (7a)$$

Under the conditions that G_I increases with field and R_T decreases with field, the left-hand side of the above expression decreases as the field increases which leads to the prediction that the breakdown field will increase as the compensation factor, N_A/N_D, decreases. The above derivations are for p-type samples; for n-type samples the compensation factor should be N_D/N_A.

Typical cryosars have been made from p-type indium-doped germanium samples compensated with antimony, in the form of wafers with ohmic contacts.[8] Figure 6(a) shows the I-V characteristics of a sample at 4.2°K. Figure 6(b) shows the breakdown field as a function of the compensation factor. We note that the breakdown field decreases with increasing compensation factor in agreement with Eq. (7a). For this sample, at low currents (e.g.,

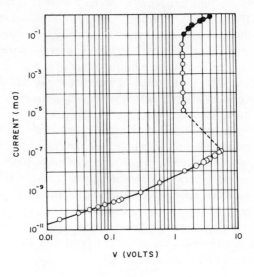

(a)

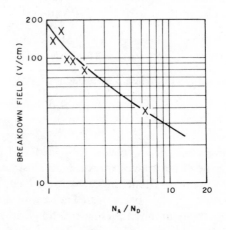

(b)

Fig. 6
(a) Current-voltage characteristics of a Ge cryosar at 4.2°K.
(b) Breakdown field versus compensation ratio.
(After McWhorter and Rediker, Ref. 8.)

10^{-5} amp), the area of the high-current filament is small—of the order of 0.1 % of the total device area. As the current increases, the area increases. Finally, for current larger than 10^{-1} amp, the filament will fill the total device area. In the above current range the sustaining voltage remains essentially the same (~ 1.5 V) corresonding to the minimum voltage, V_{min}, shown in Fig. 5(b).

One can also put two cryosars together to form a compound cryosar.[11] When the cryosar with the larger compenstaion factor is broken down by impact ionization of the impurities, the impurity conductance of the cryosar with the smaller compensation factor will limit the current. This current limiting, in addition to the differential negative resistance of the breakdown, permits one to use a matrix array of compound cryosars in a coincident-voltage memory. If a gate electrode is added to the cryosar, one has a field-effect controlled impact-ionization switch or "cryosistor" which combines the current-voltage characteristics of a junction field-effect transistor and a cryosar, and can be operated as a three-terminal bistable device.[12]

3 RIDLEY-WATKINS-HILSUM (RWH) MECHANISM[3,4]

In this section we shall consider the most important bulk voltage-controlled differential negative resistance associated with the Ridley-Watkins-Hilsum (RWH) mechanism. As mentioned in the previous section, there are various physical causes which may give rise to the voltage-controlled DNR. Since the current density in a bulk semiconductor is proportional to both the density of carriers and their drift velocity, a decrease in current density with an increasing electric field can be brought about by a decrease in either of these quantities. For example, a decrease of electron density, in the presence of an increasing field can be brought about by field-enhanced trapping.[13] This occurs in gold-doped germanium samples. Figure 7(a) shows the energy profile of an impurity state which is negatively charged. As the field increases, electrons will gain enough energy to surmount the potential barrier and subsequently "drop" into the impurity states where they are immobile. This process thus gives rise to the DNR. The trapping effects, however, are in general very slow and the device potential of this form of DNR is limited. The second possibility of reducing current density with increasing field is by decreasing the drift velocity. This is the mechanism which underlies the RWH mechanism. In addition to the above mechanisms, a bulk voltage-controlled DNR can, in principle, be obtained by transferring electrons from a positive mass state to a negative mass state[14] as shown in Fig. 7(b), or from field-induced repopulation[15] of the different (111) valleys in the conduction band of Ge.

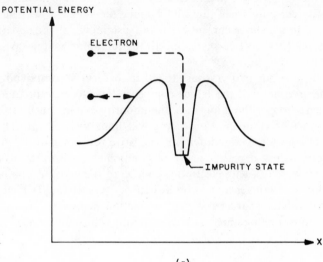

(a)

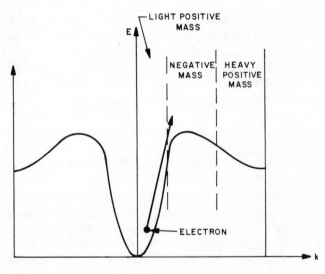

(b)

Fig. 7
(a) Energy versus distance of a negatively charged impurity state.
(b) Energy versus momentum for the conduction band of a typical semiconductor.

(1)   Qualitative Analysis

The RWH mechanism is the transfer of conduction electrons from a high-mobility energy valley to low-mobility, higher-energy satellite valleys. To understand how this effect leads to differential negative resistance, consider the simple two-valley model of GaAs shown in Fig. 8. The energy separation between the two valleys is $\Delta E = 0.36$ eV. The lower valley effective density of states is denoted by N_1, the effective mass by m_1^*, and the mobility by μ_1. The same upper valley quantities are denoted by N_2, m_2^*, and μ_2 respectively. Also, the densities of electrons in the lower and upper valleys are n_1 and n_2 respectively, and the total carrier concentration is given by $n = n_1 + n_2$. The steady-state conductivity of the semiconductor may be written as

$$\sigma = q(\mu_1 n_1 + \mu_2 n_2) = qn\bar{\mu} \tag{8}$$

where

$$\bar{\mu} \equiv (\mu_1 n_1 + \mu_2 n_2)/(n_1 + n_2)$$

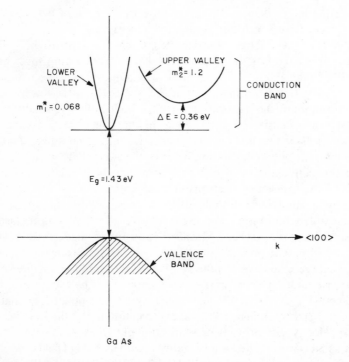

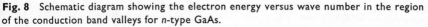

Fig. 8   Schematic diagram showing the electron energy versus wave number in the region of the conduction band valleys for n-type GaAs.

is the average mobility and q is the electronic charge. The magnitude of the current density is

$$J = qn\bar{\mu}\mathscr{E} = qnv \tag{9}$$

where v is the average electron velocity.

For simplicity the following assignments will be made for the carrier concentrations in various ranges of electric field values as illustrated in Fig. 9:

$$n_1 \sim n \quad \text{and} \quad n_2 \sim 0 \quad \text{for} \quad 0 < \mathscr{E} < \mathscr{E}_a$$

$$n_1 + n_2 = n \quad \text{for} \quad \mathscr{E}_a < \mathscr{E} < \mathscr{E}_b$$

$$n_1 \sim 0 \quad \text{and} \quad n_2 \sim n \quad \text{for} \quad \mathscr{E} > \mathscr{E}_b.$$

Using these relations the current density takes on the asymptotic values

$$J \sim qn\mu_1\mathscr{E} \quad \text{for} \quad 0 < \mathscr{E} < \mathscr{E}_a$$

$$J \sim qn\mu_2\mathscr{E} \quad \text{for} \quad \mathscr{E} > \mathscr{E}_b.$$

Now if $\mu_1\mathscr{E}_a$ is larger than $\mu_2\mathscr{E}_b$, there will exist a region of negative differential resistance between $\mathscr{E}_a$ and $\mathscr{E}_b$ as shown in Fig. 10.

In order that the electron transfer mechanism give rise to bulk DNR, certain requirements must be met: (1) the lattice temperature must be low enough that, in the absence of a bias electric field, most of the electrons are in the lower conduction band minimum, or $kT < \Delta E$; (2) in the lower conduction band minimum the electrons must have high mobility, small effective mass, and a low density of states, while in the upper satellite valleys the electrons must have low mobility, large effective mass, and a high density of states; and (3) the energy separation between the two valleys must be smaller than the semiconductor band gap so that avalanche breakdown does not set in before the transfer of electrons into the upper valleys.

Of the semiconducting materials satisfying these conditions, n-type GaAs is the most widely understood and used; however, the RWH mechanism has been observed in InP,[1] CdTe,[16–19] ZnSe,[19] and InAs[20] under hydrostatic or uniaxial pressure. The pressure was applied to InAs to decrease the energy difference between the lower conduction band valley and the satellite valleys; this difference under normal pressures is greater than the energy gap. It has been suggested that the RWH mechanism may be obtained in materials which do not strictly adhere to the above conditions.[21] Materials in which only drift velocity saturation is presently thought to exist, may in actuality possess a small value of differential negative resistivity at high fields and should be capable of supporting oscillations. In this respect RWH oscillations have been observed in Ge.[22,23]

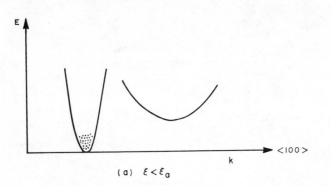

(a) $\mathcal{E} < \mathcal{E}_a$

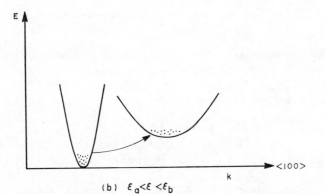

(b) $\mathcal{E}_a < \mathcal{E} < \mathcal{E}_b$

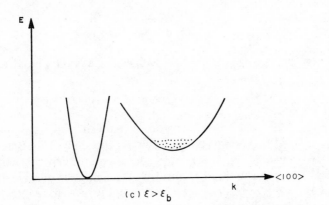

(c) $\mathcal{E} > \mathcal{E}_b$

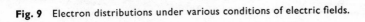

Fig. 9 Electron distributions under various conditions of electric fields.

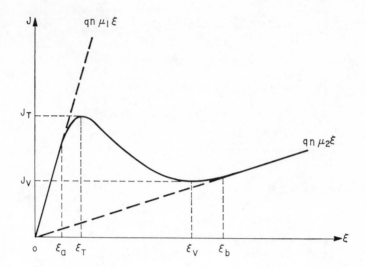

Fig. 10 A possible current versus electric field characteristic of a two-valley semiconductor.

(2)   Velocity-Field Characteristics

Since GaAs is the most widely studied and understood of the semi-conductors in which the RWH mechanism has been observed, the analysis of differential negative resistance which follows will be mainly concerned with GaAs.

Taking the derivative of Eq. (9) with respect to the electric field, one obtains

$$\frac{dJ}{d\mathscr{E}} = qn\frac{dv}{d\mathscr{E}}. \tag{10}$$

The condition for negative differential conductance can then be written as

$$\frac{dv}{d\mathscr{E}} \equiv \mu_D < 0. \tag{11}$$

To determine $dv/d\mathscr{E}$, one must carry out a detailed analysis of high-field carrier transport.[24,25] The theoretical results of the analysis by Butcher and Fawcett[24] are shown in Fig. 11 for GaAs, along with the experimental results of Ruch and Kino.[26] The threshold field, $\mathscr{E}_T$, defining the onset of DNR, is approximately 3.2 kV/cm. The low field mobility was found to be 8000 cm²/V-sec and the initial differential negative mobility was determined as -2400 cm²/V-sec. The field at which DNR ends is predicted to be about 20

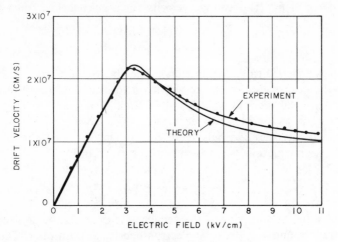

Fig. 11 Theoretical velocity-field characteristic of GaAs.
(After Butcher and Fawcett, Ref. 24.) And experimental result.
(After Ruch and Kino, Ref. 26.)

kV/cm. At higher fields the mobility approaches an asymptotic value of 175 cm²/V-sec.

In order to obtain an analytical expression for the velocity-field curve, it is assumed that the ratio of electrons between the valleys n_2/n_1 can be expressed as[27]

$$n_2/n_1 = (\mathscr{E}/\mathscr{E}_o)^k \equiv F^k \qquad (12)$$

where k is a constant and $\mathscr{E}_o$ is defined as the field such that $n_1 = n_2$ when $\mathscr{E} = \mathscr{E}_o$. We also define the ratio of upper valley mobility to lower valley mobility, μ_2/μ_1, as a constant:

$$\mu_2/\mu_1 \equiv B. \qquad (13)$$

Implicit in the model discussed here are the assumptions that the mobilities are field independent and that the local electron distribution between the valleys follows the electric field instantaneously in both space and time. In GaAs, in which the intervalley carrier transport is due to optical-phonon scattering,[28] the scattering time is of the order of 10^{-12} sec (ps). Hence for operating frequencies of the order of 10 GHz or less, the intervalley transport time can be considered as instantaneous.

Using this model the individual electron densities, n_1 and n_2, are given by

$$n_1 = n(1 + F^k)^{-1} \qquad (14a)$$

$$n_2 = nF^k(1 + F^k)^{-1} \qquad (14b)$$

where $n = n_1 + n_2$. The average velocity for a given field $\mathscr{E}$ then becomes

$$v = \frac{\mu_1 \mathscr{E}(1 + BF^k)}{1 + F^k}. \tag{15}$$

The values assigned to the parameters in the above equations that are found to give the best overall fit to the velocity-field characteristic shown in Fig. 11 are:[29] $\mu_1 = 8000$ cm^2/V-sec, $\mathscr{E}_o = 4000$ V/cm, $k = 4$, and $B = 0.05$.

We shall now consider two equations which govern the basic current-voltage characteristics of a bulk DNR device. The first one is the one-dimensional Poisson equation*

$$\frac{\partial \mathscr{E}}{\partial x} = \frac{q}{\varepsilon_s}(n - n_o) \tag{16}$$

where ε_s is the permittivity of the semiconductor, n_o is the ionized donor concentration, and n is the total electron density at position x. The second equation is the conduction current density equation

$$J = qnv(\mathscr{E}) - q\frac{\partial(Dn)}{\partial x} \tag{17}$$

or

$$J = q(n_1\mu_1 + n_2\mu_2)\mathscr{E} - q\left(D_1\frac{\partial n_1}{\partial x} + D_2\frac{\partial n_2}{\partial x}\right) \tag{17a}$$

where the first term on the right-hand side is due to carrier drift and $v(\mathscr{E})$ is the drift velocity which is a function of electric field; and the second term is due to carrier diffusion and D is the diffusion constant (assumed independent of $\mathscr{E}$).

Under steady-state conditions and for small carrier-concentration gradients, Eq. (17) reduces to $J = qnv(\mathscr{E})$. Substitution of this result in Eq. (16) yields

$$\frac{\partial \mathscr{E}}{\partial x} = -\frac{qn_o}{\varepsilon_s}\left[1 - \frac{J/qn_o}{v(\mathscr{E})}\right]. \tag{18}$$

The above first-order nonlinear equation has the boundary condition that the field $\mathscr{E}(x)$ be continuous everywhere. Equation (18) was numerically integrated for a GaAs system,[30] and the results are shown in Fig. 12 where $\mathscr{E}_T$ and J_T are the threshold electric field and current density defined in Fig. 10. It can be seen from this figure that at any point x, the electric field is a monotonically increasing function of the current density. Therefore, when one properly takes into account boundary conditions, the steady-state

---

* In order to avoid a preponderance of minus signs, a positive charge will be assigned to electrons and all operations are modified accordingly.

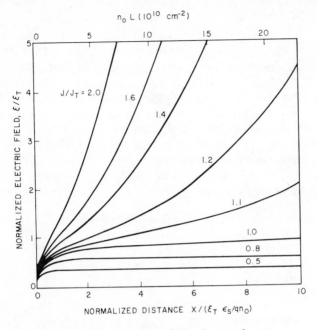

Fig. 12 Electric field versus distance x where $\mathscr{E}(0) = 0$ at $x = 0$. (After McCumber and Chynoweth, Ref. 30.)

solutions do not exhibit negative resistance. This is not surprising, however, since Shockley[31] pointed out that bulk DNR diodes must be dc stable. This is true because inside the device there exists an internal space charge (i.e., excess electrons injected from the cathode) which increases with increasing bias voltage by an amount so large that the current increases despite a decrease in electron velocity due to the negative electron mobility. However, these steady state solutions are not necessarily stable with respect to small fluctuations.

4 GUNN OSCILLATOR AND VARIOUS MODES OF OPERATION

(1) Basic Criterion

Since the first observation of microwave oscillation in GaAs and InP bulk devices by Gunn[1] in 1963, various modes of operation have been studied depending on the operating conditions and material parameters. These modes of operation are not basically different from each other, since all are derived from the RWH mechanism of intervalley carrier transport.

As mentioned previously, for a voltage-controlled DNR device under nonsteady state, either accumulation layers or dipole layers (domains) may be formed. The formation of a strong space-charge instability is dependent on the conditions that enough charge is available in the crystal and the sample is long enough that the necessary amount of space charge can be built up within the transit time of the electrons. This requirement sets up a criterion for the various modes of operation of bulk negative differential resistance devices. One can easily determine the criterion for the formation of strong space-charge instabilities from the following simple argument.[9] During the early stages of space-charge build up, the time rate of growth of the space charge layers is given by

$$Q(x, t) = Q(x - vt, 0)\exp(t/\tau_D) \tag{19}$$

where

$$\tau_D = \frac{\varepsilon_s}{\sigma} = \frac{\varepsilon_s}{q n_o |\mu_D|}$$

is the absolute value of the negative dielectric relaxation time, ε_s the permittivity, n_o the doping concentration, and μ_D is the negative mobility. If this relationship remained valid throughout the entire transit time of the space-charge layer, the maximum growth would be given by $\exp(L/v\tau_D)$ where L is the length of the semiconductor. For space-charge growth to be large, this growth factor must be greater than unity. This implies that

$$n_o L > \varepsilon_s v/q |\mu_D|. \tag{20}$$

For n-type GaAs, $\varepsilon_s v/q|\mu_D| \sim 10^{12} \text{ cm}^{-2}$. Samples with $n_o L$ products smaller than 10^{12} cm^{-2} exhibit a stable field distribution. Hence an important boundary separating the various modes of operation in GaAs is the (carrier concentration) × (sample length) product, $n_o L = 10^{12} \text{ cm}^{-2}$.

Samples with $n_o L < 10^{12} \text{ cm}^{-2}$ cannot support traveling dipole domains but can support growing space-charge waves, and they can be operated as stable linear microwave amplifiers.[33] If such a sample is connected to a resonant circuit with a sufficiently high load resistance it will oscillate in the pure accumulation-layer mode.

Samples with $n_o L > 10^{12} \text{ cm}^{-2}$ connected to a constant voltage circuit are capable of supporting fully developed traveling dipole domains and operate in the Gunn, or transit-time, mode of operation in which the frequency of oscillation is determined by the drift velocity of the carriers and the length of the sample. If a sample with $n_o L > 10^{12} \text{ cm}^{-2}$ is connected to a resonant circuit, three basic modes of operation are possible: (1) the transit-time domain mode, (2) the quenched domain mode in which fully developed dipole domains are quenched before they reach the anode; and (3) the inhibited domain mode in which the start of domain formation is delayed.

Under the condition of very small doping fluctuations across the sample or very small space-charge accumulation, the electric field can effectively remain uniform across the sample. The sample will then appear to the circuit as a true negative resistance, and this is called the limited-space-charge-accumulation (LSA) mode.

The regions of fL (frequency × length) and $n_o L$ (doping × length) for the various modes of operation are shown in Fig. 13. The mode of operation in the overlap region ($fL \sim 10^7$ cm/sec, $n_o L \sim 10^{12}$ cm^{-2}) depends strongly on bias conditions and the circuit to which the sample is connected. This section will consider the analysis of these modes of operation.

(2) Accumulation-Layer Mode ($n_o L < 10^{12}$ cm^{-2})

Lightly doped or short samples (i.e., $n_o L < 10^{12}$ cm^{-2}) exhibit a stable field distribution and a positive dc resistance but a negative resistance in a band of frequencies near the electron transit-time frequency $f_t = v/L$ and its harmonics.[32] Figure 12 shows the steady-state field distribution in such a

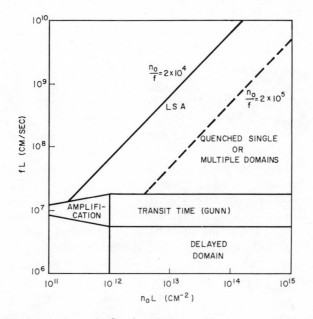

Fig. 13 Modes of operation in the fL-$n_o L$ regions.

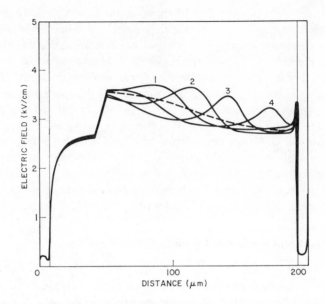

Fig. 14 Electric field distribution in a room temperature sample of GaAs biased for negative resistance operation. The dashed curve shows the steady state field distribution in the absence of modulation and the solid curves are four sequential field distributions with quarter cycle spacing during small signal modulation.
(After McCumber, Ref. 32.)

stable amplifier diode for different values of current density. Figure 14 shows four superimposed frames of a computer-generated motion picture[30,32] of the field distribution across a 200-μm GaAs sample biased for negative conductance. The sample is uniformly doped at 10^{14} cm^{-3} except for a 10-μm notch starting at 40 μm where the doping has been reduced to 0.9×10^{14} cm^{-3}. The dashed curve shows the steady-state field distribution in the absence of bias modulation. The solid curves are four sequential field distributions at quarter-cycle spacing during small-signal modulation at a frequency near which the negative conductance is greatest. This mode of operation discovered by Thim et al.[33] is called the traveling space-charge mode.

We shall now consider the effect of small sinusoidal current or voltage perturbation on the electric field and current distribution. The relevant equations for space-charge waves are Poisson's equation, Eq. (16), current density equation, Eq. (17), and the continuity equation

$$\frac{\partial n_{1,2}}{\partial t} = -\frac{1}{q}\frac{\partial J_{1,2}}{\partial x} + \left(\frac{\partial n_{1,2}}{\partial t}\right)_{\varepsilon} \qquad (21)$$

where the subscript $\mathscr{E}$ denotes the time rate of field-induced transfer of carriers from the lower conduction band valley to the satellite valleys and vice versa. Since the total carrier density between the two states is invariant, one has $(\partial n_1/\partial t)_{\mathscr{E}} = -(\partial n_2/\partial t)_{\mathscr{E}}$. Therefore, a common parameter,[34] γ, will be used to denote the rate of carrier transfer from valley 1 to valley 2.

$$\gamma \equiv (\mathscr{E}/n)(\partial n_1/\partial \mathscr{E}) = -(\mathscr{E}/n)(\partial n_2/\partial \mathscr{E}) = \frac{-kF^k}{(1+F^k)^2}. \tag{22}$$

A small-signal analysis of Eqs. (16), (17), and (21) may be performed by separating each function into a steady-state part and a space-time varying part

$$f(x, t) = f_0 + f_1(x)\exp(j\omega t)$$

where $f_0 \gg f_1$. The resulting perturbation functions form a coupled set of differential equations. The boundary conditions are that the electric field at the contacts be zero and the current at the contacts be carried predominantly by diffusion current. Also, the dc bias field within the bulk is assumed to be uniform and of such a magnitude as to be in the region of negative resistivity. It will also be assumed that $\mu_2 \ll \mu_1$ and $D_2 \ll D_1$, which is approximately true for n-type GaAs. Based on these assumptions and boundary conditions one obtains the small-signal current density $\tilde{J}(\omega)$ and field intensity $\tilde{\mathscr{E}}(x, \omega)$. The two-terminal admittance of the device is defined as

$$Y \equiv G + jB = A\tilde{J}(\omega) \bigg/ \int_0^L \tilde{\mathscr{E}}(x, \omega)\, dx \tag{23}$$

where A and L are the device area and length respectively, G is the conductance, and B is the susceptance (which is capacitive for the positive value and inductive for the negative value of B).

Figure 15 shows some computed results[34] of the admittance as a function of the transit angle for a device with a doping concentration $n_o = 3 \times 10^{13}$ cm^{-3}, $L = 40$ μm, $D = 400$ cm^2/sec, and $-1 < \gamma < -1.6$, where v_{o1} is the assumed velocity of carriers in the lower valley. It can be seen that for $\gamma < -1$ the conductance, as a function of transit angle, $(\omega L/v_{o1})$, alternates between positive and negative values, and for a large enough rate of intervalley carrier transfer (γ) there is strong enhancement of the negative and positive peaks. The magnification of the peaks is largely due to the growth of the space-charge waves and their subsequent in-phase and out-of-phase interference at certain transit angles. Several interesting features of Fig. 15 are worth pointing out. For $\gamma = -1.0$, Fig. 15(a), the conductance is always positive. This is due to the loss mechanism introduced by diffusion. As $|\gamma|$ increases in magnitude, the first negative conductance peak becomes sharper and occurs at progressively higher values of transit angle, Fig. 15(b) and (c).

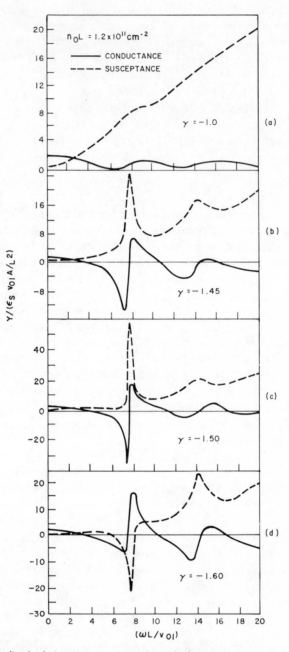

Fig. 15 Normalized admittance versus transit angle for various transfer rate γ. (After Hakki, Ref. 34.)

The capacitive-susceptance peaks occur at transit angles slightly higher than those of peak negative conductance. The transit angles of peak negative conductance are not exactly harmonically related but occur at values somewhat lower than the harmonic values. As γ exceeds some critical value, the capacitive susceptance, in the first negative conductance range, becomes inductive. As γ becomes more negative, the first negative conductance peak decreases in magnitude and occurs at progressively smaller transit angles, Fig. 15(c) and (d).

There have been several approaches to the theory of negative differential conductance amplification[29,30,35,36] and they all tend to have qualitatively similar results. It is inherent in the complexity of the problem that the analytical approximations must neglect or simplify one aspect or another of the problem. The McCumber-Chynoweth analysis[30] neglects electron diffusion effects and greatly simplifies the velocity-field characteristics. The McWhorter-Foyt analysis[35] also neglects diffusion and assumes a piecewise linear velocity-field characteristic. The analysis[34] of Hakki includes electron diffusion effects, but neglects space charge and assumes a uniform dc field, which is valid only for samples where $n_o L$ is close to 10^{12} cm^{-2}. Kroemer[29] calculated the complex admittance behavior of a bulk negative conductivity semiconductor in the limit of zero doping and zero trapping when all electrons are due to space-charge-limited emission from the negative electrode. Two different approximations were used: in the first, electron diffusion was neglected and in the second it was included. Both approximations agree at low frequencies where they result in a positive device conductance. At frequencies around the reciprocal electron transit time, both predict regions of negative conductance of slightly different magnitudes. At higher frequencies the diffusionless theory predicts slowly damped conductance oscillations, and inclusion of diffusion effects greatly increases the damping. Figure 16 shows a typical conductance and susceptance against the frequency plot showing the effects of electron diffusion on the two parameters. The important result of Kroemer's theory is that, under conditions of sufficiently strong sample bias, GaAs should show negative conductance properties even in the limit of zero doping.

The experimental gain as a function of frequency is shown in Fig. 17 for an n-type GaAs sample having a carrier concentration $n_o = 1.5 \times 10^{13}$/cm^3 and whose length is $L = 70$ μm. As the electric field is increased, the peak gain occurs at progressively higher frequencies, which agrees with the theory presented above.[34] In Fig. 18 the experimentally measured admittance as a function of frequency is shown for a sample of n-type GaAs having $n_o = 3 \times 10^{13}$/cm^3 and $L = 70$ μm. Observe that the negative conductance peaks occur at somewhere less than harmonic multiples, again in agreement with the theory. In general it has been found that amplification does not occur in

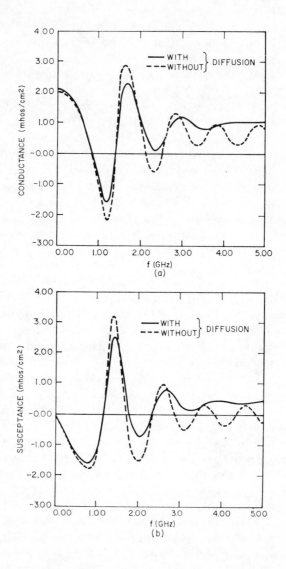

Fig. 16

(a) Conductance versus frequency, including diffusion effects, represented by the solid line, compared with the diffusion-free theory, represented by the dashed line.

(b) Susceptance versus frequency.

(After Kroemer, Ref. 29.)

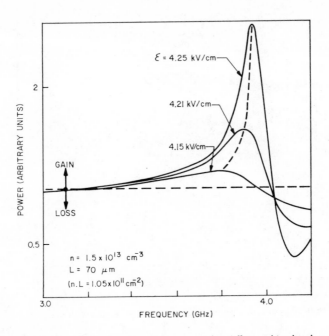

Fig. 17 Experimental gain as a function of frequency for different bias levels on a n-GaAs sample with $n_o = 1.5 \times 10^{13}$ cm^{-3} and $L = 70$ μm. (After Hakki, Ref. 34.)

samples with an $n_o L$ product below a certain value[37] ($\sim 8 \times 10^{10}$ cm^{-2}). This limit is believed to be due to the experimental difficulties, in particular, to trapping, contact, and bulk inhomogeneity problems, such that the parasitic series resistance could override the negative resistance of samples with small $n_o L$ product.[38]

(3) Dipole-Layer Mode $(n_0 L > 10^{12}$ cm$^{-2})$

When the $n_o L$ product is greater than 10^{12} cm^{-2} in GaAs, the space-charge perturbations in the material increase exponentially in space and time to form mature dipole layers which propagate to the anode. The dipole is usually formed near the cathode ohmic contact, since there exists the largest doping fluctuation and space-charge perturbation. The cyclic formation and subsequent disappearance at the anode of the fully developed dipole layers are what give rise to the experimentally observed Gunn oscillations. In Section 2 a qualitative discussion of space-charge instabilities was presented

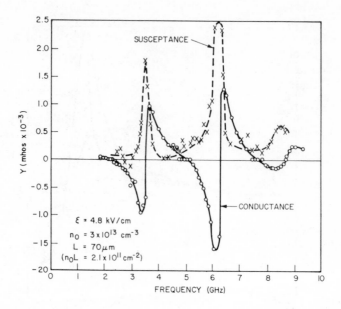

Fig. 18 Measured admittance versus frequency for a sample with $n_o = 3 \times 10^{13}$ cm^{-3} and $L = 70$ μm.
(After Hakki, Ref. 34.)

along with a short discussion of domains. In this section the theory of domains will be studied in more detail.

A.   Stable Domain Characteristics.   The characteristics of stable dipole domains will be analyzed in this section. The domains are stable in the sense that they propagate with a particular velocity but do not change in any other way with time. It will be assumed that the electron drift velocity follows the static velocity-field characteristic shown by the solid curve in Fig. 19. The equations that determine the behavior of the electron system are Poisson's equation, Eq. (16), and the total current-density equation

$$ J = qnv(\mathscr{E}) - q\,\frac{\partial D(\mathscr{E})n}{\partial x} + \varepsilon_s\,\frac{\partial \mathscr{E}}{\partial t}. \tag{24} $$

The above equation is similar to Eq. (17) except for the addition of the third term which corresponds to the displacement current component.

The type of solutions sought are those which represent a high-field domain which propagates without change of shape with a domain velocity v_{dom}. Outside the domain the carrier concentration and fields are at constant values given by $n = n_o$ and $\mathscr{E} = \mathscr{E}_r$ respectively. For this type of solution, both $\mathscr{E}$ and n should be functions of the single variable, $x' = x - v_{\text{dom}} t$, with the forms as shown in Fig. 20. It can be seen from Fig. 20 that n is a double-valued function of field. The domain consists of an accumulation layer where $n > n_o$, followed by a depletion layer where $n < n_o$. The carrier concentration, n, equals n_o at two field values, i.e., $\mathscr{E} = \mathscr{E}_r$ outside the domain and at $\mathscr{E} = \mathscr{E}_{\text{dom}}$, the peak domain field.

It will be assumed that the value of the outside field $\mathscr{E}_r$ is known. (Later it will be shown that $\mathscr{E}_r$ is easily determined.) The current outside the domain consists only of conduction current and is given by $J = q n_o v_r$, where $v_r = v(\mathscr{E}_r)$. Noting that

$$\frac{\partial \mathscr{E}}{\partial x} = \frac{d\mathscr{E}}{dx'} \quad \text{and} \quad \frac{\partial \mathscr{E}}{\partial t} = -v_{\text{dom}} \frac{\partial \mathscr{E}}{\partial x'}$$

one may rewrite Eqs. (16) and (24) as

$$\frac{\partial \mathscr{E}}{\partial x'} = \frac{q}{\varepsilon_s}(n - n_o) \tag{25}$$

and

$$\frac{d}{dx'}[D(\mathscr{E})n] = n[v(\mathscr{E}) - v_{\text{dom}}] - n_o(v_r - v_{\text{dom}}). \tag{26}$$

We can eliminate the variable x' from the above equations by dividing Eq. (26) by (25) to obtain a differential equation for $[D(\mathscr{E})n]$ as a function of the electric field:

$$\frac{q}{\varepsilon_s}\frac{d}{d\mathscr{E}}[D(\mathscr{E})n] = \{n[v(\mathscr{E}) - v_{\text{dom}}] - n_o(v_r - v_{\text{dom}})\}/(n - n_o). \tag{27}$$

In general, Eq. (27) can only be solved by numerical methods.[39-41] However, the problem may be simplified greatly by assuming that the diffusion term is independent of the electric field, $D(\mathscr{E}) = D$. Using this approximation, the solution to Eq. (27) is given by

$$\frac{n}{n_o} - \ln\frac{n}{n_o} - 1 = \frac{\varepsilon_s}{q n_o D}\int_{\mathscr{E}_r}^{\mathscr{E}}\{[v(\mathscr{E}') - v_{\text{dom}}] - \frac{n_o}{n}(v_r - v_{\text{dom}})\}\,d\mathscr{E}'. \tag{28}$$

(This may be verified by differentiation.)

Note that, when $\mathscr{E} = \mathscr{E}_r$ or $\mathscr{E}_{\text{dom}}$, one has $n = n_o$ (Fig. 20) and the left side of Eq. (28) vanishes; therefore, the integral on the right side of the equation must vanish when $\mathscr{E} = \mathscr{E}_{\text{dom}}$. However, the integration from $\mathscr{E}$ to $\mathscr{E}_{\text{dom}}$ can represent either the integration over the depletion region when $n < n_o$ or the integration over the accumulation region where $n > n_o$. Since the first term

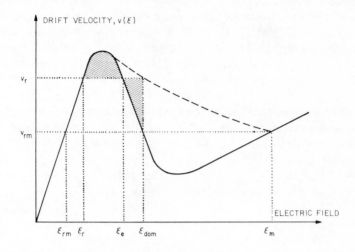

Fig. 19 Plots of drift velocity versus electric field (solid line) and peak domain field versus drift velocity outside the domain (dashed line).
(After Butcher et al., Ref. 43.)

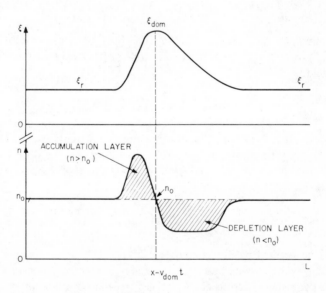

Fig. 20 Electric field and electron density distribution for a stable high-field domain propagating with velocity v_r.
(Ref. 43.)

in the integral is indpendent of n, whereas the contribution from the second term is different in the two cases, one must have $v_r = v_{dom}$ in order that the integral vanishes for both integration over the depletion region and over the accumulation region. Then for $\mathscr{E} = \mathscr{E}_{dom}$ Eq. (28) reduces to

$$\int_{\mathscr{E}_r}^{\mathscr{E}_{dom}} [v(\mathscr{E}') - v_r] \, d\mathscr{E}' = 0. \tag{29}$$

This is satisfied by requiring that the two shaded regions in Fig. 19 have equal areas. By using this, the value of the peak domain field $\mathscr{E}_{dom}$ can be determined if the value of the outside field $\mathscr{E}_r$ is known. This is the "equal areas rule" first introduced by Butcher.[42] The dashed curve in Fig. 19 is a plot of $\mathscr{E}_{dom}$ against v_r as determined by the equal areas rule. It begins at the peak of the velocity-field characteristic where the field equals the threshold field and ends at the point $(\mathscr{E}_m, v_{rm})$. For outside field values resulting in low field velocities $v(\mathscr{E}_r)$ less than v_{rm}, the equal areas rule can no longer be satisfied and stable domain propagation cannot be supported.[43]

If the field dependence of the diffusion factor in Eq. (28) is included in the equation, one must use the numerical techniques to obtain solutions. This was done by Copeland,[39,40] who used a two-valley temperature model[30] of DNR in GaAs to obtain v and D as functions of $\mathscr{E}$. These solutions showed that for a given value of outside field $\mathscr{E}_r$, there is at most one value of domain excess velocity defined as $(v_{dom} - v_r)$, for which solutions exist. In other words only one stable dipole domain configuration exists for each value of $\mathscr{E}_r$. Figure 21 shows the excess domain velocity plotted against outside drift velocity v_r for two values of n_o. Note that for $n_o = 10^{15}$ cm^{-3} one has $v_{dom} = v_r$ for a wide range of v_r values, and in this range the equal areas rule may be applied.[43]

Now consider some of the characteristics of a high-field domain. When the domain is not in contact with either of the electrodes, the device current is determined by the outside field $\mathscr{E}_r$ and is given as

$$J = qn_o v(\mathscr{E}_r). \tag{30}$$

Therefore, for a given carrier concentration, n_o, the outside field fixes the value of J. It is convenient to define the excess voltage contained by a high-field domain with outside field $\mathscr{E}_r$ by[40]

$$V_{ex} = \int_{-\infty}^{\infty} [\mathscr{E}(x) - \mathscr{E}_r] \, dx. \tag{31}$$

The computer solutions of Eq. (31) for different values of carrier concentration and outside field are shown in Fig. 22. These curves may be used to determine the outside field $\mathscr{E}_r$ in a particular diode of length L, doping concentration n_o, and bias voltage V, by noting that the following relation must

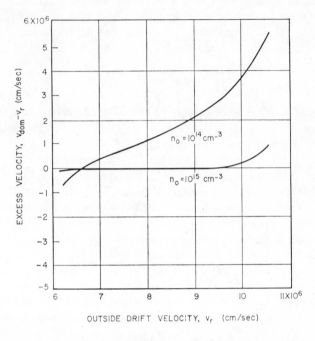

Fig. 21 The excess velocity for high-field domains versus the outside drift velocity for $n_o = 10^{14}$ cm^{-3}, and $n_o = 10^{15}$ cm^{-3}.
(After Copeland, Ref. 39.)

hold simultaneously with Eq. (31):

$$V_{ex} = V - L\mathscr{E}_r. \tag{32}$$

The straight line defined by this equation is called the device line and is shown in Fig. 22 as a dashed line for the particular values $L = 25$ μm and $V = 10$ volts. If $V/L > \mathscr{E}_T$, the threshold field, the interception of the device line and the solutions of Eq. (31) uniquely determine $\mathscr{E}_r$, which in turn specifies the current. The slope of the device line is fixed by L; however, the intercept defining $\mathscr{E}_r$ may be varied by adjusting the bias voltage V.

When the high-field domain reaches the anode, the current in the external circuit increases and the fields in the diode readjust themselves, nucleating a new domain. Then the frequency of the current oscillations depends on, among other things, the velocity of the domain across the sample, v_{dom}; if v_{dom} increases, the frequency increases and vice versa. The dependence of v_{dom} on the bias voltage can easily be determined. Increasing V in Eq. (32)

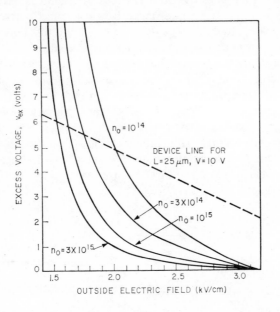

Fig. 22 Excess domain voltage versus electric field for various carrier concentrations. The dashed line is the device line for a sample $25\mu m$ long ($n_o = 10^{14} cm^{-3}$) and biased at 10 volts. (Ref. 39.)

shifts the device line up and the intercept of the device line, and the solutions of Eq. (31) occur at a lower value of $\mathscr{E}_r$. According to Eq. (30) the current therefore decreases and a differential negative resistance appears at the terminals. The decrease in $\mathscr{E}_r$ also decreases both v_r and the domain excess velocity in Figs. 19 and 21 and thereby (all other factors being equal) decreases the frequency of oscillation. The frequency of oscillation f is given by the approximate relation[40]

$$f = v_{\text{dom}}/L^* \tag{33}$$

where L^* is the effective length the domain travels from the time it is formed until the time a new domain begins to form. The length L^* may decrease as the width of the domain increases, since a new domain begins to form as soon as the leading edge of the old domain reaches the anode. This would tend to indicate that by increasing the bias voltage, and thereby increasing V_{ex} (Fig. 22), L^* would decrease and therefore increase f. The effect of a change in bias voltage on the oscillating frequency is then determined by whether the change in bias voltage has a greater effect on v_{dom} or on L^*.

Figure 23 shows a plot of domain width versus domain excess voltage. The domain width is defined as the distance between points where the electric field is greater than $\mathscr{E}_r$ by either 10 % of $\mathscr{E}_r$, or 10 % of the difference between the maximum field and the outside field, whichever is smaller.

One may determine the general shape of the high-field domain in a straight-forward manner by returning to the simple analysis of Butcher et al.[43] In the limit of zero diffusion the domain shape becomes quite simple. When $\mathscr{E}$ in Eq. (28) lies between $\mathscr{E}_r$ and $\mathscr{E}_{\text{dom}}$, the right side of the equation approaches infinity as D approaches zero; therefore, the term $[(n/n_o) - \ln(n/n_o) - 1]$ must also approach infinity. This implies that $n \to 0$ in the depletion region and $n \to \infty$ in the accumulation region. Poisson's equation can be used to show that the field builds up linearly from $\mathscr{E}_r$ to $\mathscr{E}_{\text{dom}}$ in the fully depleted region in a distance

$$d = \frac{\varepsilon_s}{q n_o} (\mathscr{E}_{\text{dom}} - \mathscr{E}_r). \tag{34}$$

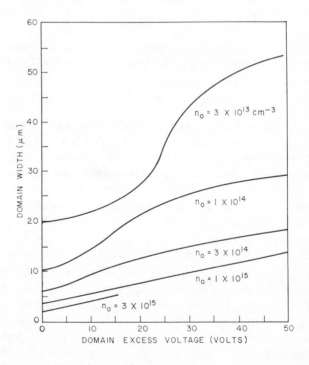

Fig. 23 Domain width versus domain excess voltage for various doping levels. (Ref. 39.)

Since $n \to \infty$ in the accumulation region, its width is zero. Then, for $\mathscr{E}_r > \mathscr{E}_{rm}$ the domain has a triangular shape as shown by the solid curve in Fig. 24. There can be no flat-topped region, in which $\mathscr{E} = \mathscr{E}_{dom}$ and $n = n_o$, separating the depletion and accumulation regions when $\mathscr{E}_r > \mathscr{E}_{rm}$ because in this region one would have $v(\mathscr{E}_{dom}) > v_r$ and the conduction current there would be more than that outside the domain, violating the conservation of total current requirement. As $\mathscr{E}_r$ approaches $\mathscr{E}_{rm}$, the peak domain field increases and both the height and width of the domain triangle increases (referring to Fig. 19). When $\mathscr{E}_r = \mathscr{E}_{rm}$ the depletion layer triangle is fully developed. Since $v(\mathscr{E}_m) = v_{rm}$ at this point, a flat-topped region of arbitrary width may now be inserted between the depletion and accumulation layers as shown by the dotted line in Fig. 24.

B. Resonant Circuit Operation. Thus far, the bulk differential negative resistance device has been examined from the point of view of isolating the device from any external circuitry and by considering only the magnitude of the bias voltage across its terminals. However, normally the RWH device is operated in some type of parallel resonant circuit, such as a high-Q microwave cavity. One would like to know how the external circuitry affects the operation of the devices. The theoretical analysis of a differential negative resistance device in a resonant circuit is complicated by the fact that the bias voltage is no longer constant but varies sinusoidally with the periodicity of the resonant circuit. This generally requires numerical solutions of the relevant equations. The resonant circuit model and the approximate velocity-field characteristic used in these studies are shown in Fig. 25. The modes of operation described in this section are divided according to (resonant frequency) × (length), fL product regions as shown in Fig. 13.

(a) *Transit Time Mode* ($fL \sim 10^7$ cm/sec). When the fL product of a device is approximately equal to the average carrier drift velocity ($\sim 10^7$ cm/sec), the sample is operating in the transit-time mode. This is the mode of operation discovered by Gunn.[1] In this mode the high-field domain is nucleated at the cathode and travels the full length of the sample to the anode. When the dipole reaches the anode, the outside field throughout the sample begins to rise through the threshold field and a new domain is nucleated at the cathode. Each time a domain is absorbed at the anode, the current in the external circuit increases; therefore, for samples in which the width of the domain is considerably smaller than the length of the sample, the current waveform tends to be spiked rather than the desired sinusoidal form. Figure 26 shows the experimentally obtained current waveform of a 100-μm sample[44] with an $n_o L$ product equal to 3×10^{13} cm^{-2}. Obviously, to obtain a more nearly sinusoidal current waveform one may either decrease the length of the sample (which increases the frequency in this mode) or increase the width

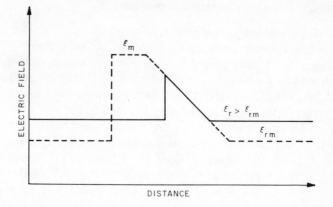

Fig. 24 Domain shapes in the zero diffusion limit when $\mathcal{E}_r > \mathcal{E}_{rm}$ (solid line) and when $\mathcal{E}_r = \mathcal{E}_{rm}$ (dashed line).
(After Butcher et al., Ref. 43.)

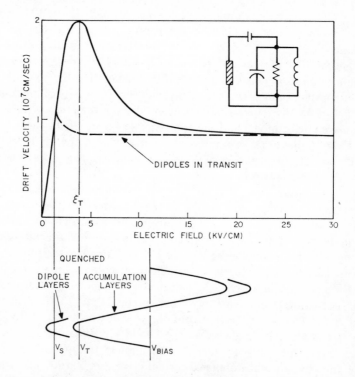

Fig. 25 Drift velocity of electrons versus electric field in *n*-GaAs. The dashed line is the dynamic *I-V* characteristic and applies when the domain is in transit. A plot of voltage versus time illustrates the principle of quenched modes.
(After Thim, Ref. 45.)

766

of the domain. It can be seen from either Eq. (34) or Fig. 23 that the domain width increases with decreasing doping level, n_o. In general, more sinusoidal waveforms may be obtained by decreasing the $n_o L$ product. Figure 27 shows a sequence of field distributions across a 35-μm sample during one rf cycle along with the doping profile and the voltage and current waveforms. These figures were obtained by the computer analysis of Thim.[45] The $n_o L$ product is 2.1×10^{12} cm^{-2} for this sample. The current waveform is much closer to sinusoidal for this device than the waveform shown in Fig. 26. Theoretical studies show that the efficiency of the transit-time mode is greatest when the $n_o L$ product is one to several times 10^{12} cm^{-2} so that the domain fills about half the sample and the current waveform is almost a sine wave. The maximum dc-to-rf conversion efficiency for this mode[45,46] is less than 10%. The variation of the dc-to-rf conversion efficiency, η, and the total output power with respect to bias voltage are shown in Fig. 28. The different curves are for different ratios of the load resistance R to the low-field resistance R_o of the device where

$$R_o = L/qn_o \mu_1 A \tag{35}$$

and μ_1 is the low-field mobility. These results are independent of device area A, length L, frequency f, and doping level n_o as long as the doping-length product, $n_o L$, and frequency-length product, fL, are constant. These results were also independent of the cavity Q as long as Q was greater than ten.

The rf power produced by a transit-time oscillator is given by[46]

$$P_{rf} = \eta P_{dc} \tag{36}$$

and

$$P_{dc} = VI = V(qvn_o A) \sim L \tag{36a}$$

where v is the average carrier velocity and V is the terminal bias voltage which is proportional to the sample L. The area A is limited by the load impedance R, since A is inversely related to R_o by Eq. (35). One can obtain the power-impedance product from Eqs. (35) and (36)

$$P_{rf} R = \eta(R/R_o)VIR_o \sim L^2 \sim 1/f^2. \tag{37}$$

Hence the power-impedance product is inversely proportional to the square of the transit-time mode frequency. This can be explained easily by noting that as the sample length decreases both the power input and the sample resistance decrease. Since the frequency is inversely proportional to the length, we thus obtain the relationship given by Eq. (37). Since the length of a device is fixed by the desired operating frequency and the doping is determined by optimizing the $n_o L$ product, only the area remains available for matching a diode to a

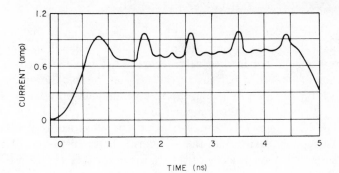

Fig. 26 Current waveform (experimental) of a 100-μm-long sample with a doping concentration of 3×10^{15} cm^{-3}.
(After Fukui, Ref. 44.)

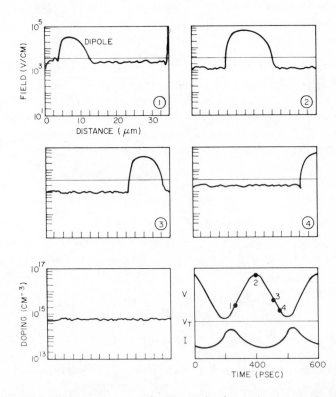

Fig. 27 Electric field versus distance at four intervals of time during one *rf* cycle (upper four plots), doping profile (lower left plot), and voltage and current waveforms (lower right plot) for a sample having $n_o L = 2.1 \times 10^{12}$ cm^{-2} and $fL = 0.9 \times 10^7$ cm/sec.
(After Thim, Ref. 45.)

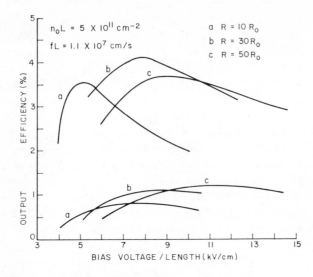

Fig. 28 Efficiency and power output versus bias voltage/length for three values of load resistance.
(After Copeland, Ref. 46.)

given impedance. The area of the device is given from Eq. (35)

$$A = (1/R)(R/R_o)(L/qn_o\,\mu_1) \tag{38}$$

after a value of R/R_o has been chosen to give proper efficiency and output power.

Because of the frequency limitations and efficiency limitations of the transit-time mode of oscillation, the bulk differential negative resistance oscillator operating in this mode has not been extensively studied. The modes of operation to be discussed in the next sections have less stringent limitations on the frequency and efficiency, and therefore have been studied more intensively.

(*b*) *Quenched Domain Mode* ($fL > 2 \times 10^7$ *cm/sec*). A bulk GaAs oscillator can oscillate at frequencies higher than the transit-time frequency if the high-field domain was quenched before it reached the anode. This mode was proposed by Carroll[47] in 1966. Dipole domain quenching occurs when the bias voltage across the sample is reduced below V_S in Fig. 25. In the dipole domain mode of operation, most of the voltage across the sample is dropped across the high-field domain itself. Therefore, as the resonant circuit reduces the bias voltage, one can see from Fig. 23 that the width of the domain is reduced. The domain width continues to reduce as the bias voltage decreases until at some point the accumulation layer and the depletion layer neutralize each other. The bias voltage at which this occurs is V_S. When the bias voltage swings back above threshold, a new domain is nucleated and the process repeats. Therefore, the oscillations occur at the frequency of the resonant circuit rather than at the transit time frequency.

In the quenched domain mode of operation it has been found, both theoretically[45] and experimentally,[48] that for samples in which the resonant frequency of the circuit is several times the transit-time frequency (i.e., $fL > 2 \times 10^7$ cm/sec), multiple high-field domains usually form, since one dipole does not have enough time to readjust and absorb the voltage of the other dipoles. Figure 29 shows multiple dipole formation in a sample operated in the quenched domain mode.[45] In this sample $n_o L = 4.2 \times 10^{12}$ cm^{-2}, $fL = 4.2 \times 10^7$ cm/sec, and $n_o/f = 10^5$ sec-cm^{-3}.

The upper frequency limit for this mode is determined by the speed of quenching, which in turn is determined by two time constants. The first is

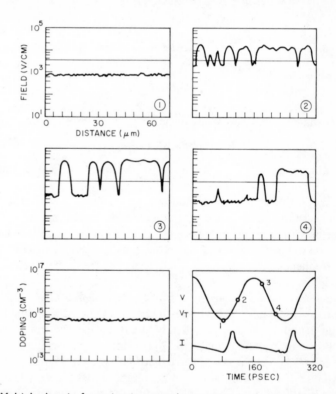

Fig. 29 Multiple domain formation in a sample operating in the quenched mode having $n_o L = 4.2 \times 10^{12}$ cm^{-2} and $fL = 4.2 \times 10^7$ cm/sec. (After Thim, Ref. 45.)

the positive dielectric relaxation time, while the second is an RC-time constant, R being the positive resistance in those regions of the diode not occupied by dipoles, and C the capacitance of all the dipoles in series. The first condition gives a minimum critical n_o/f ratio of about 10^4 sec-cm^{-3} for n-type GaAs.[49,50] The second time constant depends on the number of dipoles and sample length. The efficiency of quenched domain oscillators can theoretically reach 13%.[45,51]

(c) *Inhibited Mode.* If the total voltage across an RWH device is below threshold at the instant a dipole disappears at the anode, the formation of a new dipole is delayed until the voltage rises above threshold. If the new domain is inhibited from starting for a time equal to the domain transit time, then the waveform will be approximately a symmetrical square wave. The theoretical efficiency of this mode can reach 27% if the load resistance and the bias voltage are infinite, the voltage waveform is square, and the frequency of the resonant circuit is tuned to exactly half the transit time frequency.[45] Experimental efficiencies of 20% have been reported for this mode,[54] with pulse power output of 100 watts.

(4) Limited-Space-Charge-Accumulation (LSA) Mode[49,52]
   $(fL > 2 \times 10^7$ cm/sec)

The simplest possible mode of operation would consist of a uniformly doped semiconductor without any internal space charges. In such a case the internal electric field would be uniform and simply proportional to the applied voltage. The current, in turn, would be proportional to the drift velocity at this field level. The entire device would then have a current-voltage characteristic of the same relative shape as the velocity-field characteristic, Fig. 11. Coupled to an external resonant circuit, this device could then excite oscillations at a frequency determined by the resonant frequency of the circuit (combined with the capacitance of the device) but independent of the transit time of the electrons. This simplest "fundamental" mode of oscillation unfortunately has been given a rather obscure name, the *l*imited-*s*pace-charge-*a*ccumulation (LSA) mode of oscillation.

In the LSA mode the electric field across the diode rises from below threshold and falls back again so quickly that the space-charge distribution associated with high-field domain does not have sufficient time to form. Only the primary accumulation layer forms near the cathode, while the rest of the sample remains fairly homogeneous. Therefore, with little space-charge formation, the remainder of the sample appears as a series negative resistance which enhances the oscillations in the resonant circuit. The value to which the bias voltage must drop in order to quench the primary accumulation layer is V_T in Fig. 25 (which corresponds to the threshold field). However, in practice, due to sample inhomogeneities, high-field domains always form. The size of the domains is the quantity which distinguishes the LSA mode from the quenched (single or multiple) dipole mode.

When the electric field across a negative differential-resistance sample exceeds the threshold field, a space-charge accumulation layer begins to form at the cathode. The period of time in which this space charge grows, in the LSA mode, is only a fraction of a complete cycle. During the rest of the cycle the space charge is being quenched. Let the period of a cycle be denoted by τ and assume that the space charge is in its growth phase from time $t = 0$ to time $t = t_s$. The space-charge growth factor, G_n, which is a measure of the charge built up in time t_s, is defined as [50]

$$G_n = \exp(h/h_n) \qquad (39)$$

where $h \equiv n_0/f$, f is the frequency; and

$$h_n = \left(-\frac{q}{\varepsilon_s \tau} \int_0^{t_s} \mu \, dt\right)^{-1} \qquad (40)$$

where μ is the differential mobility. For the space-charge growth to be controlled, growth factor G_n must be sufficiently small that the space-charge fluctuations do not grow large enough to appreciably distort the electric field. During the part of the cycle during which the space charge decays, a factor denoted by G_p, the space-charge decay factor, may be defined as

$$G_p = \exp(-h/h_p) \qquad (41)$$

where

$$h_p = \left(-\frac{q}{\varepsilon_s \tau} \int_{t_s}^{\tau} \mu \, dt\right)^{-1}. \qquad (42)$$

If one separates the electric field into a dc component, $\mathscr{E}_o$, and an rf component, $\mathscr{E}_1$, at the fundamental frequency f, as

$$\mathscr{E} = \mathscr{E}_o + \mathscr{E}_1 \sin(2\pi ft) \qquad (43)$$

then Eqs. (40) and (42) may be plotted as a function of $\mathscr{E}_1$ for a fixed dc bias $\mathscr{E}_o$ as in Fig. 30 if μ as a function of the electric field is known.

Since the space-charge growth factor for an entire period, the product of G_n and G_p, must be less than unity to prevent progressive space-charge growth over many cycles, one must have

$$h_p < h_n. \qquad (44)$$

The size of (h/h_n) is somewhat arbitrary and must be chosen such that the small-signal theory is not invalidated. There is little guidance from the theory as to what its actual value should be. If we choose $h/h_n < 5$ and

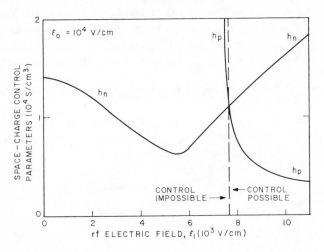

Fig. 30 Space-charge control parameters h_n and h_p versus *rf* field amplitude $\mathscr{E}_1$ for a fixed dc bias field of 10^4 V/cm.
(After Copeland, Ref. 50.)

$h/h_p > 6$, we obtain[50,53]

$$6h_p < h < 5h_n. \tag{45}$$

This in turn gives the ratio of doping to frequency in the following range for *n*-type GaAs:

$$2 \times 10^4 < (n_0/f) < 2 \times 10^5 \text{ sec-cm}^{-3}. \tag{46}$$

The results of some numerical studies suggest the maximum dc-to-*rf* conversion efficiency[50,53] is approximately 18 to 23%.

Until now, only the space-charge factors G_n and G_p have been calculated. But the magnitude of the dipoles is determined by both the growth factor and the initial doping fluctuations. In fact, it is the magnitude of these fluctuations which will determine the mode of operation, whether it is the multiple (or single) quenched dipole domain mode or the LSA mode of operation.[45] The quenched domain mode of operation is always less efficient than the LSA mode, a criterion which might be useful in distinguishing the two modes of operation.

(5) Other Devices Employing the RWH Mechanism[55]

Thus far we have dealt with the physics of differential negative resistance and related space-charge instabilities, and its application to microwave oscillators and amplifiers. However, the importance of the RWH mechanism lies not only in its ability to generate and amplify microwave signals, but also in its potentials in the conception of other new devices. This section will briefly discuss some of the other devices now employing the RWH mechanism.

A. Bulk Oscillators with Nonuniform Cross-Sectional Area.

The theory of high-field domains in one dimension can be used to analyze nonuniformly shaped oscillators if one assumes very thin high-field domains and considers phenomena in practically uniform regions in their neighborhood. These assumptions are valid if $n_o L \gg 10^{12}$ cm^{-2} and the variation of the cross-sectional area is gradual. Using the theory presented in the previous section, it can be shown that there exists a value of domain excess voltage, $V_{ex} = V_{rm}$, above which the outside electric field $\mathscr{E}_r$, remains constant. The value of the outside electric field corresponding to V_{rm} is $\mathscr{E}_{rm}$ shown in Fig. 19. The current density associated with a domain with $V_{ex} = V_{rm}$ is

$$J_{rm} = q n_o v_{rm}. \tag{47}$$

Such saturated domains move in the oscillator with a constant velocity.

Let the thickness of the bulk oscillator be s and the changing width be $b(x)$ where x is measured from the cathode. If a high-field domain is nucleated from the cathode at time $t = 0$, then at time t the domain is at $x(t) = v_{rm} t$ using the above assumptions. The current density of the device must be constant at all cross sections and, in the vicinity of the domain, is given by

$$I(t) = J_{rm} s b(v_{rm} t) + I_g(t) \tag{48}$$

where $I_g(t)$ is due to the decay of the domain and is zero except at the end of the cycle. Then from Eq. (48) the current is proportional to $b(v_{rm} t)$.

Figure 31 shows the waveform of a bulk effect oscillator for a sample shown in the insert. The solid lines in this figure show the waveforms expected from this shape. Note that the experimental current waveform is indeed similar to the shape of the sample. The symbols α, β, β', and γ correspond to the instants of time when the domain is located at points A, B, B', and C.

B. Multiterminal Devices.

Until now, only two terminal devices have been considered. The current waveform of an RWH device may be controlled by the addition of one or more electrodes located along the length of the device.[55] Figure 32(a) shows the structure of such a device with the electrode located at point B. The expected current waveform is shown in Fig. 32(b). This can be explained as follows. (The saturated domain theory described previously is used here again.) When the domain leaves the cathode at time $t = 0$, the cathode current, $I_c(t)$, is equal to the current of the

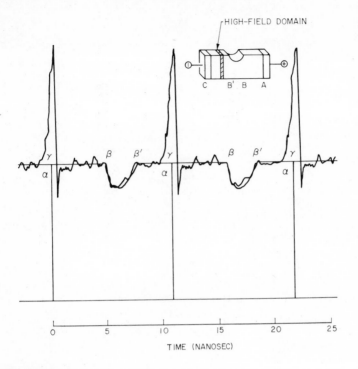

Fig. 31 Waveform generated by sample shown in the insert.
(After Shoji, Ref. 55.)

saturated domain ($Aqn_o v_{rm}$), and remains at this value until the domain reaches the electrode at B. At that time the cathode current becomes the sum of the saturated domain current and I_g, the current flowing through R_g. I_g is equal to the voltage sustained by the sample between A and B with a domain present, divided by R_g. The cathode current then remains at

$$I_c(t) = Aqn_o v_{rm} + I_g \qquad (49)$$

until the domain is absorbed at the anode, at which time the current spikes briefly.

A three-terminal device has been reported[56] which oscillated at frequencies between 60 MHz and 2.5 GHz. The structure of the device is shown in Fig. 33(a). The units were fabricated from GaAs p-n diodes by sawing into the n side with 12- to 25-μm tungsten wire to a depth close to the depletion layer. The oscillator bias is applied across the two n contacts and a variable bias is applied across the p-n junction to tune the frequency of operation. The

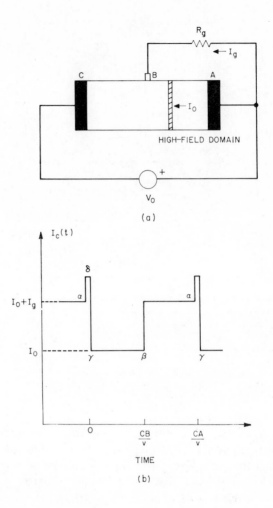

Fig. 32
(a) Controlled current step generator.
(b) Expected waveform of controlled current step generator.
(After Shoji, Ref. 55.)

frequency tunability as a function of p-n junction bias is shown in Fig. 33(b). A small-signal analysis has shown that if the space-charge waves that normally propagate in an RWH device interact with a distributed capacitance (the junction capacitance in this case), their phase velocity is substantially decreased and is a function of the total shunt capacitance. This could explain

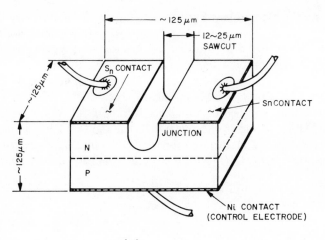

(a)

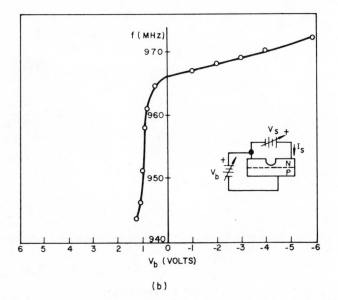

(b)

Fig. 33
(a) Three-terminal GaAs microwave CW oscillator.
(b) Tuning characteristics as a function of p-n junction bias.
(After Petzinger et al., Ref. 56.)

why the device oscillates at such low frequencies and can be tuned by the junction bias.

C. RWH-Mechanism Lasers. Radiation from bulk GaAs operated in the high-field domain mode has been reported by several workers.[57,58,59] The mechanism believed to cause this is a radiative recombination resulting from impact ionization within the high-field domain. Laser action[60] has been reported in field-ionized bulk GaAs. The devices were made from n-type Te-doped GaAs with a carrier density of about 4×10^{17} cm^{-3}. Two sides of the device were polished to a thickness of about 0.1 mm to form an optical cavity. Figure 34 shows the spectral distribution of light output from this device for both spontaneous (upper curve) and stimulated (lower curve) emissions where I_{CR} is the critical current permitting ionization to occur.

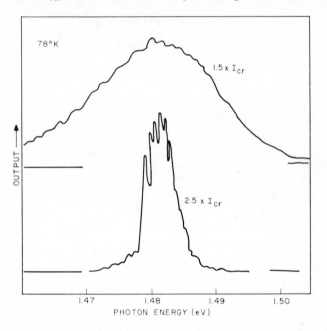

Fig. 34 Spectral distribution of light output of a bulk GaAs laser in spontaneous and stimulated emission.
(After Southgate, Ref. 60.)

5   ASSOCIATED BULK-EFFECT DEVICES

(1)   Thermistor

Thermistors, or *therm*ally sensitive res*istors*, are devices made of semiconductors whose electrical resistance varies rapidly with temperature.[61] The thermistors are useful for measurements of temperature, thermal conductivity, and radiation, and for control applications such as delay and surge suppression.

Some typical thermistors are shown in Fig. 35(a) in bead, rod, and disk forms. Figure 35(b) shows the static current-voltage characteristic for a typical thermistor. Each time the current is changed, sufficient time is allowed for the voltage to attain a new steady value. For sufficiently small currents, the power dissipation is too small to appreciably heat the thermistor, and Ohm's law is followed. However, as the current increases, the power dissipation increases, the temperature rises above ambient temperature, the resistance decreases, and hence the voltage is less than it would have been had the resistance remained constant. At some current I_m, the voltage attains a maximum value V_m. Beyond this point, as the current increases the voltage decreases, and the thermistor has a negative differential resistance.

To derive the static current-voltage relationship, we consider first the resistivity of a semiconductor which is given by $[q(\mu_n n + \mu_p p)]^{-1}$. The mobilities μ_n and μ_p and the carrier concentrations n and p are functions of temperature. The most temperature-sensitive parameter, however, is the majority carrier concentration which varies approximately as $\exp(-B/T)$ where B is a constant depending on the band gap and impurity concentrations. We can thus write the following equations for the resistance and the dissipated power:

$$R = R_0 \exp\left[B\left(\frac{1}{T} - \frac{1}{T_0}\right)\right] = V/I \qquad (50)$$

$$W = C(T - T_0) = VI \qquad (51)$$

where R_0 is the resistance at ambient tempearture T_0, and C is the dissipation constant related to the dissipated power W. By differentiating Eqs. (50) and (51) with respect to I, putting the derivative equal to zero, one obtains

$$T_m^2 = B(T_m - T_0) \qquad (52)$$

whose solution is

$$T_m = \frac{B}{2}(1 \pm \sqrt{1 - 4T_0/B}). \qquad (53)$$

The minus sign pertains to the maximum in Fig. 35(b), while the plus sign

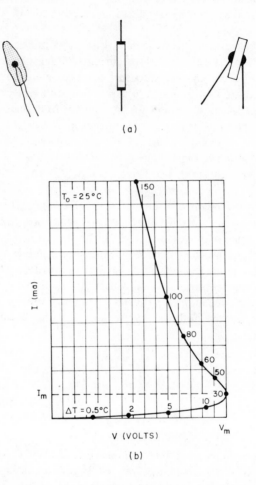

Fig. 35
(a) Typical configurations of thermistors in the bead, rod, and disk forms.
(b) Static current-voltage characteristics of a thermistor.
(After Becker et al., Ref. 61.)

pertains to the minimum (to be shown in Fig. 36). We note that T_m depends only on B and T_0, and not on R, R_0, or C. The temperature coefficient of a thermistor is defined as

$$\alpha \equiv \frac{1}{R}\frac{dR}{dT}.$$

$$(54)$$

It follows from Eq. (50) that

$$\alpha = -B/T^2. \tag{55}$$

From Eqs. (50), (51), and (55), we obtain for the quantities at the maximum of the I-V curve:

$$\alpha_m = -1/(T_m - T_0)$$

$$W_m = C(T_m - T_0)$$

$$R_m = R_0 \exp(-T_m/T_0) \tag{56}$$

$$V_m = [CR_0(T_m - T_0)\exp(-T_m/T_0)]^{1/2}$$

$$I_m = \left[\frac{C}{R_0}(T_m - T_0)\exp(-T_m/T_0)\right]^{1/2}$$

Generally the static I-V curves of thermistors are plotted in V versus I instead of I versus V as shown in Fig. 35(b). In addition, because currents and voltages for different thermistors cover such a large range of values, it is convenient to plot log V versus log I. Figure 36(a) shows such plots for three values of dissipation constant C. A line with a slope of 45 degrees represents a constant resistance, and line with a slope of -45 degrees represents constant power. Since B and T_0 are fixed in Fig. 36(a), the temperature corresponding to the maximum and minimum values of voltages as given by Eq. (53) remains the same. The values of the maximum and minimum voltages, however, decrease with decreasing value of C. Figure 36(b) shows a family of log V versus log I curves for various values of B while C, R_0, and T_0 are kept constant. For small values of $B (< 1200°K)$ the curves do not have a maximum. For large B values the maximum occurs at low powers and hence at low values of $(T - T_0)$. This follows as $W = C(T - T_0)$.

The speed of operation of a thermistor depends on its thermal time constant which is the time taken for the excess temperature to drop on cooling to $1/e$ of its initial value. The time constant τ is given by[62]

$$\tau = \text{(density)(specific heat)(volume)/(dissipation constant)}. \tag{57}$$

The time constant for typical thermistors varies from about one second to one minute, depending on the device geometry and heat sink.

(2) Hall Devices[63]

When a bulk semiconductor is placed in a magnetic field perpendicular to the direction of current flow, a voltage is developed across the semiconductor in a direction perpendicular to both the initial current direction and the

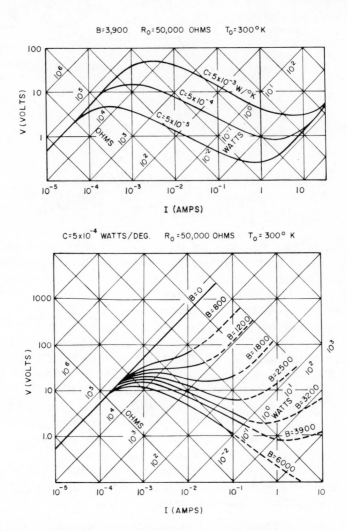

Fig. 36
(a) Logarithmic plots of voltage versus current for three values of the dissipation constant C.
(b) Logarithmic plots of voltage versus current for various values of B.
(After Becker et al., Ref. 61.)

magnetic field. This is the Hall effect, resulting from the Lorentz force on a moving charge carrier in an applied magnetic field. For an isotropic semiconductor, the electric field vector may be written as

$$\mathscr{E} = R_H(\mathbf{J} \times \mathscr{B}) \tag{58}$$

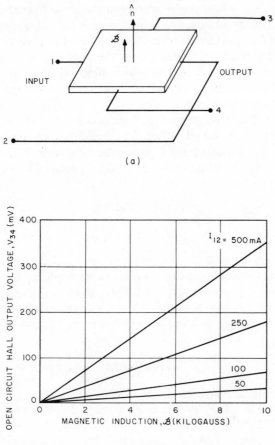

(a)

(b)

Fig. 37
(a) Hall effect probe.
(b) Characteristics of an InSb Hall probe.
(After Bulman, Ref. 63.)

where R_H is the Hall coefficient, J is the current density, and $\mathscr{B}$ is the magnetic induction.

In a four-terminal Hall device of rectangular geometry with thickness small compared with width, and length-to-width ratio sufficient to prevent field reduction due to end contacts the device equation is given as

$$V_h = \frac{R_H I}{d} \, (\hat{n} \cdot \mathscr{B}) \sim I(\hat{n} \cdot \mathscr{B}) \tag{59}$$

where V_h is the open-circuit Hall voltage, I the current, d the thickness of the active material, $\hat{n}$ is a unit vector normal to the plane formed by the length and width. A typical arrangement is shown in Fig. 37(a). From Eq. (59) we note that both I and $\mathscr{B}$ may be time-dependent, or they may be static values. For a constant current to the Hall probe, the open-circuit output voltage is linearly proportional to $\mathscr{B}$, and is also a function of angle between the normal vector and the magnetic induction vector. A set of InSb Hall probe characteristics are shown in Fig. 37(b).

When a Hall probe is operated with associated magnets, it can perform various functions. For example, when $\mathscr{B} \| \hat{n}$, Fig. 37(a). The device is called a gyrator which is an antireciprocal four-terminal device. This is because in the presence of the magnetic field the transfer impedances between two sets of terminals

$$z_{ik}(\mathscr{B}) \neq z_{ki}(\mathscr{B})$$

but

$$z_{ik}(\mathscr{B}) = -z_{ki}(-\mathscr{B}). \tag{60}$$

Thus when we apply voltage V_1 between terminal 1 and 2, we get kV_1 between terminals 3 and 4 where k is a constant. When we apply V_1 between 3 and 4 we get, however, $-kV_1$ between 1 and 2. Through the use of external shunting resistances the gyrator may be made to function as an isolator, a nonreciprocal four-terminal device in which one of the transfer impedances is zero.

REFERENCES

1. J. B. Gunn, "Microwave Oscillation of Current in III-V Semiconductors," Solid State Comm., *1*, 88 (1963), and "Instabilities of Current in III-V Semiconductors," IBM J. of Res. Develop., *8*, 141 (1964).

2. H. Kroemer, "Theory of the Gunn Effect," Proc. IEEE, *52*, 1736 (1964).

3. B. K. Ridley and T. B. Watkins, "The Possibility of Negative Resistance Effects in Semiconductors," Proc. Phys. Soc. (London), *78*, 293 (1961).

4. C. Hilsum, "Transferred Electron Amplifiers and Oscillators," Proc. IRE, *50*, 185 (1962).

5. A. R. Hutson, A. Jayaraman, A. G. Chynoweth, A. S. Coriell, and W. L. Feldmann, "Mechanism of the Gunn Effect from a Pressure Experiment," Phys. Rev. Letters, *14*, 639 (1965).

6. J. W. Allen, M. Shyam, Y. S. Chen, and G. L. Pearson, "Microwave Oscillations in GaAs$_{1-x}$P$_x$ Alloys," Appl. Phys. Letters, *7*, 78 (1965).

7. B. K. Ridley, "Specific Negative Resistance in Solids," Proc. Phys. Soc. (London), *82*, 954 (1963).

8. A. L. McWhorter and R. H. Rediker, "The Cryosar—A New Low-Temperature Computer Component," Proc. IRE, *47*, 1207 (1959).

9. H. Kroemer, "Negative Conductance in Semiconductors," IEEE Spectrum, 5, 47 (1968).

10. S. H. Koenig, R. D. Brown, and W. Schillinger, "Electrical Conduction in n-type Germanium at Low Temperature," Phys. Rev., 128, 1668 (1962).

11. R. H. Rediker and A. L. McWhorter, "Compound Cryosars for Low-Temperature Computer Memories," Solid State Electron., 2, 100 (1961).

12. L. Melngailis and A. G. Milner, "The Cryosistor—A Field-Effect Controlled Impact Ionization Switch," Proc. IRE, 49, 1616 (1961).

13. B. K. Ridley, and R. G. Pratt, "A Bulk Differential Negative Resistance Due to Electron Tunneling Through an Impurity Potential Barrier," Phys. Letters, 4, 300 (1963).

14. H. Kroemer, "Proposed Negative-Mass Microwave Amplifier," Phys. Rev., 109, 1856 (1958).

15. M. Shyam and H. Kroemer, "Transverse Negative Differential Mobilities for Hot' Electrons and Domain Formation in Germanium," Appl. Phys. Letters, (May 1968).

16. A. G. Foyt and A. L. McWhorter, "The Gunn Effect in Polar Semiconductors," IEEE Trans. Electron Devices, ED-13, 79 (1966).

17. G. W. Ludwig, "Gunn Effect in CdTe," IEEE Trans. Electron Devices, ED-13, 547 (1967).

18. M. R. Oliver and A. G. Foyt, "The Gunn Effect in n-CdTe," IEEE Trans. Electron Devices, ED-14, 617 (1965).

19. G. W. Ludwig, R. E. Halsted, and M. S. Aven, "Current Saturation and Instability in CdTe and ZnSe," IEEE Trans. Electron Devices, ED-13, 671 (1966).

20. J. W. Allen, M. Shyam, and G. L. Pearson, "Gunn Oscillations in Indium Arsenide," Appl. Phys. Letters, 9, 39 (1966).

21. R. Holmstrum, "Gunn-Type Oscillations in Ge, GaAs, and Other Materials," Proc. IEEE, 56, 331 (1968).

22. J. C. McGroddy and M. I. Nathan, "Microwave Oscillations in n-Type Ge," Solid State Device Research Conference, July 1967.

23. B. J. Elliott, J. B. Gunn, and J. C. McGroddy, "Bulk Negative Differential Conductivity and Traveling Domains in n-Type Germanium," Appl. Phys. Letters, 11, 253 (1967).

24. P. N. Butcher and W. Fawcett, "Calculation of the Velocity-Field Characteristics for Gallium Arsenide," Phys. Letters, 21, 489 (1966).

25. E. M. Conwell and M. O. Vassell, "High-Field Distribution Function in GaAs," IEEE Trans. Electron Devices, ED-13, 22 (1966).

26. J. G. Ruch and G. S. Kino, "Measurement of the Velocity-Field Characteristics of Gallium Arsenide," Appl. Phys. Letters, 10, 40 (1967).

27. H. Kroemer, "Nonlinear Space-Charge Domain Dynamics in a Semiconductor with Negative Differential Mobility," IEEE Trans. Electron Devices, ED-13, 27 (1966).

28. J. L. Birman and M. Lax, "Intervalley-Scattering Selection Rules, III-V Semiconductors," Phys. Rev., 145, 620 (1966).

29. H. Kroemer, "Detailed Theory of the Negative Conductance of Bulk Negative Mobility Amplifiers, in the Limit of Zero Ion Density," IEEE Trans. Electron Devices, ED-14, 476 (1967).

30. D. E. McCumber and A. G. Chynoweth, "Theory of Negative Conductance Amplification and Gunn Instabilities in 'Two-Valley' Semiconductors," IEEE Trans. Electron Devices, *ED-13*, 4 (1966).

31. W. Shockley, "Negative Resistance Arising from Transit Time in Semiconductor Diodes," Bell Syst. Tech. J., *33*, 799 (1954).

32. D. E. McCumber, "Numerical Studies of a Two Valley Model of the Gunn Effect," International Conference of Research on Semiconductors, Kyoto, Japan (1966).

33. H. W. Thim, M. R. Barber, B. W. Hakki, S. Knight and M. Uenohara, "Microwave Amplification in a Bulk Semiconductor," Appl. Phys. Letters, *7*, 167 (1965).

34. B. W. Hakki, "Amplification in Two-Valley Semiconductors," Appl. Phys., *38*, 808 (1967).

35. A. L. McWhorter and A. G. Foyt, "Bulk GaAs Negative Conductance Amplifiers," Appl. Phys. Letters, *9*, 303 (1966).

36. R. Holmstrom, "Small Signal Behavior of Gunn Diodes," IEEE Trans. Electron Devices, *ED-14*, 464 (1967).

37. H. W. Thim, "Temperature Effects in Bulk GaAs Amplifiers," IEEE Trans. Electron Devices, *ED-14*, 59 (1967).

38. H. Kroemer, "Effect of Parasitic Series Resistance on The Performance of Bulk Negative Conductivity Amplifiers," Proc. IEEE, *54*, 1980 (1966).

39. J. A. Copeland, "Electrostatic Domains in Two-Valley Semiconductors," IEEE Trans. Electron Devices, *ED-13*, 187 (1966).

40. J. A. Copeland, "Stable Space-Charge Layers in Two-Valley Semiconductors, "Appl. Phys., *37*, 3602 (1966).

41. P. N. Butcher, W. Fawcett, and N. R. Ogg, "Effect of Field Dependent Diffusion on Stable Domain Propagation in the Gunn Effect," Brit. J. Appl. Phys., *18*, 755 (1967).

42. P. N. Butcher, "Theory of Stable Domain Propagation in the Gunn Effect," Phys. Letters, *19*, 546 (1965).

43. P. N. Butcher, W. Fawcett, and C. Hilsum, "A Simple Analysis of Stable Domain Propagation in the Gunn Effect," Brit. J. Appl. Phys., *17*, 841 (1966).

44. H. Fukui, "New Method of Observing Current Waveforms in Bulk GaAs," Proc. IEEE, *54*, 792 (1966).

45. H. W. Thim, "Computer Study of Bulk GaAs Devices With Random One-Dimensional Doping Fluctuations," J. Appl. Phys., *39*, 3897 (1968).

46. J. A. Copeland, "Theoretical Study of a Gunn Diode in a Resonant Circuit," IEEE Trans. Electron Devices, *ED-14*, 55 (1967).

47. J. E. Carroll, "Oscillations Covering 4 GHz to 31 Ghz from a Single Gunn Diode," Electronics Letters, *2*, 141 (1966).

48. H. W. Thim and M. R. Barber, "Observation of Multiple High-Field Domains in *n*-GaAs," Proc. IEEE, *56*, 110 (1968).

49. J. A. Copeland, "A New Mode of Operation for Bulk Negative Resistance Oscillators," Proc. IEEE, *54*, 1479 (1966).

50. J. A. Copeland, "LSA Oscillator Diode Theory," J. Appl. Phys., *38*, 3096 (1967).

51. M. R. Barber, "High Power Quenched Gunn Oscillators," Proc. IEEE, *56*, 752 (1968).

52. R. W. H. Engelmann, "Circuit Performance of Field-Controlled Negative-Conductivity Devices," Presented at the Informal Conference on Active Microwave Effect in Bulk Semiconductors, New York, Feb. 3 (1966), also "Gunn Effect Devices," Technical Report ECOM-01758-1 and 2, U.S. Army Electronics Command, Fort Monmouth, N.J., March and August 1966.

53. I. B. Bott and C. Hilsum, "An Analytic Approach to the LSA Mode," IEEE Trans. Electron Devices, *ED-14*, 492 (1967).

54. D. G. Dow, C. H. Mosher, and A. B. Vane, "High Peak Power Gallium Arsenide Oscillators," IEEE Trans. Electron Devices, *ED-13*, 105 (1966).

55. M. Shoji, "Functional Bulk Semiconductor Oscillators," IEEE Trans. Electron Devices, *ED-14*, 535 (1967).

56. K. E. Petzinger, A. E. Hahn, Jr., and A. Matzelle, "CW Three-Terminal GaAs Oscillator," IEEE Trans. Electron Devices, *ED-14*, 403 (1967).

57. J. S. Heeks, "Some Properties of Moving High Field Domains in Gunn Effect Devices," IEEE Trans. Electron Devices, *ED-13*, 69 (1966).

58. S. G. Liu, "Infrared and Microwave Radiations Associated with a Current Controlled Instability in GaAs ,"Appl. Phys. Letters, *9*, 79 (1966).

59. P. Guetin, "Contribution to the Experimental Study of the Gunn Effect in Long GaAs Samples," IEEE Trans. Electron Devices, *ED-14*, 552 (1967).

60. P. D. Southgate, "Laser Action in Field-Ionized Bulk GaAs," Appl. Phys. Letters, *12*, 61 (1968).

61. J. A. Becker, C. B. Green, and G. L. Pearson, "Properties and Uses of Thermistors—The Thermally Sensitive Resistor," Trans. AIEE, *65*, 711 (1946).

62. R. W. A. Scarry and R. A. Setterington, "Thermistors, Their Theory, Manufacture and Application," Proc. IEEE, *107*, 395 (1960).

63. W. E. Bulman, "Applications of the Hall Effect," Solid State Electron., *9*, 361 (1966).

Author Index

Abraham, D., 573–575
Addiss, R. R., 641
Adler, R. B., 262, 282
Ahlstrom, E., 669
Aigrain, P., 687
Alberghini, J. E., 188
Allen, F. G., 445
Allen, J. W., 732, 744
Allen, L. C., 19
Anderson, R. L., 141, 142, 144, 612, 654, 667, 675, 676
Andrews, J. M., 389, 403
Angell, J. B., 282
Angello, S. J., 5
Angelo, E. J., 262
Arnold, K. M., 700
Ashley, J. C., 613
Atalla, M. M., 79, 505, 507, 594, 597
Augustyniak, W. M., 188
Aven, M. S., 744

Baertsch, R. D., 676, 679
Baker, R. H., 410
Baraff, G. A., 61
Barber, M. R., 752, 770, 771
Bardeen, J., 38, 261, 372
Barnes, P. A., 634, 635
Bartelink, D. J., 239, 252

Basov, N. G., 687, 723, 724
Batdrof, R. L., 60, 200, 244, 246
Bauer, R., 501
Bazin, B., 186
Becke, H., 295, 297, 506
Becker, J. A., 778, 780, 782
Beer, A. C., 13
Benda, H., 140
Bennett, W. B., Jr., 687
Berg, F., 546
Berglund, C. N., 448, 482
Bergstresser, T. K., 19, 22
Berman, R., 55, 56
Bernard, M. G. A., 687, 692
Bernard, W., 186, 188
Bethe, H. A., 378
Bevacqua, S. F., 687
Bew, G., 546
Biondi, F. J., 262
Birman, J. L., 747
Blair, J. C., 501
Blakemore, J. S., 26
Blakeslee, A. E., 295
Blue, J. L., 242–245
Bluhm, V. A., 310
Blum, S., 706
Bockemuehl, R. R., 320, 343, 348
Bode, P. E., 568

Boll, H. J., 359, 360
Boltaks, B. I., 29
Boothroyd, A. R., 262
Borkan, H., 569, 571
Bott, I. B., 773
Bowman, L. S., 248, 251, 252
Brattain, W. H., 261, 431
Bray, A. R., 269
Bricker, C. H., 244, 245
Brillouin, L., 17, 18, 19
Brockhouse, B. N., 51
Brody, T. P., 152, 570
Brondy, R. M., 188
Brotherton, S. D., 544, 545
Brown, G. A., 170, 176, 180–184
Brown, R. D., 737
Bruncke, W. C., 357
Bube, R. H., 655, 656
Buckingham, J. H., 186
Buget, U., 417, 419, 421
Bullis, W. M., 29, 49
Bulman, W. E., 781, 783
Burger, R. M., 29, 31
Burkhardt, P. J., 495
Burns, G., 687, 688, 700, 705, 708, 710, 727
Burrell, G. J., 703
Burrus, C. A., Jr., 181, 248, 251, 252
Butcher, P. N., 177, 746, 759–761, 764 766

Caldwell, J. F., 172, 188
Callaway, J., 29
Campbell, R. B., 669
Carey, G., 602, 603, 604, 607
Carley, D. R., 295, 296
Carlson, R. O., 687
Carr, W. N., 181, 636–639
Carroll, J. E., 769
Carruthers, J. A., 55
Carslaw, H. S., 29, 80, 83
Chang, H. C., 669
Chang, K. K. N., 252
Chang, L. L., 141, 143, 144, 506
Chapin, D. M., 641
Chen, Y. S., 732
Chenette, E. R., 262
Choudhury, N. K. D., 283
Chow, W. F., 150, 183
Christenson, H., 79, 261

Chu, T. L., 495
Chynoweth, A. G., 170, 172, 175, 176, 180, 184, 732, 748, 749, 752, 755, 761
Claassen, R. S., 172
Clark, L. E., 310, 313
Cohen, B. G., 200, 244, 246
Cohen, M. L., 17–19, 22
Collins, D. R., 477
Conley, J. W., 153
Conwell, E. M., 29, 30, 39, 42, 746
Copeland, J. A., 759, 761–764, 767, 769, 771–773
Corak, W. S., 537, 538
Coriell, A. S., 732
Cowley, A. M., 368, 372, 373, 376, 377
Crowell, C. R., 61, 64, 118, 119, 213, 337, 378–380, 382, 386–388, 394, 403, 404, 406–408, 548, 592, 599, 602–607, 613, 669
Cuthbert, J. D., 633
Cuttriss, D. B., 43, 44

Dacey, G. C., 320, 355
Dahlke, W. E., 487, 491
Dalton, J. V., 495
Darmagna, D., 570
D'Asaro, L. A., 675, 676, 700, 702
Dash, W. C., 52, 54, 660, 661
Davis, R. E. 175, 178
Davison, E., 641
Daw, A. N., 283
Deal, B. E., 80, 81, 426, 435, 437, 460– 462, 468, 470–472, 475, 493, 495
Dean, P. J., 627, 633
Debever, J. M., 724
Degraff, H. C., 581, 583, 587
Deloach, B. C., Jr., 200, 235, 244–246
Derrick, L., 79, 261
Dewitt, D., 262
DiDomenico, M., Jr., 657, 658, 660, 662
Dill, F. H., Jr., 687, 700
Dimmock, J. O., 486
Dolling, G., 51
Donovan, R. P., 29, 31
Doucetter, E. I., 357
Dow, D. G., 771
Duh, C. Y., 59
Dumke, W. P., 687, 700
Dunlap, W. C., 11, 78, 261

Duraffourg, G., 687, 692
Dyment, J. C., 702, 706, 716, 718, 720

Early, J. M., 235, 273, 282, 597
Eastman, P. C., 634, 635
Ebers, J. J., 303, 305, 319, 322, 324
Edwards, R., 281, 286, 291, 292
Ehrenreich, H., 39, 183, 186
Elliott, B. J., 744
Ellis, S. G., 641
Emmons, R. B., 658, 660, 667, 668
Eng, S. T., 194
Engelmann, R. W. H., 771
Epstein, A. S., 172, 188
Esaki, L., 150, 165, 183, 185, 491
Evans, W. J., 252, 253

Fan, H. Y., 629, 630
Faraday, B. J., 653
Fawcett, H., 746, 759–761, 764, 766
Feibelman, W. A., 611
Feldmann, W. L., 170–172, 175, 176, 180, 732
Fenner, G. E., 687, 720
Fischer, W., 535
Fitzgerald, D. J., 477, 509, 510, 555, 557, 560
Flatley, D., 80, 295, 297
Flinn, I., 546
Formenko, V. S., 364, 366
Forster, J. H., 133
Foster, N. F., 568
Fowler, R. H., 405, 549, 609
Fox, A. G., 699
Foyt, A. G., 744, 755
Frankl, D. R., 425, 451, 480, 513
Franks, V. M., 181
Frantz, V. L., 570
Frenkel, J., 496
Fritzsche, H., 186
Frosch, C. J., 79, 261
Fukui, H., 258, 765, 768
Fuller, C. S., 288, 641
Fulop, W., 326
Furnanage, R. A., 640

Galginaitis, S. V., 636–638
Gallagher, R. C., 537, 538
Garrett, C. G. B., 431, 480

Gartner, W. W., 39, 41, 49, 262, 266, 278, 663, 666, 667, 669
Geballe, T. H., 55
Gentry, F. E., 320, 322, 324, 326, 335, 339
Geppert, D. V., 591
Gershenzon, M., 633, 634
Gibbons, G., 82, 114–118, 121–125, 140, 175, 178, 204, 207–212, 244, 254, 255, 297, 337, 482, 492, 544, 675
Gibbons, J. F., 29, 59, 261, 262, 322, 329, 333
Gilbert, J. F., 188
Gilden, M., 215, 217, 221
Glicksman, R., 179–182
Gobeli, G. W., 445
Goetzberger, A., 120, 426, 435, 438, 445, 446, 451, 453, 454, 461, 464, 465, 473, 482, 489, 675, 676
Gold, R. D., 187
Goldberg, P., 626
Goldey, J. M., 319
Goldsmith, H., 611
Goldsmith, N., 80
Goldstein, Y., 425
Goodman, A. M., 403, 468, 477
Gordon, J. P., 687
Gray, P. E., 262
Gray, P. V., 449
Green, C. B., 778, 780, 782
Grosvalet, J., 475
Grove, A. S., 36, 71, 78, 80, 81, 426, 435, 437, 460–462, 472, 475, 477, 493, 509, 510, 513–515, 542, 543, 555, 557, 560, 562
Grover, N. B., 425
Guetin, P., 778
Guillaume, C. B., 724
Gummel, H. K., 92, 105, 106, 228, 230, 235, 236–238, 242–245, 270, 285–287, 292, 597–599, 601, 651–653
Gunn, J. B., 731, 744, 749, 765
Gurney, R. W., 421
Gutzwieler, F. W., 320, 322, 324, 335, 339

Haering, R. R., 578–581, 634, 635
Hahn, A. E., Jr., 775, 777
Haitz, R. H., 120

Hakki, B. W., 752–755, 757, 758
Hall, E. H., 44
Hall, R. N., 28, 46, 47, 78, 137, 183, 186, 261, 647, 687
Halladay, H. E., 355, 357
Halsted, R. E., 744
Handy, R. J., 650
Handy, R. M., 609, 610, 612
Hartman, T. E., 501
Hauer, W. B., 191
Hauser, J. R., 351
Haynes, J. R., 70, 71, 310
Heeks, J. S., 778
Heil, O., 505
Heiman, F. P., 505
Heine, V., 453, 465
Henisch, H. K., 364, 368, 369, 414, 415, 634
Henry, C. H., 633
Herlet, A., 326
Herman, F., 19
Hermanson, J., 29, 39
Herring, C., 19
Herriott, D. R., 687
Hielscher, F. H., 459
Hill, D. E., 52, 54, 636
Hill, E. R., 641
Hilsum, C., 13, 731, 741, 760, 761, 764, 766, 773
Hines, M. F., 215, 241, 242
Hinkley, E. D., 140
Hirshorn, H., 611
Hoerni, J. A., 79, 261
Hoffman, A., 140
Hoffman, D., 407
Hofstein, S. R., 435, 459, 463, 467, 474, 505, 545
Holland, L., 416
Holland, M. G., 55
Holmstrum, R., 744, 755
Holonyak, N. H., 319, 320, 322, 324, 335, 339
Holonyak, N., Jr., 172, 181, 687, 723
Honig, R. E., 416, 417
Hooper, W. W., 411, 562
Hopkins, J. B., 194
Howarth, L. E., 404, 613
Howson, D. P., 480
Hu, S. M., 499

Huff, G. J., 310
Hughes, A. J., 29
Hui, W., 641
Hulme, K. F., 177, 181, 186
Hurwitz, C. E., 726
Hurych, Z., 618
Hutson, A. R., 732

Iglesias, D. E., 246, 248, 252, 253
Ihanotola, H. K. J., 505, 535, 536, 545, 569
Irvin, J. C., 29, 30, 39, 40, 43, 44, 395, 400, 401
Ivey, H. F., 626, 628, 629, 631, 633
Iwersen, J. E., 269, 359, 360
Iyengar, P. K., 51

Jackson, W. H., 357
Jaeger, J. C., 29, 80, 83
Javan, A., 687
Jayaraman, A., 732
Johnson, E. O., 235, 262, 295
Johnson, H., 505, 506, 546
Johnson, K. M., 669, 675
Johnson, M. R., 723
Johnston, R. L., 200, 239, 244, 246, 252
Jones, H., 588
Jordan, A. G., 312, 667, 668
Jordan, E. L., 79
Josenhaus, J. G., 244, 248, 250
Jost, W., 29
Jund, C., 475

Kahng, D., 364, 505, 507, 550, 551, 556, 591
Kamins, T. T., 506
Kaminsky, G., 200, 244, 246
Kane, E. O., 156, 161, 163, 165–169, 181, 591, 695
Kanter, H., 607, 609, 611
Karlovsky, J., 165, 166, 196
Kawana, Y., 329
Keldysh, L. V., 167
Kendall, D. L., 29, 31, 82
Kennedy, F. J., 535
Keramidas, B. G., 641
Kerr, D. R., 495
Kessler, H. K., 188, 703
Keyes, R. J., 49, 67, 68, 687

Kibles, L. U., 674
Kingsley, J. D., 687
Kingston, R. H., 127, 129, 432
Kino, G. S., 59, 746, 747
Kirk, C. T., 285
Kittel, C., 12, 17, 18
Klaassen, F. M., 357
Kleimack, J. J., 60, 79, 261, 269
Klein, N., 298, 501
Kleinknecht, H. P., 181
Knight, S., 752
Koelmans, H., 581, 583, 587
Koenig, S. H., 737
Kontecky, J., 445
Kovel, S. R., 254, 255
Kray, W. E., 687
Kressel, H., 114, 723
Kroemer, H., 140, 283, 731, 735–737, 741, 747, 748, 750, 755–757
Krokhin, O. N., 687
Kruse, P. W., 654, 656, 657
Ku, H. Y., 621
Kunig, H. E., 570
Kunioka, A., 641

Labate, E. E., 403, 404, 602–604, 607, 613
Ladany, I., 700
Laff, R. A., 700
Lampert, M. S., 417
Landau, L. D., 161
Laraya, J. L. R., 410
Larrabee, R. D., 138, 139
Lasher, G. J., 687, 694, 695
Lasser, M. E., 658, 667, 668
Lawrence, H., 357
Lax, B., 673, 674, 687, 726
Lax, M., 747
Lebeder, P. N., 29
Lee, C. A., 60, 200, 244, 246, 261
Lee, C. H., 495
Lee, T. P., 82, 85, 90, 91, 94, 95
Leenov, D., 140
Lehner, L. L., 301
Lehorec, K., 425
Lehrer, W. I., 411
Leistiko, O., 513–515, 562
LeMee, J. M., 355

Lepselter, M. P., 261, 280, 389, 401–403, 412, 413, 546–550, 561, 562, 669, 680–683
Lesk, I. A., 310
Leuenberger, F., 544
Levine, A. K., 689
Levinstein, H., 656
Levitt, R. S., 708, 710
Li, T., 699
Lifshitz, E. M., 161
Ligenza, J. R., 79, 472
Lilienfeld, J. E., 505
Lindner, R., 425, 480
Liu, S. G., 778
Loar, H. H., 79, 261
Loferski, J. J., 641
Logan, R. A., 60, 63, 170–172, 180, 184, 188, 495
Longini, R. L., 262
Lucas, R. C., 613
Lucovsky, G., 140, 658, 660, 667, 668
Ludwig, G. W., 744
Lukes, Z., 537, 541
Lynch, W. T., 640, 660, 662, 668, 669, 676–678, 680

McCumber, D. E., 748, 749, 751, 752, 755, 761
McDonald, B., 120
MacDonald, R. W., 261, 282, 412, 413
McGill, T. C., 151, 407
McGlauchlin, L. D., 654, 656, 657
McGroddy, J. C., 744
McIntyre, R. J., 678, 679, 681
McMahon, M. E., 134
McMullin, P. G., 675, 676
McMurtry, B. J., 654, 667, 675
McWhorter, A. L., 687, 733, 738–741, 744, 755
Madelung, O., 11, 13
Magalhaes, F. M., 258
Mahan, G. D., 153
Maiman, T. H., 687
Maita, J. P., 28
Many, A., 425
Marinaccio, L. P., 244, 246, 247, 249, 254, 256, 258
Marinace, J. C., 700
Mathis, V. P., 310

Matizuki, K., 588
Matzelle, A., 775, 777
Mayburg, S., 700
Mead, C. A., 151, 376, 377, 395, 404,
    407, 409–411, 468, 470, 471, 501, 567,
    591, 594
Melchior, H., 298, 300, 459, 461, 660,
    662, 669, 675–678, 680–683
Melngailis, I., 719, 722, 723
Melngailis, L., 741
Meyerhofer, D., 176, 177, 180–184
Michel, W., 79
Middlebrook, R. D., 350
Mierop, H., 723
Miksic, M. G., 580, 581
Miller, H. S., 506
Miller, R. C., 719, 721
Miller, S. L., 60
Milner, A. G., 741
Milnes, A. G., 141, 144, 145, 667, 668
Minton, R. M., 179–182
Misawa, T., 200, 201, 224–226, 228,
    229, 231, 233, 244–250, 254, 256, 258,
    329
Mitchell, J. P., 477, 479
Mitra, R. N., 283
Miyahara, Y., 181, 183, 185
Moll, J. L., 11, 50, 56, 59, 77, 103, 105,
    109, 164, 167, 268, 282, 302, 303,
    305–308, 319, 425, 480, 505, 535, 536,
    545, 569, 592, 594, 613
Mooradian, A., 629, 630
Morant, M. J., 273–275, 277
Morato De Andrade, C. A., 173, 181
Morgan, J. V., 161, 177, 181, 186
Morin, F. J., 28
Morris, C. E., 618, 620
Mosher, C. H., 771
Mott, N. F., 363, 414, 415, 421, 588, 605

Nanavati, R. P., 173, 181
Nathan, M. I., 687, 688, 691, 705, 708,
    710, 727, 744
Nelson, O. L., 612, 703
Neuberger, M., 19, 24, 28, 49
Neustadter, S. F., 432
Newman, R., 52, 54, 660, 661
Newmark, G. F., 545

Nicollian, E. H., 426, 435, 445, 446, 451,
    453, 454, 459, 461, 464, 465, 473,
    482, 489
Nielson, E. G., 293
Nishizawa, J. I., 687
Noel, G., 641
Nordheim, L., 549, 609
Norris, C. B., 59
Norwood, M. H., 133, 134, 136
Noyce, R. N., 46, 47, 77, 272

O'Dwyer, J. J., 498, 501
Ogg, N. R., 759
Oldham, W. G., 141, 144, 145
Oliver, M. R., 744
Olsen, H. M., 138, 139
O'Reilly, T. J., 576
Owen, B., 480

Padovani, F. A., 387
Page, D. J., 584–586
Pankove, J. I., 153
Pao, H. C., 524, 530, 532–534
Parker, G. H., 407
Paul, W., 24, 186
Pearson, G. L., 310, 445, 505, 641, 732,
    744, 778, 780, 782
Pearson, W. B., 14
Pederson, D. O., 262
Pell, E. M., 137
Penfield, P. Jr., 133, 135, 136
Pennebaker, W. B., 570
Perkins, D. M., 641
Perry, E. W., 359, 360
Petzinger, K. E., 775, 777
Pfann, W. G., 425, 480
Pfann, W. H., 261
Phelan, R. J., 486, 723, 724
Philipp, H. R., 52, 54
Phillips, A. B., 262
Phillips, J. C., 19, 29, 39
Pilkuhn, M. H., 700, 706–709
Pittman, G. E., 636, 637
Pliskin, W. A., 495
Pochobradsky, J., 618
Poechler, T. O., 573–575
Pollack, S. R., 618, 620
Poon, H. C., 285–287
Popov, Y. M., 687, 696, 725, 726
Prager, H. J., 252
Pratt, R. G., 741

Prince, M. B., 39, 40, 137, 138, 642, 643, 649, 650
Pritchard, R. L., 262, 282
Pritchett, R. L., 286
Pugh, D., 378, 445

Queisser, H. J., 645, 649
Quinn, J. J., 588, 593, 613
Quist, T. M., 673, 674, 687

Racette, J. H., 28, 183, 186
Rafuse, R. P., 133, 135, 136
Rappaport, P., 641, 644, 645, 648, 649
Read, W. T., 200, 215, 647
Reddi, V. G. K., 527, 529, 530
Rediker, R. H., 140, 667, 673, 674, 687, 719, 722–724, 733, 738–741
Reizman, F., 495, 500
Reynand, J., 570
Richer, I., 350
Ridley, B. K., 731–733, 738, 741
Rietz, R. P., 663, 664
Rinder, W., 186, 188
Ripper, J. E., 711, 713–716, 719, 720
Ritchie, R. H., 613
Roehr, W. D., 302, 303
Rose, A., 591
Rose-Innes, A. C., 13
Rosenberg, H. M., 55
Rosenzweig, R., 651–653
Ross, I. M., 268, 320, 355
Ross, M., 654
Roth, H., 186, 189
Ruch, J. G., 59, 746, 747
Rulison, R. L., 244
Rupprecht, H. S., 700, 706–709
Ryan, F. M., 719, 721
Ryder, R. M., 195

Sah, C. T., 46, 47, 77, 272, 357, 426, 435, 437, 459, 461, 462, 475, 477, 493, 505, 513–515, 522–524, 527, 529, 530, 532–534
Sakai, Y., 641
Salama, C. A. T., 576, 577
Salzberg, C. D., 367
Sarace, J. C., 394, 403, 404, 406
Sawyer, D. E., 667
Scarlet, R. M., 120
Scarry, R. W. A., 781

Schafft, H. A., 298, 299
Scharfetter, D. L., 59, 92, 228, 230, 235–239, 252, 285–287, 312, 390, 392
Schawlow, A. L., 687, 699
Schiff, L. J., 110
Schillinger, W., 737
Schlig, E. S., 580, 581
Schlosser, W. O., 258
Schmidt, P. F., 79
Schmidt, R., 467
Schneider, M. V., 669, 670, 672
Schottky, W., 363, 378, 381
Schrieffer, J. R., 513
Schwartz, R. F., 140, 591
Scott, J., 80
Searle, C. L., 262
Sears, F. W., 626, 627
Seidel, T. E., 59
Seitz, F., 17
Senhouse, L. S., 138, 139, 310, 311
Setterington, R. A., 781
Sevick, J., 294, 309
Sevin, L. J., 320, 353, 356
Shallcross, F. V., 570
Sharpless, W. M., 674
Shatz, E., 133, 134, 136
Shewchun, J., 140
Shewmon, P. G., 29
Shive, J. N., 262
Shockley, W., 34, 35, 38, 46, 47, 65, 70, 71, 77, 96, 98, 100, 140, 194, 200, 214, 261, 263, 272, 310, 319, 320, 341, 403, 445, 505, 506, 645, 647, 749
Shoji, M., 546, 773–776
Shyam, M., 732, 741, 744
Simmons, C. D., 298
Simmons, J. G., 613, 614, 616
Sirkis, M. D., 723
Skalski, J. F., 570
Sklar, M., 460, 472
Slobodskoy, A., 425
Smith, A. C., 262
Smith, R. A., 11, 24, 25, 27, 29, 35, 36, 40, 45, 46, 52, 53
Smith, W. V., 688
Smits, F. M., 42, 651–653
Snow, E. H., 426, 435, 437, 460–462, 468, 470–472, 475, 477, 493, 495
Soltys, T. J., 687

Sommer, H. S., Jr., 176, 177, 180–184
Sorokin, P. P., 687, 688
Soshea, R. W., 594, 597, 613
Southgate, P. D., 778, 779
Spakowski, A. E., 641
Sparks, M., 261, 588
Spencer, H. C., 5
Spenke, E., 140
Spicer, W. E., 613
Spitzer, W. G., 376, 377, 404, 407–409, 613
Spratt, J. P., 591
Srelto, O., 657, 658, 660, 662
Stanley, A. G., 544
Statler, R. L., 653
Statz, H., 295
Stern, F., 694–697, 707
Stevenson, D. T., 49, 67, 68
Stevenson, M. J., 687
Stiles, P. J., 491
Stillman, G. E., 723
Stolnitz, D., 295, 297
Stone, H. A., 357
Stopek, S., 140
Stratton, R., 387
Straube, G. F., 134
Strutt, M. J. O., 109, 110, 112, 298, 300
Stuart, R., 613
Sugano, T., 613
Sundresh, T. S., 340
Swan, C. B., 244, 245, 254, 256, 258
Sylvan, T. P., 310, 314, 315
Sze, S. M., 29, 30, 39, 40, 43, 44, 60, 61, 63, 64, 82, 85, 90, 91, 94, 95, 114–119, 121–125, 140, 195, 204, 207, 214, 292, 297, 337, 367, 368, 372, 373, 376–378, 382, 386–388, 394, 401–404, 406–408, 412, 413, 487, 491, 492, 498, 499, 544, 546–551, 556, 561, 562, 592, 597–599, 601–607, 613, 614, 669, 675, 680
Szedon, J. R., 495

Tada, K., 410
Taft, E. A., 52, 54
Tamm, I., 445
Tanenbaum, M., 79, 261, 319
Tanke, R. V., 653
Teal, G. K., 261
Terman, L. M., 425, 646, 647

Theuerer, H. C., 79, 261
Thim, H. W., 752, 757, 766–768, 770, 771, 773
Thomas, D. E., 79, 184, 261
Thomton, C. G., 298
Thornber, K. K., 151
Thornton, R. D., 262
Thorpe, D., 302, 303
Tiemann, J. J., 186
Torrey, H. C., 194
Townes, C. H., 687, 699
Trofimenkoff, F. N., 310
Trumbore, F. A., 32

Uenohara, M., 752
Ullmann, F. G., 621

Vadasz, L., 542, 543
Valdes, L. B., 42
Van der Pauw, L. J., 44
Vanderwal, N. C., 395, 400, 401
Van Der Ziel, A., 130, 355, 357, 659
Vane, A. B., 771
Van Gelder, W., 495, 500
Vassell, M. O., 746
Veloric, H. S., 137, 138
Villa, G. G., 367
Vohl, P., 641
Von Munch, W., 295
Von Zastrow, E. E., 320, 322, 324, 335, 339
Voulgaris, N. C., 328, 329

Wallmark, J. T., 505, 506, 546
Wang, S., 629
Ward, J. H. R., 140
Warfield, G., 435, 448, 459, 475, 545
Warner, R. M., 357
Warschauer, D. M., 24, 186
Watanabe, Y., 687
Waterman, T., 611
Watkins, T. B., 731, 741
Watson, H. A., 395, 400, 401
Wu, S. Y., 546
Waugh, J. L. T., 51
Webster, W. M., 272, 273, 282
Wedlock, B. D., 350, 352
Wei, L. Y., 140
Weimer, P. K., 506, 567, 569–572
Weisberg, L. R., 187

Weisbrod, S., 252
Weiser, K., 708, 710, 726
Weisskopf, V. F., 39
White, G. K., 55, 56
White, H. G., 60
White, J., 506
White, M. H., 506
Whitmer, C. A., 194
Wiegman, W., 200, 244, 246
Willardson, R. K., 13
Williams, S., 468
Wilson, A. H., 605
Winogradoff, N. N., 188, 703
Wolfstirn, K. B., 34, 40, 288
Woodall, J., 708, 709
Woods, J. F., 726

Wooten, F., 613
Wright, G. T., 417, 419, 421, 480, 591
Wysocki, J. J., 641, 644, 645, 648, 649

Yang, E. S., 328, 329, 340
Yariv, A., 692
Young, C. F., 435
Young, L., 576, 577
Yu, H. N., 506

Zachos, T. H., 711, 713–717, 719–721
Zaininger, K. H., 448, 477
Zeiger, H. J., 687
Zemansky, M. W., 626, 627
Ziman, J. M., 17, 18, 22, 23, 55

Subject Index

Abrupt approximation, 82
Abrupt junction, 84–92, 204
Absorption, free carrier, 52
Absorption coefficient, 52, 54, 661
  temperature dependence, 53
Absorption process, 689
Acceptor, 25
  level, 29
  surface states, 445
Accumulation, 429, 430, 432
  layer, 736
  -layer mode (IGFET), 751
Acoustic branch, 50
Acoustic mode, 50
Acoustic phonon, 38
Activation energy, 27
Active region, transistor, 302
Admittance, ac, 108, 753
  equivalent circuit, 109
Al–Al$_2$O$_3$–Al–Al$_2$O$_3$–Al structure, 612
Al–Al$_2$O$_3$–Au, 611
Alloy-junction, 261
  method, 78
Al$_2$O$_3$, 607, 611
Alpha cutoff frequency, 283
Aluminum, 78
Aluminum antimonide, crystal structure,
  15 (T)

properties, 20 (T)
Ambipolar diffusion coefficient, 99
Ambipolar mobility, 99
Anodization, 79
Arsenic, 32
Atoms per cm³, of Ge, Si, and GaAs, 57
  (T)
Auger process, 46
Available power, avalanche photodiode,
  677
  photoconductor, 659
  photodiode, 650
Avalanche breakdown, curvature effect,
  119
  field, 117, 118, 122, 210, 481
  method, 634
  temperature effect, 119
  voltage, 115, 116, 121, 124, 126, 207,
    208
Avalanche mode, 302
Avalanche multiplication, 111
  breakdown condition, 113, 726
  temperature effect, 119
Avalanche photodiode, 676
Avalanche region, 201, 202, 210, 217–
  220, 320
Axial mode, 697, 713

Backward diode, 193–197
Ball alloy, 174
Ballistic mean free path, 599, 603
Band diagram, *see* Energy band diagram
Band-edge tailing, 151, 155
Band gap, 19
  doping effect, 151–156
  pressure effect, 24
  of semiconductors, 20 (T)
  temperature effect, 23
Bands, frequency range, 249 (T)
Band-tail tunneling, 170
Band-to-band recombination, 46, 48
Band-to-band transitions, direct, 409
  indirect, 409
Band-to-band tunnel current, 156
Baraff theory, 61, 118
  plot, 63
Barrier height for metal-semiconductor
    diodes, experimental value, 376,
    377, 397–399 (T)
  general expression, 372
  temperature dependence, 407
Base layer charging time, 282
Base resistance, 597
Base thickness, 281
Basic characteristics of Shockley diode,
    320
Basic equations, device operation, 65–72
Basic equivalent circuit, 446
Beam-lead technology, 261, 280
Beam-of-light transistor, 673
Bipolar transistor, 261; *see also* Junction,
    transistor
Blackbody radiation, 692
Bloch's theorem, 18
Body-centered cubic, 17
Bohr radius, 588
Boltzmann approximation, 96
Boltzmann factor, 693
Boltzmann statistics, 27
Bond picture, 25
Boron, 32, 79
Boron nitride, crystal structure, 15 (T)
  properties, 20 (T)
Boron phosphide, crystal structure, 15
    (T)
  properties, 20 (T)
Breakdown, junction, 109–126
  avalanche multiplication, 111

condition, 113
  thermal instability, 109
  tunneling effect, 110
Breakdown surface potential, 481
Breakdown voltage, abrupt junction,
    114, 115
  composite junction, 124
  curvature effect, 121
  linearly graded junction, 116
  maximum field, 117, 118, 122, 210
  MIS diode, 483
  p-i-n diodes, 140, 207, 208
  punch-through, 125
  surface effect, 562
  temperature effect, 119
  universal expression, 114
Breakover voltage, 335, 338
Brillouin zone, 18, 19
Built-in field, 267
Built-in potential, abrupt junctions, 87,
    88
  linearly graded junctions, 92, 94
Bulk differential negative resistance,
    732–741

Cadmium selenide, crystal structure, 15
    (T)
  properties, 21 (T)
Cadmium sulfide, crystal structure, 15
    (T)
  properties, 21 (T)
Cantilever beam, 602
Capacitance, abrupt junction, 90
  cylindrical junction, 90
  depletion layer, 90
  diffusion capacitance, 107
  exact calculation, 92
  at flat-band condition, 435
  linearly graded junction, 107
  metal-insulator-semiconductor diode,
    436
  metal-semiconductor diode, 404
  spherical junction, 91
  surface effect, 555
Capacitance method, 447
Capacitance-voltage measurement, 403
Capture cross section, 48, 103, 454
  of surface traps, 576
Carbon, crystal structure, 15 (T)
  properties, 20 (T)

Carrier concentration, 33
  gradient, 66
  thermal equilibrium, 25
Carrier storage time, 308
Carrier transport phenomena, 38–50
  in insulating films, 492–501
Carrier velocity, 59
Cathodoluminescence, 626
CdS, 376
CdSe thin-film transistor, 574, 583
Channel, 507
  conductance, 345, 346, 351, 512–515, 520
  width, 340
Characteristic frequency, 201
Characteristic penetration length in the metal, 621
Charge-storage diode, 137
Clamped transistor, 410
Classification, of semiconductor devices, 3
  of states and charges, 444
Cleavage plane, 17
Collection efficiency, 599
Collector, capacitance, 597
  characteristic, 603
  depletion-layer transit time, 283
  saturation current, 274
Color comparison of SiO_2 and Si_3N_4 films, 495 (T)
Common-base configuration, 262, 263, 272–276, 308
  input characteristics, 275
  output characteristics, 275
Common-base current gain, 270, 597
Common-collector configuration, 262
Common-emitter configuration, 262, 301, 308
  input characteristics, 276
  output characteristics, 276
Common-gate configuration, 537
Compensation factor, 739
Composite junction, 122
Composite systems, 495
Concentration, intrinsic carrier, 27
Conductance method (MIS), 451
Conduction, processes in insulators, 496 (T)
Conduction band, effective density of states, 26

subbands, 23
Contact potential, 369
Continuity equations, 66
Conversion efficiency, 767
Coplanar-electrode structure, 568
Copper, heat sink, 257
  thermal conductivity, 56
Covalent bond, 25
Cryosar, 738
Cryosistor, 741
Crystal orientation effect, 471
Crystal structure, 12
  semiconductor, 15 (T)
Current-controlled DNR, 732
Current density equation, 66
Current-gain, 270–272, 604
Current limiter, 357–361
Current mode, switching transistor, 302
Current-voltage characteristic, 156, 172–190
  doping effect, 175
  electron bombardment effect, 184
  pressure effect, 184
  temperature effect, 181–184
Current transport theory, 378–393
Current-voltage characteristics, avalanche photodiode, 677
  backward diode, 194
  cryosar, 740
  hot-electron transistor, 603
  IGFET, 519, 524
    with Schottky contacts, 548, 549
  junction FET, 343, 349, 353
  junction transistor, 262
  metal-semiconductor diode, 393, 402
  MIS structures, 489, 497
  p-n junction, 96
  p-n-p-n devices, 323
  solar cell, 642
  space-charge-limited diode, 421
  thermistor, 780
  thin-film transistor, 571
  tunnel diode, 157, 192
  unijunction transistor, 315
  voltage-controlled differential negative resistance, 733
Current-voltage measurement, 393
Curvature coefficient, 194, 196
Curvature effect, avalanche breakdown, 119, 563

Cutoff frequency, IMPATT diode, 222
  junction FET, 355
  junction transistor, 282, 288 (T)
  MIS surface varactor, 480
Cutoff region, switching transistor, 302
Cylindrical junction, 82

d-Band shielding effect, 613
Decay time, 307
Deep level, 29
Degenerate, 19, 151
Degradation, 187
Delay time, 201
Density of states, 32, 155
  distribution, 696
  effective mass, for electrons, 25
    for holes, 27
Depletion, 429, 430, 432
  capacitance, 84
Depletion layer, 370
  capacitance, 89, 95
  region, 84
Depletion-type IGFET, 536
Depletion-type photodiode, 659
Depletion-type TFT, 570
Depletion width, abrupt junction, 89
Detectivity, 654, 656
Deutron irradiation, 49
Device, "bulk property," 2
  optoelectronic, 2
Device geometry, 246, 279–282
Device line, 762
Diamond heat sink, 257
  thermal conductivity, 56
Diamond lattice, 12
Dielectric relaxation time, 750
Dielectric strength, 500
Differential capacitance, 432
Differentiation procedure, 447
Diffused-base transistor, 280
Diffused mesa junction, 79
Diffusion, 261
  capacitance, 107, 109
  coefficient, of impurities, 29, 31
    of carriers, 40
  conductance, 108
  constant surface concentration, 82
  length, 50, 69, 266
  limited source condition, 82
  method, 700, 702

noise, 131
potential, 84; see also Built-in
    potential
process, 80
profile, 82
solid state, 29, 79
theory, 381
Diode geometry, 244–246
Dipole-layer mode, 737, 757
Direct band-gap semiconductor, 690
Direct lattice, 17
Dispersion, 698
  solubility, solid, 32
Domain velocity, 758
Domain width, 764
Donor, 25, 34
  level, 28
  surface state, 445
Doping distribution, generalized, 134
Double-base diode, 310; see also Uni-
    junction transistor
Double-diffused transistor, 280
Double-injection, 420
Drain (channel) conductance, IGFET,
    520
  junction FET, 346, 348
  thin-film transistor, 371
Drain current, IGFET, 520
  junction FET, 346, 347
  thin-film transistor, 569
Drift, effect, 462
  region, 201, 202, 210, 220
  transistor, 267, 283
  velocity, 38
Dynamic characteristics, 215–221, 355

Early effect, 273
Ebers-Moll model, 303
Effective concentration, 176
Effective density of surface traps, 576
Effective diffusion velocity, 385
Effective mass, 12, 19, 177, 178, 379
Effective mobility, 513
Effective recombination velocity, 384
Effective Richardson constant, 385, 389
Effective temperature, 56
Effects, of mobility, 355
  of temperature, 355
  of traps and surface state, 573
  of unswept layer, 254

Efficiency, 235
  electroluminescent, 636
  photodiodes, 662
  solar cell, 644, 648
Einstein relationship, 42
Electric abrupt junction, 88
  surface, 431
Electric displacement, 3 (T)
Electroluminescent devices, 626–640
  diode, 7 (T), 9 (T)
Electromagnetic spectrum, 627
Electron, affinity, 369, 427
  capture cross section, 48
  -electron mean free path, 588
  energy, 19, 587
  energy distribution, 610, 611
  -impurity scattering process, 604
  imref, 532
  irradiation, 49
  mean free path, 607 (T)
  -phonon, mean free path, 589 (T)
    scattering process, 386, 604
  velocity at the Fermi level, 588
Emission rate, 695
Emission spectra, 627
  GaAs diode, 639
Emitter, characteristic, 312, 600 (T)
  charging time, 261
  conductance, 593, 598
  depletion-layer charging time, 282
  efficiency, 270
  stripe width, 281, 597
  -to-collector signal delay time, 597
Energy band diagram, avalanche photo-
    diode, 679
  avalanche p-n junction, 634
  backward diode, 195
  Fermi level, 37
  GaP p-n junction, 633
  heterojunction, 142, 144, 145, 673
  hot-electron transistors, 594, 595
  IGFET, 509, 510, 526
    with floating gate, 551
    with Schottky contacts, 550
  impurity states, 742
  junction transistor, 265, 268
  laser diode, 704
  luminescent efficiency, 631
  metal-insulator-metal structure, 615–
    620

metal-semiconductor contacts, 369,
  371, 373, 383, 390, 405, 671
MIS diodes, 427–429, 434, 463, 468,
  471, 485
Mott barrier, 415
n-type GaAs, 743
photodiodes, 667
p-n junction, 86, 93, 98
recombination processes, 47
semiconductor, 22
space-charge-limited diode, 419
TFT, 574, 582
transition processes, 628
tunnel diode, 158–160, 171, 174
tunnel emission structures, 608, 610
tunneling MIS diode, 487
tunnel metal-semiconductor contact,
  635
Energy distribution, 609
  emitter electrons, 611
  half width, 609
Energy gap, 3 (T), 20 (T)
Energy-momentum relationship, 17
Enhancement-type TFT, 570
Epitaxial substrate, 79
Epitaxial technology, 80, 261
Equal areas rule, 761
Equivalent circuit, avalanche photo-
    diode, 678
  IGFET, 535
  IMPATT (Read) diode, 216
  junction FET, 358
  junction transistor, 290, 305
  MIS structures, 446, 458, 464, 492
  photoconductor, 658
  photodiode, 662
  tunnel diode, 190
  unijunction transistor, 313
Equivalent minima, 25
Error function complement, 82
Esaki diode, 150; *see also* Tunnel diode
Ethyl orthosilicate, 80
Eutectic temperature, 78
Excess current, 169–172
Excess velocity, 761
Excess voltage, 763
Exchange effect, 613
Exciton, 52, 634
Experiments, IMPATT diodes, 244–258

Exponential excess current, 157, 169, 170
Extrinsic transitions, 629

Fabrication methods of lasers, 174
Fabry-Perot cavity, 700
Face-centered cubic structure, 17
Far-field pattern, 716
Fast-recovery diode, 136
Fast states, 445
Fermi-Dirac distribution, 26, 32, 691
Fermi-Dirac integral, 26
Fermi energy level, 3 (T), 27, 32, 35, 37, 589 (T)
Fick's equation, 80
Field distribution, abrupt junction, 88
  linearly graded junction, 92
Field-effect diode, 7 (T), 9, 357
Field-enhanced trapping, 741
Field-induced repopulation, 741
Filamentary transistor, 310; see also Unijunction transistor
Film thickness, 495
Flat-band capacitance, 438, 440
  condition, 428, 430, 432
  voltage, 450
Flicker noise, 130
Floating-gate memory device, 550
Frenkel-Poole emission, 496 (T), 552
Forward bias cutoff frequency, 395
Forward blocking, p-n-p-n devices, 321, 336
Four-point probe method, 42
Fowler-Nordheim tunneling, 552
Free electron density, 589 (T)
Frequency of oscillation, 763
Fundamental equation of the JFET, 346
Fundamental IMPATT mode, 235–238

GaAs, devices, beam-of-light transistor, 674
  bulk-effect devices, 731–787
  field-effect transistor, 410
  heterojunction, 672
  hot-electron transistor, 604
  IGFET, 506
  infrared source, 636
  junction transistor, 288, 291, 294, 295, 297
  metal-semiconductor diodes, 397 (T)

MIS devices, 481
  semiconductor laser, 687–730
  solar cell, 641
  tunnel diode, 181, 187
Gain, 692, 696, 705
Gain-bandwidth product, 573, 584
Gallium arsenide (GaAs), 57, 58
  absorption coefficient, 52
  crystal structure, 15 (T)
  diffusion coefficient, 31
  diffusion potential, 87, 92
  drift velocity, 59
  energy band, 19
  intrinsic concentration, 28
  ionization rate, 62
  maximum diffusion length, 70
  mobility, 39
  phonon spectra, 51
  properties, 20 (T), 57 (T)
  resistivity, 44
  thermal conductivity, 55
  velocity-field curve, 746
Gallium antimonide, 181, 193
  crystal structure, 15 (T)
  properties, 20 (T)
  tunnel diode, 181
Gallium nitride, crystal structure, 15 (T)
  properties, 20 (T)
Gallium phosphide, 183, 376, 633, 634
  crystal structure, 15 (T)
  ionization rates, 60
  properties, 20 (T)
Gamma function, 45
γ-ray, 477, 479
Gate, 340
  capacitance, 530
  electrode, 506
Gate-controlled diode, 555
Gaussian function, 82
Gaussian lineshape, 694
Gauss' law, 431
Ge, 57, 58
  absorption coefficient, 52, 661
  crystal structure, 15 (T)
  depletion layer width, 437
  diffusion coefficient, 29
  diffusion length, 70
  diffusion potential, 87
  drift velocity, 59

energy band, 19
intrinsic concentration, 28
ionization rates, 60
mobility, 39, 40
phonon spectra, 51
properties, 20 (T), 57 (T)
resistivity, 44
thermal conductivity, 55
Ge *p-i-n* diode, 207
transistor, 288, 291, 294, 295, 297
tunnel diode, 180, 181, 186, 193
Germanium oxide, 80
General characteristics of IGFET, 524–546
General expression, breakdown due to curvature effect, 121
Generalized current-voltage characteristics, *p-n-p-n* devices, 329
Generalized small-signal analysis, IMPATT diode, 221–234
General solutions, IMPATT diode, 225
Generation-recombination current, 102
Generation-recombination noise, 131
Geometry, electroluminesent diodes, 636
laser diode, 700, 715
transistor, 279, 597
Glass substrate, 586
Gold, 49
film, 599, 607
Gradual-channel approach, IGFET, 525
approximation, 517
Graphical method, Fermi level, 34
junction FET, 351
p-n-p-n devices, 329–333
Grey tin, crystal structure, 15 (T)
properties, 20 (T)
Ground state degeneracy, 34, 445
Grown-junction, 261
Growth factor, 222
Gummel number, 270
Gunn oscillator, 749, 765
Gyrator, 784

Half-channel depth, 340
Half-width of total energy distribution, 609
Hall coefficient, 45, 783
Hall devices, 9 (T), 781
Hall effect, 42, 44

Hall generator, 6 (T)
Hall mobility, 45
Hall probe, 784
Haynes-Shockley experiment, 70
Heat dissipation, 254
Heavy-hole band, 19
Hermite-Gaussian function, 712, 716
Heterojunction, 140
energy band model, 141
interface states, 141, 145
photodiode, 672
Hexagonal close-packed lattice, 13
Hexagonal structure, Wigner-Seitz cell, 17
High-current filament, 734, 741
High doping, 151-156
High-efficiency IMPATT diode, 252
High-efficiency IMPATT mode, 238–240
High-energy radiation, 49
High-field carrier transport, 746
High-field domain, 734
High-field property, 56
High-frequency capacitance curve, 435, 437
characteristics, junction transistor, 289
High-injection condition, 104, 106
Hole capture cross section, 48
Hot-electron devices, 567
Hot-electron lifetime, 587, 588, 590
in aluminum, 593
in gold, 592, 613
Hot-electron transistors, 587–613
Hot-hole lifetime, 590
"Hump" current, 172
Hybrid connection, 258
Hydrogen-atom model, 29
Hyperabrupt junction, varactor, 134

Ideal asymmetrical MIM structure, 616
Ideal diode law 96, 100; *see also* Shockley equation
Ideal metal-insulator-semiconductor diode, 426-444
Ideal MIS curves, 433, 438
IGFET, 505
basic structure, 506
with a floating gate, 550
general characteristics, 524–546
noise, 546

IGFET (cont.), with Schottky contacts, 546
Illumination effect, 475
Image force, 364
  dielectric constant, 367
Impact ionization, 59, 739
  generation rate, 59
  ionization rate, 64
  see also Avalanche multiplication
IMPATT diodes, 200–260
  measurement setup, 246
  Schottky contact, 412
  small-signal admittance measurement, 247
  small-signal property, 231
Impedance, avalanche region, 219
  drift region, 220
Impurity band, 152
Impurity distribution, varactor, 134
Impurity energy level, 29, 30
Impurity gradient, 268
Imref, 96
InAs, 183
  crystal structure, 15 (T)
  properties, 20 (T)
  tunnel diode, 183
Indirect band-gap semiconductor, 690
Indium antimonide, 183
  crystal structure 15 (T)
  properties, 20 (T)
Indium phosphide, crystal structure, 15 (T)
  properties, 20 (T)
Infrared region, 626
Inhibited mode, 771
Injection laser, see Junction, laser
Injection method, 633
InP tunnel diode, 181
Input, characteristics, 273
Input impedance, 191
Input resistance, 354
Instability in TFT, 580
Insulated-gate thin-film, transistor (TFT), 567, 568–586
Insulating layer, 79
Integrated arrays, EL diodes, 640
Integrated circuits, 10, 79
Integration procedure, 448
Interband transitions, 626

Interface, 2
  states, 444; see also Surface states
Intervalley-transfer mechanism, 2, 59
Intervalley transport time, 747
Intraband transitions, 626
Intravalley scattering, 39
Intrinsic barrier lowering, 403
Intrinsic carrier concentration, 28
Intrinsic Fermi level, 428
Intrinsic method, 633
Intrinsic semiconductor, 25
Intrinsic stand-off ratio, unijunction transistor, 312
Intrinsic transitions, 628
Inversion, 429, 430, 432
Ion drift instability, 467
Ion-drift method, 137
Ionic conduction process, 496 (T)
Ion implantation technology, 261
Ionization, coefficients, 60–65, 225
  energy of impurities, 30
  integral, 112
  integrand, 202
Ionized acceptors, 34
Ionized donors, 34
Ionized impurities, 38, 39
Ionized impurity scattering, 45
Irradiation, electron, neutron, deutron, 49
Isolator, 784

Junction, curvature, 85, 562
  field-effect transistor, 340–357
  laser, 699–723
  multiple, 5
  resistance, 395
  single p-n, 5
  transistor, 261–318

Kirk effect, 285
KPR (Kodak photoresist) process, 79

Large-signal analysis, IMPATT diode, 234–240
Laser, avalanche breakdown, 726
  degradation, 721
  electron beam pumping, 724
  fabrication technology, 700
  magnetic effect, 719
  materials, 727 (T)

optical pumping, 723
physics, 688–699
stress effect, 717
time delay, 720
Lasing, requirement, 699
Lateral alternating current flow, MIS
        diode, 464
Lattice constant, 14
semiconductors, 15 (T)
Lattice spacing, 3 (T)
Lead sulfide, crystal structure, 15 (T)
properties, 21 (T)
Lead telluride, crystal structure, 15 (T)
properties, 21 (T)
Light-activated semiconductor controlled
        rectifiers (LASCR), 335
Light-hole band, 19
Limited-space-charge-accumulation
        (LSA) mode, 771
Limiting current, 357
Limiting-velocity diode, 357, 359
Linearly graded approximation, 82
Linearly graded junction, 92, 205
Lineshape, 693
Lorentz force, 44, 782
Lorentzian lineshape, 694
Low-injection condition, 48
Low-temperature-coefficient regulator,
        133
LSA oscillator, 771
Luminescence, 626
Luminescent efficiency, 629, 631
Luminescent emission spectrum, 630

Magnetic induction, 3 (T)
Magnetoresistance effect, 46
Magnetoresistor, 6 (T)
Majority carrier, 38
Maximum attainable drain voltage, 544
Maximum available gain (MAG), 291
Maximum dielectric strength, 500
Maximum field, abrupt junction, 88
linearly graded junction, 92
Maximum junction temperature, 297
Maximum operating frequency, 535
Maximum oscillation frequency, 293,
        598
Maximum power rectangle, solar cell,
        643

Maximum resistive cutoff frequencies,
        193
Maximum width, 436
Maxwell equation, 65, 711
Mean free path, optical phonon, 61
Mean free time, 45
Measured Schottky barrier heights, 393–
        409
Measurement setup, IMPATT diode, 246
Metal-insulator-metal-insulator-metal
        structure, 591, 594
Metal-insulator-metal structure, 614
Metal-insulator-semiconductor devices,
        425–504, 505–566
Metal-semiconductor, contacts, 634, 635
diode (Schottky), 137, 369, 371, 564
field-effect transistor, 410
IMPATT diode, 412
photodiode, 669
with diffused guard ring, 561
Metal work function, 366, 469 (T)
effect, 467
Methods of excitation, optoelectronic
        devices, 632
Microplasmas, 675
effect, 125
Microwave, measurement, 246–254
oscillations, 248
power, 200, 245
switch, 140
transistor, 279–294
Miller indices, 12, 16, 17
MIM tunneling diode, 487
Minimum capacitance, 435, 438
Minimum noise figure, 602
Minimum reactive cutoff frequency, 193
Minimum voltage, 435, 438
Minority carrier, 38
lifetime, 48, 49
injection ratio, 390
MIS optical detectors, 486
structure, 466
surface varactor, 480
tunnel diode, 487
Mobility, 38, 40, 41, 514
conductivity, 42
drift, 42
Hall, 42
Modes of operation, bulk-effect devices,
        751

Modulation, frequency response, 666
index, 657, 658
noise, 131
MOSFET, 505; *see also* IGFET
MOST, 505; *see also* IGFET
Mott barrier, 363, 414
Mott-Gurney law, 421
Multiple-level traps, 48
recombination, 46
Multiplication factor, 331, 675

n-Channel, depletion IGFET, 536
enhancement IGFET, 537
Negative differential conductance amplification, 755
Negative mobility, 750
Neutrality, 34
Neutral region, 99
Neutron irradiation, 49
nL products, 750, 757
Noise, 130–131, 240–244, 357
equivalent power (NEP), 654
figure, 193, 240, 242, 293, 294
in IGFET, 546
$1/f$-type, 459
measure, 240, 245
in photoconductor, 659
power, avalanche photodiode, 678
voltage, 242, 459
Nomenclatures of transistors, 262, 263
Nominal current density, 695, 697
Nondegenerate, 27
Nondirectional laser cavity, 700
Nonideal I-V characteristics, 105
n-p-n Junction transistor, 594
np Product, 36

Ohmic contact, 78, 416
process, 496 (T)
One-sided abrupt p-n junction, 202
Orthogonalized plane wave method, 19
Optical branch, 50
Optical filter, 6 (T)
Optical mode, 50
Optical phonon energy, 57
Optical-phonon scattering, 386
Optical pumping, laser, 724
Optical transitions, 53
Optimum frequency, 223
Optoelectronic diodes, 625–686

Oscillation, frequency, 50
Out-diffusion, 79
Output characteristics, 274, 546
Overlap integral, 167
Overlap structure, 296
Oxide thickness, 80, 81

Parabolic barriers, 162, 164
Parallel connection, 258
Pauli Exclusion Principle, 587
p-Channel, depletion IGFET, 537
enhancement IGFET, 537
Pd film, 607
Peak current, 156, 178
Peak current to valley-current ratio, 179
Peltier effect, 7 (T)
Peltier refrigerator, 7 (T)
Penetration of the electric field, 621
Phase plot, 236
Phonon, 184
Phonon-assisted indirect tunneling, 166, 167
Phonon-assisted tunneling process, 183
Phonon scattering, 45
Phonon spectra, 51
Phosphorus, 32
Photoconduction effect, 49
Photoconductivity gain, 655, 656
Photoconductor, 654, 657
Photocoupled isolator, 639
Photodetectors, 653–683
Photoelectric measurement, 367
of barrier height, 404–409
Photoelectromagnetic effect, 49
Photoluminescence, 626
Photomultiplication factor, 675
Photoresist technique, 79
Piezoresistance effect, 6 (T)
properties, 568
Pinch-off condition, 341
Pinch-off point, 515
p-i-n Diode, 137, 206, 213, 214, 226
p-i-n Photodiode, 663
Planar process, 79, 175
Planar technology, 261
Plasma, frequency, 588
reaction, 79
Plasmons, 588
energy, 591 (T)

p-n Junction, 77–147
  photodiode, 667
p-n-p-n Devices, triple-junction, 319
Point-contact, photodiode, 674
  rectifier, 363, 415
Poisson's equation, 65, 431
Population inversion, 690
Power, gain, 290
  impedance product, 767
  limitation, 546, 584
Power dissipation limitation, 586
Power-frequency, limitation, 234
  relationship, 244
Primitive basis vector, 12
Properties of Si crystal planes, 473 (T)
Proton bombardment, 653
Pseudopotential method, 19
Pulse bond, 175
Punch-through, 336, 545
  barrier, 414
  effect, 125

Quality factor, 222
Quantum-mechanical reflection, 387, 599
Quasi-Fermi level, 96; *see also* Imref
Quenched domain mode, 769

Radiation dose, 477
Radiation effect, 477
  on solar cell, 651
Radiation pattern of EL diode, 638
Radiative process, 46
Radiative recombination process, 46,
  629
Radiative transitions, 626
Radioluminescence, 626
Rate equations, 631
Reactive cutoff frequency tunnel diode,
  191
Read diode, 200, 202, 203
  breakdown voltage, 205
  maximum field, 208, 215, 244
Reciprocal lattice, 14, 16
Recombination current, 647
Recombination-generation current, 272
Recombination processes, 46, 47
Rectification ratio, 132
Rectifier, 132
Reflection coefficient, 52

Refractive index, 52, 711
  equivalent, 713
Resistive cutoff frequency, tunnel diode,
  191
Resistivity, 42, 43
Resonant circuit operation, 765
Resonant frequency, 219, 222
Resonant mode, (laser) frequency sepa-
  ration, 713
Response time, photodetectors, 656 (T)
Reverse blocking characteristics, 336
Richardson constant, 379, 389
Ridley-Watkins-Hilsum (RWH) mecha-
  nism, 741–749
RWH device, 776
RWH-mechanism lasers, 778

Sapphire substrate, 586
Satellite frequency, 714
  valley, GaAs, 59, 743
Saturated drain current, 571
Saturated mode, switching transistor, 302
Saturation, current density, 101, 396
  region, 302
  temperature dependence, 102
  voltage, 357
Schottky barrier, 2, 195, 363
  avalanche photodiode, 680
  effect, 364–368
  emission, 496 (T)
  gate FET, 410
  IMPATT diode, 412
  lowering, 365
Schrodinger equation, 17
Secondary peak, tunnel diode, 172
Second breakdown, 298
Seeback effect, 6 (T), 7 (T)
Self-consistent dielectric constant approach, 588
Semiconductor, 1, 11
  degenerate, 151–156
  homogeneous, 5
  nonpolar, 38
  optical, thermal and high-field prop-
    erties, 50–65
  partially compensated, 35
  polar, 39
Semiconductor controlled rectifier
  (SCR), 319, 320–340

Semiconductor-metal-semiconductor
transistor, 594, 599
Series resistance, 107, 354, 395
of solar cell, 649
Shallow level, 29
Shockley diode, 319, 320–340
Shockley equation, 96, 100
Shot noise, 130
Si, $p^+ nin^+$, 207
Signal-to-noise ratio, (S/N), 654, 659
photodiode, 663
Silicon (Si), 57, 58
absorption coefficient, 52, 661
crystal plane, 473 (T)
crystal structure, 15 (T)
depletion layer width, 437
diffusion coefficient, 29
diffusion length, 70
drift velocity, 59
energy gap, 49
Fermi level, 36
intrinsic, 28
ionization rate, 60
mobility, 39
phonon spectra, 51
properties, 20 (T), 57 (T)
resistivity, 44
solid solubility, 32
thermal conductivity, 55
tunnel diode, 181
Silicon carbide, crystal structure, 15 (T)
properties, 20 (T)
Silicon dioxide, 79
Silicon nitride, 467, 495
films, 495, 500
Silicon oxide, 501
Read diode, 212
TFT, 577
transistor, 288, 291, 294, 295, 297
Silver, 607
Single-level recombination, 46
Slow states, 445
Small-signal, ac transconductance, TFT,
579
admittance measurement, IMPATT
diode, 247
alphas of transistor, 270, 326, 329
assumptions, IMPATT diode, 219
current gain, 289
properties, IMPATT diode, 231

Snapback diode, 137; *see also* Step-
recovery diode
Sodium ion, 461, 467
Solar cell, 640–653
Solar power density, 645
Solid solubility, 32
Solid-state diffusion, 49
Solution regrowth method, 703
Space-charge, build up, 750
decay factor, 772
density, 432
effect, 214
growth factor, 772
instability, 734
insulator, 463
Space-charge-limited, 496 (T)
diode, 417–421
transistor, 594
Space-charge-region widening, 527
Space-charge resistance, 214
Spatial coherence, 688
Spatial distribution of laser output, 715,
718, 719
Specific charge distributions, JFET, 346
Spectral distribution of laser output, 708,
719
Spectral response, solar cell, 645
Spectrum of oscillation, 246
Speed index, tunnel diode, 178
Speed of operation of a thermistor, 781
Spherical p-n junction, 82
Spin, 19
Spin-orbit interaction, 19
Spontaneous emission, 690
Stability factor, 291
Stable domain characteristics, 758
Staggered-electrode structure, 568
Static characteristics, of IMPATT diode,
202–215
of junction transistor, 262–279
Static common-emitter current gain, 270
Static I-V curves of thermistors, 781
Statistical distribution of surface charges,
456
Step-recovery diode, 137
Stevenson-Keyes method, 67
Stimulated emission, 690, 692
function, 694, 695
Strain gauge, 6 (T)
Stress effect, tunnel diode, 189

Stripe geometry contact, 700
Strong inversion, 436
Suhl effect, 6 (T)
Surface charge, 460
  effect, 102
  electron concentration, 532
  field effect, 555
  instability, 467
  potential, 429, 532
  recombination, 71
Surface-space-charge region, 429,
  506–512
Surface states, 372, 444, 445
  continuous, 456
  density, 452, 454
  lifetime, 446
  measured surface states, 452
Surface traps, 576
  trapping center, 72
Switching, modes, 302
  of p-n-p-n device, 326
  time, 302, 309
  transistor, 302–309
Symbol, backward diode, 194
  IGFET, 538, 539
  junction FET, 358
  junction transistor, 263
  p-n junction diode, 640, 641
  tunnel diode, 190
  unijunction transistor, 310
Symmetrical MIM structure, 615

Tantalum oxide, 501
Technology, device, 78–84
Temperature, coefficient, 111, 780
  dependence of barrier height, 407
  effect, 213, 473
    on avalanche breakdown, 119
  procedure, MIS diode, 449
Temporal coherence, 688
Terminal function, p-n junction diodes,
  131–140
Tetrahedral phases, 12
Thermal conductivity, 54, 55, 585
Thermal current, 157
Thermal diffusion current, 174
Thermal instability, 109
Thermal limitation, 301
Thermal noise, 130
Thermal oxidation, 79

Thermal property, 53
Thermal resistance, 254
Thermal sink, 56
Thermal velocity, 4
Thermionic emission, 146
Thermionic emission-diffusion theory,
  382
Thermionic emission theory, 378
Thermistor, 778
  speed of operation, 781
  static I-V curves, 781
Thermocouple, 7 (T)
Thermoelectric generator, 7 (T)
Thermoelectric power, 53
Thin-film devices, 567–624
  optical detectors, 568
  transducer, 568
Thomson heat, 6 (T)
Threshold, current density, 704, 705,
    707, 721, 722, 748
  electric field, 748
  voltage, 519, 524, 540, 570, 576
Thyristor, 319, 332
Time-constant dispersion, 456
Total impedance, 221
Transconductance, 348, 520, 522, 531,
    571, 578, 580
Transfer characteristics, 350, 522, 537,
    538
Transient behavior, 127–130
Transient operations, 337
Transient time, 128
Transistor, insulated-gate field effect,
    505–546
  junction, 261–318
    field-effect, 340–362
    thin-film, 568–613
    unijunction, 310–315
Transit angle, 222
Transition, allowed direct, 52
  exciton states, 52
  forbidden direct, 52
  indirect, 52
  processes, 688
  in semiconductors, 628
Transit time, 201, 222, 536, 545, 750
  -delay photodetector, 665
  -mode, 750, 765
Transmission probability, 110
Transport factor, 271

Transverse electric field, 533
Trap density, 48
Trap energy level, 48
Traveling, mode, 752
  space-charge accumulation, 736
Triangular barrier, 162
Triggering, methods, 334
  temperature, 298
  time, 298
Tunnel diode, 150–193
  transistor, 594
Tunnel emission, 496 (T)
  device, 608, 610
Tunneling, 485
  breakdown, 112
  current, 161, 165
  effect, 110
  in insulator, 487
  method, 634
  probability, 161, 549, 575, 614
  process, 156–169, 487
  time, 150
Turn-off time, 340
Turn-on time, 307, 337
Turn-on voltage, 437, 521
Turnover voltage, 110
Two-valley model, 743

Ultraviolet detector, 669, 682
Ultraviolet region, 626
Uniformly avalanching, 201
Unijunction transistor, 310–315
Unilateral gain, 291, 294 (T)
Unionized acceptors, 44
Unipolar transistor, 320, 340
Unit cells, 12–14
Universal expression, breakdown voltage,
  114

Valence band, effective density of states,
  27
  subbands, 19
Valley current, 157, 169
Vapor, phase reaction, 79
  pressure, 417, 418
  regrowth method, 703
Varactor, 133–136, 400
Variation of alphas with current, 327
Variolosser, 140
Varistor, 133
Velocity, field characteristics, 746, 765
  of GaAs, 746
  scattering limited, 57
  of sound, 56
Visible region, 626
Voltage, regulator, 133
  variable capacitor, 425
Voltage-controlled DNR, 732

Wave equation, 711
Wavelength response, 660
Webster effect, 272
Wigner-Seitz cell, 17, 19
WKB approximation, 161
Work function, of elements, 366
Work function difference, 427
Wurtzite lattice, 13

X-ray, 477

Zincblende lattice, 12
Zinc oxide, crystal structure, 15 (T)
  properties, 21 (T)
Zinc sulfide (ZnS), 633
  crystal structure, 15 (T)
  properties, 21 (T)
Zone-refining, 261